T0205335

Advances in Intelligent Systems and Computing

Volume 823

Series editor

Janusz Kacprzyk, Polish Academy of Sciences, Warsaw, Poland
e-mail: kacprzyk@ibspan.waw.pl

The series "Advances in Intelligent Systems and Computing" contains publications on theory, applications, and design methods of Intelligent Systems and Intelligent Computing. Virtually all disciplines such as engineering, natural sciences, computer and information science, ICT, economics, business, e-commerce, environment, healthcare, life science are covered. The list of topics spans all the areas of modern intelligent systems and computing such as: computational intelligence, soft computing including neural networks, fuzzy systems, evolutionary computing and the fusion of these paradigms, social intelligence, ambient intelligence, computational neuroscience, artificial life, virtual worlds and society, cognitive science and systems, Perception and Vision, DNA and immune based systems, self-organizing and adaptive systems, e-Learning and teaching, human-centered and human-centric computing, recommender systems, intelligent control, robotics and mechatronics including human-machine teaming, knowledge-based paradigms, learning paradigms, machine ethics, intelligent data analysis, knowledge management, intelligent agents, intelligent decision making and support, intelligent network security, trust management, interactive entertainment, Web intelligence and multimedia.

The publications within "Advances in Intelligent Systems and Computing" are primarily proceedings of important conferences, symposia and congresses. They cover significant recent developments in the field, both of a foundational and applicable character. An important characteristic feature of the series is the short publication time and world-wide distribution. This permits a rapid and broad dissemination of research results.

More information about this series at http://www.springer.com/series/11156

Sebastiano Bagnara · Riccardo Tartaglia
Sara Albolino · Thomas Alexander
Yushi Fujita
Editors

Proceedings of the 20th Congress of the International Ergonomics Association (IEA 2018)

Volume VI: Transport Ergonomics and Human Factors (TEHF), Aerospace Human Factors and Ergonomics

Editors
Sebastiano Bagnara
University of the Republic of San Marino
San Marino, San Marino

Riccardo Tartaglia
Centre for Clinical Risk Management and Patient Safety, Tuscany Region
Florence, Italy

Sara Albolino
Centre for Clinical Risk Management and Patient Safety, Tuscany Region
Florence, Italy

Thomas Alexander
Fraunhofer FKIE
Bonn, Nordrhein-Westfalen
Germany

Yushi Fujita
International Ergonomics Association
Tokyo, Japan

ISSN 2194-5357 ISSN 2194-5365 (electronic)
Advances in Intelligent Systems and Computing
ISBN 978-3-319-96073-9 ISBN 978-3-319-96074-6 (eBook)
https://doi.org/10.1007/978-3-319-96074-6

Library of Congress Control Number: 2018950646

This Springer imprint is published by the registered company Springer Nature Switzerland AG
The registered company address is: Gewerbestrasse 11, 6330 Cham, Switzerland

Preface

The Triennial Congress of the International Ergonomics Association is where and when a large community of scientists and practitioners interested in the fields of ergonomics/human factors meet to exchange research results and good practices, discuss them, raise questions about the state and the future of the community, and about the context where the community lives: the planet. The ergonomics/human factors community is concerned not only about its own conditions and perspectives, but also with those of people at large and the place we all live, as Neville Moray (Tatcher et al. 2018) taught us in a memorable address at the IEA Congress in Toronto more than twenty years, in 1994.

The Proceedings of an IEA Congress describes, then, the actual state of the art of the field of ergonomics/human factors and its context every three years.

In Florence, where the XX IEA Congress is taking place, there have been more than sixteen hundred (1643) abstract proposals from eighty countries from all the five continents. The accepted proposal has been about one thousand (1010), roughly, half from Europe and half from the other continents, being Asia the most numerous, followed by South America, North America, Oceania, and Africa. This Proceedings is indeed a very detailed and complete state of the art of human factors/ergonomics research and practice in about every place in the world.

All the accepted contributions are collected in the Congress Proceedings, distributed in ten volumes along with the themes in which ergonomics/human factors field is traditionally articulated and IEA Technical Committees are named:

I. Healthcare Ergonomics (ISBN 978-3-319-96097-5).
II. Safety and Health and Slips, Trips and Falls (ISBN 978-3-319-96088-3).
III. Musculoskeletal Disorders (ISBN 978-3-319-96082-1).
IV. Organizational Design and Management (ODAM), Professional Affairs, Forensic (ISBN 978-3-319-96079-1).
V. Human Simulation and Virtual Environments, Work with Computing Systems (WWCS), Process control (ISBN 978-3-319-96076-0).

VI. Transport Ergonomics and Human Factors (TEHF), Aerospace Human Factors and Ergonomics (ISBN 978-3-319-96073-9).
VII. Ergonomics in Design, Design for All, Activity Theories for Work Analysis and Design, Affective Design (ISBN 978-3-319-96070-8).
VIII. Ergonomics and Human Factors in Manufacturing, Agriculture, Building and Construction, Sustainable Development and Mining (ISBN 978-3-319-96067-8).
IX. Aging, Gender and Work, Anthropometry, Ergonomics for Children and Educational Environments (ISBN 978-3-319-96064-7).
X. Auditory and Vocal Ergonomics, Visual Ergonomics, Psychophysiology in Ergonomics, Ergonomics in Advanced Imaging (ISBN 978-3-319-96058-6).

Altogether, the contributions make apparent the diversities in culture and in the socioeconomic conditions the authors belong to. The notion of well-being, which the reference value for ergonomics/human factors is not monolithic, instead varies along with the cultural and societal differences each contributor share. Diversity is a necessary condition for a fruitful discussion and exchange of experiences, not to say for creativity, which is the "theme" of the congress.

In an era of profound transformation, called either digital (Zisman & Kenney, 2018) or the second machine age (Bnynjolfsson & McAfee, 2014), when the very notions of work, fatigue, and well-being are changing in depth, ergonomics/human factors need to be creative in order to meet the new, ever-encountered challenges. Not every contribution in the ten volumes of the Proceedings explicitly faces the problem: the need for creativity to be able to confront the new challenges. However, even the more traditional, classical papers are influenced by the new conditions.

The reader of whichever volume enters an atmosphere where there are not many well-established certainties, but instead an abundance of doubts and open questions: again, the conditions for creativity and innovative solutions.

We hope that, notwithstanding the titles of the volumes that mimic the IEA Technical Committees, some of them created about half a century ago, the XX Triennial IEA Congress Proceedings may bring readers into an atmosphere where doubts are more common than certainties, challenge to answer ever-heard questions is continuously present, and creative solutions can be often encountered.

Acknowledgment

A heartfelt thanks to Elena Beleffi, in charge of the organization committee. Her technical and scientific contribution to the organization of the conference was crucial to its success.

References

Brynjolfsson E., A, McAfee A. (2014) The second machine age. New York: Norton.
Tatcher A., Waterson P., Todd A., and Moray N. (2018) State of science: Ergonomics and global issues. Ergonomics, 61 (2), 197–213.
Zisman J., Kenney M. (2018) The next phase in digital revolution: Intelligent tools, platforms, growth, employment. Communications of ACM, 61 (2), 54–63.

Sebastiano Bagnara
Chair of the Scientific Committee, XX IEA Triennial World Congress
Riccardo Tartaglia
Chair XX IEA Triennial World Congress
Sara Albolino
Co-chair XX IEA Triennial World Congress

Organization

Organizing Committee

Riccardo Tartaglia (Chair IEA 2018)	Tuscany Region
Sara Albolino (Co-chair IEA 2018)	Tuscany Region
Giulio Arcangeli	University of Florence
Elena Beleffi	Tuscany Region
Tommaso Bellandi	Tuscany Region
Michele Bellani	Humanfactor[x]
Giuliano Benelli	University of Siena
Lina Bonapace	Macadamian Technologies, Canada
Sergio Bovenga	FNOMCeO
Antonio Chialastri	Alitalia
Vasco Giannotti	Fondazione Sicurezza in Sanità
Nicola Mucci	University of Florence
Enrico Occhipinti	University of Milan
Simone Pozzi	Deep Blue
Stavros Prineas	ErrorMed
Francesco Ranzani	Tuscany Region
Alessandra Rinaldi	University of Florence
Isabella Steffan	Design for all
Fabio Strambi	Etui Advisor for Ergonomics
Michela Tanzini	Tuscany Region
Giulio Toccafondi	Tuscany Region
Antonella Toffetti	CRF, Italy
Francesca Tosi	University of Florence
Andrea Vannucci	Agenzia Regionale di Sanità Toscana
Francesco Venneri	Azienda Sanitaria Centro Firenze

Scientific Committee

Sebastiano Bagnara (President of IEA2018 Scientific Committee)	University of San Marino, San Marino
Thomas Alexander (IEA STPC Chair)	Fraunhofer-FKIE, Germany
Walter Amado	Asociación de Ergonomía Argentina (ADEA), Argentina
Massimo Bergamasco	Scuola Superiore Sant'Anna di Pisa, Italy
Nancy Black	Association of Canadian Ergonomics (ACE), Canada
Guy André Boy	Human Systems Integration Working Group (INCOSE), France
Emilio Cadavid Guzmán	Sociedad Colombiana de Ergonomia (SCE), Colombia
Pascale Carayon	University of Wisconsin-Madison, USA
Daniela Colombini	EPM, Italy
Giovanni Costa	Clinica del Lavoro "L. Devoto," University of Milan, Italy
Teresa Cotrim	Associação Portuguesa de Ergonomia (APERGO), University of Lisbon, Portugal
Marco Depolo	University of Bologna, Italy
Takeshi Ebara	Japan Ergonomics Society (JES)/Nagoya City University Graduate School of Medical Sciences, Japan
Pierre Falzon	CNAM, France
Daniel Gopher	Israel Institute of Technology, Israel
Paulina Hernandez	ULAERGO, Chile/Sud America
Sue Hignett	Loughborough University, Design School, UK
Erik Hollnagel	University of Southern Denmark and Chief Consultant at the Centre for Quality Improvement, Denmark
Sergio Iavicoli	INAIL, Italy
Chiu-Siang Joe Lin	Ergonomics Society of Taiwan (EST), Taiwan
Waldemar Karwowski	University of Central Florida, USA
Peter Lachman	CEO ISQUA, UK
Javier Llaneza Álvarez	Asociación Española de Ergonomia (AEE), Spain
Francisco Octavio Lopez Millán	Sociedad de Ergonomistas de México, Mexico

Donald Norman	University of California, USA
José Orlando Gomes	Federal University of Rio de Janeiro, Brazil
Oronzo Parlangeli	University of Siena, Italy
Janusz Pokorski	Jagiellonian University, Cracovia, Poland
Gustavo Adolfo Rosal Lopez	Asociación Española de Ergonomia (AEE), Spain
John Rosecrance	State University of Colorado, USA
Davide Scotti	SAIPEM, Italy
Stefania Spada	EurErg, FCA, Italy
Helmut Strasser	University of Siegen, Germany
Gyula Szabò	Hungarian Ergonomics Society (MET), Hungary
Andrew Thatcher	University of Witwatersrand, South Africa
Andrew Todd	ERGO Africa, Rhodes University, South Africa
Francesca Tosi	Ergonomics Society of Italy (SIE); University of Florence, Italy
Charles Vincent	University of Oxford, UK
Aleksandar Zunjic	Ergonomics Society of Serbia (ESS), Serbia

Contents

Transport Ergonomics and Human Factors (TEHF)

Life on the Road: Exposing Drivers' Tendency to Anthropomorphise In-Vehicle Technology

David R. Large[(✉)] and Gary Burnett

Human Factors Research Group, University of Nottingham, Nottingham, UK
david.r.large@nottingham.ac.uk

Abstract. Anthropomorphism is often used in the design of products and technology, with the aim of enhancing the user experience. However, 'human' elements may also be employed for practical reasons, e.g. using speech as an interaction mechanism to minimise visual/manual distraction while driving. A self-report questionnaire survey (attracting 285 respondents from the UK), enriched by over thirteen hours of ethnographic-style observations involving 14 participants, explored drivers' tendency to anthropomorphise a routine in-vehicle navigation device (employing speech to deliver instructions). While the self-reported behaviour of drivers revealed only limited evidence of anthropomorphism, the observations clearly demonstrated that such behaviour was abundant during everyday use, with plentiful examples of drivers and passengers assigning gender, names and personality to the device. Drivers also attempted to engage the device in conversation, apparently endowing it with independent thought, and blamed it for mistakes. The results raise important considerations for the design and development of future in-vehicle technology (where speech is employed as an interaction mechanism), and speech-based systems more widely.

Keywords: Anthropomorphism · Navigation device · Survey
Ethnography

1 Introduction

Drivers are increasingly presented with novel technological solutions to augment or replace manual aspects of the driving task. The willingness to engage with such technology is not only contingent on objective factors, such as the reliability and performance of such systems, but is also shaped by subjective perceptions [1]. This means that a driver's preferences and usage patterns are likely to be based on individual attitudes, expectations and experience [2], although these factors need not be directly related to functionality. Given recent trends towards the 'humanising' of interactions with technology (with exemplars such as Siri, Alexa and Cortana), an area that is increasingly of interest is the impact of anthropomorphism on users' attitudes and behaviour.

Anthropomorphism is defined as the "attribution of human motivation, characteristics, or behaviour to inanimate objects, animals, or natural phenomena" [3]. According to social and psychological theory, humans engage in anthropomorphism for various reasons, including familiarity, comfort, as a 'best-bet' solution, or to make sense of,

S. Bagnara et al. (Eds.): IEA 2018, AISC 823, pp. 3–12, 2019.
https://doi.org/10.1007/978-3-319-96074-6_1

or exert authority over an object [4]. The term is often used pejoratively in science, when behaviours considered exclusive to humans are attributed to non-human entities. Nevertheless, people's desire to anthropomorphise appears to be highly compelling, and there is an abundance of rich evidence across cultures and time, with contemporary examples including: interpreting the behaviour of pets sympathetically using human emotions, declaring the personality of a car based on its physical appearance, and coaxing or threatening a recalcitrant computer [5, 6]. In these examples, exponents apparently use their experiences of 'being human' to help explain and guide their interactions with the nonhuman entity. Moreover, this occurs in spite of objective knowledge to the contrary [7]. Nevertheless, directing anthropomorphic attention towards certain subjects (e.g. pet animals, cartoon characters etc.) is generally considered to be more acceptable than attention towards other entities, such as cars and technology [8].

Anthropomorphism has been used extensively in design, with examples ranging from physical adornments to expressions of behavioural qualities [9]. Anthropomorphic designs are said to comfort users by providing clues about the product's function and mode of use, avoiding uncertainty and ambiguity, and facilitating certain social modes of interaction [10], but can also invite people to attach different personality traits to the host, or associate personal and social significance with it [4]. Anthropomorphism can therefore play a significant role in shaping user preferences [11]. However, using anthropomorphism in design can be problematic: it not only emphasizes the similarities but also hides the differences, and this can confuse or mislead users and create unrealistic expectations [12].

1.1 Speech

The inclusion of 'human' elements (e.g. speech) for pragmatic reasons may also result in anthropometric interpretations. One of the quintessential markers of humanness, speech is the primary means of social identification amongst humans [13]. From early infancy, humans are able to differentiate speech-like sounds from other sounds in their environment, and use this to maintain a sense of presence, even when their visual field is absent or obscured [14]. As humans develop and grow, they rapidly acquire the ability to extract salient, socially-relevant, paralinguistic cues from speech [13] based on vocal characteristics such as pitch, cadence, speech rate and volume, and use these to provide systematic and highly compelling guidance for determining gender, personality and emotion-specific actions [15]. Extensive research has demonstrated that humans appear to lack the wherewithal to overcome these evolutionary instincts, and behave towards vocal utterances from a computer – under experimental conditions at least – in a similar manner, ascribing humanlike characteristics and attending to talking machines as if they were interacting with another human [15, 16]. Manipulating speech interfaces has therefore enabled researchers to exploit these automatic responses. For example, different digital 'personalities', created by varying the vocal characteristics and language content of spoken language interfaces, have been shown to influence trust, performance, learning and even consumers' buying habits during research studies [15].

From a pragmatic point of view, however, speech-based interfaces also afford 'hand-free' and 'eyes-free' interaction, and are therefore often favoured by designers of computers or interfaces intended for deployment in situations already posing high levels of visual and/or manual demand, such as driving [17]. There is a substantial body of evidence which shows that using speech-based interfaces in vehicles (rather than traditional visual/manual methods of interaction) improves driving performance, reduces workload and passive task-related fatigue, and keeps drivers' eyes on the road [17, 18]. However, as speech-based interfaces become more 'natural' (due to steady, sustained improvements in speech synthesis and natural language delivery), drivers' attitudes towards the technology (and its impact on their behaviour) may be influenced by subtle paralinguistic cues that define factors such as gender and personality, and create the illusion of human presence.

1.2 Overview of Study

To explore driver's tendency to attribute human motivation, characteristics, or behaviour to in-vehicle technology, based on the presence of a voice, and begin to consider the implications that this may have on their interactions with the host device, we selected a ubiquitous example of an in-vehicle speech-based interface – the navigation device ('satnav'). We devised a bespoke questionnaire survey, and invited respondents to self-report their behaviour when using a satnav. We then conducted ethnographic-style observations documenting the routine use of satnav – from this we hoped to expose illustrative episodes of anthropomorphism, informed by the earlier questionnaire. The paper therefore reports two studies: the method and results are summarised independently for each, whereas the discussion and conclusions are common.

2 Questionnaire Survey

The questionnaire survey was designed to explore how people interpret and define their own anthropomorphic behaviour when driving and using a navigation device, and document their attitudes towards this type of behaviour when displayed by others. In addition, it aimed to explore the factors that may influence this behaviour (e.g. the nature of the voice delivering the navigation messages, the style of delivery, the level of trust that drivers placed in the device, usage occasion and duration etc.). Given this broad scope, a bespoke questionnaire was devised, comprising items selected from relevant scales and questionnaires, including the trust in automation scale [19], the anthropomorphic tendencies scale [8], the 'Godspeed' questionnaire [20] and human personality constructs [21]. All questions were presented as seven-point Likert scales, comprising a stem statement, such as *"I can trust the satnav"*, followed by numerically assigned scale points anchored with verbal response descriptors – typically, 'strongly agree' and 'strongly disagree', or 'not at all' and 'completely'. Respondents were asked to indicate their response by selecting a number on the Likert scale that best indicated their level of agreement with each statement. In addition, several open questions inviting written responses were included (e.g. asking respondents to elucidate their selection of voice or name for the device). Ratings underwent multivariate analysis to reveal any significant relationships. Only using selected items from the aforementioned scales may have contravened the analytic techniques specified by each questionnaire, however, the intention was to gain a general understanding of the incidence of anthropomorphism, and the attitudes displayed by respondents, to inform subsequent research activities.

2.1 Method

The survey was created online, with details posted on a number of online discussion forums and consumer groups. Details were also distributed via email to students and staff at the University of Nottingham. As an incentive to take part, and to encourage respondents to complete the survey (taking approximately 10 and 15-min), a donation of £1 (GBP) was made to a UK registered charity, selected from a shortlist by the respondent, for each completed questionnaire. Only complete datasets were considered in the analysis. Two hundred and eight-five respondents completed the online survey (228 male, 57 female; mean age 44.3 years, range 18–74 years). All participants were experienced UK drivers (mean years with licence: 24.0; mean annual mileage: 16,150), and regular users of satnavs (mean number of years: 3.45). Almost 90% of respondents used portable devices, comprising dedicated units (79.9%) and smartphone applications (9.7%); the remainder used factory-fitted devices. All respondents received directions using voice messages, with some supplementing these with visual iconography.

2.2 Results

Drivers generally placed high levels of trust in their satnavs. Multivariate analysis revealed that the level of trust was influenced by the length of time that drivers had used the device (Beta = .20, $p = .001$) and their annual mileage (Beta = –.13, $p = .03$), with the highest trust indicated by drivers who had used the devices for the longest. There was evidence to suggest that respondents associated 'human' qualities with their satnav, although some of the findings were inconclusive. For example, it is interesting to note that while respondents generally stated that they would not 'praise' their satnav when it performed well, responses suggest a greater likelihood that they would reprimand the device if it 'made a mistake'.

Almost a quarter of respondents (n = 66) stated that they would give their satnav a name (it was evident from the use of 'he' or 'she' in the written responses that respondents also tended to assign a gender to the device). Many of the names were inspired by human names and characteristics (e.g. 'Bossy Betty', 'Naggy Nora', 'Silicon Sal', 'Sally Satnav' and 'Suzy Satnav'), while some names were inspired by the device manufacturer or the term 'Satnav', with several mildly derogatory derivatives (e.g. 'Sat Nag', 'Nagman'). Drivers who gave their satnav a name also tended to rate statements relating to their own engagement in anthropomorphic behaviour more highly. However, multivariate analysis revealed that the most significant predictor of anthropomorphic tendency was gender, with females rating their own engagement in anthropomorphism more positively than male respondents (Beta = .29, $p < .05$). Even so, females also made lower ratings regarding the 'acceptability' of anthropomorphic behaviour displayed by others (Beta = –.17, $p = .008$). Driving experience was also a predictor of anthropomorphic tendency (Beta = –.36, $p = .04$), with responses suggesting that more experienced drivers were less inclined to engage in anthropomorphic behaviour directed towards their satnav. There was also evidence of anthropomorphism revealed through respondents' comments: *"it was only doing this that I realised how often I talk to it"*; another stated: *"I often shout at my satnav when it goes wrong"*.

Almost eighty percent of respondents (n = 224) chose a female voice to guide them (although this may have been influenced by the system default); only 13 female drivers indicating a preference towards a male voice. Several respondents (n = 18) indicated that they used a voice based on a famous character or celebrity. In these situations, the most popular celebrity voice was John Cleese. Other celebrity voices included Joanna Lumley, Kim Cattrall, Homer Simpson, Mr T and the Queen. Drivers who used a celebrity voice often tended to also refer to the device with that name.

Most respondents identified aspects of human personality associated with their navigation voice, based on both the 'dominance' and 'affiliation' dimensions. Additionally, navigation voices were generally considered to be 'natural' and 'humanlike'. However, higher ratings for these qualities may simply reflect designers' ability to reproduce authentic human voices within the technology. Perhaps most revealing were the written responses when drivers were asked if they would change the navigation voice they used. Most respondents (n = 268) indicated that they would not change the voice. While many citied practical reasons such as, *"clarity"*, or felt that their current navigation voice presented, *"minimal distraction"*, other responses were more revealing. For example: *"Joyce is now a family member – you wouldn't change the voice of a family member"*, *"she sounds nice"*, *"[it's a] more human female voice"*, and *"the familiarity with that voice provides reassurance and makes it feel more natural somehow"*. However, several respondents indicated that they would change the voice depending on context or occasion of use, stating that they downloaded character or celebrity voices to guide them. A variety of reasons were provided for such changes: *"after a while the voice gets annoying or boring and I find it better to keep it changing to keep me noticing it"*, *"depends on my mood"*, *"it is fun to change the voices"* and *"when on journeys with my children I often select a funny voice, dalek, pirate, Darth Vader etc."*

3 Ethnographic Observations

Ethnography emerged as a method for social inquiry from anthropological roots in the early 1920s, and is popularised today as a method to observe everyday people, undertaking everyday activities, in everyday settings [22]. Within the field of HCI, ethnography has been used to uncover the sort of practical ways that people overcome problems when interacting with a computer, and to reveal how they configure and adapt technology for specific tasks and contexts of use [23]. Findings have been consequently used to inform the design of future technology to ensure that it is applicable to the real world. In its simplest form, undertaking ethnography work involves observing people in a naturalistic setting.

3.1 Method

During the current study, observations took place in the UK, France and Southern Ireland, although all drivers and passengers were from the UK. Participants were contacted personally and asked if they were interested in taking part in the study. Fourteen people were subsequently recruited. These comprised a number of different social groups and use cases, including: two work colleagues travelling to a business meeting, a family of four embarking on a shopping trip in the UK, and journeys associated with two family holidays (with each family unit comprising four members) – holiday travel included travelling to the destination by car (in France and Southern Ireland) and daily excursions while there. All observations took place within the drivers' own vehicles and in most cases involved navigation devices owned by the participants (for one observation, a navigation system was provided, although the participants already had extensive experience using such devices). Only experienced drivers took part (mean years with driving licence: 22.8), with ages ranging from 38 to 45 years (mean age: 41 years). Interactions were documented using an audio recorder operated by each participant during their journeys in which a satnav was used as a matter of course. In total, over thirteen hours of dialogue were recorded. All interactions were transcribed using simplified CA conventions to enable subsequent thematic analysis.

3.2 Results

Transcripts were coded using thematic analysis, with the aim of identifying, analysing and reporting patterns (or themes) within the data. A common theme that emerged during the analysis was that drivers and passengers responded to and interacted with the satnav as if it had human qualities. Indeed, people talked to the device, shouted at it, gave it a name, and attempted to evoke it in conversational dialogue. They were also courteous during interactions, offering salutations (*"Bye, Jane"*) and even apologising to the device (*"Sorry love, we're having a rest"*). People also discussed the satnav as if it were somebody else present, referring to it as 'he' or 'she' (*"Where is he taking us? Yeah, alright. Oh dear, he's got a bit chatty now"*). In fact, people generally attended to the satnav in a similar manner to how they responded to each other in the car, and applied the same social etiquette, for example, employing conversational turn-taking. Moreover, this behaviour was commonplace and unremarkable – other car occupants did not respond to its peculiarity.

People also appeared to assign more complex human characteristics, behaviours and motivations to the devices, praising it when they successfully reached their destination (*Good job, Jane*), or reprimanding it if it failed to support them ("*She was nagging us through the forest. That was a bad thing, Jane"*), as if it intended to act as it did. People even appeared to endow the device with the power of independent thought ("*It still thinks we're in Beeston"*), human emotions (*"She hates it when you come off at the services"*), and an awareness of the environment that far exceeded the device's technical capability (*"He obviously heard what I said. See, that's what I wanted the first time"*).

There was evidence throughout the observations that it was the voice and the style of delivery that encouraged drivers and passengers to respond to the device in this manner. The voice was not only seen as a medium for delivering navigational instructions – in keeping with human-human interactions, the presence of a voice also appeared to encourage social interaction. The characteristics of the voice were used to assign attributes, such as 'gender' and 'personality' that subsequently defined and determined responses (similar to attributions made during human-human interactions). Some voices therefore encouraged stereo-typical responses from drivers and passengers, for example affecting trust (*"Well, I had a female voice on it...It was extremely posh...And just didn't feel like, well, just didn't feel like I could believe her"*).

The presence of a voice clearly influenced what participants thought about the device. For instance, participants appeared to believe that they were interacting with an intelligent device whose 'motivations' were the same as theirs. This may explain why some people appeared willing to place their trust in what was being said and accept routing errors as genuine mistakes. The voice may have also influenced how drivers used and relied on the support the device provided. Indeed, there were instances when drivers appeared to contemplate changing their normal behaviour, for example by contravening safe driving practices, making illegal manoeuvres, or selecting a road that would normally be deemed to be inappropriate (*"Breaking the law... I could just park here and we could walk?... No. No. Go on...Put your foot down. Follow the [satnav]"*).

4 Discussion

Although only limited evidence of anthropomorphism could be gleaned from the questionnaire (possibly due to limitations and biases in this approach [24]), there were abundant examples from the observations of participants interacting with and referring to the satnav as if it had humanlike qualities, and endowing it with human traits, such as a name, gender and personality. To help explain and interpret this behaviour, a number of theses can be offered. The familiarity thesis, for example, offers a cognitive motivation for anthropomorphism, and posits that anthropomorphism allows humans to explain things that they do not understand in terms of things that they do understand (i.e. themselves) [25]. An alternative explanation is offered by the comfort thesis, which proposes that anthropomorphism derives from an emotional motivation, arguing that people anthropomorphise because they are uncomfortable with things that are not like them and 'making' things more like them reduces that discomfort [25]. Moreover, in the face of chronic uncertainty about the nature of the world, guessing that something is humanlike, or has a human cause, constitutes a 'good bet'. If this assumption is found to be true, there is much to be gained, whereas if it is wrong, usually little is lost [25].

While these theories may go some way to explaining the observed behaviours, a more likely explanation is that drivers and passengers were using their experience of being human to define the social structure of the interaction. The Species-Specific Group-Level Coordination System [26], also referred to as the 'Social' Thesis [4], suggests that attributing human characteristics to non-human artefacts changes the value we place on them, and thus defines how we can behave towards them. The social thesis therefore claims that the act of anthropomorphising possesses the potential for social consequence [26]. Indeed, evidence from the observations, in particular, suggests that the satnav played an important role in the social fabric in which it was situated. Moreover, drivers (and indeed, passengers) were prepared to afford it high social status, and interacted with it as if it were another human entity (as evidenced by the adoption of social norms and etiquette, such as politeness and turn-taking, when interacting with the device). This might explain why drivers and passengers were quick to praise, or indeed, reprimand the satnav, depending on its performance.

This behaviour is likely to have been inspired by the presence of a 'human' voice, which activated evolutionary instincts [16], inspiring ascriptions of gender and personality, and leading to expectations of 'intelligence' and 'awareness'. However, this naturally raises concerns. The motivation behind using a voice to direct drivers is to ensure that instructions are clear and understandable, and do not distract drivers. The evidence from the study suggests that it also has the potential to inspire unrealistic expectations and inappropriate allocations of trust and reliance. Indeed, during the observations, participants were quick to trust the satnav (sometimes apparently against their better judgement), and there have been many similar anecdotal accounts of poorly calibrated trust in satnavs reported elsewhere (whereby drivers have incorrectly followed instructions, often with dire results) – it is possible that these have occurred due to inappropriate and unrealistic expectations of 'intelligence' inspired by the perceived 'human' qualities and capabilities.

Nevertheless, a fundamental question posed by the research is whether the participants genuinely and sincerely believed that the satnav had human qualities – the language they used to refer to the device, for example, may simply have been chosen for convenience, in the same way that people routinely use language and terminology to refer to their daily interactions with computers and technology that are taken directly from humans (e.g. becoming 'infected with a virus', being user-'friendly' or 'thinking' while a particularly complex program is being executed [27]). However, anthropomorphism makes no claims that it is the same as relating to other people: "we do not impute human personality in all its subtle complexity; we paint with broad strokes, thinking only of those traits that are useful to us in the particular context" [28]. Therefore, rather than dismissing anthropomorphism as a probable explanation for the observed behaviours and responses, it is suggested that the findings instead reinforce this view, drawing particular attention to the peculiarities and character of modern day anthropomorphism, as well as revealing some of the potential pitfalls in this context.

Finally, it is worth noting that, although responses were actually collected during 2010, the structure and delivery of navigational voice messages, and the front-end of satnavs, remain largely unchanged today, despite technological advancements in the underpinning technology. Moreover, with the current resurgence in speech interfaces, and the expectation that speech is likely to play an important role in supporting drivers through increasing levels of automated driving in the near-future [29, 30], findings are likely to be highly relevant today. The results can be used to inform the design and development of future in-vehicle speech-based technology, and suggest that, on the one hand, designers must be mindful of drivers' automatic attributions associated with 'natural' methods of interaction, such as speech, and the potential impact that this may have on their behaviour (particularly in terms of inappropriate allocation of trust and over-reliance on the technology). On the other hand, knowledge on this topic can be used to enhance the user experience and acceptance/uptake of speech-based technology in the automotive domain (and indeed, elsewhere).

5 Conclusion

A questionnaire survey enriched by ethnographic-style observations explored the propensity of drivers (and passengers) to engage in anthropomorphic behaviour during the routine use of an in-vehicle navigation device. Although, on face value, some of the self-reported behaviour of the questionnaire respondents was unremarkable, further interrogation of the data revealed subtle behaviours traditionally considered exclusive to humans, such as naming the device, talking to it and blaming it when mistakes were made. The observations were far more revealing, with evidence of anthropomorphism common and widespread, including making social reference to the device (using 'he' or 'she'), attempting to engage it in discussions, reprimanding it (following erroneous instructions or navigational errors) and praising it after successfully reaching a destination. Moreover, this behaviour was evident during routine interactions, was not dependent on occupancy or occasion use, and was not highlighted by other occupants as unusual or abnormal. The results can be used to inform the design and development of future in-vehicle speech-based technology.

References

1. Ghazizadeh M, Lee JD, Boyle LN (2012) Extending the technology acceptance model to assess automation. Cogn Technol Work 14(1):39–49
2. Schade J, Baum M (2007) Reactance or acceptance? Reactions towards the introduction of road pricing. Transp Res Part A Policy Pract 41(1):41–48
3. The Free Dictionary. http://www.thefreedictionary.com/anthropomorphism. Accessed 03 Apr 2017
4. DiSalvo C, Gemperle F (2003) From seduction to fulfillment: the use of anthropomorphic form in design. In: Proceedings of the 2003 international conference on Designing pleasurable products and interfaces. ACM, pp 67–72
5. Caporael LR (1986) Anthropomorphism and mechanomorphism: two faces of the human machine. Comput Hum Behav 2(3):215–234
6. Daston L, Mitman G (eds) (2005) Thinking with animals: new perspectives on anthropomorphism. Columbia University Press, New York
7. Robert F, Robert J (2000) Faces. Chronicle Books. New York
8. Chin MG, Sims VK, Clark B, Lopez GR (2004) Measuring individual differences in anthropomorphism toward machines and animals. In: Proceedings of the human factors and ergonomics society annual meeting, vol 48, no 11. SAGE Publications, Los Angeles, pp 1252–1255
9. DiSalvo C, Gemperle F, Forlizzi J (2005) Imitating the human form: four kinds of anthropomorphic form. Accessed Apr. (Unpublished manuscript)
10. Mitchell RW, Thompson NS, Miles HL (eds) (1997) Anthropomorphism, anecdotes, and animals. SUNY Press, Albany

11. Choi J, Kim M (2009) Anthropomorphic design: projecting human characteristics to products. In: International Association of Societies of Design Research Conference (IASDR2009), Seoul
12. Shneiderman B, Plaisant C (2004) Designing the user interface: strategies for effective human-computer interaction, 4th edn. Pearson Addison Wesley, London
13. Barthes R (1977) Music image text. Hill and Wang, New York
14. Chion M (1999) The voice in cinema. Columbia University Press, New York
15. Nass C, Brave S (2005) Wired for speech: how voice activates and advances the human-computer relationship. MIT press, Cambridge
16. Large DR, Burnett GE (2013) Drivers' preferences and emotional responses to satellite navigation voices. Int J Veh Noise Vib 9(1–2):28–46
17. Barón A, Green P (2006) Safety and usability of speech interfaces for in-vehicle tasks while driving: a brief literature review (No. UMTRI-2006-5). University of Michigan, Transportation Research Institute
18. Large DR, Burnett G, Antrobus V, Skrypchuk L (2018) Driven to discussion: engaging drivers in conversation with a digital assistant as a countermeasure to passive task-related fatigue. IET Intelligent Transport Systems (in press)
19. Jian JY, Bisantz AM, Drury CG (1998) Towards an empirically determined scale of trust in computerized systems: distinguishing concepts and types of trust. In: Proceedings of the human factors and ergonomics society annual meeting, vol 42, no 5. SAGE Publications, Los Angeles, pp 501–505
20. Bartneck C, Croft E, Kulic D (2008) Measuring the anthropomorphism, animacy, likeability, perceived intelligence and perceived safety of robots. In: Metrics for HRI workshop, technical report, vol 471, pp 37–44
21. Kiesler DJ (1983) The 1982 interpersonal circle: a taxonomy for complementarity in human transactions. Psychol Rev 90(3):185
22. Crabtree A (2006) Designing collaborative systems: a practical guide to ethnography. Springer, London
23. Schensul SL, Schensul JJ, LeCompte MD (1999) Essential ethnographic methods: observations, interviews, and questionnaires, vol 2. Rowman Altamira
24. Couch A, Keniston K (1960) Yeasayers and naysayers: agreeing response set as a personality variable. J Abnorm Soc Psychol 60(2):151
25. Guthrie SE, Guthrie S (1993) Faces in the clouds: a new theory of religion. Oxford University Press on Demand
26. Caporael LR, Heyes CM (1997) Why anthropomorphize? Folk psychology and other stories. In: Anthropomorphism, anecdotes, and animals, pp 59–73
27. Branscomb LM (1979) The human side of the computer. Paper presented at the Symposium on Computer, Man and Society, Haifa
28. Laurel B (1997) Interface agents: metaphors with character. In: Human values and the design of computer technology, pp 207–219
29. Large DR, Clark L, Quandt A, Burnett G, Skrypchuk L (2017) Steering the conversation: a linguistic exploration of natural language interactions with a digital assistant during simulated driving. Appl Ergon 63:53–61
30. Eriksson A, Stanton NA (2017) The chatty co-driver: a linguistics approach applying lessons learnt from aviation incidents. Saf Sci 99:94–101

Theoretical Considerations and Development of a Questionnaire to Measure Trust in Automation

Moritz Körber(✉)

Chair of Ergonomics, Technical University of Munich,
Boltzmannstraße 15, 85747 Garching, Germany
moritz.koerber@tum.de

Abstract. The increasing number of interactions with automated systems has sparked the interest of researchers in trust in automation because it predicts not only whether but also how an operator interacts with an automation. In this work, a theoretical model of trust in automation is established and the development and evaluation of a corresponding questionnaire (*Trust in Automation*, TiA) are described.

Building on the model of organizational trust by Mayer et al. (1995) and the theoretical account by Lee and See (2004), a model for trust in automation containing six underlying dimensions was established. Following a deductive approach, an initial set of 57 items was generated. In a first online study, these items were analyzed and based on the criteria item difficulty, standard deviation, item-total correlation, internal consistency, overlap with other items in content, and response quote, 40 items were eliminated and two scales were merged, leaving six scales (*Reliability/Competence*, *Understandability/Predictability*, *Propensity to Trust*, *Intention of Developers*, *Familiarity*, and *Trust in Automation*) containing a total of 19 items.

The internal structure of the resulting questionnaire was analyzed in a subsequent second online study by means of an exploratory factor analysis. The results show sufficient preliminary evidence for the proposed factor structure and demonstrate that further pursuit of the model is reasonable but certain revisions may be necessary. The calculated omega coefficients indicated good to excellent reliability for all scales. The results also provide evidence for the questionnaire's criterion validity: Consistent with the expectations, an unreliable automated driving system received lower trust ratings as a reliably functioning system. In a subsequent empirical driving simulator study, trust ratings could predict reliance on an automated driving system and monitoring in form of gaze behavior. Possible steps for revisions are discussed and recommendations for the application of the questionnaire are given.

Keywords: Trust in automation · Automated driving · Questionnaire

S. Bagnara et al. (Eds.): IEA 2018, AISC 823, pp. 13–30, 2019.
https://doi.org/10.1007/978-3-319-96074-6_2

1 Introduction

It has become impossible to evade automation: Thanks to the technological progress made, many functions that were previously carried out by humans can now be fully or partially replaced by machines (Parasuraman et al. 2000). As a consequence, they are taking over more and more functions in work and leisure environments of all kinds in our day-to-day lives. The resulting increase in the number of interactions with automated systems has sparked the interest of human factors researchers to investigate trust in automation with the overall goal to ensure safe and efficient joint system performance in mind (Drnec et al. 2016). An empirical investigation of trust in automation necessitates a measurement of trust in automation. Trust in automation is a latent construct, which is not directly observable; thereby, researchers rely on indicators such as neuroscientific methods (Drnec et al. 2016), behavioral measures (e.g., eye tracking; (Hergeth et al. 2016), or questionnaires (Jian et al. 2000; Madsen and Gregor 2000).

Trust in automation and reliance on automation are closely related: "People tend to rely on automation they trust and tend to reject automation they do not" (Lee and See 2004). Yet, trust in automation and reliance on automation are at the same time distinct constructs. In their theory of reasoned action, Ajzen and Fishbein (1980) argue that behavior, such as reliance, results from an intention and that this intention is a function of attitudes, which in turn are an affective evaluation of beliefs. Trust in automation as an attitude, thus, stands between the belief about the characteristics of an automated system, such as its reliability, and the intention to rely on it. Attitude, intention, and actual behavior are not in a deterministic but in a probabilistic relationship (Ajzen and Fishbein 1980). Whether trust translates into reliance behavior depends on a dynamic interaction of operator, automation, situational factors, and interface (Lee and See 2004). As a result, other factors, such as the effort to engage or self-confidence, also affect the intention to rely on an automated system (Kirlik 1993; Bisantz and Seong 2001; Dzindolet et al. 2001; Lee and See 2004; Meyer 2004). Environmental and cognitive constraints, such as time pressure, then determine whether a formed intention translates into actual reliance on automation. Even if trust is at a high level and the automated system is perceived as capable, reliance does not necessarily follow (Kirlik 1993). That means, to measure trust as an attitude itself, a questionnaire or another similar methodology that is distinct from observable risk taking is necessary (Mayer et al. 1995). Furthermore, the conceptualizations of trust in automation refer to the construct as an attitude (Lee and See 2004), a mainly affective response closely related to beliefs and expectations. Affective responses are not always accompanied by overt behavior. For example, students with and without math anxiety may behave the same way during a math test even though their internal state differs (McCoach et al. 2013). An affective response is, thereby, probably only completely accessible through self-report (Paulhus and Vazire 2007). A questionnaire, therefore, is an attractive method to measure trust in automation.

2 Theoretical Model

A literature review of available questionnaires on trust in automation revealed that the questionnaires comprise single-item as well as multi-item scales. Single-item scales allow a quick, uncomplicated measurement such as a dynamic assessment during an experiment. However, these instruments also have some drawbacks. Dimensions and models of trust have been extensively discussed, resulting in a variety of facets and concepts (Lee and See 2004). It is questionable whether the broadness and depth of this construct can be captured by a single questionnaire item. In contrast, multiple, heterogeneous indicators (= questionnaire items) enhance construct validity by increasing the probability of adequately identifying the construct (Eisinga et al. 2013). Consequently, Fuchs and Diamantopoulos (2009) do not recommend single-item scales if the construct in question is abstract. Likewise, a single item does not allow for a detailed analysis of the underlying reasons for a favorable or non-favorable trust score. Is the machine perceived as unreliable? Does a participant simply not trust a certain brand? It is not possible to give an answer with a single item scale. Using multiple items also helps to cancel out errors due to specificities inherent in single items, which lowers measurement error and, thereby, increases reliability (Robins et al. 2001; Diamantopoulos et al. 2012; Moosbrugger and Kelava 2012). Single-item scales are correspondingly more susceptible to unknown biases in meaning and interpretation (Hoeppner et al. 2011). For a detailed discussion on the choice between single-item and multi-item scales, readers may consider Körber (2018). Because of these drawbacks, it was decided to develop a multi-item questionnaire.

The measurement of a latent construct such as trust requires the process of construct validation (Flake et al. 2017). In the substantive phase, the literature is reviewed, the construct is defined and conceptualized, and its dimensions, boundaries, and structure are identified. For this purpose, theoretical discourses on trust in automation were screened along with empirical articles and articles with a stronger focus on interpersonal trust[1]. The most widespread and most cited model of trust is the dyadic model of organizational trust by Mayer et al. (1995). Integrating previous theoretical accounts on trust, the parsimonious model differentiates trust from its contributing factors and its outcome, risk taking in a relationship. The authors argue that trust is only necessary in a risky situation or when having something invested. In this context, they define trust as the willingness of a party to be vulnerable to the actions of another party based on the expectation that the other will perform a particular action important to the trustor, irrespective of the ability to monitor or control that other party. (Mayer et al. 1995)

According to their model, a person's trust depends on two components, a person's *individual propensity* or general willingness to trust others and the *trustworthiness* of the party to be trusted (trustee). A person's trust propensity results from different

[1] In this literature review, the following work was considered: Barber (1983), Blomqvist (1997), Butler and Cantrell (1984), Butler (1991), Deutsch (1958), Deutsch (1960), Dzindolet et al. (2001), Hoff and Bashir (2015), Hoffman et al. (2013), Jian et al. (2000), Lee and Moray (1992), Lee and See (2004), Madhavan and Wiegmann (2007), Madsen and Gregor (2000), Mayer et al. (1995), McKnight and Chervany (1996), McKnight and Chervany (2001), Muir (1987), Muir (1994), Muir and Moray (1996), Rempel et al. (1985), Rotter (1971).

developmental experiences, personality type, and cultural background and determines how much a person trusts a trustee prior to any knowledge of that particular party being available. The second component, the perceived trustworthiness, is determined by three relevant attributes of the trustee: (1) *Ability*: The level of skills, competencies, and characteristics that the trustee possesses and that enables him to have influence within a specific domain. (2) *Benevolence*: The extent to which a trustee is perceived to want to do good to the trustor and avoids egocentric motives. (3) *Integrity*: The extent to which the trustee consistently adheres to a set of principles that the trustor finds acceptable. Risk taking is then the behavioral manifestation of the willingness to be vulnerable, i.e. the outcome of trust.

Since interpersonal trust and trust in automation exhibit fundamental differences (Körber 2018), the model from Mayer et al. (1995) does not completely apply to trust in automation. Taking this into account, Lee and See (2004) follow the model of trust by Mayer et al. (1995) but fit their dimensions to the context of trust in automation. They argue that previously found bases for trust in automation can be summarized into three dimensions, *performance*, *process*, and *purpose*, which correspond to the dimensions of trustworthiness in the model by Mayer et al. (1995), as illustrated in Fig. 1. *Performance* refers to the current and previous operation of the automated system and comprises characteristics such as reliability, competency, and ability. Performance information describes what the automated system can do reliably and matches the attribute ability in Mayer et al. (1995). *Process* describes how the automated system operates and if this modus operandi is appropriate for the situation and the operator's goals. It subsumes characteristics such as understandability and matches integrity in Mayer et al. (1995). *Purpose* describes the intention in the automated system's design, the perception that the designers possess a positive orientation towards the operator, and the degree to which automation is used as intended by the designer. It corresponds to benevolence in Mayer et al. (1995). We follow the model from Lee and See (2004) but divide the three components into more detailed facets for item generation. Three underlying dimensions of trust in automation were postulated: *Reliability/Competence*, *Understandability/Predictability*, and *Intention of Developers*. Trust exhibits a stable individual component (Körber 2018). Individuals consistently vary in their general propensity to trust, depending on their developmental experiences, personality type, and cultural backgrounds. Additionally, not objective characteristics but a person's subjective perception of a system's characteristics determines trust in automation in the end (Lee and See 2004; Merritt and Ilgen 2008). We, therefore, added the individual component, *Propensity to Trust*, from the model of Mayer et al. (1995) as a moderator but also as a direct determinant of trust in automation.

The model of Mayer et al. (1995) addresses interpersonal trust. While other human individuals may be perceived more or less as individuals, different driving automation systems seem to be perceived as a single technology (Schoettle and Sivak 2014). This increases the importance of prior familiarity because trust is thereby probably not evaluated again for each driving automation system. Familiarity is assumed to have an indirect influence on trust in automation. With increasing familiarity, operators form expectations, calibrate their trust, and eventually, their confidence in the evaluation of the attributes increases (Hergeth et al. 2017). For example, if no unexpected failures occur, the confidence in the system's reliability increases. As experience with a system

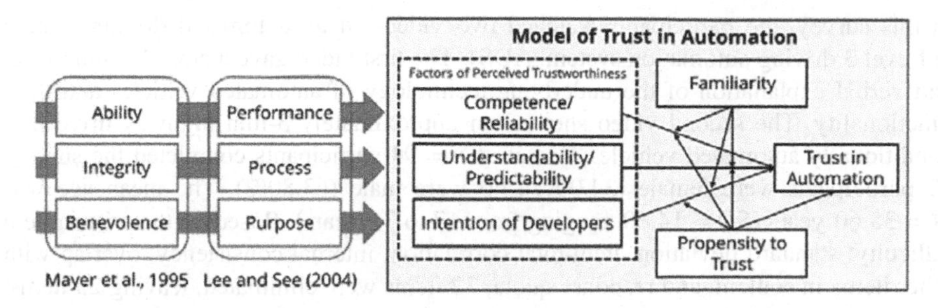

Fig. 1. Model of trust in automation based on the postulated dimensions by Mayer et al. (1995) and Lee and See (2004).

grows, trust builds up until a certain level is reached (Beggiato et al. 2015). Taking this into account, *Familiarity* with an automated system was included as a moderator in the theoretical model. Figure 1 illustrates the complete model structure. Based on Mayer et al. (1995), we define trust in automation as the attitude of a user to be willing to be vulnerable to the actions of an automated system based on the expectation that it will perform a particular action important to the user, irrespective of the ability to monitor or to intervene.

3 Item Generation and Analysis

Likert-scales are used as means of measurement. Measurement by Likert-scales is based on summative scaling, where respondents use a ranked scale to indicate their agreement with statements. The goal is to combine the single item responses of an individual to obtain a total score that represents a reliable measurement – multiple Likert-type items form one coherent Likert scale (Uebersax 2006; Hubley and Zumbo 2013). A 5-point rating scale ranging between 1 (= strongly disagree) to 5 (= strongly agree) was chosen as the response format. Rating scales with a very fine-grained range, for example from 1 to 100 as in Brown and Galster (2004), offer a resolution that might be inadequate for the provided precision of the measurement, resulting in merely artificial precision. Furthermore, the self-report of trust is based on introspection. It is questionable whether the participants are able to access their trust by introspection with such a granularity as provided by the scale. Such a fine-grained scale might map an empirical structure, which does not exist in this resolution, onto numbers with limited meaning. If such scales provide no anchor points, measurement at interval scale level is also even more problematic since equidistance between the rating scale points becomes even more questionable.

We followed a deductive approach for the generation of items (Burisch 1978) and constructed the questionnaire based on classical test theory (Moosbrugger and Kelava 2012). An initial set of 57 items was generated. Approximately one third of the items was inversely formulated to reduce response bias (e.g., acquiescence bias) and based on Likert's notion that someone with a positive attitude about the object should also disagree with negative statements. An online survey was conducted for item analysis.

In this survey, the participants watched two videos of an automated driving system (a Level 3 driving automation system; ADS). The first video gave a circa 10-min visual and verbal explanation of the underlying technology of automated vehicles and their functionality. The second video showed an approximately 3-min highway drive in a conditionally automated vehicle. A total of $n = 94$ participants completed the survey, 32 participants were female (34.00%), 60 were male (63.80%). The mean age was $M = 35.60$ years ($SD = 14.60$, ranging from 17 to 71 years). Based on the criteria item difficulty, standard deviation, item-total correlation, internal consistency, overlap with other items in content, and response quote, 32 items were eliminated, leaving 25 items.

The first validation was carried out in a subsequent online study. In a between-subjects design, a sample of $n = 58$ participants (age range 17 to 72, mean age $M = 34.00$ years, $SD = 15.10$, 58.60% male, 37.90% female) watched a video of a conditionally automated highway drive. Participants were randomly assigned to a *reliable* condition, where the video showed a perfectly functioning automation, or a *non-reliable* condition, where participants watched an extended version including a take-over request. As expected, participants of the reliable condition rated the ADS more reliable ($t(41.32) = 3.76$, $p < .001$, $d = 1.05$). Additionally, participants rated their trust directly by answering the item "I trust this system" on a 5-point rating scale ranging between 1 (= strongly disagree) to 5 (= strongly agree). All scales correlated positively with different strength with this rating (lowest: *Familiarity*: $r = .33$; highest: *Reliability*: $r = .85$). Although the total questionnaire correlated strongly with this item ($r = .81$), we found no significant difference between the two conditions ($t(46.92) = 1.21$, $p = .23$, $d = 0.33$), on the contrary for the direct question ($t(45.63) = 2.58$, $p = .01$, $d = 0.71$). Because of their high correlation, the scales competence and reliability were merged, leading to a reduction to 17 items. The internal consistency of the scales ranged from acceptable ($\alpha = .75$; *Propensity to trust*) to excellent ($\alpha = .92$; *Reliability/Competence*).

McCoach et al. (2013) recommend utilizing an exploratory factor analysis (EFA) to evaluate the structure in the very first pilot study because it allows for the highest flexibility of potential solutions. An exploratory factor analysis was conducted to assess whether the structure of the covariation among items is consistent with the proposed factor structure of the trust model. The analysis was performed in JASP (Love et al. 2015). The dataset showed a sufficient basis to conduct an initial exploratory factor analysis (KMO = .80, Bartlett-Test: $\chi^2(136) = 418.81$, $p < .001$). Following the recommendations of Sakaluk and Short (2017) and McCoach et al. (2013), we chose principal axis factoring as the extraction method and oblique rotation (oblimin) to make the factor solution more interpretable. Parallel analysis by Horn (1965) as well as multiple item factor loadings >.40 on only one single factor determined the extracted factors (Fig. 2). Results of the analysis provide initial support for the assumed factorial structure. The resulting pattern matrix (Table 1) shows a clear structure of four factors with high over-determination, "the degree to which each factor is clearly represented by a sufficient number of variables" (MacCallum et al. 1999). Each factor exhibits high pattern coefficients (>.50) by multiple variables while each of the items does not load substantially (>.35) onto other factors, a requirement for a stable solution. Medium to high communalities were observed. Tables 2 and 3 provide further information on the resulting solution.

Table 1. Pattern matrix generated by principal axis factoring; loadings <.35 have been omitted.

	Factor 1	Factor 2	Factor 3	Factor 4	Uniqueness
Familiarity 1			.81		.31
Familiarity 2			.80		.34
Intention of Developers 1		.74			.46
Intention of Developers 2		.49			.45
Propensity to Trust 1				.58	.60
Propensity to Trust 2				.55	.36
Propensity to Trust 3				.59	.55
Reliability/Competence 1	.88				.15
Reliability/Competence 2	.70				.34
Reliability/Competence 3	.79				.23
Reliability/Competence 4	.82				.30
Reliability/Competence 5	.86				.28
Reliability/Competence 6	.70				.44
Understanding/Predictability 1		.65			.36
Understanding/Predictability 2		.60			.44
Understanding/Predictability 3	.64				.24
Understanding/Predictability 4		.62			.50

Table 2. Inter-correlations matrix of the extracted factors.

	Factor 1	Factor 2	Factor 3	Factor 4
Factor 1				
Factor 2	.65			
Factor 3	.25	.24		
Factor 4	.19	.31	.04	

Table 3. Fit indices of the resulting model.

Chi-squared test			Additional fit indices	
Value	*df*	*p*	RMSEA	TLI
112.14	74	.003	0.09 [0.05, 0.11]	.85

McNeish (2017) advises against using Cronbach's alpha as a reliability index because its rigid assumptions are routinely violated. He suggests using the omega coefficient, which is conceptually similar to Cronbach's alpha but makes less strict assumptions. In fact, omega total is a more general version of Cronbach's alpha: It also assumes unidimensionality, but the items are allowed to vary in how strongly they are related to the measured construct. Revelle's omega differs from omega total in its more sophisticated variance decomposition. Given that the items each implement a 5-point rating scale, relying on Pearson covariance matrices is reasonable (Rhemtulla et al. 2012). All scales exhibited good to excellent internal consistency (Table 4).

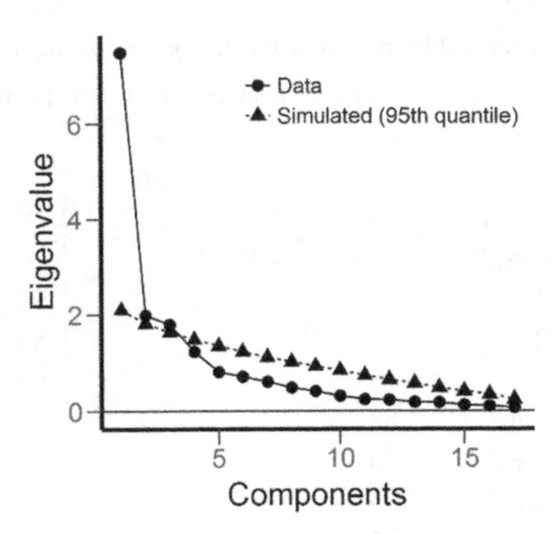

Fig. 2. Scree plot of the extracted factors with a parallel analysis by Horn (1965) superimposed.

Table 4. Indices of the internal consistency of each scale; [a] since Omega total and Revelle's omega cannot be calculated for scales with fewer than three items, the Spearman-Brown coefficient according to Eisinga et al. (2013) was calculated.

	Omega Total	Revelle's Omega
Familiarity	.83[a]	-
Intention of Developers	.79[a]	-
Propensity to Trust	.78	.77
Reliability/Competence	.92	.95
Understanding	.81	.88

The factor Reliability/Competence was the first extracted factor and, therefore, explained a very major part of the variance, which may be expected given the design of the study, i.e. automation reliability was manipulated between the conditions. However, no factor for *Intention of the Developers* could be extracted. The reason for this may lie in the domain of automated driving. A driving automation system is an expensive, highly sophisticated system whose development was motivated by the increase in safety and comfort. The developers of the system are known to be professional car manufacturers. Thus, it is hard to imagine that a driving automation system's developers did not act in a benevolent manner. A revised version of the questionnaire may eliminate this dimension, at least in the domain of automated driving. Item 3 of Understanding ("The system state was always clear to me") seems to exhibit a certain degree of multidimensionality and may also be eliminated if this again is the case in future analyses.

Although the aim was to conduct an EFA, fit indices for the model, known from confirmatory factor analysis (CFA), are also reported (Table 3). Fit indices indicate how well the empirical data of the study actually conform to the proposed model. A CFA, therefore, is a more stringent test if the pattern of relationships among the items can be explained by the proposed model/factor structure (McCoach et al. 2013).

The chi-squared test evaluates the null hypothesis that the proposed model exactly reproduces the population covariance matrix implied by the data (McCoach et al. 2013). This null hypothesis has to be rejected for the four-factor model. Besides the chi-squared test is generally too liberal at small samples sizes, as in this study, the informative value of this rejection is limited by the fact that a model is always a simplification of a process in reality that never intends to exactly recreate it (McCoach 2003). The root mean square error of approximation (RMSEA) is an index of absolute fit that compensates for the effect of model complexity (Hu and Bentler 1999) and can be considered an estimate of the misfit of the model per degree of freedom in the population (Preacher et al. 2013). Cut-offs for small sample sizes ($N \leq 250$) are .08 for a mediocre fit whereas .10 and larger indicates a poor fit (MacCallum et al. 1996; Hu and Bentler 1999; Heene et al. 2011), indicating a mediocre fit for the four-factor structure of the trust model. However, the estimate is positively biased and the amount of the bias depends on the smallness of the sample and the degrees of freedom (Kenny et al. 2014). The Tucker-Lewis Index (TLI) indicates an incremental fit and also compensates for model complexity. A TLI value at or above .95 indicates a good fit, TLI values below .90 are generally considered less than satisfactory (McCoach 2003). The four-factor model does not fulfill this criterion. However, once again, the TLI is biased in small samples, i.e. it is underestimated in samples with fewer than 100 participants. Heene et al. (2011) echoe previous critique on the application of fixed cut-off rules for model fit because of the multiple dependencies of the fit indices on the conditions (e.g., the achieved factor loadings) and on sample size. After establishing the trust model, two items for measurement of trust in automation itself ("I trust the system" and "I can rely on the system") forming the subscale *Trust in Automation* were added.

The EFA gathered sufficient preliminary evidence of the factor structure and shows that further pursuit of the model is reasonable. Nevertheless, this analysis of construct validity is certainly not sufficient. Firstly, the sample size of $n = 58$ participants results in a case/item ratio of approximately 3:1, which reflects the absolute minimum for a sensible analysis and may be too small to produce a stable solution. However, the minimum required ratio is not constant across studies but rather depends on aspects of the variables and study design (MacCallum et al. 1999). Given a clear factor structure, a high degree of over-determination and high communalities (constantly >.60, as in this study), it is nevertheless possible to reach a stable factor solution even with a sample size smaller than 100 participants (McCoach et al. 2013). Secondly, the participants did not experience driving automation themselves but watched videos of it. Thirdly, the participants only got a short, probably first impression of a driving automation system. This may promote a single-factor structure because the participants might not have had enough insight into the driving automation system to form themselves a detailed, multifaceted impression.

The results of the initial exploratory factor analysis established sufficient initial evidence for the factor structure, affirming that further work is sensible but also needed. Thus, the development process for the questionnaire has certainly not yet come to its end. Future studies have to investigate and ensure the construct validity in greater detail and need to investigate the structure in an applied setting with an adequate sample size. Future work should also follow up this analysis with a CFA to put the established structure to a more rigorous test. In a structural equation model, the claimed paths and relationships of the model can be directly tested and different models can be compared. The questionnaire's criterion validity was examined in its first use in a driving simulator study in Körber et al. (2018).

4 Predictive Validity

In the study by Körber et al. (2018), the developed questionnaire to measure trust in automation was used in an applied setting for the first time. In this driving simulator study, 40 participants encountered three critical situations while driving in a conditionally automated vehicle (SAE Level 3) on a highway while being engaged in a non-driving-related task. Eye tracking was used to assess how much the participants rely on driving automation. Furthermore, the instruction for the ADS was varied between two groups with participants receiving either trust-promoting (*Trust promoted* group) or trust-lowering (*Trust lowered* group) introductory information. The trust questionnaire was administered three times: (1) after an introductory video, (2) after an introductory drive, (3) after the experimental drive. It was expected that, firstly, self-reported trust will correlate positively with reliance on automation and, secondly, that participants of the *Trust promoted* group will report higher trust than the *Trust lowered* group.

The analysis comprised the whole *Trust in Automation Questionnaire* (TiA; 19 items) as well as just the subscale *Trust in Automation* and the subscale *Competence*. Regarding the reliability of the *Trust in Automation* subscale, the drawbacks of short scales become eminent. The scale exhibits a low reliability of $\alpha = .63$ after the video and of $\alpha = .70$ after the introductory drive, while it achieved a high reliability of $\alpha = .85$ after the experimental drive. This reflects the problems mentioned earlier with single-item scales: They are more vulnerable to random measurement errors and more susceptible to unknown biases in meaning and interpretation (Emons et al. 2007; Hoeppner et al. 2011). Nevertheless, the subscale *Trust in Automation* was the scale that showed the largest difference ($M_{\text{diff}} = 0.45$, $d = 0.59$, $BF_{-0} = 4.35$) between the two groups after the introductory drive. The subscale might be more sensitive than the whole questionnaire, but this does not guarantee that its predictive performance regarding trust in other systems is superior – predictive quality might vary in different situations and context. The experiment included two situations (Situation 1: overtaking maneuver; Situation 2: adapting speed to a headway vehicle) that were solved by the automated vehicle, but a take-over was a reasonable action if one does not trust automation. In both situations, participants who intervened showed lower trust than participants who did not intervene. The effect size was comparable between the full TiA questionnaire (Situation 1: $d = 0.41$, Situation 2: $d = 0.51$) and the subscale *Trust in Automation* (Situation 1: $d = 0.50$, Situation 2: $d = 0.45$). The same results were obtained for the take-over situation, where participants who crashed reported higher trust than collision-free participants (Full TiA: $d = 0.51$; subscale *Trust in Automation*: $d = 0.58$). Both scales correlated moderately with take-over time (Full TiA: $r = .27$; subscale *Trust in Automation*: $r = .33$) and minimum time-to-collision (Full TiA: $r = -.29$; subscale *Trust in Automation*: $r = -.35$). Both full questionnaire and subscale *Trust in Automation* correlated with the participants' gaze behavior with the expected sign and at approximately the same magnitude (medium effect) in all three measurement intervals.

In summary, participants with higher trust scores consistently showed stronger reliance in all behavioral measurements compared to participants with a lower trust score. Consequently, the study confirms the predictive validity of the questionnaire. Furthermore, the medium-sized correlation between the TiA questionnaire score and the affinity for technology questionnaire (Feuerberg et al. 2005) of $r = .47$ (BF = 18.85) shows that trust is related to affinity for technology, yet it represents a distinct construct, supporting its construct validity.

5 Is a Single Item Enough to Measure Trust in Automation?

The two-item subscale *Trust in Automation* showed lower reliability but was more sensitive regarding group differences and performed equally as well as the full TiA questionnaire regarding all other measures. This provokes the question of whether a single-item scale may be sufficient for a valid measurement of trust. The benefits of using single-item measures have been listed by several researchers (Fuchs and Diamantopoulos 2009; Hoeppner et al. 2011): Single-item scales are less monotonous and time-consuming. They can also be administered during an experiment for a momentary assessment, for example while driving. The aforementioned advantages of multi-item scales are also accompanied by drawbacks, such as boredom caused by redundant items and fatigue in lengthy questionnaires (Burisch 1984). Nevertheless, for a detailed assessment of a multidimensional construct such as trust in automation, a multi-item measure is typically necessary (Nunnally and Bernstein 1994).

Yet, Fuchs and Diamantopoulos (2009) argue that the use of a single-item scale may still be appropriate in certain cases. For example, Sloan et al. (2002), while discussing the quality of life measurement, claim that "there comes a point where the construct becomes so complex that a single question may be the best approach" (p. 481). Hence, when measuring overall job satisfaction, the best measurement may be a question like "Overall, how satisfied are you with your job?" (Scarpello and Campbell 1983; Fuchs and Diamantopoulos 2009). A single item on trust in automation reflects the conceptualization of trust in automation as a mainly affective response with influences from analytic and analogical processes. Lee and See (2004) suggest that because of the complexity of automation technology, operators probably rely less on analytic calculations to guide their behavior but rather apply heuristics to accommodate the limits of the human bounded rationality (Gigerenzer and Selten 2002). A situation might occur where operators cannot form a complete mental model of an automated system as it is too complex to perfectly predict its behavior. Emotions can then guide behavior when rules are not effective or when cognitive resources are too limited for a calculated rational choice (Damasio 1996; Lee and See 2004). In the validation study, 78% of the participants have had no contact with conditionally automated driving before. Thus, it might not have been possible for the participants to rate each dimension of the trust questionnaire adequately because of a lack of knowledge or experience. Differences in the ability to accurately rate a system have been pointed out by Annett (2002) who gives the example of expert test drivers who learn by experience to identify and rate the subtle dynamic features of a vehicle. It is conceivable that the participants' trust rating was a rather global impression or rating, which can be captured accurately by a single item. It is unclear if participants would also provide a global rating if they had more experience with an automated vehicle.

Yet, such a simplification of the construct trust in automation comes with a cost: The *Trust in Automation* scale consists of two items, one of them with the content "I can rely on the system". It is not surprising that such a measure highly correlates with behavioral reliance measures such as eye tracking and intervention frequency. For such a narrow conceptualization of trust, the high validity may justify the use of a single-item measure (Flake et al. 2017). The construct trust in automation, which possesses a detailed underlying theory (Mayer et al. 1995; Lee and See 2004) would then, at the same time, become one with its measure and loses any theoretical meaning beyond that measure (Bagozzi 1982). This measurement would then be in conflict with the definition of what it intends to measure. Indeed, as already mentioned, trust is an attitude that stands between the belief about characteristics of an automated system and the intention to rely on it. Attitude, intention, and actual behavior are not in a deterministic but in a probabilistic relationship (Ajzen and Fishbein 1980). Whether trust translates into actual reliance on an automated system is also influenced by other factors such as self-confidence or time constraints (Dzindolet et al. 2001; Lee and See 2004; Meyer 2004).

Besides the psychometric drawbacks mentioned in Körber (2018), the use of a single-item measure is also problematic in longitudinal studies: If the observed value changes, it is not possible to differentiate between a true change in the construct and a change caused by imperfect reliability of the measurement (Fuchs and Diamantopoulos 2009). Here, researchers may fall back on the multi-item questionnaire. If a single-item is administered to obtain a global assessment, it has to be taken into account that the respondents each consider an individual set of aspects of trust and of the automated system, weighted by their own individual preferences, providing a tailor-made impression (Nagy 2002). Hence, respondents may not consider the same aspects or may not even think of a relevant aspect at all. It, thereby, remains unknown how the assessment is constituted. To ensure that each participant assesses the same construct, i.e. that a common understanding of trust exists, an accurate definition of trust in automation has to be provided in this case (Fuchs and Diamantopoulos 2009). On the other hand, multi-item scales are less individual but more comparable. A preset of aspects, formed by the questionnaire's scales, also helps and guides the participants to rate the system.

Multiple scales also provide the possibility to express the trust rating in greater detail. With a single-item scale, should the trust score turn out to be low, the researchers then have no indication for the reason. Contrarily, multiple scales may enable researchers to find the cause in a certain characteristic of the automated system. For example, it could be perceived as reliable, but participants did not understand its functioning. Thus, it is reasonable to use a multiple-item scale such as the TiA if the aim is a thorough, multi-faceted assessment.

In conclusion, if the research objective is a global assessment, an overall feeling, or impression by the participants, then a single-item may provide all the desired information (Fuchs and Diamantopoulos 2009). It represents a useful supplement that might be sufficient for a single and quick, yet valid assessment and "can provide an acceptable balance between practical needs and psychometric concerns" (Robins et al. 2001). This is particularly true if trust is merely used as a moderator or as a control variable (Fuchs and Diamantopoulos 2009). If the goal is a detailed assessment of trust in automation or if a longitudinal design is implemented, then the multi-item questionnaire may be preferred.

6 Objectives for Revision and Further Development

The questionnaire's further development certainly needs to address its psychometric qualities. The low internal consistency of the subscale *Trust in Automation* at the beginning of the study raises the question of whether a short scale of two rather direct items is sufficient as a measurement of trust itself. Mayer and Davis (1999) provide a questionnaire for their model of interpersonal trust, which includes a four-item scale to assess trust. The items are less direct than the two trust items of the TiA questionnaire and rather aim at the willingness to be vulnerable, corresponding to their definition of trust (Mayer et al. 1995). Thus, a revised version of the TiA questionnaire may adopt this approach and offer a four-item scale (besides the original scales) for trust in automation that is closer to its definition by Körber et al. (2018). Items from Mayer and Davis (1999) adapted to the domain of automation could, for example, read "I would be comfortable handing over the driving task to the driving automation system without monitoring it" or "If I had my way, I wouldn't let a driving automation system have any influence on the driving task". A single item for assessing trust in automation such as "I trust this driving automation system" then may function as the aforementioned pragmatic variant alongside the multi-item questionnaire. In addition, information on the questionnaire's discriminant validity is still missing. Also, further data on the questionnaire's predictive performance have to be gathered.

A revision may also reconsider the inclusion of the scale *Familiarity*. Familiarity itself is not an element of trust in automation but indirectly influences it as a moderator. With increasing familiarity, operators form expectations and the confidence in their evaluation of the attributes increases. If this moderating role is of no interest in a study, the scale could be eliminated to shorten the questionnaire. A core questionnaire only containing the factors that directly influence trust then may be more appropriate. Beyond this, familiarity could also induce response bias: Low familiarity with an automated system could induce a tendency towards a global evaluation of the system due to a lack of in-depth knowledge. It would, therefore, be interesting to administer the questionnaire to participants who are already very familiar with a driving automation system. This is especially of interest regarding the difference between the predictive performance of a single-item measure and the multi-item TiA questionnaire.

In closing, it has to be considered that the measurement of trust in automation by means of a questionnaire certainly has to be viewed in perspective of its position in measurement theory. There have been concerns doubting the possibility of measurement of psychological constructs and their quantitative nature in general (Michell 1997). However, using rating scales for the measurement of psychological constructs, such as trust, does not exclusively have to be regarded as a form of measurement in the strict sense of the term, i.e. in terms of the representational theory of measurement, where a homomorphic representation of physical empirical relations is mapped to numerical relations (Annett 2002; Krantz et al. 2007). Instead, following a model-based account of measurement, measurement of trust can rely on an abstract model that is valid for the prediction of an individual's performance during a certain task (Tal 2017). As Tal (2017) argues, such a model is defined by theoretical and statistical assumptions about the measured psychological construct and its relation to the measurement task. Inference from the indication of a measurement instrument (e.g., a rating scale) to the measurement outcome is non-trivially derived from the model. Measurement is then the coherent and consistent assignment of values to parameters in this model, based on instrument indications. The model defines the content of the measurement outcome, which does not have to hold a counterpart in the observable world – a construct, in the end, is a concept, model, or schematic idea (McCoach et al. 2013). As for the measurement of intelligence, the values do not represent physical properties but empirical relationships between theoretical constructs and other constructs or behavior (Annett 2002). Trust measurement, thus, may not deliver meaningful, absolute values per se but values that are meaningful in the context of a model of trust, which is defined by theoretical and statistical assumptions such as confirmed construct validity. In this way, the measurement outcome can be used to predict and explain behavior, decisions, or performance. For this reason, it is unreasonable to apply the same standards to the measurement of trust as to measurements such as take-over time. Nevertheless, the results of Körber et al. (2018) show that the questionnaire produces meaningful measures with relation to observable and safety-relevant behavior.

References

Ajzen I, Fishbein M (1980) Understanding attitudes and predicting social behavior. Prentice Hall, Englewood Cliffs

Annett J (2002) Subjective rating scales: science or art? Ergonomics 45:966–987. https://doi.org/10.1080/00140130210166951

Bagozzi RP (1982) The role of measurement in theory construction and hypothesis testing: toward a holistic model. In: Fornell C (ed) A second generation of multivariate analysis. Praeger, New York, pp 5–23

Barber B (1983) The logic and limits of trust. Rutgers University Press, New Brunswick

Beggiato M, Pereira M, Petzoldt T, Krems JF (2015) Learning and development of trust, acceptance and the mental model of ACC. A longitudinal on-road study. Transp Res Part F Traffic Psychol Behav 35:75–84. https://doi.org/10.1016/j.trf.2015.10.005

Bisantz AM, Seong Y (2001) Assessment of operator trust in and utilization of automated decision-aids under different framing conditions. Int J Ind Ergon 28:85–97. https://doi.org/10.1016/S0169-8141(01)00015-4

Blomqvist K (1997) The many faces of trust. Scand J Manag 13:271–286. https://doi.org/10.1016/S0956-5221(97)84644-1

Brown RD, Galster SM (2004) Effects of reliable and unreliable automation on subjective measures of mental workload, situation awareness, trust and confidence in a dynamic flight task. In: Proceedings of the human factors and ergonomics society annual meeting 2004, pp 147–151

Burisch M (1978) Construction strategies for multiscale personality inventories. Appl Psychol Measur 2:97–111. https://doi.org/10.1177/014662167800200110
Burisch M (1984) Approaches to personality inventory construction: a comparison of merits. Am Psychol 39:214–227. https://doi.org/10.1037/0003-066X.39.3.214
Butler JK, Cantrell RS (1984) A behavioral decision theory approach to modeling dyadic trust in superiors and subordinates. Psychol Rep 55:19–28. https://doi.org/10.2466/pr0.1984.55.1.19
Butler JK (1991) Toward understanding and measuring conditions of trust: evolution of a conditions of trust inventory. J Manag 17:643–663. https://doi.org/10.1177/014920639101700307
Damasio AR (1996) The somatic marker hypothesis and the possible functions of the prefrontal cortex. Philos Trans R Soc Lond B Biol Sci 351:1413–1420. https://doi.org/10.1098/rstb.1996.0125
Deutsch M (1958) Trust and suspicion. J Conflict Resolut 2:265–279. https://doi.org/10.1177/002200275800200401
Deutsch M (1960) The effect of motivational orientation upon trust and suspicion. Hum Relations 13:123–139. https://doi.org/10.1177/001872676001300202
Diamantopoulos A, Sarstedt M, Fuchs C, Wilczynski P, Kaiser S (2012) Guidelines for choosing between multi-item and single-item scales for construct measurement: a predictive validity perspective. J Acad Mark Sci 40:434–449. https://doi.org/10.1007/s11747-011-0300-3
Drnec K, Marathe AR, Lukos JR, Metcalfe JS (2016) From trust in automation to decision neuroscience: applying cognitive neuroscience methods to understand and improve interaction decisions involved in human automation interaction. Front Hum Neurosci 10:54. https://doi.org/10.3389/fnhum.2016.00290
Eisinga R, Grotenhuis MT, Pelzer B (2013) The reliability of a two-item scale: Pearson, Cronbach, or Spearman-Brown? Int J Public Health 58:637–642. https://doi.org/10.1007/s00038-012-0416-3
Emons WHM, Sijtsma K, Meijer RR (2007) On the consistency of individual classification using short scales. Psychol Methods 12:105–120. https://doi.org/10.1037/1082-989X.12.1.105
Feuerberg BV, Bahner JE, Manzey D (2005) Interindividuelle unterschiede im umgang mit automation – entwicklung eines fragebogens zur erfassung des complacency-potentials. In: Urbas L, Steffens C (eds) Zustandserkennung und systemgestaltung. 6. Berliner werkstatt mensch-maschine-systeme., Als Ms. gedr. VDI-Verlag, Düsseldorf, pp 199–202
Flake JK, Pek J, Hehman E (2017) Construct validation in social and personality research. Soc Psychol Pers Sci 8:370–378. https://doi.org/10.1177/1948550617693063
Fuchs C, Diamantopoulos A (2009) Using single-item measures for construct measurement in management research: conceptual issues and application guidelines. Die Betriebswirtschaft 69:195
Gigerenzer G, Selten R (eds) (2002) Bounded rationality: the adaptive toolbox. MIT Press, Cambridge
Heene M, Hilbert S, Draxler C, Ziegler M, Bühner M (2011) Masking misfit in confirmatory factor analysis by increasing unique variances: a cautionary note on the usefulness of cutoff values of fit indices. Psychol Methods 16:319–336. https://doi.org/10.1037/a0024917
Hergeth S, Lorenz L, Vilimek R, Krems JF (2016) Keep your scanners peeled: gaze behavior as a measure of automation trust during highly automated driving. Hum Factors 58:509–519. https://doi.org/10.1177/0018720815625744
Hergeth S, Lorenz L, Krems JF (2017) Prior familiarization with takeover requests affects drivers' takeover performance and automation trust. Hum Factors 59:457–470. https://doi.org/10.1177/0018720816678714

Hoeppner BB, Kelly JF, Urbanoski KA, Slaymaker V (2011) Comparative utility of a single-item versus multiple-item measure of self-efficacy in predicting relapse among young adults. J Subst Abuse Treat 41:305–312. https://doi.org/10.1016/j.jsat.2011.04.005
Hoff KA, Bashir M (2015) Trust in automation: integrating empirical evidence on factors that influence trust. Hum Factors 57:407–434. https://doi.org/10.1177/0018720814547570
Hoffman RR, Johnson M, Bradshaw JM, Underbrink A (2013) Trust in automation. IEEE Intell Syst 28:84–88. https://doi.org/10.1109/MIS.2013.24
Horn JL (1965) A rationale and test for the number of factors in factor analysis. Psychometrika 30:179–185. https://doi.org/10.1007/BF02289447
Hu L-T, Bentler PM (1999) Cutoff criteria for fit indexes in covariance structure analysis: conventional criteria versus new alternatives. Struct Eq Model Multidiscip J 6:1–55. https://doi.org/10.1080/10705519909540118
Hubley AM, Zumbo BD (2013) Psychometric characteristics of assessment procedures: An overview. In: Geisinger KF (ed) Test theory and testing and assessment in industrial and organizational psychology, 1st edn. American Psychological Association, Washington, pp 3–20
Jian J-Y, Bisantz AM, Drury CG (2000) Foundations for an empirically determined scale of trust in automated systems. Int J Cogn Ergon 4:53–71. https://doi.org/10.1207/S15327566IJCE0401_04
Kenny DA, Kaniskan B, McCoach DB (2014) The performance of RMSEA in models with small degrees of freedom. Soc Methods Res 44:486–507. https://doi.org/10.1177/0049124114543236
Kirlik A (1993) Modeling strategic behavior in human-automation interaction: why an "aid" can (and should) go unused. Hum Factors 35:221–242. https://doi.org/10.1177/001872089303500203
Körber M (2018) Individual differences in human-automation interaction: a driver-centered perspective on the introduction of automated vehicles. Dissertation, Technical University of Munich
Körber M, Baseler E, Bengler K (2018) Introduction matters: manipulating trust in automation and reliance in automated driving. Appl Ergon 66:18–31. https://doi.org/10.1016/j.apergo.2017.07.006
Krantz DH, Luce RD, Suppes P, Tversky A (2007) Additive and polynomial representations. Foundations of measurement, vol 1. Dover Publisher, Mineola
Lee JD, Moray N (1992) Trust, control strategies and allocation of function in human-machine systems. Ergonomics 35:1243–1270. https://doi.org/10.1080/00140139208967392
Lee JD, See KA (2004) Trust in automation: designing for appropriate reliance. Hum Factors 46:50–80. https://doi.org/10.1518/hfes.46.1.50_30392
Love J, Selker R, Verhagen J, Marsman M, Gronau QF, Jamil T, Šmíra M, Epskamp S, Wild A, Ly A, Matzke D, Wagenmakers E-J, Morey RD, Rouder JN (2015) Software to sharpen your stats. APS Obs 28:27–29
MacCallum RC, Browne MW, Sugawara HM (1996) Power analysis and determination of sample size for covariance structure modeling. Psychol Methods 1:130–149. https://doi.org/10.1037/1082-989X.1.2.130
MacCallum RC, Widaman KF, Zhang S, Hong S (1999) Sample size in factor analysis. Psychol Methods 4:84–99. https://doi.org/10.1037/1082-989X.4.1.84
Madhavan P, Wiegmann DA (2007) Similarities and differences between human–human and human–automation trust: an integrative review. Theor Issues Ergon Sci 8:277–301. https://doi.org/10.1080/14639220500337708
Madsen M, Gregor S (2000) Measuring human-computer trust. In: Proceedings of the 11th Australasian conference on information systems, pp 6–8

Mayer RC, Davis JH, Schoorman FD (1995) An integrative model of organizational trust. Acad Manag Rev 20:709–734. https://doi.org/10.5465/AMR.1995.9508080335
Mayer RC, Davis JH (1999) The effect of the performance appraisal system on trust for management: a field quasi-experiment. J Appl Psychol 84:123–136. https://doi.org/10.1037//0021-9010.84.1.123
McCoach DB (2003) SEM isn't just the Schoolwide enrichment model anymore: structural equation modeling (SEM) in gifted education. J Educ Gifted 27:36–61. https://doi.org/10.1177/016235320302700104
McCoach DB, Gable RK, Madura JP (2013) Instrument development in the affective domain: School and corporate applications, 3rd edn. Springer, New York
McKnight DH, Chervany NL (2001) Trust and distrust definitions: one bite at a time. In: Falcone R, Singh M, Tan Y-H (eds) Trust in cyber-societies: integrating the human and artificial perspectives. Springer, Heidelberg, pp 27–54
McNeish D (2017) Thanks coefficient alpha, we'll take it from here. Psychol Methods. https://doi.org/10.1037/met0000144
Merritt SM, Ilgen DR (2008) Not all trust is created equal: dispositional and history-based trust in human-automation interactions. Hum Factors 50:194–210. https://doi.org/10.1518/001872008X288574
Meyer J (2004) Conceptual issues in the study of dynamic hazard warnings. Hum Factors 46:196–204
Michell J (1997) Quantitative science and the definition of measurement in psychology. Br J Psychol 88:355–383. https://doi.org/10.1111/j.2044-8295.1997.tb02641.x
Moosbrugger H, Kelava A (eds) (2012) Testtheorie und fragebogenkonstruktion, 2nd edn. Springer, Heidelberg
Muir BM (1987) Trust between humans and machines, and the design of decision aids. Int J Man Mach Stud 27:527–539. https://doi.org/10.1016/S0020-7373(87)80013-5
Muir BM (1994) Trust in automation: Part I. Theoretical issues in the study of trust and human intervention in automated systems. Ergonomics 37:1905–1922. https://doi.org/10.1080/00140139408964957
Muir BM, Moray N (1996) Trust in automation. Part II. Experimental studies of trust and human intervention in a process control simulation. Ergonomics 39:429–460. https://doi.org/10.1080/00140139608964474
Nagy MS (2002) Using a single-item approach to measure facet job satisfaction. J Occup Organ Psychol 75:77–86. https://doi.org/10.1348/096317902167658
Nunnally JC, Bernstein IH (1994) Psychometric theory, 3rd edn. McGraw-Hill, New York
Parasuraman R, Sheridan TB, Wickens CD (2000) A model for types and levels of human interaction with automation. IEEE Trans Syst Man Cybern 30:286–297. https://doi.org/10.1109/3468.844354
Paulhus DL, Vazire S (2007) The self-report method. In: Robins RW, Fraley RC, Krueger RF (eds) Handbook of research methods in personality psychology. The Guilford Press, New York, pp 224–239
Preacher KJ, Zhang G, Kim C, Mels G (2013) Choosing the optimal number of factors in exploratory factor analysis: a model selection perspective. Multivar Behav Res 48:28–56. https://doi.org/10.1080/00273171.2012.710386
Rempel JK, Holmes JG, Zanna MP (1985) Trust in close relationships. J Pers Soc Psychol 49:95–112. https://doi.org/10.1037/0022-3514.49.1.95
Rhemtulla M, Brosseau-Liard PÉ, Savalei V (2012) When can categorical variables be treated as continuous? A comparison of robust continuous and categorical SEM estimation methods under suboptimal conditions. Psychol Methods 17:354–373. https://doi.org/10.1037/a0029315

Robins RW, Hendin HM, Trzesniewski KH (2001) Measuring global self-esteem: construct validation of a single-item measure and the Rosenberg self-esteem scale. Pers Soc Psychol Bull 27:151–161. https://doi.org/10.1177/0146167201272002

Rotter JB (1971) Generalized expectancies for interpersonal trust. Am Psychol 26:443–452. https://doi.org/10.1037/h0031464

Sakaluk JK, Short SD (2017) A methodological review of exploratory factor analysis in sexuality research: used practices, best practices, and data analysis resources. J Sex Res 54:1–9. https://doi.org/10.1080/00224499.2015.1137538

Scarpello V, Campbell JP (1983) Job satisfaction: are all the parts there? Pers Psychol 36:577–600. https://doi.org/10.1111/j.1744-6570.1983.tb02236.x

Sloan JA, Aaronson N, Cappelleri JC, Fairclough DL, Varricchio C (2002) Assessing the clinical significance of single items relative to summated scores. Mayo Clin Proc 77:479–487. https://doi.org/10.4065/77.5.479

Tal E (2017) Measurement in science. In: Zalta EN (ed) The Stanford encyclopedia of philosophy. Metaphysics Research Lab, Stanford University

Uebersax JS (2006) Likert scales: dispelling the confusion. http://www.john-uebersax.com/stat/likert.htm. Accessed 8 Feb 2018

Analysis of the Possible Savings of Cycle Time on Tram Line No. 17 in Zagreb

Davor Sumpor[1(✉)], Tanja Jurčević Lulić[2], Marko Slavulj[1], and Sandro Tokić[1]

[1] Faculty of Transport and Traffic Sciences, University of Zagreb, Vukelićeva 4, 10000 Zagreb, Croatia
{davor.sumpor,marko.slavulj}@fpz.hr, sandro.tokic@gmail.com

[2] Faculty of Mechanical Engineering and Naval Architecture, University of Zagreb, Ivana Lučića 5, 10002 Zagreb, Croatia
tanja.jurcevic@fsb.hr

Abstract. The cycle time on tram line T is theoretically defined as a sum of the total operating (travel) time on line T_o and the total time at terminals t_t. In order to ensure the increasing of the operating and cycle speeds of trams in Zagreb, as well as increasing the gross transport work of a single line with the same number of trams and drivers in one shift, this paper has explored a scenario for simulating cycle time savings by reducing the total operating time, both achieved by the possible implementation of tram priority via traffic lights based on the local automation. Possible realistic saving of the total operating time ΔT_o on line No. 17 in Zagreb is based on the assumption that all signals at the traffic lights for trams should be green due to tram priority via traffic lights. Real operating time savings ΔT_o was measured by using the stopwatch - pencil - record list method. The measurements were conducted in the middle shift during the afternoon peak load, as well as in the early shift during the morning peak load, for all of the seven days in one week, and for the new tram NT 2200 which accounts for 51% of the ZET tram fleet, on line No. 17. This paper has explored the ideal and maximum positive cycle time scenario, based on the measurement of possible maximum operating time savings ΔT_o as the sum of the duration of all red signals at traffic lights for trams in one cycle in real traffic situations in Zagreb.

Keywords: Tram priority · Local automation · Green traffic lights · Cycle time savings

1 Introduction

For the purpose of simulating cycle time savings ΔT on line No. 17 by reducing the total operating (travel) time on line T_o a possible realistic saving of the total operating time ΔT_o in Zagreb was recorded (method of stopwatch - pencil - record list). The measurements were implemented during the morning and afternoon peak loads, for all the days of the week including the weekend, and for the new tram NT 2200, on line No. 17 which is one of the four statistically most dangerous lines operated by the new tram NT 2200 [1].

S. Bagnara et al. (Eds.): IEA 2018, AISC 823, pp. 31–39, 2019.
https://doi.org/10.1007/978-3-319-96074-6_3

One of the key parameters that determine the quality of the service offered to tram users is the operating (travel) time T_o [2]. For operators the most important cycle time on line T (min) is defined by Eq. (1) where ΣT_o (min) is the total operating (travel) time on the line and Σt_t (min) is the total time at terminals, calculated for both directions:

$$T = \sum T_o + \sum t_t \quad (1)$$

The cycle time on line T (min) is defined by the basic Equation of the transport process (2) on the public transport line which connects the headway of vehicles on line h (min/veh) and the number of vehicles on line N.

$$T = h * N \quad (2)$$

Furthermore, from Eq. (2) follows that if the same number of vehicles N are retained, and the cycle time on line T decreases, headway h (min/veh) must also be reduced.

The cycle time on line T can be reduced if the operating (travel) time on line T_o is reduced, in the way that all traffic lights are set to the green light for trams (the tram priority), assuming the real total time at terminals t_t.

The operating (travel) speed V_o (km/h) on the same line length $2L$ (km) will increase if the operating (travel) time on line T_o is reduced, which is defined by Eq. (3).

$$V_o = 2L/T_o \quad (3)$$

Consequently, the cycle speed V_c (km/h) on the same line length $2L$ (km) will increase if the cycle time on line T is reduced, which is defined by Eq. (4).

$$V_c = 2L/T \quad (4)$$

It will also result in an increase in the transport work of line w (sps•km/h) defined by Eq. (5)

$$w = C * L \quad (5)$$

because it has increased the dynamic line capacity C (sps/h) defined by Eq. (6).

$$C = C_v * f \quad (6)$$

So, with the same vehicle capacity C_v (sps/veh) and for the known number of standing and sitting places in the new tram NT 2200, and the same number of vehicles N on line No. 17 the frequency of vehicles on line f (veh/h) is higher because headway h (min/veh) is smaller.

The information on absolute savings of real operating time ΔT_o is expected, according to the assumption that all tram lights should be green due to tram priority via the traffic lights on open sections as well as on the stations which include traffic lights. The results of similar model research in which priority is given to trams and restricting other vehicles entering the “yellow line” confirm the increased cycle speed V_c as well as the increased dynamic capacity [3].

2 Research

Thus, with the aim of studying a real reduction of cycle time on line ΔT two time scenarios have been compared in this paper:

- the objectively measured scenario for line No. 17 (contains all stop times because of red signals at traffic lights for the tram while driving as well as at stations),
- ideal and maximum positive cycle scenario with the possible time savings ΔT of cycle time on line No. 17, based on the measurement of possible time savings ΔT_o of operating (travel) time on line No.17 as the sum of the duration of all red signals at traffic lights while driving as well as at stations in one cycle.

It should be noted that the total length of line No. 17 [4] in both directions (Borongaj – Prečko, Prečko – Borongaj) is $2L$ = 25,536 m.

Tables 1, 2 and 3 show objective individual and average amounts of the total operating (travel) times T_o, total times at terminal t_t, operating speeds V_o and cycle speeds V_c, all on line No. 17, for both directions, which represents the first (real and measured) scenario for line No. 17.

Table 1. Objective operating (travel) time T_o' and time at terminal t_t' on line No. 17 for Borongaj - Prečko direction

Day in the week	Date	Driving mark	Borongaj - Prečko direction				
			1 Start of retention time	2 Time of departure from terminal Borongaj	3 Time of arrival at terminal Prečko	t_t' 2–1 (min)	T_o' 3–2 (min)
Sat	March 17	NJ	13:10	13:17	14:09	7	52
Mon	March 19	A	6:41	6:49	7:36	8	47
Mon	March 19	B	14:31	14:39	15:24	8	45
Tue	March 20	C	6:58	7:02	7:53	4	51
Tue	March 20	D	13:51	13:55	14:50	4	55
Wed	March 21	E	7:05	7:10	7:58	5	48
Wed	March 21	F	13:00	13:06	13:58	6	52
Thu	March 22	G	6:43	6:49	7:38	6	49
Thu	March 22	H	13:08	13:15	14:04	7	49
Fri	March 23	I	7:06	7:11	8:01	5	50
Fri	March 23	J	13:09	13:15	13:59	6	44
Sat	March 24	K	7:17	7:28	8:13	11	45
Sat	March 24	L	12:57	13:05	13:56	8	51
Sun	March 25	M	7:53	8:03	8:49	10	46
Sun	March 25	N	13:01	13:07	13:57	6	50
Average amounts						6.73	48.93

Table 2. Objective operating (travel) time To″ and time at terminal tt″ on line No. 17 for Prečko - Borongaj direction

Day in the week	Date	Driving mark	Prečko - Borongaj direction				
			6	4	5	t_t''	T_o''
			Start of retention time	Time of departure from terminal Prečko	Time of arrival at terminal Borongaj	4–3 * 4–6 (min)	5–4 (min)
Sat	March 17	NJ		14:16	15:01	7	45
Mon	March 19	A		7:44	8:42	8	58
Mon	March 19	B		15:38	16:32	14	54
Tue	March 20	C		8:00	9:05	7	65
Tue	March 20	D		14:56	15:53	6	57
Wed	March 21	E		8:02	8:56	4	54
Wed	March 21	F		14:05	14:53	7	48
Thu	March 22	G		7:46	8:49	8	63
Thu	March 22	H		14:19	15:13	15	54
Fri	March 23	I		8:09	9:11	8	62
Fri	March 23	J		14:05	14:54	6	49
Sat	March 24	K	6:23	6:31	7:17	8*	46
Sat	March 24	L		14:01	14:50	5	49
Sun	March 25	M	6:58	7:10	7:53	12*	43
Sun	March 25	N		13:59	14:43	2	44
Average amounts						7.80	52.73

Table 3. Real amounts of total operating (travel) times T_o and total cycle times T, operating speeds V_o and cycle speeds V_c, all on line No.17, for both directions

Day in the week	Date	Driving mark	T_o	T	V_o	V_c
			(min)	(min)	(km/h)	(km/h)
Sat	March 17	NJ	97	111	15.68	13.71
Mon	March 19	A	105	121	14.49	12.57
Mon	March 19	B	99	121	15.37	12.57
Tue	March 20	C	116	127	13.12	11.98
Tue	March 20	D	112	122	13.58	12.47
Wed	March 21	E	102	111	14.92	13.71
Wed	March 21	F	100	113	15.21	13.46
Thu	March 22	G	112	126	13.58	12.07
Thu	March 22	H	103	125	14.77	12.17
Fri	March 23	I	112	125	13.58	12.17
Fri	March 23	J	93	105	16.36	14.49
Sat	March 24	K	91	110	16.72	13.83
Sat	March 24	L	100	113	15.21	13.46
Sun	March 25	M	89	111	17.09	13.71
Sun	March 25	N	94	102	16.18	14.92
Average amounts			101.67	116.20	15.06	13.15

Tables 4, 5 and 6 show the reduced amounts of total operating (travel) times T_{or} and total cycle times T_r (which include an average amounts of the real times at terminal t_t from first real scenario), increased operating speeds V_{oi} as well as increased cycle speeds V_{ci}, all on line No. 17, for both directions, which represents the second ideal and maximum positive cycle scenario for line No. 17.

Table 4. Reduced operating (travel) times T'_{or} on line No. 17 for Borongaj - Prečko direction

Day in the week	Date	Driving mark	Borongaj - Prečko direction				
			1	2	3	1 − (2 + 3)	t'_t (min)
			T'_o (min)	$\sum T'_{otl}$ (sec)	$\sum T'_{oss}$ (sec)	T'_{or} (sec)	
Sat	March 17	NJ	52	501	94	2,525	6.7
Mon	March 19	A	47	442	122	2,256	6.7
Mon	March 19	B	45	219	107	2,374	6.7
Tue	March 20	C	51	206	178	2,676	6.7
Tue	March 20	D	55	274	233	2,793	6.7
Wed	March 21	E	48	283	87	2,510	6.7
Wed	March 21	F	52	359	148	2,613	6.7
Thu	March 22	G	49	173	42	2,725	6.7
Thu	March 22	H	49	246	144	2,550	6.7
Fri	March 23	I	50	460	150	2,390	6.7
Fri	March 23	J	44	297	60	2,283	6.7
Sat	March 24	K	45	199	82	2,419	6.7
Sat	March 24	L	51	315	94	2,651	6.7
Sun	March 25	M	46	245	111	2,404	6.7
Sun	March 25	N	50	393	78	2,529	6.7
Average amounts			48.93	307.47	115.33	2,513.20	/

Possible savings of operating (travel) times $\Delta T'_o$ and $\Delta T''_o$ in one direction can be realized by omitting the following times:

- $\sum T'_{otl}$ and $\sum T''_{otl}$ from Tables 4 and 5 which represent total times of all red traffic signals at traffic lights in one direction on open sections while driving
- $\sum T'_{oss}$ and $\sum T''_{oss}$ from Tables 4 and 5 which represent the sum of differences between particular times of all red traffic signals at traffic lights at stations and the average amount of all green traffic signals at traffic lights at stations in one direction (only for all stations which contain the traffic lights).

Table 5. Reduced operating (travel) times T''_{or} on line No. 17 for Prečko – Borongaj direction

Day in the week	Date	Driving mark	Prečko - Borongaj direction				
			4	5	6	4 − (5 + 6)	t''_t
			T''_o (min)	$\sum T'_{otl}$ (sec)	$\sum T'_{oss}$ (sec)	T''_{or} (sec)	(min)
Sat	March 17	NJ	45	190	312	2,198	7.8
Mon	March 19	A	58	311	304	2,865	7.8
Mon	March 19	B	54	291	240	2,709	7.8
Tue	March 20	C	65	520	355	3,025	7.8
Tue	March 20	D	57	386	300	2,734	7.8
Wed	March 21	E	53	335	267	2,578	7.8
Wed	March 21	F	48	222	286	2,372	7.8
Thu	March 22	G	63	287	302	3,191	7.8
Thu	March 22	H	54	458	237	2,454	7.8
Fri	March 23	I	62	387	373	2,960	7.8
Fri	March 23	J	49	373	183	2,384	7.8
Sat	March 24	K	46	371	259	2,130	7.8
Sat	March 24	L	49	350	241	2,349	7.8
Sun	March 25	M	43	255	173	2,152	7.8
Sun	March 25	N	44	204	196	2,240	7.8
Average amounts			52.67	329.33	268.53	2,562.13	/

Table 6. Reduced amounts of total operating (travel) times T_{or} and total cycle times T_r, increased amounts of operating speeds V_{oi} and cycle speeds V_{ci}, all on line No.17, for both directions

Day in the week	Date	Driving mark	T_{or} (min)	T_r (min)	V_{oi} (km/h)	V_{ci} (km/h)
Sat	March 17	NJ	78.72	93.22	19.33	16.32
Mon	March 19	A	85.35	99.85	17.82	15.24
Mon	March 19	B	84.72	99.22	17.96	15.33
Tue	March 20	C	95.02	109.52	16.01	13.89
Tue	March 20	D	92.12	106.62	16.52	14.27
Wed	March 21	E	84.80	99.30	17.94	15.32
Wed	March 21	F	83.08	97.58	18.31	15.59
Thu	March 22	G	98.60	113.10	15.43	13.45
Thu	March 22	H	84.92	99.42	17.92	15.30
Fri	March 23	I	89.17	103.67	17.06	14.68
Fri	March 23	J	77.78	92.28	19.56	16.49
Sat	March 24	K	75.82	90.32	20.07	16.84
Sat	March 24	L	83.33	97.83	18.26	15.55
Sun	March 25	M	75.93	90.43	20.04	16.82
Sun	March 25	N	79.48	93.98	19.14	16.19
Average amounts			84.59	99.09	18.09	15.42

Possible savings of cycle times ΔT and operating times ΔT_o on line No. 17 are shown in Table 7; these are total amounts for both directions.

Table 7. Possible savings of cycle times ΔT and operating times ΔT_o on line No. 17

Day in the week	Date	Driving mark	T_o (min)	T_{or} (min)	ΔT_o %	T (min)	T_r (min)	ΔT %
Sat	March 17	NJ	97	78.72	−23.22	111	93.22	−19.07
Mon	March 19	A	105	85.35	−23.02	121	99.85	−21.18
Mon	March 19	B	99	84.72	−16.86	121	99.22	−21.95
Tue	March 20	C	116	95.02	−22.08	127	109.52	−15.96
Tue	March 20	D	112	92.12	−21.58	122	106.62	−14.43
Wed	March 21	E	102	84.80	−20.28	111	99.30	−11.78
Wed	March 21	F	100	83.08	−20.37	113	97.58	−15.80
Thu	March 22	G	112	98.60	−13.59	126	113.10	−11.41
Thu	March 22	H	103	84.92	−21.29	125	99.42	−25.73
Fri	March 23	I	112	89.17	−25.60	125	103.67	−20.57
Fri	March 23	J	93	77.78	−19.57	105	92.28	−13.78
Sat	March 24	K	91	75.82	−20.02	110	90.32	−21.79
Sat	March 24	L	100	83.33	−20.00	113	97.83	−15.51
Sun	March 25	M	89	75.93	−17.21	111	90.43	−22.75
Sun	March 25	N	94	79.48	−18.27	102	93.98	-8.53
Average amounts			101.7	84.59	−20.20	116.2	99.09	−17.35

Average amounts of results from Table 7 point to the possibility for the cycle time savings $\Delta T = -17.35\%$ as well as the possibility for the operating time savings $\Delta T_o = -20.20\%$, during the morning and afternoon peak loads of passengers, for all days of the week including the weekend (Saturday and Sunday).

Possible increase of the individual and average amounts of cycle speeds ΔV_c as well as possible increases of the individual and average amounts of operating speeds ΔV_o on line No. 17 are shown in Table 8; these are the total amounts for both directions in one cycle.

In the City of Zagreb, public transport attractiveness is possible to improve significantly by increasing the quality of service. In the current network state, operating speed increase related to the quality of service could bring significant passenger travel time decrease, and this can be made by introducing public transport priority in the network [5].

Table 8. Possible increase of the cycle speeds ΔV_c and operating speeds ΔV_o on line No. 17

Day in the week	Date	Driving mark	V_o (km/h)	V_{oi} (km/h)	ΔV_o %	V_c (km/h)	V_{ci} (km/h)	ΔV_c %
Sat	March 17	NJ	15.68	19.33	+18.88	13.71	16.32	+15.99
Mon	March 19	A	14.49	17.82	+18.69	12.57	15.24	+17.52
Mon	March 19	B	15.37	17.96	+14.42	12.57	15.33	+18.00
Tue	March 20	C	13.12	16.01	+18.05	11.98	13.89	+13.75
Tue	March 20	D	13.58	16.52	+17.80	12.47	14.27	+12.61
Wed	March 21	E	14.92	17.94	+16.83	13.71	15.32	+10.51
Wed	March 21	F	15.21	18.31	+16.93	13.46	15.59	+13.66
Thu	March 22	G	13.58	15.43	+11.99	12.07	13.45	+10.26
Thu	March 22	H	14.77	17.92	+17.58	12.17	15.30	+20.46
Fri	March 23	I	13.58	17.06	+20.40	12.17	14.68	+17.78
Fri	March 23	J	16.36	19.56	+16.36	14.49	16.49	+12.13
Sat	March 24	K	16.72	20.07	+16.69	13.83	16.84	+17.87
Sat	March 24	L	15.21	18.26	+16.70	13.46	15.55	+13.44
Sun	March 25	M	17.09	20.04	+14.72	13.71	16.82	+18.49
Sun	March 25	N	16.18	19.14	+15.46	14.92	16.19	+7.84
Average amounts			15.06	18.09	+16.77	13.15	15.42	+14.69

3 Conclusion

The dominant mode of public passenger transport in the center of Zagreb are the trams. This paper has proven the possibility of significant simulating cycle time savings ΔT by reducing the operating time ΔT_o, both achieved by the possible implementation of tram priority via green traffic signals at traffic lights based on the local automation. Based on the proved cycle time savings ΔT realized by the operating time savings ΔT_o on line No. 17, the average amount of the increased cycle speed V_{ci} = 15.42 km/h which contains speed increase of 14.69% as well as the average amount of increased operating speed V_{oi} = 18.09 km/h which contains speed increase of 16.77%, indicate the need for further research of potential increase of cycle speed and operating speed on other tram lines in Zagreb. The technical realization of the proposed solution is possible because of successful implementation in several urban areas in the world. The application of the tram priority in urban areas by using the local automatic setting the points and signals for trams on the junctions can provide the following benefits to some extent: reduction of tram driver's workload (by reducing the number of the manually served commands on the control panel) as well as increase of tram driver's performance level.

References

1. ZET (2013) Izvješće o stanju sigurnosti JJP za 2013. godinu, Zagreb (Report on safety condition of public transport for 2013)
2. Pyrgidis C, Chatziparaskeva M (2012) The impact of the implementation of green wave in the traffic light system of a tramway line - the case of Athens tramway. In: 2nd international conference on road and rail infrastructure (CETRA), Croatia, pp 891–897
3. Brčić D, Slavulj M, Šojat D (2012) Analysis of tram priority in the City of Zagreb. In: 32nd conference of transportation systems with international participation, Zagreb, pp 173–176
4. ZET (2010) Eksploatacijski pokazatelji tramvajskih linija, Zagreb. (Exploitation Indicators of Tram Lines)
5. Šojat D, Brčić D, Slavulj M (2017) Analysis of public transport service improvements on tram network in the city of Zagreb. Tech Gaz Sci Prof J Tech Fac Josip Juraj Strossmayer Univ Osijek 24(1):217–223

Intelligent In-Car Health Monitoring System for Elderly Drivers in Connected Car

Se Jin Park[1,2,3](✉), Seunghee Hong[1,2], Damee Kim[1,2], Iqram Hussain[1,2,3], and Young Seo[1]

[1] Korea Research Institute of Standards and Science, Daejeon, South Korea
[2] Electronics Telecommunication Research Institute, Daejeon, South Korea
[3] University of Science and Technology, Daejeon, South Korea

Abstract. Introduction: Health has become a major concern nowadays. People pass significant amount of time of daily life on driving seat. Some health complexity happens during driving like heart problem, stroke etc. Driver's health abnormality may also effect safety of other vehicles. So, automotive manufacturers and users are interested to include real-time health monitoring in car system.

Intelligent in-car health monitoring is considered most innovative technology which is able to measure real-time physiological parameters of drivers, feed data to web cloud, analysis using machine learning, artificial intelligence and big data. Brain stroke is most deadly diseases and effected persons lose conscience and ability to contact emergency services or hospital. Emergency medical assistance is necessary in order to survive from any kind of disability due to stroke.

Purpose: The aim of our study is to develop a health monitoring system for elderly drivers using air cushion car seat and embedded IoT (Internet of Things) devices in order to detect stroke onset during driving.

Method: Real-time monitoring is desired to detect stroke onset during regular activities like driving. Abnormal physiological signals, face pattern generated during stroke onset can be traced by real-time monitoring using sensors. Here, we have suggested a framework of stroke onset detection using sensors and developed a system suitable for elderly drivers. This system can measure and analyze data of ECG, EEG, heart rate, seat pressure balance data, face/eye tracking etc. using IoT sensors. Physiological data will be feed to cloud and compared with reference normal person data.

Findings: If any health abnormality such as stroke is found in real-time monitoring, system will predict type and severity of stroke and suggest possible steps. System may switch car control to autonomous driving mode if available and move the car to safe place. System may also generate alarm and send message with available information such as position to relatives and emergency services to provide emergency assistance so that effected driver can be transferred to hospital/clinic.

Keywords: Internet of Things · Elderly healthcare · Brain stroke
Real-time monitoring

S. Bagnara et al. (Eds.): IEA 2018, AISC 823, pp. 40–44, 2019.
https://doi.org/10.1007/978-3-319-96074-6_4

1 Introduction

The Internet of Things (IoT) performs a significant role in the development of smart vehicles, which offers smart transportation, cloud connectivity, vehicle-to-vehicle interaction, smartphone integration, safety, security, and e-healthcare services. Recent development trends show that auto industries are already paying attention to develop IoT cars that could integrate driver's health status and driving safety. Both auto industry and key global original equipment manufacturers are integrating healthcare services into their next-generation products [1, 2]. Aging originates from increasing longevity, and results in deteriorating fertility [3]. Population aging is taking place in nearly all the countries of the world. As age increases, older drivers become more conservative on the road. Age-related decline in cognitive function hampers safety and quality of life for an elder. As the aged population in the developed world is increasing, so the number of older drivers is becoming higher. Research on age-related driving has shown that an increased risk of being involved in a vehicle crash is more at around the driver's age of 65. Among of all health complexity during driving, stroke is the top one. Stroke is the sudden collapse of brain cells due to lack of oxygen, caused by blockage of blood flow to the brain or breakdown of blood vessels. Stroke is the second top reason of death above the age of 60 years, and its proportion is rising [4]. Many health abnormality happens after stroke. Postural disorders is observed as one of the most common disabilities after stroke [5].

This paper focused on briefly explaining the design and framework of the elderly drivers' health monitoring services in connected car using IoT devices. The purpose of this study is to develop a real-time health monitoring system for elderly drivers' to detect health abnormality like stroke using air cushion and IoT devices and to successfully detect health complications and generate messages & alarms to responsible ones such as family members, emergency services or hospitals about drivers' stroke onset while driving in order to transfer victim to hospital or clinic.

2 Model and Methodology

A sensor-based integrated health monitoring system has been proposed in order to measure physiological measurement of car driver (Fig. 1). A model of air cushion is designed for monitoring health status, specially sitting balance of driver. Stroke results unbalance body pressure of a person. For measuring pressure in each cell, each air chamber is equipped with air pressure sensor. This air cushion is inserted to inside of air seat and covered with seat cover. For better comfortability, air cushion top surface is placed in same level of car seat flat surface. EEG, EMG, ECG sensors will monitor corresponding physiological signals of elderly adult car drivers. Face tracking and eye tracking camera are employed in this system in order to detect abnormality in drivers' appearance due to stroke or other heart diseases. All described sensors are able to measure basic physiological parameters for predicting health status of an elderly adult driver. Physiological data analysis procedures are described in Fig. 2.

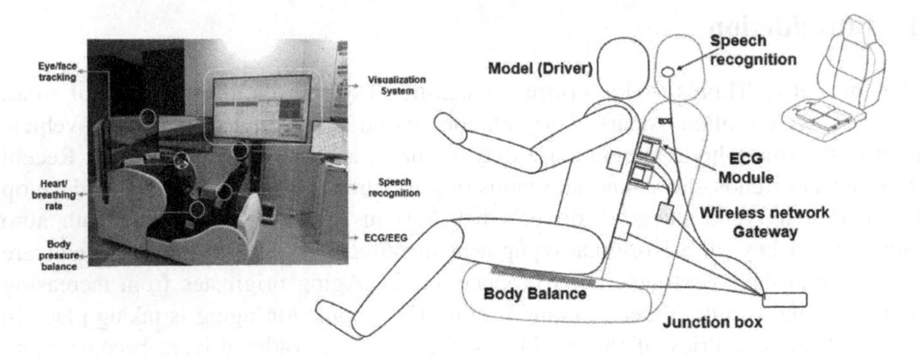

Fig. 1. Model of intelligent in-car health monitoring system.

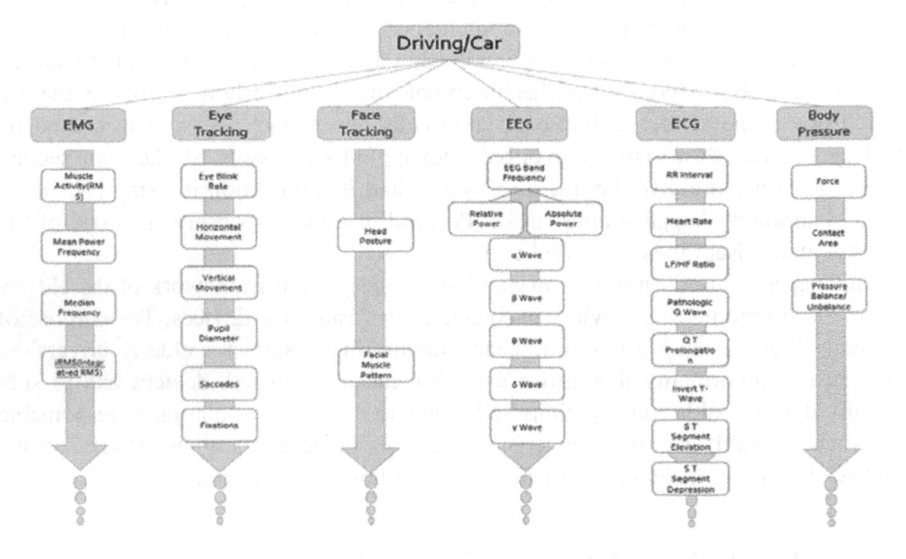

Fig. 2. Physiological data analysis procedures for real-time health monitoring.

3 Architecture and Framework of In-Car Health Monitoring System

Framework of biosensor based drivers' health monitoring system is presented in Fig. 3. Real-time data also train-up big data and capture ECG/EMG/heart rate/face pattern that boost up Self-learning engine. For real-time IoT sensor data, MapR Streams is used for scalable data collection, Spark streaming is used for data processing. Processed data is stored using MapR-DB(HBase). In cloud, drivers' health record, normal physiological data have to be stored first as reference data. Real-time ECG/EMG/heart rate will be compared with reference normal data in order to find out health abnormality during driving. Complete system will feed drivers' physiological data to cloud engine for comparison of real-time data and already stored reference data in order to detect stroke onset of elderly drivers.

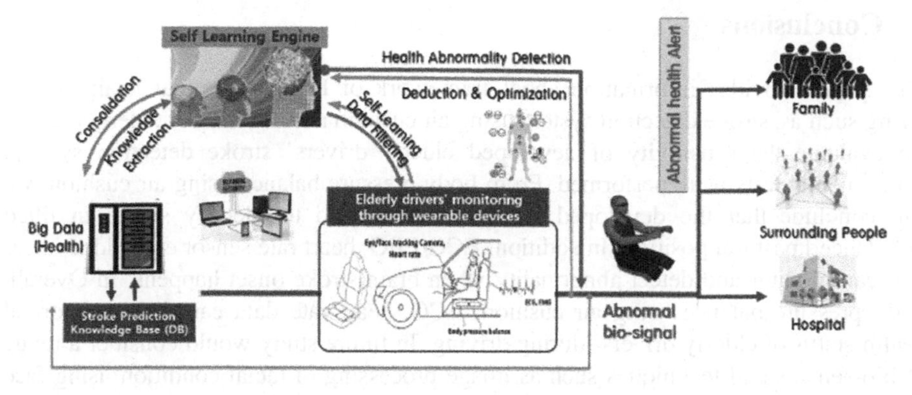

Fig. 3. Architecture and framework of in-car health monitoring system.

4 Result and Discussion

ECG signal has been analyzed and key features (RRI, HR, RH, PH, PRG, QT, QTC, ST segments) has been extracted for both stroke patients and normal control subjects. Then statistical test, ANOVA (Analysis of Variance) has been conducted. Outcome of ANOVA test has been presented in Fig. 4. Heart rate also showed significant difference between elderly normal adult and elderly stroke patient by measuring with in-car sensor system. System will also predict variation of physiological signal pattern in order to generate an alarm and deliver messages to emergency services, family of the victim, people around the victim, and healthcare professionals in order to ensure the timely medical assistance. Each sensor stroke prediction result contributes own probability in large set of IoT sensor network. The more sensor in health monitoring system, the more reliability of monitoring system.

ANOVA Test: ECG features (Stroke patient & Normal controls)		*Sum of Squares*	*df*	*Mean Square*	*F*	*Significance*
RRI	Between Groups	4.548	1	4.548	195.552	.000
	Within Groups	139.047	5979	.023		
	Total	143.594	5980			
HR	Between Groups	72924.495	1	72924.495	281.603	.000
	Within Groups	1548334.632	5979	258.962		
	Total	1621259.128	5980			
RH	Between Groups	134.040	1	134.040	11.204	.001
	Within Groups	71531.427	5979	11.964		
	Total	71665.467	5980			
PH	Between Groups	71.580	1	71.580	5.398	.020
	Within Groups	72305.865	5453	13.260		
	Total	72377.445	5454			
PRQ	Between Groups	4.222	1	4.222	6.238	.013
	Within Groups	3644.577	5385	.677		
	Total	3648.799	5386			
QT	Between Groups	1.507	1	1.507	346.656	.000
	Within Groups	24.703	5682	.004		
	Total	26.210	5683			
QTC	Between Groups	.451	1	.451	131.564	.000
	Within Groups	19.472	5682	.003		
	Total	19.923	5683			
ST	Between Groups	.943	1	.943	35.464	.000
	Within Groups	149.153	5611	.027		
	Total	150.096	5612			

* RRI: R-R interval, HR, RH, PH=Height of the waves from the iso-electric level, QRS=Time interval of the QRS complex, PRQ=Time interval of the QRS complex, QT,ST= Q-T/ST interval.

Fig. 4. ANOVA test of ECG features between stroke patients and normal control subjects.

5 Conclusions

This paper provides information about framework of Real-time in-car health monitoring such as, stroke detection system using air cushion and sensors for elderly drivers. To evaluate the sensitivity of developed elderly drivers' stroke detection system, experimental tests were performed. From body pressure balance using air cushion, we can conclude that the developed car seat is expected to identify stroke in tilted unbalanced postural position. In addition, ECG/EEG, heart rate sensor embedded in car seat can monitor and detect abnormality when brain stroke onset happens. In Overall, body pressure balance using air cushion, ECG, heart rate data can detect abnormal health status of elderly drivers' during driving. In future study would consider a range of bio-sensors and techniques such as image processing of facial condition using face tracking and eye tracking, thermal imaging of elderly drivers' body in order to detect stroke onset during driving.

References

1. Park SJ, Hong S, Kim D, Seo Y et al. (2018) Development of a real-time stroke detection system for elderly drivers using quad-chamber air cushion and IoT devices, SAE Technical Paper 2018-01-0046. https://doi.org/10.4271/2018-01-0046
2. Park SJ, Subramaniyam M, Hong S, Kim D, Yu J (2017) Conceptual design of the elderly healthcare services in-vehicle using IoT. SAE Technical paper (No. 2017-01-1647)
3. Park SJ, Subramaniyam M, Kim SE, Hong SH, Lee JH, Jo CM (2017) Older driver's physiological response under risky driving conditions–overtaking, unprotected left turn. In: Duffy V (ed) Advances in applied digital human modeling and simulation. AISC, vol 481. Springer, Heidelberg, pp 107–114. https://doi.org/10.1007/978-3-319-41627-4_11
4. Park SJ, Min SN, Lee H, Subramaniyam M (2015) A driving simulator study: elderly and younger driver's physiological, visual and driving behavior on intersection. In: IEA 2015, Melbourne, Australia
5. Pérennou D (2006) Postural disorders and spatial neglect in stroke patients: a strong association. Restor Neurol Neurosci 24:319–334

Air Travel Accessibility: Interaction Between Different Social Actors

Jerusa Barbosa Guarda de Souza[1(✉)], Talita Naiara Rossi da Silva[2], and Nilton Luiz Menegon[1]

[1] Federal University of São Carlos, São Carlos, Brazil
jerusaguarda@gmail.com
[2] University of São Paulo, Belo Horizonte, Brazil

Abstract. Despite the efforts of regulatory agencies to set out requirements to ensure accessibility conditions in air travels, studies show that there are assistance inadequacies, because the services offered to passengers who require special assistance result from interactions between aircraft manufacturers, regulatory agencies, operators, and air travelers. They have different interests which remain little explored. The objective of this study was to discuss the social actors' interactions that determine the services offered to these passengers. An exploratory research including 7 regulatory agencies, 377 passengers, and 20 traveling companions was carried out using semi structured interviews and questionnaires to collect their specific interests. A checklist based on current Brazilian laws was used to investigate the current air transport accessibility conditions in three airlines. The results showed that accessibility problems are mainly related to the inefficiency of services provided by operators, ineffectiveness of aircrafts in terms of passenger needs, or inefficiency of regulations.

Keywords: Accessibility · Transport · Aircraft

1 Introduction

The growth in air transport and global phenomena, such as population aging [20, 25] and the increasing number of people with disabilities [23] and of obese people [24] have changed the profile of air travelers. Therefore, knowing air travel users also means knowing passengers requiring special assistance (PNAEs, in Portuguese), a group that has an increasingly important role in air transportation [9, 10, 22].

Passengers requiring special assistance are people with disabilities, people aged 60 years or more, pregnant women, nursing mothers, people with young children, people with reduced mobility, or anyone who has a specific condition that limits his/her autonomy as a passenger [1]. The present study focuses only on people with disabilities and reduced mobility, including the elderly (people > 60 years) and obese people (Body Mass Index > 30).

Due to the increased number of these passengers and the concerns about their safety in air transport, regulatory agencies have been making an effort to set out specific requirements to ensure the provision of adequate assistance to this specific traveling public, protect their physical and moral integrity, and ensure they have equal rights.

S. Bagnara et al. (Eds.): IEA 2018, AISC 823, pp. 45–54, 2019.
https://doi.org/10.1007/978-3-319-96074-6_5

Despite some actions in terms of laws and regulations, studies show that there are gaps and inadequacies in the assistance provided to these passengers [3–5, 12, 19, 21]). The main difficulties are related to lack of trained staff to help people who require special assistance; lack of adequate equipment for boarding; unavailability of priority seats; lack of movable or folding armrests, making it even more difficult to transfer the passenger; difficult access and limited space inside the lavatory; lack and inadequacy of on-board wheelchairs; damage of baggage, wheelchair, or other assistive devices.

These gaps and inadequacies can be explained by the fact that the service offered to PNAEs result from interactions between different social actors, aircraft manufacturers, regulatory agencies, airline and airport operators, and air travelers who need special assistance. These social actors have different interests and concerns, which remain little explored and unknown. Thus, the present study aimed to discuss the interactions between these social actors that determine the service offered to passengers requiring special assistance.

2 Theoretical Framework

The present study was based on the institutional theory [14] and on the activity-centered ergonomics, which guided methodological procedures adopted.

According to Scott [14] the institutions challenge and replace previous institutions. The author introduced two different perspectives of institutionalization. The first is the naturalistic view, in which institutions are not created by the objective actions of interested-based actors, but rather emerge from the collective sense-making and problem-solving behavior of actors. On the other hand, the agent-based view, stresses the importance of identifying particular actors as causal agents, which involve individual and collective actors who participate in the construction of new institutional forms and exercise influence on formal processes.

Nation State actors represent the first type of those actors, who, together with the legal professionals, can define the nature, capacity, and rights of economic and political actors, including collective actors. States exert significant influence not only on individual organization's structures and behaviors but also on the structure of organizational fields [14].

Professions also create new institutional structures and are divided into three categories in the contemporary society: cultural-cognitive agents, normative agents, and regulatory agents [14, 15].

Other institutional agents include the associations, which exercise authority in the cultural-cognitive, normative, and regulatory domains. Therefore, their agents are the individual and collective actors that operate with different cognitive-cultural, normative, or regulatory tools that support their construction efforts [14].

The created institutions influence organizations by formulating the rules of the game, while the organizations are the players [14]. Powell [13] states that institutional effects compel organizations to conform to the expectations of the fields in which they were members. Similarly to Powell's [13] perspective, Scott [14] defined the concept of isomorphism, which suggests that units subjected to the same environmental conditions acquire similar forms of organizations. In contrast, although institutional pressures under the same conditions lead to organizational isomorphism (structures and practices), there are examples of identical institutional forces that lead to different outcomes.

Comparing to the accessibility regulations in Brazil, Costa, Maior, Lima [6] argue that although there has been increased society mobilization, the implementation of these regulations depends on cultural changes. Silva [16] states that nowadays there is a need for joint actions between the State and civil society to ensure the inclusion of people with disabilities in the socioeconomic and cultural scenario.

The advances in accessibility laws resulting from restrictions on the participation of people with disabilities demonstrates the challenge of understanding the distinction between prescribed and real work or task, as described by ergonomics. Prescribed task involves goals, resolutions, norms that represent some determinants of a situation. Real work or task is related to performing the activities as prescribed, which are unique object-oriented activities involving particular subjects and contexts and their variability. In this perspective, in addition to the prescribed work, the activity includes purpose, life experiences, and environmental and social determinants, highlighting the need to understand the situations considering the different representations and experiences of the actors involved [7, 11].

The research problem demonstrates that air transport accessibility issues arise from the interaction between different social actors who have peculiarities and limitations. The institutions represented by the regulatory agencies, together with the users who help establish them, characterize the service offered by organizations since they set out the rules to be followed by the airport, airline and airport operators, and aircraft manufacturers (Fig. 1).

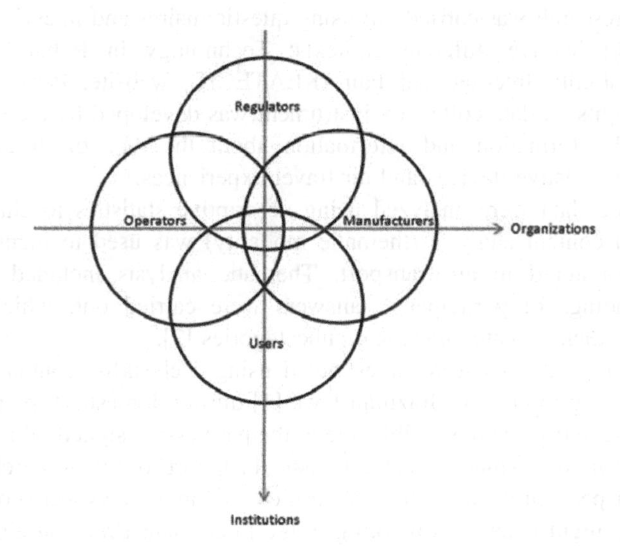

Fig. 1. Interaction between the agents that establish institutions and organizations (Source: SOUZA [17])

Since the model proposed in this study suggests an interaction between social actors and considering that a fundamental characteristic of ergonomics is the involvement of the different actors in the development processes [8], this study was based on different perspectives to understand their particularities and to discuss their interactions.

3 Materials and Methods

The present study was approved by the Ethics Committee for research on humans (CAAE: 18017613.7.0000.5504, August 1st, 2013). Participation was voluntary, and all participants signed an Informed Consent Form prior to participation.

Interviews with representatives of Brazilian regulatory agencies were conducted during visits, and questionnaires were administered to people who fall into the category of "passengers requiring special assistance" [1]. The current Brazilian air transport accessibility conditions were verified.

The visits to Brazilian regulatory agencies aimed to understand how they are organized regarding the introduction of accessibility regulations in Brazil. Preeminent regulatory agencies involved with air transport or the rights of people with disabilities were selected. Semi-structured interviews were conducted using a checklist that included questions about structure of the regulatory agencies to address accessibility issues, research on this topic, and future trends in this sector. The interviews were recorded and the transcribed interviews were presented to the respondents to check their accuracy. Data were analyzed descriptively.

In order to collect data about the needs of passengers requiring special assistance, an exploratory research was carried out using questionnaires and interviews. This step was carried out in five different contexts: Technology in Rehabilitation, Inclusion and Accessibility International Fair (REATECH), website, Paralympic events, airports, and flights. A data collection instrument was developed by the researchers to obtain personal information and information about the type of disability, use of mobility aids or assistive devices and air travel experiences.

Questionnaire data were analyzed using descriptive statistics to characterize the population, and content analysis (thematic modality) was used to identify the main difficulties encountered in air transport. Thematic analysis included floating and exploratory readings of participants' answers were carried out, which led to the identification of themes, categories, and subcategories [2].

Air transport conditions were investigated using a checklist containing 26 items drawn up according to current Brazilian laws [1] during domestic trips in the second half of 2013. These trips were possible due to the partnership signed with the Brazilian Paralympic Committee, which provided logistical support for this research. During the trips, different aspects of the aircraft were verified, and interviews were conducted with airport staff and flight attendants working in the three main Brazilian airlines.

The three airlines investigated were denominated A, B, and C in order to protect their identity since this study is not intended to improve or denigrate the image of the companies but rather to investigate their adherence to existing regulations. The researchers flew twice with airline A, twice with airline B, and once with airline C.

4 Results

4.1 Interviews with Regulatory Agencies

The first step of the research consisted of the interviews with representatives of Brazilian regulatory agencies, as described: Agency A (National aviation regulatory agency); Agency B (National secretariat for protecting the rights of persons with disabilities); Agency C (State secretariat for protecting the rights of persons with disabilities); Agency D (National agency for elder rights protection); Agency E (National institute of tourism promotion and management); Agency F (National agency responsible for actions and measures concerning means of transport); Agency G (National agency responsible for actions and measures concerning cities and urban areas);

The results obtained show that all of these agencies, with exception of agencies A and F, have an adequate internal structure to deal with accessibility matters. Some agencies, such as agency B, reported that they have an active interaction with society, government, and entities that represent people with disabilities. Others mentioned having control functions, such as agencies C and G, and actions to improve lives of people with disabilities. Agency D mentioned having a role in both developing and oversighting guidelines for the National Policy for the Elderly through research carried out in partnership with universities. The main function of agency E is to guarantee inclusion of people with disability in tourism.

The demographic data of people with disabilities are provided to these agencies by the Brazilian Population Census, airline information, and research results. All agencies, with the exception of agency F, mentioned participating in studies or discussion groups about accessibility although not all of these discussions are specifically related to air transport. With regard to the difficulties of PNAEs in air transport, the participant regulatory agencies mentioned the lack of trained staff, boarding and disembarking without adequate equipment, inadequate signage in airports and aircraft, damage to mobility aids, use of the cabin lavatory, lack of sign language interpreters, and lack of appropriate supervision of the compliance with existing regulations.

Based on the aforementioned considerations, it can be concluded that the agencies interviewed have participated in discussions about enforcement and adoption of accessibility regulations in Brazil, but this can be considered a scenario undergoing transformation.

4.2 Exploratory Research Involving Passengers Requiring Special Assistance

A total of 377 individuals were considered in this study, being: 25,46% people with physical disability and wheelchair users, 16,18% people with physical disability and not wheelchair users, 12,73% obese people, 11,67% elderly people, 7,69% obese people with disability, 6,1% people of visual impairment, 5,57% people with hearing impairment, 5,84% people with dwarfism, 3,18% elderly people with disability, 2,92% obese and elderly people, 1,5% obese and elderly people with disability and 1,06% people with multiple disabilities.

A total of 20 travelling companions were considered, being 35% companion of people with disability, 15% companion of people with hearing impairment, 10% companion of elderly people, 5% companion of people with multiple disabilities and 35% did not specify.

The main difficulties encountered by the participants in the airports, during boarding, during the flight, and during disembarking as well the number of times each difficulty was mentioned are summarized by categories in Table 1.

Table 1. Main difficulties mentioned by the participants

Difficulties	Total
Boarding and disembarking from remote gates or without boarding bridges: need to be hand-carried up the stairs	174
Lack of adequate equipment for boarding (boarding bridges and ambulift)	117
Difficult access and limited space inside the lavatory (inside aircraft)	117
Limited legroom	61
Seats are narrow, small, and uncomfortable	59
Lack of trained staff to help people who require special assistance (inadequate training and assistance)	56
Long distances from one location to another	49
Narrow aisles (which do not allow full wheelchair accessibility to the seat and transfer to the seat)	43
Disembarking delay (non-priority; disembarking after all other passengers)	42
Front row seats are not always available (disrespect for designated priority seats)	35
Long wait on checked bags to arrive at the carousel	25
Damage of baggage, wheelchair, or other assistive devices	24
Communicating and asking for help since people do not know sign language: lack of interpreters	23
Lack of movable or folding armrests making it even more difficult to transfer the passenger	23
Boarding from remote gates or without boarding bridges: need to take a shuttle bus	18
Limited space inside the cabin	16
Understanding instructions and information given in auditory announcements	16
Delay in getting the wheelchair after exiting the cabin of the aircraft	15
Inadequate signage	12

Other difficulties mentioned were: check-in counters are too high (10), not all aircrafts have on-board wheelchairs (10), boarding with a wheelchair with caterpillar tracks (9), few elevators available inside airports (9), lack of accessible restrooms inside airports (9), lack of accessibility in airports (8), inadequate on-board wheelchairs (7), seats without footrest (7), long lines and delay at the check-in counter (6), going up and down stairs (6), stowing and retrieving carry-on items in and from the overhead compartment (6), moving around and way finding (5), there are no designated priority storage areas for the assistive devices (5), small size fonts and poor contrast in flight information display boards (4), distance and traffic to the airport (4), lack of or inaccurate information (4), finding baggage claim (4), shuttle bus is always crowded (3), retrieving luggage from the carousel (3), bad quality of airport wheelchairs (2), boarding gate changes and auditory announcements (2), information is not always up-to-date in displays (2), the pre-flight safety briefing is performed in the form of a video (2).

4.3 Air Transport Conditions

With regard to air transport conditions, it was found that the three companies investigated did not keep records of PNAEs that have permanent disability, limit the number of PNAEs per flight, do not offer escort for passengers who request this service, and do not have a quality management system. Moreover, in these companies, boarding and disembarkation are done without assistance equipment. These aspects indicate non-compliance with the procedures established in the Resolution 280 that addresses the air transport of PNAEs [1]. Company B does not offer a discounted ticket for a companion travelling with PNAE, and Company C does not offer assistance during flight connections.

There was not enough information to reach a conclusion on whether the three companies investigated allow passengers to use their own assistive devices during the flight and whether they offer training to their staff, according to the regulations. With regard to companies A and B, it was not possible to concluded whether they have a professional responsible for ensuring accessibility; as for company C, this requirement is not complied. The results of such investigation are described in the study carried out by Souza [18].

5 Discussion

Based on the interviews conducted, it can be said that the institutions that are concerned with accessibility matters in Brazil are guided by agents. All of the regulatory agencies interviewed were linked to the State, but in some of them it was possible to observe a connection with other agents of society, such as associations. The regulatory agencies interviewed exercise their authority by rulemaking, guiding, evaluating, supervising, and fostering awareness to consolidate institutions that are involved with accessibility matters in Brazil.

However, the respondents' answers show that there is an ongoing debate about this topic and the accessibility laws, which corroborates the findings of Costa, Maior, Lima [6] and Silva [16], who pointed out the need for joint actions between the State and civil society to ensure the inclusion of people with disabilities in the socioeconomic and cultural scenario and to help raising social actors' awareness to promote the adoption of mechanisms for ensuring effective implementation of legislations.

Based on the interviews with the regulatory agencies, the data obtained through the questionnaires and the analyses of air transport conditions it was possible to fulfill the model. Figure 2 illustrates the model and Table 2 explains the user's complaints related to regulators, operators and manufacturers.

The model proposed shows that air transport accessibility problems are mainly due to the deficiency of the services offered by airline and airport operators, ineffectiveness

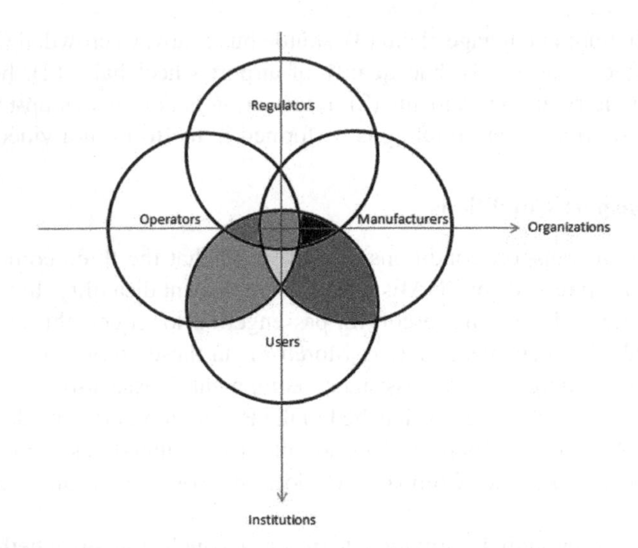

Fig. 2. User's complaints related to operators, regulators and manufacturers (Color figure online)

Table 2. User's complaints related to operators, regulators and manufacturers

	Problems related to airport infrastructure (physical barriers and signage); Inadequate staff training; Waiting in lines; Need to be hand-carried up the stairs; Boarding without assistance equipment; Difficulty going up stairs; Difficulty understanding pre-flight safety procedures; Disembarking delay.
	Lack of sign language interpreters; Long distances from one location to another inside the airports; Delay in getting the wheelchair after landing; Boarding with a wheelchair with caterpillar tracks; Finding baggage claim; Wayfinding in the airport; Long wait on checked bags to arrive at the carousel; Retrieving luggage from the carousel; Frequent boarding gate changes; Use of shuttle buses when boarding from remote gates; Inadequate on-board wheelchairs; Flight information display boards is not always up-to-date.
	Difficulty going up and down the stairs (smaller aircrafts); Difficulty understanding auditory announcements during the flight; Small cabin seats and limited space between them.
	Front row seats are not always available; Damage of luggage and assistive devices; Not all aircrafts have on-board wheelchairs.
	Lack of movable armrests; Narrow aisles.
	Lack of accessible lavatories; Difficulty finding storage areas for the assistive devices; Seats without footrest; Difficulty stowing and getting items from the overhead compartment; Inadequate on-board wheelchair inadequate

of aircrafts in terms of passenger needs, or inefficiency of regulations regarding the real needs of the users; although there are regulations, they often do not help to improve the services provided to PNAEs.

6 Conclusion

Although the difficulties encountered by air travelers mentioned by the participants were similar to those reported in the studies included in the theoretical framework, *the present study differs from others* in the literature in that it investigates accessibility issues from a holistic point of view, showing that, in order to solve these problems it is necessary to promote the effective integration among these four social actors.

On the other hand, this study investigated accessibility issues in the Brazilian context only, which suggests a valuable opportunity for further research extending the scope of this investigation to include other countries that have different regulations, users, and operators. Furthermore, another aspect emphasized in the present study is that in addition to enforcing laws and regulations, regulatory agencies should supervise their compliance.

References

1. ANAC homepage (2018). http://www.anac.gov.br/assuntos/legislacao/legislacao-1/resolucoes/resolucoes-2013/resolucao-no-280-de-11-07-2013/@@display-file/arquivo_norma/RA2013-0280.pdf. Accessed 08 Apr 2018
2. Bardin L (2011) Análise de conteúdo. Editora Edições 70, São Paulo
3. Chang YC, Chen CF (2011) Identifying mobility service needs for disabled air passengers. Tourism Manag 32:1214–1217. https://doi.org/10.1016/j.tourman.2010.11.001
4. Chang YC, Chen CF (2012) Meeting the needs of disabled air passengers: factors that facilitate help from airlines and airports. Tourism Manag 33:529–536. https://doi.org/10.1016/j.tourman.2011.06.002
5. Chang YC, Chen CF (2012) Service needs of elderly air passengers. J Air Transp Manag 18:26–29. https://doi.org/10.1016/j.jairtraman.2011.07.002
6. Costa GRV, Maior IMM, Lima NM (2005) Prodam webpage. http://www.prodam.sp.gov.br/multimidia/midia/cd_atiid/conteudo/ATIID2005/MR1/01/AcessibilidadeNoBrasilHistorico.pdf. Accessed 08 Apr 2018
7. Daniellou F, Rabardel P (2005) Activity-oriented approaches to ergonomics: some traditions and communities. Theor Issues Ergon Sci 6(5):353–357. https://doi.org/10.1080/14639220500078351
8. Dul J (2012) A strategy for human factors/ergonomics: developing the discipline and profession. Ergonomics 55(4):377–395. https://doi.org/10.1080/00140139.2012.661087
9. Henley Centre Headlightvision (2008) Amadeus Homepage. http://www.amadeus.com/amadeus/documents/corporate/TravellerTribes.pdf. Accessed 08 Apr 2018
10. Henley Centre Headlightvision (2015) Amadeus Homepage. http://www.amadeus.com/documents/future-traveller-tribes-2030/travel-report-future-traveller-tribes-2030.pdf. Accessed 08 Apr 2018

11. Leplat J (1990) Relations between task and activity: elements for elaborating a framework for error analysis. Ergonomics 33(10–11):1389–1402. https://doi.org/10.1080/00140139008925340
12. Poria Y, Reichel A, Brandt Y (2010) The flight experiences of people with disabilities: an exploratory study. J Travel Res 49(2):216–227. https://doi.org/10.1177/0047287509336477
13. Powell WW (2007) The international encyclopedia of organization studies. Sage Publishers, Sydney
14. Scott WR (2008) institutions and organizations –ideas and interests. Sage Publications, Stanford
15. Scott WR (2008) Lords of the dance: professionals as institutional agents. Organ Stud 29(2):219–238. https://doi.org/10.1177/0170840607088151
16. Silva ERP (2010). http://www.unifieo.br/pdfs/marketing/dissertacoes_mestrado_2010/ELIANA.pdf. Accessed 08 Apr 2018
17. Souza JBG (2014) UFSCAR webpage. https://repositorio.ufscar.br/bitstream/handle/ufscar/3448/6273.pdf?sequence=1&isAllowed=y. Accessed 08 Apr 2018
18. Souza JBG, Silva TNR, Lunardon L, Menegon NL (2015) Differences between established and existing: verification of service provision to passengers who need special assistance in air transport from the perspective of the new Brazilian Resolution. In: XIX international ergonomics association proceedings, Melbourne
19. Suen SL, Wolfe HP (2006). http://www.elderairtravel.com/NewDelhi.pdf. Accessed 03 Mar 2013
20. United Nations webpage (2012). http://esa.un.org/unpd/wpp/Documentation/pdf/WPP2010_Volume-II_Demographic-Profiles.pdf. Accessed 20 Nov 2012
21. Walton J (2013). http://blog.apex.aero/cabin-interior/passengers-size-face-big-challenges-aircraft-seats-smaller/. Accessed 10 Mar 2013
22. Wolfe HP (2012) http://www.elderairtravel.com/eldertravel.htm. Accessed 20 Dec 2012
23. World Health Organization (2011) http://www.who.int/disabilities/world_report/2011/en/. Accessed 10 Mar 2018
24. World Health Organization (2014). http://www.who.int/nmh/publications/ncd-status-report-2014/en/. Accessed 20 Feb 2018
25. World Health Organization (2015). http://www.who.int/ageing/events/world-report-2015-launch/en/. Accessed 20 Feb 2018

Suppressed Articulatory Rehearsal Mechanism and Driving Errors

Sajad Najar and Premjit Khanganba Sanjram(✉)

Human Factors and Applied Cognition Lab, Indian Institute of Technology Indore, Simrol, Indore 453552, India
sajadanajar@gmail.com, sanjrampk@iiti.ac.in

Abstract. Drivers get driving related information mainly through visual, auditory, and haptic sensory channels but it is predominantly based on the information received through visual senses. In working memory visual information fades away faster than the auditory information and in order to retain the visual information for a longer duration it gets recoded into phonological information through Articulatory Rehearsal Mechanism (ARM) [1]. After every 2 s, ARM recites and rehearses the phonological information making it to re-enter into the phonological store, where it starts to decay again immediately [1, 2]. Individuals when engaged in processing visual information in order to perform driving and if there is suppression of ARM chances are high that visual information processing will be compromised. This distraction is ought to suppress the visual information from being rehearsed and remembered acoustically. The present study investigates the effect of suppression of ARM on driving performance in terms of driving errors. 30 drivers voluntarily participated in the study. They drove an instrumented vehicle and were required to follow certain directions displayed on signboards. The signboards were installed along a two-lane track. Drivers were randomly assigned to one of the three conditions of suppression of ARM namely non-suppression, simple suppression, and complex suppression. Driving errors were analyzed in terms of slips and lapses. The results indicate that there are significantly more driving errors under complex suppression as compared to other two conditions of suppression (i.e., non-suppression and simple suppression). Further analysis reveals that there are significantly more cases of slips than lapses.

Keywords: Suppressed articulatory rehearsal mechanism (ARM)
Distracted driving · Driving errors · Slips · Lapses

1 Introduction

Distraction is an important factor concerning compromise of driving performance. During driving drivers could face various potential distractors (e.g., audiovisual entertainment systems, navigation systems, information and communication systems, eating, drinking, and talking to co-passenger). Driving is predominantly visual in nature and visual information fades away as fast as 0.5–1.0 s unless rehearsed by Articulatory Rehearsal Mechanism (ARM). ARM in working memory (WM) which recites and recodes visual information into phonological information and rehearses it in

S. Bagnara et al. (Eds.): IEA 2018, AISC 823, pp. 55–61, 2019.
https://doi.org/10.1007/978-3-319-96074-6_6

order to prevent it from decaying [3–5]. It does not only recite phonological information but also recodes that visual information which has a verbal label (e.g., speed limit 30 km/h written on the traffic sign board has a verbal label of thirty) into phonological information. Information presented in a visual form is named or labeled and the phonological information produced from this labeling process then gets access to the phonological store where it is recited again and again to stay for a longer duration. This recitation process of ARM makes the phonological information re-enter into the phonological store where it starts to decay again. The process of reciting and refreshing the information is called as articulatory rehearsal. Phonological coding helps in situations where the delay is more between encountering the information and executing it in terms of actions [6], one such situation is driving. There is an increase in the use of wireless nomadic devices inside the vehicles from the last decade which gives rise to many new sources of distractions, it is highly possible that visual information could be suppressed from entering into phonological store for articulatory rehearsal. In this view, the current study is undertaken to investigate the role of suppression of ARM on driving errors.

2 Driving Error

Errors are treated to have occurred when the planned actions fail to achieve their intended consequences [7]. When the action is carried out incorrectly, the error is termed as slip and when the action is simply omitted or not carried out at all, the error is termed as lapse. Since driving error has been a major concern for road safety researchers, a number of methods have been developed for measuring driving error including the Driver Behaviour Questionnaire (DBQ) [7] and the Wiener Fahrprobe method [8]. Studies on distracted driving and driving errors have focused on issues related to mobile phone use [9]; sending texts and e-mails while driving [10]; and roadside advertising and information billboards [11]. A synthesis of specific driver distractions – like speaking on phone (both hand-held and hands-free), sending and receiving texts, using GPRS – research reveals that driver distractions reduces longitudinal control [12]; reduces lateral control [13]; increases reaction time [14]; reduces situation awareness [15]; and impairs hazard detection and response [16]. However, the link between distraction and different error types is often not clear, because distraction has been viewed as a driving error rather than a causal factor by a number of driving error taxonomies [7, 17, 18]. According to these taxonomies, due to some distraction if a driver is not able to stop at a red traffic signal it would be listed as a driving error. Other taxonomies consider distraction as a causal factor for driving errors but do not indicate the mechanisms by which it contributes [19, 20]. Wierwille et al. [20] for example, list internal and external distraction as one of the factors contributing to recognition errors, but do not indicate how distraction contributes to these errors. In this perspective the current paper makes an attempt, in terms of suppressing ARM, for establishing a causal relationship between distracted driving and driving errors. In this study, driving errors are analyzed in terms of Reason's [7] slips and lapses. Within this framework when slips and lapses occur the individual's intentions (plans) are correct, but the execution is flawed. When the action is carried out incorrectly, the error is termed as slip (comprised of- changing lanes wrongly, doing lane excursions, and over/under speeding in the current study) and when the action is simply omitted or not carried out at all, the error is termed as lapse (comprised of- forgetting to indicate lane change, forgetting to execute lane change, and forgetting to turn off the indicator after changing the lane in the current study).

3 Methodology

An instrumented vehicle was driven by 30 participants on a two lane track with a total length of 1 km. 4 pairs of direction signboards (with a distance of 250 m away from each other) were installed on both sides along the track, where the direction shown on one side was a copy of the direction shown on the other side of the track. Each pair of directions were marked by two signs "↑" and "X". "↑" sign indicates that the driver has to take the corresponding lane, while the other lane was marked by a "X" sign, which indicates that the driver has to avoid driving in the corresponding lane. The starting lane of the drive was decided on random basis which came out to be the left lane. The height of the direction signboards, color of the figure and background, and the size of the "↑" and "X" were decided by strictly following the Code of Practice for Road Signs (Third Revision) published by Indian Roads Congress [21]. Drivers were required to drive within the speed range of 20–45 km/h.

There were three levels of suppression of ARM, non-suppression (NS), simple suppression (SS), and complex suppression (CS). Equal number of drivers were randomly assigned to each level. Participants in NS group were not exposed to any kind of suppression of ARM so it served as a control group. The other two groups of participants were exposed to CS of ARM and SS of ARM. Participants in CS were instructed to start count down rapidly from 50 as soon as they first time look at the signboard and in the case of SS, they were instructed to count rapidly from 1–30 (contrasted with counting down from 50 in case of CS) as soon as they first time look at the signboard. This process continued after crossing every direction signboard for both CS as well as for SS.

3.1 Design

This experiment employs a 3 Suppression of ARM (NS vs. SS vs. CS) between-groups design. In the current framework, the levels of suppression of ARM are relative to each other where SS is an intermediate level of suppression (NS is a control, i.e., no suppression).

Data related to driving performance were collected through an instrumented vehicle (Volkswagon POLO, 2015 Model, Hatchback 1.2 petrol highline, right-hand drive), and Video Velocity Box (20 Hz GPS video data logger with a 4 camera video system). The data was collected in terms of 6 parameters- lane excursions, over/under speeding, changing lanes wrongly, forgetting to indicate lane change, forgetting to execute lane change, and forgetting to turn off the indicator after changing the lane. For capturing the front view of driving, one camera was attached to the top of the dashboard near the windshield of the car; to capture the usage of direction indicators, one camera was attached to the steering column of the car; the third and fourth cameras were attached on the left and right side of the bonnet of the car in order to capture the lane excursions to the left and right side respectively.

3.2 Participants

Participation to this naturalistic 'distracted driving study' was open to only those who possessed valid driving licenses and had normal vision. A set of data collected from 30 male drivers were analyzed (*M* = 30.16 years; *SD* = 7.66; *age range* = 19–43 years). The participants received 250 INR each for their voluntary participation in the study.

3.3 Procedure

Participants were briefly explained about the task they had to undergo so that they could arrive at an informed decision about their participation in the study. Each participant first went through a practice trail so that he/she can get acquainted with the instrumented vehicle. The participants were taken to the experimental track (for actual data recording) only when they were confident about all the functions of the instrumented vehicle. The experimenter was seated in the front passenger seat (next to driver) so that he could start and stop the recording devices, to instruct the participant to start and stop driving, and also for experimental observation and planning for consequent appropriate debriefing. Post data collection, each participant was asked some probing questions about his/her experience while driving.

4 Results

In order to analyze the effect of suppression of ARM on driving performance, this paper incorporates Reason's [7] slips and lapses. Statistical significance was tested at an alpha level of 0.05. An ANOVA was carried out in order to determine whether the three levels of suppression differ from each other with respect to their effect on slips and lapses separately. It was found that the three levels of suppression significantly differ from each other with respect to their effect on slips (NS, $M = 1.16$; SS, $M = 2.79$; and CS, $M = 3.29$), $F(2, 27) = 7.84$, $p = 0.002$, $\eta p2 = 0.36$. LSD analysis also reveals a highly significant difference between CS and NS, $p = 0.001$; whereas the differences between NS and SS, and SS and CS are insignificant. With respect to lapses also the three levels of suppression differ significantly (NS, $M = 0.26$; SS, $M = 0.96$; and CS, $M = 1.63$), $F(2, 27) = 10.00$, $p = 0.001$, $\eta p2 = 0.42$. Like slips, LSD analysis for lapses also reveals that CS and NS are significantly different, $p = 0.001$ whereas the differences between NS and SS, and SS and CS are insignificant (Fig. 1).

The analysis of overall driving error with respect to the three levels of suppression of ARM was performed. ANOVA revealed that there is a statistically significant difference among the three levels of suppression with respect to their effect on driving performance (driving errors), (NS, $M = 1.433$; SS, $M = 3.766$; and CS, $M = 4.933$), $F(2, 27) = 12.138$, $p = 0.001$, $\eta p2 = 0.473$. LSD analysis reveals a significant difference between CS and NS, $p = 0.001$; and NS and SS also differ significantly $p = 0.003$.

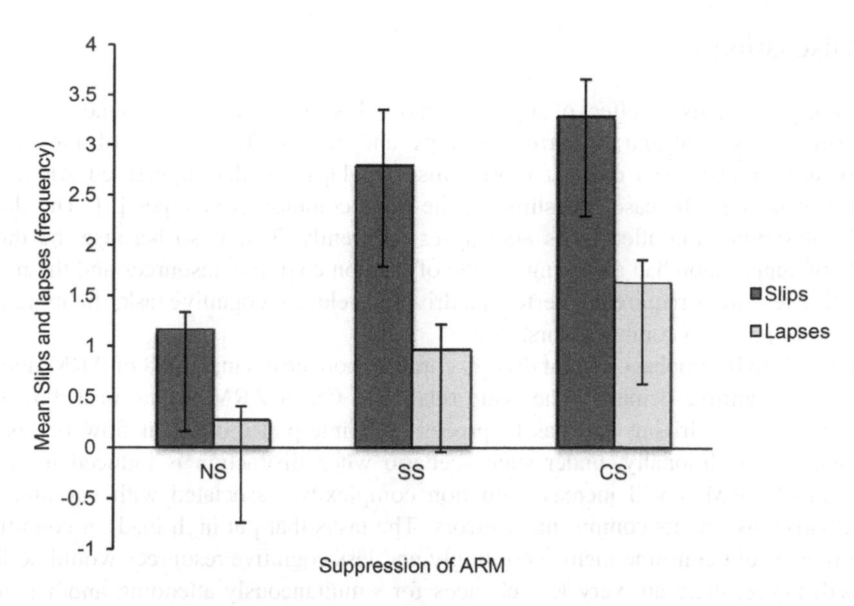

Fig. 1. Comparison of slips and lapses (frequency) (Color figure online)

LSD analysis also indicated that there is no significant difference between SS and CS. It was also found that in comparison to lapses (M = .955), it was the slips (M = 2.422) that are committed more by the drivers. Figure 2 shows the contribution in slips and lapses as a result of three levels of suppression of ARM.

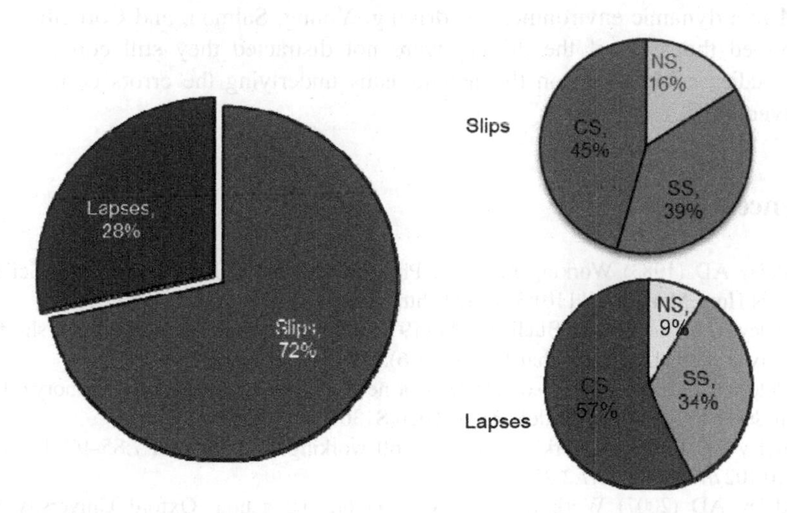

Fig. 2. Contribution in slips and lapses as a result of three levels of suppression of ARM (Color figure online)

5 Discussion

This study examines the effect of suppression of ARM on driving performance. Driving performance is analyzed in terms of slips and lapses. The results of the study demonstrate that drivers commit more slips than lapses under suppressed ARM. In general it is often the case that slips are the most common error types [7]. The three levels of suppression affect slips and lapses differently. This is so because the three levels of suppression had a varying degree of load on cognitive resources and the more cognitive resources required to perform a driving irrelevant cognitive task, the more are the drivers prone to commit errors.

It needs to be emphasized that drivers commit more errors under CS of ARM due to additional cognitive demand. The issue related to CS of ARM is that in a dynamic environment like driving one has to process and interpret a constant flow of visual information. Additionally, under such scenario when distraction is induced by suppression of ARM it will increase situation complexity (associated with variation in workload) thus drivers commit more errors. The tasks that put high load on cognitive resources would consume them accordingly and less cognitive resources would be left unused, hence, there are very less chances for simultaneously attending another task without any compromise in either of the task [22]. The current study was designed in such a manner that the second task (counting backward and forward under CS and SS respectively) was to be performed necessarily while driving. Driving itself is a complex task performed in a dynamic environment and it accordingly consumes cognitive resources, when an additional cognitively challenging task (e.g., counting backward) is simultaneously performed, driving performance is significantly compromised.

An aspect of error analysis is that of errors committed in the absence of suppression of ARM in a dynamic environment of driving. Young, Salmon, and Cornellison [23] also reported that even if the drivers were not distracted they still commit errors. Further studies could focus on the mechanisms underlying the errors committed by such drivers.

References

1. Baddeley AD (1983) Working memory. Philos Trans R Soc Lond Ser B Biol Sci Funct Aspects Hum Mem 302(1110):311–324. https://doi.org/10.1098/rstb.1983.0057
2. Baddeley AD, Thomson N, Buchanan M (1975) Word length and the structure of short-term memory. J Verbal Learn Verbal Behav 14(6):575–589
3. Baddeley AD (2000) The episodic buffer: a new component of working memory? Trends Cogn Sci 4:417–423. https://doi.org/10.1016/S1364-6613(00)01538-2
4. Baddeley AD (2002) Is working memory still working? Eur Psychol 7:85–97. https://doi.org/10.1027//1016-9040.7.2.85
5. Baddeley AD (2007) Working memory, thought, and action. Oxford University Press, Oxford
6. Henry L (2011) The development of working memory in children: discoveries and expansions in child development. Sage Publications, London
7. Reason J (1990) Human error. Cambridge University Press, Cambridge

8. Risser R, Brandstätter C (1985) Die Wiener Fahrprobe: Freie Beobachtung, Kleine Fachbuchreihe des Kuratoriums für Verkehrssicherheit, Band 21
9. Horrey WJ, Wickens CD (2006) Examining the impact of cell phone conversations on driving using meta-analytic techniques. Hum Factors 48:196–205. https://doi.org/10.1518/001872006776412135
10. Young KL, Rudin-Brown CM, Patten C, Ceci R, Lenne MG (2014) Effects of phone type on driving and eye glance behavior while text messaging. Saf Sci 68:47–54. https://doi.org/10.1016/j.ssci.2014.02.018
11. Herrstedt L, Greibe P, Andersson P (2013) Roadside advertising affects driver attention and road safety. In: Proceedings of the 3rd international conference on driver distraction and inattention, Gothenburg, Sweden (No. 05-P), pp 1–14
12. Hosking SG, Young KL, Regan MA (2009) The effects of text messaging on young drivers. Hum Factors 51(4):582–592. https://doi.org/10.1177/0018720809341575
13. Engström J, Johansson E, Östlund J (2005) Effects of visual and cognitive load in simulated motorway driving. Transp Res Part F Traffic Psychol Behav 8(2):97–120. https://doi.org/10.1016/j.trf.2005.04.012
14. Maciej J, Vollrath M (2009) Comparison of manual vs. speech-based interaction with in-vehicle information system. Accid Anal Prev 41:924–930. https://doi.org/10.1016/j.aap.2009.05.007
15. Kass SJ, Cole KS, Stanny CJ (2007) Effects of distraction and experience on situation awareness and simulated driving. Transp Res Part F Traffic Psychol Behav 10:321–329. https://doi.org/10.1016/j.trf.2006.12.002
16. Smiley A, Smahel T, Donderi DC, Caird JK, Chisholm S, Lockhart J, Teteris E (2007) The effects of cellphone use and CD use on novice and experienced driver performance. Insurance Bureau of Canada, Ottawa
17. Sabey BE, Staughton GC (1975) Interacting roles of road environment, vehicle and road user in accidents. In: Proceedings of the fifth international conference of the international association for accident and traffic medicine (IAATM). IAATM, London, pp 1–17
18. Sabey BE, Taylor H (1980) The known risks we run: the highway. In: Schwing RC, Albers WA Jr (eds) Societal risk assessment: how safe is safe enough? Plenum Press, New York
19. Staubach M (2009) Factors correlated with traffic accidents as a basis for evaluating advanced driver assistance systems. Accid Anal Prev 41(5):1025–1033. https://doi.org/10.1016/j.aap.2009.06.014
20. Wierwille WW, Hanowski RJ, Hankey JM, Kieliszewski CA, Lee SE, Medina A, Keisler AS, Dingus TA (2002) Identification and evaluation of driver errors: overview and recommendations. U.S. Department of Transportation
21. Indian Roads Congress (2012) Code of Practice for Road Signs (3rd ed) IRC, New Delhi
22. Lavie N, Hirst A, de Fockert JW, Viding E (2004) Load theory of selective attention and cognitive control. J Exp Psychol Gen 133(3):339–354. https://doi.org/10.1037/0096-3445.133.3.339
23. Young Kl, Salmon PM, Cornelissen M (2013) Distraction-induced driving error: an on-road examination of the errors made by distracted and undistracted drivers. Accid Anal Prev 58:218–225. https://doi.org/10.1016/j.aap.2012.06.001

Impact of Waiting Times on Risky Driver Behaviour at Railway Level Crossings

Grégoire S. Larue[1,2(✉)], Ross Blackman[1], and James Freeman[1]

[1] Centre for Accident Research and Road Safety – Queensland, Queensland University of Technology (QUT), Brisbane, Australia
g.larue@qut.edu.au

[2] Australasian Centre for Rail Innovation (ACRI), Canberra, Australia

Abstract. Increased road and rail traffic in Australia results in actively protected crossings being closed for extended periods of time during peak hours. This results in road congestion. It is known that extended periods of warning/waiting times at level crossings have impacts on drivers' decision making in regards to violating crossing rules. Excessive waiting times could lead to non-compliant behaviour by motorists, resulting in incidents, including injuries and fatalities. However, the correlation between waiting time and rule violation is not well documented, although it is known that a range of personal and environmental factors influence rule non-compliance. This leads to the question of whether longer waiting times affect motorists' assessment of risk and how long motorists are prepared to wait at level crossings before undertaking risky behaviour. A driving simulator study was used to obtain objective measures of railway level crossing (RLX) rule violations. Sixty participants completed six driving tasks each, with the tasks varying in terms of waiting times. Compliance with road rules at the level crossing during the simulated drives was examined. Main results include that increased waiting times result in increased likelihood of risky driving behaviour, particularly for waiting times longer than three minutes. Risky driving behaviours included entering the activated crossing before boom gates are down; entering the crossing after the train passage but before signals are deactivated; and stopping/reversing on the crossing. The results suggest that, where possible, waiting times should be standardized at values lower than three minutes in order to reduce the likelihood of risky road user behaviour.

1 Introduction

Crashes at level crossings are rare but catastrophic events. In Australia between 2002 and 2012 [1], 31% of rail-related injuries and fatalities (excluding suicides) occurred at railway level crossings. The cost of level crossing crashes in terms of lives and serious injuries is over-represented in the statistics [2]. While numbers of collisions have decreased, the most recently available statistics show that the number of fatalities and serious injuries at level crossings are in fact stagnating [1].

Previous research has shown road user behaviour to be the most prominent factor in collisions between trains and road users at railway level (at-grade) crossings [3, 4]. Among the factors influencing non-compliance with railway level crossing (RLX) rules

S. Bagnara et al. (Eds.): IEA 2018, AISC 823, pp. 62–69, 2019.
https://doi.org/10.1007/978-3-319-96074-6_7

are errors of perception, errors of judgement and deliberate attempts to circumvent safety-oriented countermeasures (including regulatory and engineering measures) [5–7]. Within the category of deliberate circumvention, road user frustration and reluctance to wait as directed has been identified as one of the underlying factors in RLX violations [8].

Opportunities to circumvent RLX rules are seen to vary according to the specific RLX conditions. As such, research has examined driver and pedestrian responses and compliance in relation to some of these conditions, including the common measures designed to prevent rail and road user collisions. Such research has compared driver responses to passive and active controls [4, 9, 10] as well as to different types of active controls [11], with combinations of boom gates and active signals generally found to be most effective [12]. To a lesser extent, research has also addressed issues around road traffic volumes and congestion in regard to driver behaviour and RLX safety [13, 14]. The finding that drivers are more likely to be killed in a level crossing crash during peak periods than at other times [13] suggests that road traffic congestion creates a particularly hazardous RLX environment, but the specific mechanisms involved have received relatively little research attention.

Increased road and rail traffic in Australia results in actively protected crossings being closed for extended periods of time during peak hours [8, 15]. This results in road congestion. It is known that extended periods of warning/waiting times at level crossings have impacts on drivers' decision making in regards to violating crossing rules. Excessive waiting times lead to non-compliant behaviour by motorists [8, 15], and could result in incidents, including injuries and fatalities. However, the correlation between waiting time and rule violation is not well documented, although it is known that a range of personal and environmental factors influence rule non-compliance. This leads to the question of whether longer waiting times affect motorists' assessment of risk and how long motorists are prepared to wait at level crossings before undertaking risky behaviour.

The current paper seeks to address an identified gap in research regarding the relationship between RLX waiting times and risky driver behaviour. A clearer understanding of this relationship is needed to enhance the depth and reliability of evidence on which to base refined measures for improving RLX safety. This study therefore examines whether drivers' behaviour at level crossings becomes more risky with longer waiting times.

For the purpose of the current paper, 'waiting time' can be defined as the time it takes to traverse the crossing, from when first stopped before the crossing to when the crossing has been passed completely. 'Warning time' can be defined as the time between initial RLX signal activation to the time when the train arrives at the crossing. Waiting time effectively includes a proportion of the warning time, the amount of which depends on how soon a driver stops after activation of the RLX signals.

2 Methods

2.1 Design and Measures

An Advanced Driving Simulator (Fig. 1) was used to obtain measurements of the frequency of violations of relevant RLX rules, specifically: entering the activated crossing before the boom gates are down; entering the crossing after the train passage but before the flashing lights are deactivated; stopping and/or reversing on the crossing.

Fig. 1. View of the driving simulator from the control room. The level crossing was just deactivated, and is congested.

The experiment used a mixed design with two factors: warning time and waiting time. 'Waiting time' was a within-subject factor with six levels, operationalised in the current study as the time from when the driver first stops before the crossing to the moment the driver has passed the crossing. It includes the time needed to reach the crossing in case of congestion, the time stopped at the crossing and the time to drive through the crossing beyond the stop line on the other side (for the traffic in the other direction). 'Warning time' was a between-groups factor with three levels (Groups A, B and C), being the time between activation of the flashing lights at the crossing and the arrival of the (first) train at the crossing.

Realistic warning time values were selected: 28 s (Group A, minimum warning time); 75 s (Group C, long warning time); and 50 s (Group B, approximate average of the previous values). Six levels of waiting times were selected for this study: 1.5, 2, 3, 5, 8 and 10 min, obtained by varying the number of consecutive trains going through the crossing, as well as the number of road vehicles at the crossing.

A third of participants (selected randomly within each group) experienced increasing waiting times, another third experienced decreasing waiting times, while the remainder experienced only shorter waiting times (<3 min) and were the control group (baseline).

Participants first undertook a familiarisation drive to acquaint themselves with the driving simulator and the task they had to perform. The task was to drive in a suburban environment to an office building for an appointment (job interview) for which they should try to arrive on time, following an itinerary where the participant encountered one active level crossing with flashing lights, boom gates and bells. The itinerary took around three minutes to complete if the crossing was inactive (no trains present). For all six driving tasks participants were requested to drive as they normally would if driving on a public road.

A monetary incentive/disincentive scheme was used to induce time pressure and a risk/reward component in the simulated driving tasks, similar to the methods used successfully by Fairclough and Spiridon [16] and Lee [17]. All participants were offered $50 as compensation for their time and effort at the initial recruitment stage, but were not informed about any other monetary incentives/disincentives until arriving at the testing station. For the incentive, participants were informed that they would be given an extra $5 in each instance that they arrived at the destination within six minutes of commencing the drives (i.e., 'on time'). A digital clock was positioned on the centre display console of the vehicle so that participants could track time as they progressed. For the disincentive, participants were told that they may be penalized the amount of $5 for committing a traffic violation, if detected in the scenario by police, including 'speeding, running a red light, or violating the level crossing rules'. They were told that this penalty would also apply if they crashed the car. The incentive/disincentive scheme constituted a 'minor deception' in that, contrary to the information provided before the test drives, they did not actually receive any extra payments for arriving 'on time' nor receive any penalties for violations.

A questionnaire administered before the simulation gathered information on socio-demographic characteristics of the participants.

2.2 Analysis

The analysis focused on compliance and violations for the different scenarios using generalized linear mixed models (GLMMs), as they are adapted for repeated measures design. Waiting times, warning times, drive number, and sequence of scenarios were used as independent factors that influence decision making and rule compliance, as well as their potential interactions.

2.3 Participants

A total of 62 male participants with a valid driver's licence were recruited for the study. Only male drivers were recruited as they are over-represented in RLX violation and fatal crash statistics in Australia and all parts of the world [5, 12, 18] and also more likely to be involved in violations at level crossings [2]. Therefore it was considered that such a sample would provide more relevant data relating to violations than a male/female sample. All participants who completed the six simulator test drives and the Time 1 and 2 questionnaires (n = 60) received $50 for their time and effort.

Excluding the two participants who did not complete the test drives, the mean age of participants was 29.4 years (SD 10.4), ranging from 18 to 63 years. Participants reported a mean period of 11.5 years since they first obtained a driver's licence (SD 10.1), with a range of 0.33 to 39.0 years.

3 Results

Analysis showed no meaningful effect of warning times on the probabilities of the driver (a) entering the crossing while activated (before boom gates down), (b) entering the crossing before it was completely deactivated (gates up, lights still flashing), nor (c) stopping on the level crossing (behind congested traffic). The following presentation and discussion of results therefore focuses on the effect of waiting times in the simulated driving tasks.

The most common violation observed was entering the level crossing while it was activated (lights flashing) but the gates were not down yet, with 23 participants (38%) entering the crossing under such conditions. With 50 such violations in total, these participants committed this violation on average 2.2 times over the 4 drives in which they had the opportunity to do so (generally this was not possible in Scenarios 5 and 6).

The next most common transgressions were to stop on the yellow marking at the crossing when there was congestion on the other side of the crossing, and to enter the crossing before signals were deactivated. Fifteen participants (25%) over 18 occurrences stopped on the yellow marking, apparently not realizing they would be unable to proceed completely through the crossing when they decided to undertake the manoeuvre. A further two participants stopped momentarily on the crossing, but reversed back when they realised that they would become stuck on the crossing behind congested traffic. Participants stopping on the crossing committed this violation on average 1.2 times out of their two opportunities to stop on the crossing behind stationary traffic, suggesting that most learnt from their initial mistake.

The odds ratio of entering the crossing early were higher by an overall factor of 16.2 for participants who encountered the longer waiting times first ($t = 2.75$, $p = .008$). The odds ratio was reduced by a factor 2.5 for each drive performed for this group of participants ($t = -2.83$, $p = .005$). The effect of waiting time was also significant for this group of participants, with a reduction by a factor 1.61 times the amount of time waited at the crossing ($t = -2.30$, $p = .022$).

For entering the crossing after the train passed but before the lights were deactivated, the odds ratio of violating increased by a factor .62 ($t = 3.49$, $p < .001$) for each drive of the participant, showing an effect of driving the same road six times in a row. Waiting times also increased this ratio by a factor 1.28 ($t = 2.35$, $p < .001$), but this was compensated through its interaction with the drive number, which decreased the ratio by a factor 1.28 for each drive ($t = -4.97$, $p < .001$). The effect of driving the scenarios with increasing waiting times resulted in an overall reduction of this type of violation as compared to the baseline by a factor of −3.86 ($t = -3.01$, $p = .004$). This effect was complemented with an interaction with the waiting time of factor 1.25 ($t = 4.38$, $p < .001$).

For baseline participants, the probability of entering the crossing before it was deactivated tended to decrease as the waiting time increased. For the participants experiencing the scenarios with increasing waiting times, the violation rate increased from scenario 1 to 4, with a larger increase as waiting times increased. It must be noted that participants in this group tended to be more respectful of the rules as compared to the baseline participants, as can be seen with their initial violation rates for scenario 1. The increasing trend stopped when participants experienced congested level crossing in scenarios 5 and 6. For participants experiencing decreasing waiting times, the violation rate increased as waiting times decreased.

On the probability of stopping on the level crossing (yellow grid), for the participants experiencing increasing waiting times the odds ratio was lower by a factor 5.06 ($t = -4.29$, $p < .001$) for scenario 6 as compared to scenario 5. The ratio was lower for participants experiencing the decreasing waiting times by a factor 1.32 ($t = -4.61$, $p < .001$) for scenario 5, and by a factor .26 for scenario 6 ($t = -4.69$, $p < .001$). This shows that participants tended to not make the same mistake twice, and were more likely to stop on the crossing the first time they arrived at the congested crossing. Violation rate was higher for the participants who experienced the increasing waiting times, as these scenarios were their last two driving scenarios, while they were the first two scenarios for the group of participants experiencing decreasing waiting times.

4 Discussion

With the effects of waiting times disentangled from the effects of task repetition, the total waiting time was found to have an effect on participants' engagement in RLX violations. Considering all types of observed risky behaviour, participants in the current study engaged in more RLX violations when they had to wait at the crossing longer than three minutes. However, this time threshold varied according to the specific violation in question. Additionally, reducing the waiting time below the three minute threshold did not result in an immediate reduction in violations for drivers who experienced decreasing waiting times. For these participants, waiting times had to be below two minutes to observe an improvement of their driving behaviour.

While there was no increase in the likelihood of entering the activated crossing while lights were flashing but before the boom gates closed for the baseline group and the group experiencing ascending waiting times, this likelihood tended to increase for participants experiencing the decreasing waiting times. These results suggest that once participants are accustomed to a level crossing with high level of congestion and extended waiting times, it is difficult for them to return to the baseline behaviour, even when waiting times are decreasing. The effects of decreasing waiting times are not immediate, suggesting a delay for participants in realising that the situation is improving at the crossing. The study also suggests that waiting times up to four minutes were not sufficient to result in an increased engagement in this risky behaviour at the crossing. Waiting times longer than four minutes and involving high levels of road congestion at the crossing (resulting in eight to ten minute wait times in this study) appear to stimulate engagement in this risky behaviour. To return to a baseline situation, waiting times shorter than two minutes were required.

Entering the RLX after a train had passed but before the lights were deactivated was found to be associated with longer waiting times. While the likelihood of engagement in this behaviour did not change for the baseline group, engagement increased for the two other groups when waiting times were longer than three minutes (both ascending and descending waiting times). The likelihood of engaging in this behaviour increased with waiting time for the participants experiencing increasing waiting times, until road congestion was present at the exit of the crossing. When congestion was present at the exit of the crossing, there was less incentive to go through the crossing before complete deactivation, as the road traffic would block participants after or on the crossing. In the case of the participants experiencing decreasing waiting times, the probability of engagement in this violation increased while waiting times were decreasing. As for the RLX violation discussed in the preceding paragraph, it appears that once a level of frustration or impatience has stimulated engagement in risky behaviour, reduced waiting times thereafter do not produce an immediate or certain return to the original driving behaviour.

Waiting times did not have an effect on stopping on the level crossing. The main effect leading to this risky behaviour appeared to be a lack of experience of this situation. In other words, drivers were more likely to make an error the first time that they were blocked on the crossing due to the stationary traffic ahead. Participants were not likely to be blocked on the crossing a second time, suggesting that the observed behaviours were due to errors rather than deliberate violations and that participants rapidly learnt to be alert to the risk of this occurrence. This study also found that the group experiencing increasing waiting times was more likely to engage in this level crossing violation. This is likely to be due to higher levels of cumulative frustration among participants who drove the longer scenarios last, and the repetition of the scenarios was found to be related to increased engagement in this behaviour.

5 Conclusions

The current study has demonstrated through an Advanced Driving Simulator experiment that increased RLX waiting times result in an increased likelihood of risky driving behaviour, particularly for waiting times longer than three minutes. The analysis has shown that while waiting times are of high importance for increasing the likelihood of risky driver behaviour at railway level crossings, the presence of congestion and the need to stop at crossings for multiple cycles are additional factors likely to compound driver frustration and impatience. Overall, the results further reinforce the recognised need to minimize road congestion around railway level crossings, and provide the rail industry with a recommended maximum waiting time to ensure road users' violations at level crossings remain unlikely. This threshold may depend between countries, and the current study applies mainly to the Australian context. The study has also highlighted the need for treating congestion early in its development, as reducing congestion at level crossings does not immediately result in improved road user behaviour.

Acknowledgements. The authors gratefully acknowledge the Australasian Centre for Rail Innovation (ACRI) for funding this research (Project LC/7-8).

References

1. Australian Transport Safety Bureau, Australian Rail Safety Occurrence Data 1 July 2002 to 30 June 2012 (2012) Australian Transport Safety Bureau, Canberra
2. Searle A, Di Milia L, Dawson D (2012) An investigation of risk-takers at railway level crossings. CRC for Rail Innov, Rep 2:79
3. McCollister GM, Pflaum CC (2007) A model to predict the probability of highway rail crossing accidents. Proc Inst Mech Eng, Part F: J Rail Rapid Transit 221(3):321–329
4. Tey L-S, Ferreira L, Wallace A (2011) Measuring driver responses at railway level crossings. Accid Anal Prev 43(6):2134–2141
5. Abraham J, Datta TK, Datta S (1998) Driver Behaviour at Rail-Highway Crossings. Transp Res Rec 1648:28–34
6. Freeman J, McMaster M, Rakotonirainy A (2015) An exploration into younger and older pedestrians' risky behaviours at train level crossings. Safety 1(1):16–27
7. Mulvihil CM et al (2016) Using the decision ladder to understand road user decision making at actively controlled rail level crossings. Appl Ergon 56:1–10
8. Naweed A et al (2016) Level with me: Human factors in pedestrian and road-user violations at a notorious Victorian railway level crossing. Road Transp Res 25(2):40–47
9. Lenné MG et al (2011) Driver Behaviour at Rail Level Crossings: Responses to Flashing Lights, Traffic signals and Stop Signs in Simulated Rural Driving. Appl Ergon 42:548–554
10. Liu J et al (2015) What are the differences in driver injury outcomes at highway-rail grade crossings? Untangling the role of pre-crash behaviors. Accid Anal Prev 85:157–169
11. Rudin-Brown CM et al (2012) Effectiveness of traffic light vs. boom barrier controls at road-rail level crossings: A simulator study. Accid Anal Prev 45:187–194
12. Raub R (2009) Examination of highway-rail grade crossing collisions nationally from 1998 to 2007. Transp Res Rec J Transp Res Board 2122(1):63–71
13. Hao W, Daniel J (2013) Severity of Injuries to Motor Vehicle Drivers at Highway-Rail Grade Crossings in the United States. Transp Res Rec J Transp Res Board 2384:102–108
14. Oh J, Washington SP, Nam D (2006) Accident prediction model for railway-highway interfaces. Accid Anal Prev 38(2):346–356
15. Larue GS, Naweed A, Rodwell D (2018) The road user, the pedestrian, and me: Investigating the interactions, errors and escalating risks of users of fully protected level crossings. Saf Sci
16. Fairclough SH, Spiridon E (2012) Cardiovascular and electrocortical markers of anger and motivation during a simulated driving task. Int J Psychophysiol 84(2):188–193
17. Lee Y-C (2010) Measuring Drivers' Frustration in a Driving Simulator. Proc Hum Factors Ergonom Soc Ann Meet 54(19):1531–1535
18. Bureau of Infrastructure Transport and Regional Economics (2013) Road deaths Australia - 2012 statistical summary. Department of Infrastructure and Transport, Editor. Australian Government, Canberra, Australia

Keeping the Driver in the Loop: The 'Other' Ethics of Automation

Victoria Banks[1], Emily Shaw[2], and David R. Large[2(✉)]

[1] Human Factors Engineering,
Transportation Research Group, Southampton, UK
[2] Human Factors Research Group, University of Nottingham, Nottingham, UK
david.r.large@nottingham.ac.uk

Abstract. Automated vehicles are expected to revolutionise everyday travel with anticipated benefits of improved road safety, comfort and mobility. However, they also raise complex ethical challenges. Ethical debates have primarily centred around moral judgements that must be made by autonomous vehicles in safety-critical situations, with proposed solutions typically based on deontological principles or consequentialism. However, ethics should also be acknowledged in the design, development and deployment of partially-automated systems that invariably rely upon the human driver to monitor and intervene when required, even though they may be ill-prepared to do so. In this literature review, we explore the lesser-discussed ethics associated with the role of, and expectations placed upon, the human driver in partially-automated vehicles, discussing factors such as the marketing and deployment of these vehicles, and the impact upon the human driver's development of trust and complacency in automated functionality, concluding that the human driver must be kept 'in the loop' at all times.

Keywords: Ethics · Automated driving · Literature review

1 Introduction

Automated driving is expected to enhance road safety by reducing the number of accidents attributable to driver error, improve comfort by reducing driver workload and enabling drivers to engage in secondary tasks and activities, and deliver 'mobility for all' [1]. These are undeniably bold and admirable claims, which on face value suggest that the current drive towards the 'driverless car' favours the 'greater good' and should be applauded and encouraged. However, the visionary hype surrounding the concepts of 'self-driving', 'driverless' and 'autonomous cars' is masking a disconnect between public expectations and current technological capabilities [2, 3]. Journalists in the popular media are failing to clarify important differences in terminology that define the varying levels of automation within the automotive industry. Vehicle manufacturers are also using choice words to describe product capabilities that carry a multitude of interpretations [4]. For example, it has been suggested that Tesla's choice of the name 'Autopilot' for its partially automated driving feature suggests that it has fully autonomous capability and therefore, promises more than it can deliver [5, 6]. Indeed, Tesla

S. Bagnara et al. (Eds.): IEA 2018, AISC 823, pp. 70–79, 2019.
https://doi.org/10.1007/978-3-319-96074-6_8

faced intense media scrutiny following the first fatal collision involving one of its vehicles being operated in Autopilot mode in May 2016. The driver of a Tesla Model S was fatally injured when their vehicle collided with a tractor trailer that was crossing an intersection on a highway west of Williston, Florida. Data taken directly from the Tesla Model S vehicle in question confirmed that, at the time of the incident, the vehicle was being operated in Autopilot mode. The Autonomous Emergency Brake (AEB) system did not provide any warning nor initiated automated braking and there was no attempt made by the driver to take evasive action. There was speculation that the radar and camera technology used on Tesla's AEB system failed to detect the trailer against a brightly lit sky, or the trailer was misclassified as an overhead sign by the software. However, the National Highway Traffic Safety Administration [7] concluded that "human error" was the primary cause of the fatal collision and speculated that the driver must have been distracted from the driving task for an 'extended period'. Further investigation by the National Transportation Safety Board [8] reported that the drivers' hands were only detected for 25 s over a thirty-seven minute journey. It is therefore reasonable to assume that the driver was on the whole operating the vehicle completely "hands and feet free", despite Tesla's own Human-Machine Interface (HMI) alert that states "Always keep your hands on the wheel, be prepared to take over at any time" when Autopilot is initially activated. According to Norman [9, 10], automation is at its most dangerous when it behaves in a consistent and reliable manner for most of the time. This is because of an increased risk of out-of-the-loop performance problems including complacency [11], over-trust [12], and loss of situation awareness [13].

Instead, the full benefits of automation can only be realised if the driver is completely removed from the driving task. In these situations, the vehicle would then need to make moral and ethical judgements in otherwise conflicting safety-critical situations. The development of control algorithms and artificial intelligence to enable such decisions is therefore the subject of many ongoing works. Studies propose the use of either: deontological principles, such as the laws of robotics [14, 15]; or consequentialism, which uses the classic 'trolley dilemma', a thought-scenario based on unavoidable harm, to determine moral right and wrong in order to inform computational algorithms [16, 17]. However, we currently reside in an era whereby only some driving tasks are automated. This means that the human driver still plays a vital role within the wider driving task.

1.1 The Role of Terminology

One of the greatest challenges surrounding automobile automation is the inconsistent use of terminologies, in particular, two key concepts: autonomous and automation. Understanding the differences between these terms is extremely important because the terminologies used to describe a feature can lead to inferences over what the role of the human driver is within the wider system. According to the Oxford English Dictionary [18], to be 'autonomous' is to mean "having the freedom to act independently". An autonomous vehicle therefore can act independently and control its own behaviour without human intervention. One example of an 'autonomous' vehicle system is the Electronic Stability Control (ESC) system that can improve vehicle stability by detecting and then reducing a loss of traction that can lead to skidding.

In contrast, 'automation' refers to the "use or introduction of automation equipment" [18]. Thus, an automated vehicle is not capable of acting independently from the driver, and any 'automated control' must be carefully engineered to ensure seamless transition from and back to manual control. The driver therefore has overarching control of vehicle operation meaning that automated systems can be disengaged whenever the driver feels it is necessary. Automated systems can therefore be considered as 'driver-initiated' because it is the human that has ultimate authority of the primary task.

There are also various levels of automated functionality that further complicate our understanding of automation. The Society of Automotive Engineers (SAE) [19] defined five levels of automation specific to the driving domain, outlining the functional capabilities of systems operating within each of these levels. SAE propose that automation ranges from Level 0 (Fully Manually) to Level 5 (Fully Autonomous). However, this can be criticised because automation is not considered to be a dichotomous concept. Instead, automation operates on a continuum [20] making it difficult to classify systems whereby different component parts operate at different levels of automation [2].

There are, however, a number of taxonomies that have sought to better define the role of the human within automated systems. The oldest and most widely cited taxonomy was developed by Sheridan and Verplanck [21], who offer a ten-level taxonomy specifying which functions are the responsibility of the human operator and which are the responsibility of the computer system. Endsley and Kaber [22] sought to better define the intermediate levels by identifying "who" was doing "what" in terms of system monitoring, strategy generation, decision making and response execution processes. Defining the allocation of system function in this way can better help us understand the explicit roles and responsibilities of the driver at each level of automation.

1.2 The Role of the Driver

Kaber and Endsley [23] captured the idea of the changing driver role eloquently by describing a shift from 'active operation' to 'passive monitoring' as the level of automation increases. However, the expectation on the human driver to passively monitor operations and rapidly resume control in emergency situations, or when systems go beyond their parameters, is posited as the main human factors challenge facing the deployment of automated vehicles [2]. It is argued that the designers of automated systems have a moral and ethical responsibility to consider how the operational characteristics and capabilities of the systems impact upon the human driver. Poulin et al. [24] go as far to suggest that drivers will end up with none of the control but all of the accountability for a system that has not been designed in recognition of any potential performance issues. This is termed "the responsibility gap", and is often an inevitable outcome relating to the complexity of a system [25]. A responsibility gap would be the consequence of technologies being deployed with little knowledge as to how they will behave in context. In the case of autonomy, this would be a result of autonomous vehicles being programmed to learn as they operate. This would also mean that not even the human initially responsible for programming these vehicles, would be able to understand or predict the processes underpinning artificial agent decision making in future scenarios [26].

Too much automation can take the human out of the loop, deskill them and lower morale [27, 28]. In addition, automated assistance can lead to decrements in situation awareness and in some instances cause erratic changes to driver mental workload [29]. This is because any sudden transfer of control may result in mode confusion [30] and/or startle effects [31]. Finding ways to encourage and motivate drivers to stay in-the-loop is an enduring challenge, one that could rely heavily upon the role of the driver monitoring. Hancock [32], however, suggests that there is a need to consider motivation theory in relation to human-computer interaction associated with the automotive domain. The general consensus is that mental workload optimisation is crucial in maintaining effective task performance (e.g. [33]). Thus, being able to recognise different driver states at higher levels of automation will enable researchers to not only explore the most efficient strategies of keeping the driver 'in-the-loop' but always the most efficient means to transfer control.

The application of automation by designers without due consideration of the consequences to human performance is something Parasuraman and Riley [34] referred to as 'abuse' of automation and arguably violates the ethical principle of respect for persons [35, p. 1845]. Without a human-centred focus, the benefits of automation for individual drivers are overlooked in pursuit of higher levels of functionality [36]. We also miss opportunities to learn from driver experiences in using such systems that could help inform future generations of technology that can improve the safety of both its users and wider public [6].

At the core of the Human Factors and Ergonomics discipline is the goal to enable the development of innovative and beneficial technologies that can optimise the performance and safety of the human that interact with them [36]. Systems engineering uses the principles of complementarity to help inform the design of complex sociotechnical systems such as those seen in driving automation. This approach is based upon the idea that the allocation of tasks should serve to maintain control whilst retaining human skill [37]. Such an approach aims to encourage shared situation awareness between driver and automated systems [38]. 'Team cognition' appears to be the binding mechanism that goes on to produce coordinated behaviour [39]. According to Cuevas et al. [39], a human-automation team can be defined as the coupling of both human and automated systems working both collaboratively and in coordination with one another to successfully complete a task. Hoc [40] originally applied the concept of cooperation as a means to identify, analyse, implement and support cooperative activities between humans and machines. This essentially describes the essence of partially automated driving solutions, but does not necessarily reflect the reality of such systems.

1.3 The Role of Marketing

The drive for vehicle automation is likely to continue at pace given the potential economic benefits available from competitive advantage within a mass-market commodity [32, 41]. However, this enthusiasm and push for technological innovation may encourage more focus on the functional capabilities and characteristics of a system rather than the impacts that it may have on driver behaviour. It is after all the former that dominates discussions within the public arena [32]. The marketing and deployment of automated technologies may therefore negatively impact upon driver expectations [6, 42], risk perception [43] and trust [11]. All of these factors can contribute to automation 'misuse' (i.e. over reliance) and/or 'disuse' (i.e. underutilisation) [34].

The marketing of automated systems is also likely to influence how drivers interpret and predict what may happen within the driving environment (i.e. their mental model of the system) [3, 27], particularly if these are reinforced by positive interactions with technology (i.e. systems appears to be reliable, accurate and dependable). Of course, any deviation in expected behaviour has the potential to challenge existing models and affect user trust in system functionality [44].

There seems to be a lack of transparency regarding the true technological capabilities of autonomous systems. For example, Endsley [45] recalls that a service representative from Tesla initially told her the automated systems were 100% reliable. However, when it comes down to a matter of liability, Tesla were clear to state that the Autopilot in use around the time of the fatal accident in May 2016 was in the beta testing phase [46]. Mobileye, who make the image recognition software for assisted driving, condemned Tesla's claims that their Autopilot was safer than a human driver on highways by citing the dangers involved with causing mistrust through consumer confusion of technological capabilities [47]. In fact, Smith [48] calculated that in order to say with 99% confidence that automated vehicles crash less frequently than vehicles with human drivers, it would need to drive 725,000 mile on representative roadways without incident. If only fatal crashes were considered, this figure rises dramatically to 300 million miles. To date, no automated vehicle has yet to travel such distances unassisted [49] and therefore it is not possible to validate the claim that automated vehicles are safer than vehicles driven by human drivers. In reality, even with perfect sensing, it is unrealistic to suggest that a truly crash free environment can exist [49]. This is because the driving environment is characterised by unpredictable obstacles and dynamically changing situations that are difficult to design for. We also cannot overlook the fact that inevitable failures will occur. Software and/or hardware failures remain an inevitable threat to the safety of automated systems. In an effort to minimise such failures, vehicles will need multiple redundancies, extensive testing and likely need mandatory maintenance [49].

1.4 The Role of Technology

Given that software and/or hardware failures appear to remain an inevitable threat to drivers [49], research into failure-induced transfer of control has been extensively studied [50, 51]. Early research concentrated on determining appropriate thresholds relating to how long a driver would need to know in advance about an upcoming control transition. Damböck et al. [52] utilised three different lead times – 4, 6 and 8 s – and found that performance did not differ significantly from manual driving when an 8 s lead time was used. Similarly, Gold et al. [53] suggested that drivers needed a lead time of 7 s to ensure a safe resumption of control. For partially automated driving, the literature suggests that drivers can take anywhere between 1.2 s [54] and 15 s to respond to emergency situations [55]. Further, for self-paced control transitions (i.e.

non-critical), response times vary between 1.9 and 25.7 s depending upon task engagement and criticality [56]. This has significant implications for the design of warning systems as they must provide drivers with adequate warning about takeover situations. Whilst it is common practice to design for the 90th percentile (taking into the consideration the range between the 5th percentile female and 95th percentile male populations [57]), it is important that manufacturers are given more information about the 'extreme' ends of the population (i.e. range in performance). Using only mean values as a basis to design systems is troublesome because we do not understand the range or spread of distributions. Thus, Eriksson and Stanton [56] argue that a more inclusive design approach is needed. Median response times are simply not sufficient enough when it comes to designing control transitions for automated driving.

Even so, Louw [58] argues that, in addition to providing adequate warning to drivers, a more pressing need is to develop an objective measure relating to handover quality and safety. Thomas et al. [59] suggest that it is in the seconds immediately after takeover when most errors occur. For the case of partial automation, drivers can take between 35 and 40 s to stabilise lateral control of the vehicle following a transfer of control from automated to manual driving [55]. Transfer of control and the provision of adequate warning remains an ongoing challenge.

2 Conclusions

The utilitarian justification for automated vehicles is that they can improve safety, comfort and deliver greater mobility for both the user and wider population [1, 2]. However, there is a need to balance these justifications with the current state of the art. In reality, fully autonomous vehicles are a long way off [10]. Partially automated driving solutions are already available today, but can give the impression of higher level functionality. Systems that can automate both longitudinal and lateral control, as well as automate aspects of traditional driver decision-making enable drivers to become "hands and feet free". However, they still require the driver to supervise, maintain and potentially take over from the system [2]. Unfortunately, humans are notoriously inefficient at completely sustained vigilance tasks (e.g. [50, 60]). Combine that with the fact that drivers can quickly become complacent in situations where automated systems behave in a consistent manner for extended periods, it appears inevitable that drivers will over-trust systems offering limited automated assistance (e.g. [11, 61, 62]). This behaviour will heavily impact upon the capabilities of the driver to respond appropriately when required.

Finally, whilst the concepts of "automation" and "autonomy" are clearly distinct, it is worth noting that some of the more sophisticated *automated* technologies have the potential to become *autonomous* in the future [32]. This is of course a natural progression for technological development. For example, at a basic level 'cruise assist' technologies have evolved from systems that simply maintain a pre-set speed (Cruise Control), to systems that can adapt their speed profile depending upon the speed of the vehicle ahead, although notably will not exceed a pre-set maximum (Adaptive Cruise Control). It therefore seems likely that continued extension of pre-existing automated systems could pave the way for future autonomous functionality. Presently, however, it appears that we do not fully understand or appreciate the complexities of human-machine interaction in vehicles offering enhanced automated driving solutions [63]. This is further exacerbated by the current approach to accident investigation following an incident involving an 'automated' vehicle. Rather than adopting a holistic 'system' view, individual subsystems tend to be looked at in isolation; if they are deemed to be operationally sound, blame will often be attributed to 'human error' [64]. Such an approach is problematic because it fails to consider how multiple subsystems operating together can synergistically change the way in which the vehicle performs and the impact it has on the driver. It also fails to consider contextual factors that could significantly influence the operator's behaviour within the system (e.g. how automated systems are named and marketed). If the circle of investigation widens to include such factors, it is likely that the conclusions drawn from the investigation process will provide a much more accurate and informative depiction of accident causation. After all, Dekker [65] argues that human error should be the starting point of an accident rather than the end, and that any case involving human error "demands an explanation" [65, p. 68].

References

1. Stanton NA, Marsden P (1996) From fly-by-wire to drive-by-wire: safety implications of automation in vehicles. Saf Sci 24(1):35–49
2. Kyriakidis M, De Winter JCF, Stanton N., Bellet T, Van Arem B, Brookhuis K, Martens MH, Bengler K, Andersson J, Merat N, Reed N, Flament M, Hagenzieker M, Happee R (2016). A human factors perspective on automated driving. Theoretical Issues in Ergonomics Science, pp 1–27
3. Neilsen J (2010) Mental models. https://www.nngroup.com/articles/mental-models/. Accessed 15 Aug 2017
4. Shladover SE (2016) The truth about self-driving cars. Sci Am 314(6):52–57
5. Abraham H, Seppelt B, Mehler B, Reimer B (2017) What's in a name: vehicle technology branding consumer expectation for automation. In: Proceedings of Automotive UI 2017, Oldenburg, Germany
6. Stilgoe J (2017) Tesla crash report blames human error – this is a missed opportunity. The Guardian. https://www.theguardian.com/science/political-science/2017/jan/21/tesla-crash-report-blames-human-error-this-is-a-missed-opportunity. Accessed 14 Aug 2017
7. National Highway Traffic Safety Administration (2017) ODI Resume. https://static.nhtsa.gov/odi/inv/2016/INCLA-PE16007-7876.PDF. Accessed 15 Aug 2017
8. National Transportation Safety Board (2017) Driver assistance systems specialists factual report. https://dms.ntsb.gov/pubdms/. Accessed 09 Aug 2017
9. Norman DA (1990) The "problem" with automation: inappropriate feedback and interaction, not "over- automation". Phil Trans R Soc Lond - Ser B Biol Sci, l327(1241), 585–593
10. Norman DA (2015) The human side of automation. In Road Vehicle Automation 2. Springer, Cham, pp 73–79
11. Lee JD, See KA (2004) Trust in automation: designing for appropriate reliance. Hum Factors 46:50–80
12. Walker GH, Stanton NA, Salmon PM (2016) Trust in vehicle technology. Int J Veh Des 70(2):157–182

13. Endsley MR (1995) Toward a theory of situation awareness in dynamic systems. Hum Factors 37(1):32–64
14. Asimov I (1942) I Robot. Gnome Press, New York
15. Murphy RR, Woods DD (2009) Beyond Asimov: the three laws of responsible robotics. IEEE Intell Syst 24(4):14–20
16. Sütfeld LR, Gast R, König P, Pipa G (2017) Using virtual reality to assess ethical decisions in road traffic scenarios: applicability of value-of-life-based models and influences of time pressure. Front Behav Neurosci 11:122
17. Skulmowski A, Bunge A, Kaspar K, Pipa G (2014) Forced-choice decision-making in modified trolley dilemma situations: a virtual reality and eye tracking study. Front Behav Neurosci 8:426
18. Oxford English Dictionary (2017). https://en.oxforddictionaries.com/definition/autonomous. Accessed 15 Aug 2017
19. Society of Automotive Engineers (2016) Taxonomy and Definitions for Terms Related to On-Road Motor Vehicle Automated Driving Systems. http://standards.sae.org/j3016_201401/. Accessed 12 Oct 2015
20. Parasuraman R, Sheridan TB, Wickens CD (2000) A model for types and levels of human interaction with automation. IEEE Trans Syst, Man, Cybern Part A, Syst Hum. A publication of the IEEE Systems, Man, and Cybernetics Society, 30(3), 286–297
21. Sheridan TB, Verplanck WL (1978) Human and computer control of undersea teleoperators. MIT Man-Machine Laboratory, Cambridge
22. Endsley MR, Kaber DB (1999) Level of automation effects on performance, situation awareness and workload in a dynamic control task. Ergonomics 42(3):462–492
23. Kaber DB, Endsley MR (2004) The effects of level of automation and adaptive automation on human performance, situation awareness and workload in a dynamic control task. Theor Issues Ergon Sci 5(2):113–153
24. Poulin C, Stanton NA, Cebon D, Epple W (2015) Responses to autonomous vehicles. Ingenia 62:8–11
25. Matthias A (2004) The responsibility gap: ascribing responsibility for the actions of learning automata. Ethics Inf Technol 6(3):175–183
26. Johnson DG, Norman M (2014) Recommendations for future development of artificial agents. IEEE Technol Soc Mag, Winter 2014:22–28
27. Stanton NA, Young MS, Walker GH (2007) The psychology of driving automation: a discussion with Professor Don Norman. Int J Veh Des 45(3):289–306
28. Bainbridge L (1983) Ironies of automation. Automatica 19(6):775–779
29. Young MS, Stanton NA (2002) Malleable attentional resources theory: a new explanation for the effects of mental underload on performance. Hum Factors 44(3):365–375
30. Sarter NB, Woods DD (1995) How in the world did we ever get into that mode? mode error and awareness in supervisory control. Hum Factors 37:5–19
31. Sarter NB, Woods DD, Billings CE (1997) Automation surprises. Handb Hum Factors Ergonom 2:1926–1943
32. Hancock PA (2016) Imposing limits on autonomous systems. Ergonomics 60(2):284–291
33. Wilson JR, Rajan JA (1995) Human-machine interfaces for systems control. In: Wilson JR, Corlett EN (Eds.), Evaluation of human work: a practical ergonomics methodology. Taylor Francis, London, pp 357–405
34. Parasuraman R, Riley V (1997) Humans and automation: use, misuse, disuse, abuse. Hum Factors 39(2):230–253
35. Kumfer WJ, Levulis SJ, Olson MD, Burgess RA (2016) A human factors perspective on ethical concerns of vehicle automation. Hum Factors 60(1):1844–1848
36. Hancock PA (2014) Automation: how much is too much? Ergonomics 57(3):449–454

37. Grote G, Weik S, Wafler T, Zolch M (1995) Criteria for the complementary allocation of functions in automated work systems and their use in simultaneous engineering projects. Int J Ind Ergon 16:326–382
38. Stanton NA, Stewart R, Harris D, Houghton RJ, Baber C, McMaster R, Salmon PM, Hoyle G, Walker G, Young MS, Linsell M, Dymott R, Green D (2006) Distributed situation awareness in dynamic systems: theoretical development and application of an ergonomics methodology. Ergonomics 49(12–13):1288–1311
39. Cuevas HM, Fiore SM, Caldwell BS, Strater L (2007) Augmenting team cognition in human– automation teams performing in complex operational environments. Aviat Space Environ Med 78:B63–B70
40. Hoc JM (2000) From human-machine interaction to human-machine cooperation. Ergonomics 43(7):833–843
41. Parasuraman R, Wickens CD (2008) Humans: still vital after all these years of automation. Hum Factors 50:511–520
42. Merat N, Lee JD (2012) Preface to the special section on human factors and automation in vehicles designing highly automated vehicles with the driver in mind. Hum Factors 54(5): 681–686
43. Brill JC, Bliss JP, Hancock PA, Manzey D, Meyer J, Vredenburgh A (2016) Matters of ethics, trust, and potential liability for autonomous systems. Hum Factors 60(1):308–312
44. Banks VA, Stanton NA (2015) Discovering driver-vehicle coordination problems in future automated control systems: Evidence from verbal commentaries. Procedia Manuf 3:2497–2504
45. Endsley MR (2017) Autonomous driving systems: a preliminary naturalistic study of the Tesla models. J Cogn Eng Decis Making
46. Tesla Motors (2016). A Tragic Loss. Tesla (press release). https://www.tesla.com/blog/tragic-loss. Accessed 15 Aug 2017
47. Brown B, Laurier E (2017) The trouble with autopilots: assisted and autonomous driving on the social road. In: Proceedings of Human Factors in Computing Systems
48. Smith BW (2012) Driving at perfection. The Center for Internet and Society at Stanford Law School. http://cyberlaw.stanford.edu/blog/2012/03/driving-perfection. Accessed 15 Aug 2017
49. Goodall NJ (2014) Ethical decision making during automated vehicle crashes. Transp Res Board 2424:58–65
50. Molloy R, Parasuraman R (1996) Monitoring an automated system for a single failure: vigilance and task complexity effects. Hum Factors 38:311–322
51. Strand N, Nilsson J, Karlsson ICM, Nilsson L (2014) Semi-automated versus highly automated driving in critical situations caused by automation failures. Transp Res Part F Traffic Psychol Behav 27(Part B):218–228
52. Damböck D, Bengler K, Farid M, Tönert L (2012) Übernahmezeiten beim hochautomatisierten Fahren [Takeover times for highly automated driving]. Tagung Fahrerassistenz 15:16–28
53. Gold C, Damböck D, Lorenz L, Bengler K (2013) 'Take over!' How long does it take to get the driver back into the loop? In: Proceedings of the Human Factors and Ergonomics Society 57th Annual Meeting. Santa Monica, CA: Human Factors and Ergonomics Society, pp 1938–1942
54. Zeeb K, Buchner A, Schrauf M (2015) What determines the take-over time? an integrated model approach of driver takeover after automated driving. Accid Anal Prev 78:212–221
55. Merat N, Jamson AH, Lai FCH, Daly M, Carsten OMJ (2014) Transition to manual: driver behaviour when resuming control from a highly automated vehicle. Transportation Research Part F: Traffic Psychology and Behaviour 26(Part A):1–9

56. Eriksson A, Stanton NA (2017) Take-over time in highly automated vehicles: non-critical transitions to and from manual control. Hum Factors 59(4):689–705
57. Porter JM, Case K, Marshall R, Gyi D, neé Oliver RS (2004) 'Beyond Jack and Jill': designing for individuals using HADRIAN. Int J Ind Ergonom 33(3):249–264
58. Louw T, Merat N, Jamson H (June 2015) Engaging with highly automated driving: to be or not to be in the loop? In: Eighth International Driving Symposium on Human Factors in Driver Assessment, Training, and Vehicle Design, Salt Lake City, Utah
59. Thomas MJ, Schultz TJ, Hannaford N, Runciman WB (2013) Failures in transition: learning from incidents relating to clinical handover in acute care. J Healthc Qual 35(3):49–56
60. Casner SM, Schooler JW (2015) Vigilance impossible: diligence, distraction, and daydreaming all lead to failures in a practical monitoring task. Conscious Cogn 35:33–41
61. Parasuraman R, Molloy R, Singh IL (1993) Performance consequences of automation-induced 'complacency'. Int J Aviat Psychole 3:1–23
62. Hollnagel E, Woods DD (2005) Joint Cognitive Systems Foundations of Cognitive Systems Engineering. CRC Press, Boca Raton
63. Weyer J, Fink D, Adelt F (2015) Human-machine cooperation in smart cars. An empirical investigation of the loss-of-control thesis. Saf Sci 72:199–208
64. Rasmussen J (1990) Human error and the problem of causality in analysis of accidents. Philos Trans R Soc Lond 327:449–462
65. Dekker SW (2006) The field guide to understanding human error. Ashgate, Aldershot

Revealing the Complexity of Road Transport with Accimaps

Rich C. McIlroy[(✉)], Katherine L. Plant, and Neville A. Stanton

Human Factors Engineering, Transportation Research Group,
Faculty of Engineering and the Environment,
University of Southampton, Southampton, UK
r.mcilroy@soton.ac.uk

Abstract. The UK has one of the safest road systems of any country, yet road traffic accidents still represent the 12th leading cause of death. Although casualty and fatality rates have dropped dramatically since the 1980s, there has been little change in the past five years or so, suggesting that roads safety initiatives have plateaued in their effectiveness. Following calls for a new approach to the challenge we adopt a sociotechnical systems viewpoint. Traditionally, road safety has been addressed through the three 'E's of engineering, education, and enforcement; we have added to these with an additional four 'E's, namely economics, emergency response, enablement, and ergonomics. We use the Actor Map representation, the first step in the Accimap approach to accident analysis, to model the road transport system, with the resulting diagram giving an indication of the level of complexity we must face when designing road safety interventions. The research presented in this article represents the first step in a broader project that takes a sociotechnical approach to global road safety, involving partners in five geographically dispersed, and economically, developmentally, and culturally distinct nations.

Keywords: Accimaps · Road safety · Sociotechnical systems

1 Introduction

In 2015, road traffic incidents claimed the lives 1.34 million people (WHO 2017a). Globally, it is the 10th leading cause of death (ibid.), and the number one leading cause of death for those aged between 15 and 24 (WHO 2015). Across England and Wales, road incidents represent the 12th leading cause of death (ONS 2017). In 2016, 181,384 road incidents involving injury were reported to the UK police (DfT 2017); of these, 24,101 resulted in serious injury, and 1,794 in loss of life. These road collisions are estimated to have cost the UK economy £16 billion in 2015, a figure that rises to £36 billion when including incidents that went unreported (estimated from National Travel Survey Data; see DfT 2016). Furthermore, although the UK's roads have become far safer over the past thirty years, with a 68% drop in fatalities despite a population increase of 15%, there has been little change in road fatality numbers since 2010 (DfT 2017), suggesting that current road safety efforts have plateaued in their effectiveness.

S. Bagnara et al. (Eds.): IEA 2018, AISC 823, pp. 80–89, 2019.
https://doi.org/10.1007/978-3-319-96074-6_9

The traditional three 'E's approach to road safety, i.e., engineering, enforcement, and education, has had significant and positive lasting effects on road safety. Nevertheless, to make further improvements we will need to consider more than only those three facets of the complex road safety picture. Hence our discussion of four additional 'E's, namely economics, emergency response, enablement, and the overarching term that brings efforts together, ergonomics. It is from a viewpoint that has its basis in the socio technical approach to understanding systems and their behaviours that the additional 'E's are justified. It is by no means novel to suggest a systems view of road safety (see, for example, Scott-Parker et al. 2015; Parnell et al. 2017); however, an exploration of how human factors and ergonomics analysis and modelling techniques can apply to, and help expand the more traditional views of road safety, is yet to receive sufficient attention in the extant literature. Using Rasmussen's (1997) Risk Management Framework, and the associated Accimap approach, we aim to address this gap.

2 The Seven 'E's of Road Safety

2.1 The Traditional 'E's: Engineering, Enforcement, Education

As aforementioned, engineering, enforcement, and education have been critical in improving road safety outcomes in the past, and will continue to be so into the future. Engineering can be broadly separated into two categories through which safety is enhanced; road environment engineering and vehicle engineering. Many of the engineering solutions, particularly in the vehicle, have now become legal requirements in the UK, for example seat belts, crumple zones, anti-lock braking systems, etc. Such is one aspect of the enforcement approach to road safety; enforcing design or engineering standards. Another aspect is the use of law to influence driver behaviour; this can also be split into two broad groups. Firstly, general deterrence of certain behaviours via the fostering of the belief that performing those behaviours will result in a negative outcome (e.g. fine or loss of license). Secondly, specific deterrence through a person being caught performing an illegal action, stopped performing said action (where it is still in process), and punished for the act. Education, the last of the three traditional 'E's, can again be considered in two ways; training the driver and educating the public. Driver training is necessary for drivers to acquire licenses, while educating the public can happen in myriad ways, from teaching school children the green cross code to using television advertising to highlight the negative effects of drink driving.

2.2 Economics

As aforementioned, the value of prevention of all *reported* road casualties (not including unreported ones) was estimated at £16 billion for the year 2015 (DfT 2016). This cost, however, is far less tangible than the immediate cost of building and fixing road infrastructure, or the funding of road safety campaigns; hence, it is perhaps unsurprising that investment in road safety is nowhere near this figure. Improving road safety is not always a case of expenditure, however; income generated through fining drivers caught speeding can more than outweigh the cost of the police patrols required to catch offenders (as was seen in Uganda; Bishai et al. 2008). In the UK, there has been considerable controversy surrounding speed cameras as source of income rather than a means for improving road safety (e.g., Pilkington 2003), with some news outlets drawing attention to the large revenues garnered in particular areas of the country (e.g. Lydell 2017). Despite the UK parliament ensuring that the placement of speed cameras is based on safety considerations rather than financial ones (House of Commons Transport Committee, House of Commons Transport Committee 2017), cost benefit analyses still suggest a net financial gain for the economy (e.g. Gains et al. 2005).

2.3 Emergency Response

The quality of emergency care available undoubtedly affects the health outcomes of a particular patient (e.g. Razzak and Kellermann 2002; Balikuddembe et al. 2017). Regarding speed of response to traffic incidents, it has been demonstrated that timely transport to a hospital is significantly linked with high survival rates (e.g., Clark et al. 2013; Sánchez-Mangas et al. 2010). In England specifically, a review by the Trauma Audit and Research Network (TARN; see The University of Manchester 2015) showed that the survival rates of major trauma patients (many of which are casualties of road traffic incidents) increased by 63% over the seven-year period in which twenty-seven new Major Trauma Centres were constructed (e.g. NHS 2016). Previous to this, the fall in casualty rates in the UK from road traffic incidents across the years 1978 to 1998 has been partly explained by lower in-patient lengths of stay, higher per-capita levels of NHS staffing, and lower numbers of people (per-capita) waiting for hospital treatment (three proxies for improving medical care standards; see Noland and Quddus 2004). There is now wide acceptance of the role emergency response plays in road safety. Not only was post-impact care the focus of a recent European Commission report on road safety (European Commission 2016), but it represents a 'pillar of action' in the UN's Global Plan for the Decade of Action (on road safety; see WHO 2017b), as well as a key strategy in the World Health Organisation's world report on road traffic injury prevention (Peden et al. 2004).

2.4 Enablement

We use the term enablement here to refer to the activities that make advances in other areas possible. Funding is of course part of this, and is important enough to merit its own section (above), but also of significant importance is the enablement of research through a cultural and societal environment that supports such work, and through the availability of data to allow the research to be undertaken. The UK is home to a plethora of road safety organisations and charities. Historically, these kinds of organisations have had significant impacts on road safety, providing support and funding to a huge variety of road safety projects over the years, and their continued existence is a testament to the societal support that the vast majority of road safety initiatives receive. Additionally, the government itself backs a consortium of research institutes and councils, each of which supports research in different subjects and disciplines. In terms of data availability and openness, the UK is a world leader. Since its creation in 2013, the Open Data Barometer (run by the World Wide Web Foundation; see opendatabarometer.org), has ranked the

UK number one (out of 115 countries) in terms of the readiness of a government's open data initiatives, implementation of open data programmes, and impact of open data on society. Complex problems (such as road safety) require rich, complex data sets to guide solutions (Arzberger et al. 2004), with open access to publicly funded data providing greater returns from public investment (Janssen et al. 2012).

2.5 The Seventh 'E': Ergonomics

The traditional three 'E's approach takes a view of the road safety system that puts the driver squarely in the centre; the roads are engineered around the driver, who is educated to drive safely, and is punished for transgressions. This focus is beginning to change, however, with increased adoption of the 'safe systems' philosophy. This road safety perspective has its roots in Sweden in the 1990s (see visionzeroinitiative.com) and one of its central themes is that humans are fallible and that responsibility for road safety should be moved away from road users and placed on road system designers. Although the philosophy is supported in the UK by a variety of charitable organisations, (e.g., PACTS 2017; Brake 2017), and Bristol City Council (2015), it has not yet been implemented nationwide across the UK.

Like any approach, the 'safe systems' perspective (and the Vision Zero project in which it has its roots) has evolved over time, and now, in many descriptions of it, includes additional considerations such as emergency response, funding, and legislature (aspects included in the UK Parliamentary Advisory Council for Transport Safety's description of the approach; PACTS 2017). Nevertheless, inclusion of these considerations of higher system levels is not always seen. Focus is often still limited to the driver, with emphasis on designing a road system that tolerates the errors of the user, rather than designing a road system that takes into account the interaction of the system components that influence system behaviours. We wholeheartedly agree that we should move away from placing all blame on the road user; however, this should not be simply shifted to the system designer. Indeed, nor should it be shared between only those two actors; complex systems have numerous hierarchical levels to them (more on this below), and blame is shared across all of those levels, with multiple actors and organisations responsible for system outcomes. The language used in some of the safe systems publications, i.e., that humans "make mistakes", suggests a persistence of the idea that the human is the faulty element in the system, the component that is likely to fail. We would argue that a fundamental shift in viewpoint is required, to one that takes the system in its entirety; it is not the error-prone human, but the nature and design of the system as a whole from which failures emerge.

To this end, we advocate a socio technical systems approach, rather than the safe systems philosophy of the Vision Zero project. Although similar in name, these approaches are not the same. The 'safe systems' philosophy has its foundation in road safety efforts in Sweden in the 1970s; the 'systems approach' can be traced back as far as the early 1930s (Heinrich 1931). We use the term 'sociotechnical' (a largely synonymous term that first appeared in the early fifties; e.g. Trist 1953) as this not only neatly alludes to the fact that the system in question involves both social and technical elements, but also is more easily distinguishable from 'safe systems'.

2.6 Accimaps

There are a wide variety of sociotechnical models and frameworks that can be used to hierarchically describe a given system; however, it is Jens Rasmussen's (1997) Risk Management Framework that has received the most recent attention in the road safety domain (e.g. Salmon et al. 2013; Newnam and Goode 2015; Scott-Parker et al. 2015; Parnell et al. 2017). Rasmussen's (1997) original framework contained six levels, from the equipment at the bottom of the hierarchy up to the central government at the top. This has since been expanded upon to include two additional, higher levels; national and international committees (Parnell et al. 2017). The Accimap approach to accident analysis uses this framework as its basis, considering accidents as emergent properties that arise from the interplay of technology, individuals, and organisations at various levels of system abstraction. The first step of this type of analysis is the creation of the Actor Map. It is the Actor Map that this article focuses on.

The Actor Map. To construct the Actor Map for the UK road transport system recourse was first made to the existing literature. Work from Salmon et al. (2013), Scott-Parker et al. (2015), Newnam and Goode (2015), and Parnell et al. (2017) provided the starting point; however, although all provide Actor Maps of the road transport system, they each focus on a particular aspect (e.g., distraction, road-rail intersections, road freight, young drivers). It was therefore necessary to expand the document review beyond the academic literature. As habitual road users and road safety researchers and practitioners ourselves, we were able to use our own experience of the road system to guide initial document searches. Government and local council documents and statistical releases, as well as charity and non-governmental organisation websites represented the primary information sources, alongside the UK's parliament website (parliament.uk). Informal consultations with other members of the University of Southampton's Transportation Research Group served to supplement the model, and further additions and alterations were made following informal conversations with subject matter experts in the motor industry, public health, emergency medical response, and emergency police response domains. Several iterations were performed, resulting in the final Actor Map that can be seen in Fig. 1.

The top most levels of Fig. 1 show the International Committees (e.g. European Union, International Organisation for Standardisation) and National Committees (e.g. Transport Select Committee); they are independent of the government and oversee/review policies and directives. These are followed by Central Government, i.e., the departments that run the country and define its laws and policies (e.g. Department for Transport, Department of Health, etc.). In the Regulators and Associations level are the actors concerned with the implementation and management of mandates passed down by higher government (e.g., Highways England, Police, National Health Service), and the associations that give voice to business and industry (e.g. Institution of Civil Engineers, Association of British Insurers). Industry and Local Government, the next level down, includes actors with more narrowly defined remits (e.g. specific companies, such as Taxi Companies, Haulage and Delivery Companies) as well as those that work more locally under the national bodies that sit above them (e.g. Local Fire & Rescue, Local Police Constabularies, Traffic Control Centres). The specific

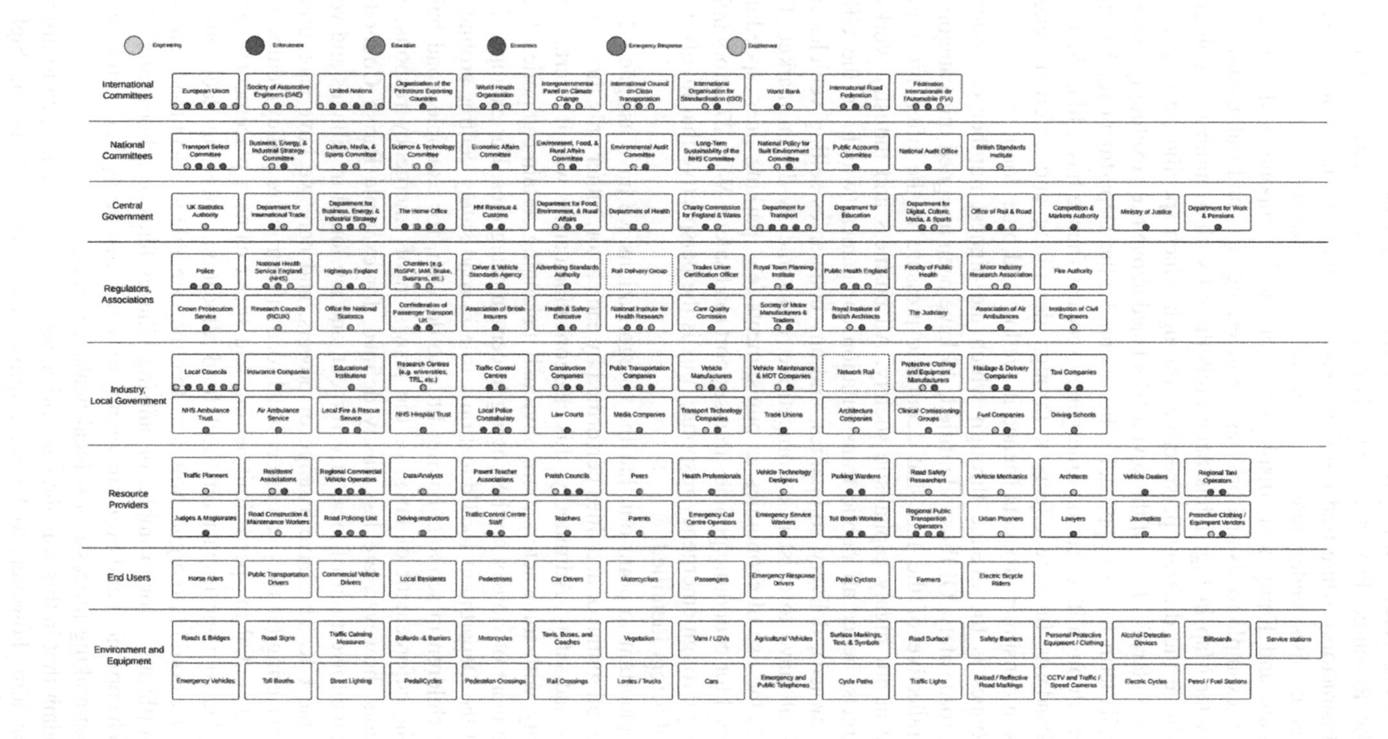

Fig. 1. Actor Map of the road transport system, with coloured dots indicating the relevant road safety perspective described above. (Color figure online)

functions and services of these actors are typically implemented by those in the level below, the Resource Providers (e.g. Traffic Police, Emergency Service Workers, Driving Instructors), while End Users represent the potential users of the system (e.g. car drivers, cyclists, pedestrians, commercial vehicle drivers, etc.). The lowest level, Environment and Equipment, displays the physical components that make up the system, for example roads and bridges, vehicles, signage, and traffic calming devices.

The Actor Map in Fig. 1 has been embellished with a number of colourful dots. These indicate the different perspectives through which specific actors can influence road safety, in terms of six of the seven 'E's listed above (i.e., excluding 'ergonomics', the umbrella term for our approach). For example, the Department for Health affects road safety through its effect on emergency services (hence has a lilac dot), while the British Standards Institute affects it through describing and enforcing engineering standards in consumer products (hence its yellow and blue dots).

Economics is the avenue through which the highest number of system actors influence road safety. Of those identified, 45 have been assigned this category. This is perhaps reflective of the fact that economic pressure is the often most influential constraint in our highly capitalistic society. All business and industry works to economic targets, hence all public transportation companies (which, in the UK, are all privately owned), delivery companies, and any companies selling vehicles or equipment will all have to exist in a competitive, economically-driven market. This is in addition to the use of economics as punishments (e.g. parking fines, speeding fines), and the need for economic policies that support road safety. With 27 actors influencing road safety through emergency response, this is the perspective that involves the least number of actors identified.

It is important to point out that the categorisations we have assigned do not represent the full gamut of activities performed by each actor, rather only those specifically related to road safety. Admittedly, this categorisation carries with it a certain degree of subjectivity, with some actors more obviously belonging certain categories than others, and some more obviously *not* belonging to certain categories. For example, it is fairly clear that the Department for Education's role in road safety centres around the education of children in schools, having little to do with, for example, road building or design, the enforcement of laws, or the provision of emergency response services. Other actors, however, are less easily defined, for example Parish Councils. These represent the lowest tier of local government, and are found in villages and very small towns in the UK. The have a range of powers, and here we have categorised them under engineering (as they can affect local road infrastructure policy), enforcement (as they are influential in speed limit setting and speed camera placement), and in economics (due to their impact on local fund allocation, spending decisions, and tax raising). One could also argue, however, that they are able to affect road safety through education (through, for example, organising villages fêtes or school fairs where road safety is championed) and even emergency response, insofar as they could have a voice in decision-making processes for local health services.

We admit that in this way the categorisation used in Fig. 1 is open to discussion. It is not our aim, however, to be prescriptive, rather to highlight the high level of complexity inherent to the road transport system. Indeed, there are actors identified in Fig. 1 that influence road safety in ways *not* captured in the six 'E's described above. In particular, the observant reader will notice that there are two nodes (other than those in the lower two levels) that have not been assigned a coloured dot at all, and that have dotted (rather than solid) outlines; 'Rail Delivery Group' and 'Network Rail'. This is because their effect on road safety is somewhat indirect. A well-managed rail system that is organised, clean, good value, and appealing to users will attract people away from their cars and onto the train. This can also be said for road public transport. In this way, the public transport industry, including the rail system, affect road safety; they play a role in reducing traffic volumes, one of the factors most affecting on-road fatality rates (e.g. Ahangari et al. 2017). Such an effect is not captured by our colour coding.

3 Conclusions and Future Work

The road transport picture is highly complex when taken as a whole, as we have attempted to do here. Of course, with increasing breadth one must sacrifice depth (there exist whole bodies of literature devoted to single components of the model presented above); however, it is important to bear in mind that in an interconnected sociotechnical system successful strategies and interventions not only provide benefit in one domain, but across a whole raft of them. The challenge is great, but the opportunity equally so. Moreover, the road safety challenge is far greater in Low- and Middle-Income Countries (LMICs) than it is in the UK. The analysis above presents the first step in the Socio Technical Approach to Road Safety (STARS) project, an undertaking of the National Institute for Health Research's newly formed Global Health Research Group in Road Safety. In addition to UK researchers, the project involves partners in Bangladesh, China, Kenya, and Vietnam. The approach we have applied to analysing the UK system will also be applied in all project partner contexts such that consistent, structured comparisons can be made. This will allow us to recommend interventions, couched in one or more of the 'E's presented above, in a way consistent with the sociotechnical systems philosophy.

Funding Statement and Disclaimer. This research was commissioned by the National Institute for Health Research using Official Development Assistance (ODA) funding. The views expressed are those of the authors and not necessarily those of the NHS, the NIHR, or the Department of Health and Social Care.

References

Ahangari H, Atkinson-Palombo C, Garrick NW (2017) Automobile-dependency as a barrier to vision zero, evidence from the states in the USA. Accid Anal Prev 107:77–85

Amos L, Davies D, Fosdick T (2015) Road Safety Since 2010. PACTS & RAC Foundation, London

Arzberger P, Schroeder P, Beaulieu A, Bowker G, Casey K, Laaksonen L, Moorman D, Uhlir P, Wouter P (2004) An international framework to promote access to data. Science 303: 1777–1778

Balikuddembe JK, Ardalan A, Khorasani-Zavareh D, Nejati A, Kasiima S (2017) Factors affecting the exposure, vulnerability and emergency medical service capacity for victims of road traffic incidents in Kampala metropolitan area: a Delphi study. BMC Emergency Medicine, 17, no pagination

Bishai D, Asiimwe B, Abbas S, Hyder AA, Bazeyo W (2008) Cost-effectiveness of traffic enforcement: case study from Uganda. Injury Prev 14:223–227

Brake (2017) Annual Report 2015. Brake, Huddersfield, UK

Bristol City Council (2015) A safe systems approach to road safety in Bristol. Bristol City Council, Bristol, UK

Broughton J, Knowles J (2010) Providing the numerical context for British casualty reduction targets. Saf Sci 48:1134–1141

Ciaburro T, Spencer J (2017) UK road safety seizing the opportunities safe roads. Parliamentary Advisory Council for Transport Safety, London

Clark DE, Winchell RJ, Betensky RA (2013) Estrimating the effect of emergency care on early survival after traffic crashes. Accid Anal Prev 60:141–147

DfT (2016) Reported Road Casualties Great Britain: 2015 Annual Report. https://www.gov.uk/government/uploads/system/uploads/attachment_data/file/568484/rrcgb-2015.pdf. Accessed 16 Mar 2018

DfT (2017) Reported Road Casualties Great Britain: 2016 Annual Report. https://www.gov.uk/government/uploads/system/uploads/attachment_data/file/668504/reported-road-casualties-great-britain-2016-complete-report.pdf. Accessed 16 Mar 2018

European Commisison (2016) Post-impact care. Summary 2016. European Road Safety Observatory. https://ec.europa.eu/transport/road_safety/sites/roadsafety/files/ersosynthesis2016-summary-postimpactcare5_en.pdf. Accessed 28 Sep 2017

Heinrich HW (1931) Industrial accident prevention: a scientific approach. McGraw-Hill, New York

House of Commons Transport Committee (2017) Road traffic law enforcement. Second Report of Session 2015–16. The Stationary Office, London

Gains A, Noerdstrum N, Heydecker B, Shrewsbury J (2005) The national safety camera programme. Four-Year evaluation report. Department for Transport, London

Janssen M, Charalabidis Y, Zuiderwijk A (2012) Benefits, adoption barriers, and myths of open government. Inf Syst Manag 29:258–268

Lydell R (2017) Exposed: London's most prolific speed camera which has raised £1.5 m in six months. Evening Standard, Tuesday the 3rd of January. https://www.standard.co.uk/news/transport/exposed-londons-most-prolific-speed-camera-which-has-raised-15m-in-six-months-a3431156.html. Accessed 27 Sep 2017

Newnam S, Goode N (2015) Do not blame the driver: a systems analysis of the causes of road freight crashes. Accid Anal Prev 76:141–151

NHS (2016) Major Trauma Canters in England October 2016. https://www.nhs.uk/NHSEngland/AboutNHSservices/Emergencyandurgentcareservices/Documents/2016/MTS-map.pdf. Accessed 30 Oct 2017

Noland RB, Quddus MA (2004) Improvements in medical care and technology and reductions in traffic-related fatalities in Great Britain. Accid Anal Prev 36:103–113

ONS (2017) Death registration and summary tables – England and Wales. https://www.ons.gov.uk/peoplepopulationandcommunity/birthsdeathsandmarriages/deaths/datasets/deathregistrationssummarytablesenglandandwalesreferencetables. Accessed 25 Sep 2017

PACTS (2017) Safe System. http://www.pacts.org.uk/safe-system/. Accessed 6 Sep 2017

Parnell KJ, Stanton NA, Plant KL (2017) What's the law got to do with it? Legislation regarding in-vehicle technology use and its impact on driver distraction. Accid Anal Prev 100:1–14

Peden M, Scurfield R, Sleet D, Mohan D, Hyder AA, Jarawan E, Mather C (2004) World report on road traffic injury prevention. World Health Organisation, Geneva
Pilkington P (2003) Speed cameras under attack in the United Kingdom. Injury Prev 9:293–294
Rasmussen J (1997) Risk management in a dynamic society: A modelling problem. Saf Sci 27:183–213
Razzak JA, Kellermann AL (2002) Emergency medical care in developing countries: is it worthwhile? Bull World Health Organ 80:900–905
Salmon PM, Read GJM, Stanton NA, Lenné MG (2013) The crash at Kerang: investigating systemic and psychological factors leading to unintentional non-compliance at rail level crossings. Accid Anal Prev 50:1278–1288
Sánchez-Mangas R, García-Ferrera A, de Juan A, Martín Arroyo A (2010) The probability of death in road traffic accidents. How important is a quick medical response? Accid Anal Prev 42:1048–1056
Scott-Parker B, Goode N, Salmon P (2015) The driver, the road, the rules… and the rest? A system-based approach to young driver road safety. Accid Anal Prev 75:297–305
Trist EL (1953) Some observations on the machine face as a socio-technical system. Tavistock Documents Series, London
The University of Manchester (2015) Review of major trauma networks reveals increase in patient survival rates. http://www.manchester.ac.uk/discover/news/article/?id=14891. Accessed 30 Oct 2017
WHO (2015) Global Status Report on Road Safety 2015. World Health Organisation, Geneva
WHO (2017a) The top 10 causes of death. http://www.who.int/mediacentre/factsheets/fs310/en/. Accessed 25 Sep 2017
WHO (2017b) Decade of action. http://www.who.int/roadsafety/decade_of_action/. Accessed 29 Sep 2017

Analysis of Discomfort During a 4-Hour Shift in Quay Crane Operators Objectively Assessed Through In-Chair Movements

Bruno Leban[1(✉)], Federico Arippa[1], Gianfranco Fancello[2,3], Paolo Fadda[2,3], and Massimiliano Pau[1]

[1] Department of Mechanical, Chemical and Materials Engineering, University of Cagliari, Cagliari, Italy
bruno.leban@dimcm.unica.it

[2] CENTRALABS Sardinian Center of Competence for Transportation, Sardinia, Italy

[3] Department of Civil Engineering, Environment and Architecture, University of Cagliari, Cagliari, Italy

Abstract. This study aims to investigate the existence of possible changes in postural strategies adopted by quay crane operators during a 4-h shift performed in a simulated environment. In particular, the analysis is carried out by analyzing the trend of in-chair-movement (ICM) as indicator of discomfort and fatigue. Using a pressure sensitive mat placed on the seat pan, average and peak body-seat pressure and trunk center-of-pressure (COP) time series were acquired and processed to calculate ICMs with two methods based on pressure changes and one which considers the COP shifts. The results show a well-defined linear trend for ICM, which was detected by all the tested approaches, with significant increases occurring after 45–60 min from the beginning of the shift. However, the method which employs COP data appears potentially more adequate to accurately identify ICM due to its relative insensitivity to external factors associated with individual's anthropometry and body composition and presence of external vibrations. Future developments of the study will be focused on establishing the suitability of the method as non-invasive early predictor of fatigue.

Keywords: Discomfort · In chair movement · Sitting posture

1 Introduction

Although the word "comfort" express a well-established theoretical concept, it remains quite difficult to provide a precise definition for it, as comfort can be either a physical sensation, a psychological state or both simultaneously [23]. Thus, to date there is no universal definition, and it is beyond dispute that comfort and discomfort are feelings or emotions subjective in nature [8]. Contrary to what one might think, comfort and discomfort are not two distinct and separate entities, but should be treated as different and complementary entities in ergonomic evaluations [27]. It is therefore not possible to completely separate these two aspects, which combine together with the general definition of well-being sensation. In working contexts, comfort has a strong influence

S. Bagnara et al. (Eds.): IEA 2018, AISC 823, pp. 90–100, 2019.
https://doi.org/10.1007/978-3-319-96074-6_10

on performance [19] and is also strictly related to both physical and environmental conditions in which the individual is required to operate. In particular, workers who are forced to adopt sitting postures for long time may experience a reduction in comfort as the work shift progresses and, moreover, they are likely to be more exposed to musculoskeletal disorders due to prolonged static muscle exertions [28]. For instance, it has been demonstrated that workers who drive in excess of 20 h a week are reportedly 6 times more likely to be absent from work due to back pain than those who drive less than 10 h per week [24]. Similarly, quay crane operators work in a really challenging environment, that forces them to adopt constrained sitting postures for prolonged periods (typically 4–6 consecutive hours), while controlling the container position with short hand movements using joysticks and receiving feedback on correct operation through direct visualization through the transparent walls and floor of the control station [18, 22]. Thus they are required to adopt prolonged static, non-neutral postures, particularly of the neck and trunk, which could be a risk factor of musculoskeletal problems [22]. In these cases conditions, the need for change in position (i.e. "postural shifts") may represent an effective indicator of postural comfort [16], as suggested in several previous studies [9, 19, 23, 29, 30]. However, although quite clear in principle, the practical definition of postural change still remains quite challenging. Since '60s, such postural shifts are defined as "in-chair-movements" (ICM) indicating with this term the quantity composed by the sum of required task movements, movements that a person performs in excess of task demands [11] and extraneous movements [7].

In recent times, the measurement of ICM [26] has been demonstrated to be a successful tool in determining the effect of seated position on the body. In fact, several previous studies reported that both discomfort and seated movements increase linearly over time, also with similar steep slope [3, 4, 10, 14], thus indicating that people change position more frequently when feeling uncomfortable. In practice, as time progresses body continuously attempts to alleviate the discomfort through micro- and macro-movements [17]. To date, various methods have been proposed for identification of such movements during sitting posture, and among these, the most common are based on the analysis of body-seat pan interface average and peak pressure data [17, 21], on qualitative assessment of video acquisitions [10], on data deriving from motion capture systems [1] and, more recently, on Kinect depth camera images [2]. Nevertheless, given their simplicity of acquisition even outside the laboratory, data related to body-seat pan contact pressure distribution are today among the most widespread and used.

On the basis of the above-mentioned considerations this study aims to investigate the trend of ICM in a cohort of professional quay crane operators while performing a 4-h work shift task in a dedicated simulator. We calculated ICM with two of the most employed methods (e.g. those proposed by Na et al. [21], and Le et al. [17]) and with an innovative approach based on the analysis of the position of the center-of-pressure (COP), namely the point of application of the resultant of the forces exchanged between body and seat.

2 Methods

2.1 The Quay Crane Simulator

Data presented in this study were obtained from trials carried out on a custom quay crane simulator (the "Chameleon Simulation Team Portainer", [6]). The simulator, installed in a 40-foot container, is equipped with a standard Brieda DYCS (Brieda Cabins, Italy) control station, which is composed by a seat and a console with two joysticks located in the armrests that allow operators to control all the engagement/disengagement of the spreader/container coupling when, for example, container has to be hoisted or put in place. The control station is placed over a three-axis platform with electric actuators controlled by a high-performing computer, which provides for the application of random vibrations, rotations and translations compatible with the task executed by the operator, according to patterns previously acquired under realistic conditions. The control station is surrounded by large screens positioned in front, on the sides and under the operator's feet, in order to faithfully reproduce the visibility conditions of an actual quay crane cabin. The software is able to simulate all weather conditions by controlling wind intensity and direction, rain, sea conditions and sky entities (sun, moon, clouds). World locations are entirely controlled by a supervisor station, from which it is possible to decide the upload/unload schedule to/from a container ship and record subjects' performance (i.e. number of successfully moved containers and collisions).

2.2 Participants

Sixteen male professional quay crane operators (age 37.3 ± 5.1 years, stature 178.5 ± 8.2 cm, body mass 79.2 ± 14.7 kg, working experience 6.2 ± 3.6 years) were recruited for this study. All participants provided written informed consent, after a detailed explanation of the purposes of the study and a description of the experimental methodology. All of them were familiar with the control station previously described, as they routinely employ it in their workplaces.

2.3 Interface Pressure Data Acquisition and Post-processing

Data on the contact pressure at the body-seat interface were obtained by means of a pressure-sensitive mat (Tekscan 5330E 471.4 × 471.4 mm active area, 1024 sensing elements arranged in a 32 × 32 matrix). The sensor was connected to a two port hub (Tekscan Versatek) using RJ-45 cables and then to a PC via USB connection. Before starting the test, the mat was calibrated according to the manufacturer's instructions and operators were allowed to briefly familiarize with the simulator, transferring as many containers as possible following a predetermined schedule. Each working session was composed by a combination of several movements. The operators moved the spreader over the selected container and hoisted it to the appropriate height, then moved the container along the bridge rails to the container-stacking bay, placing the container over a truck. Such cycle is typically repeated at least 20 times per hour [5].

Seat-body pressure was continuously recorded for 4 h (which represents the usual duration of the work shift) at 10 Hz sampling frequency. Raw data were then post-processed using the Tekscan Conformat Research Software v.7.10 to extract time-series for mean and peak pressure, and COP position. First and last 7.5 min were discarded as they include the phases of sit and stand up from the control station seat.

Starting from the acquired pressure data, ICMs were calculated using three different approaches:

(1) The first method, based on Na et al. [21] algorithm is based on the mean value of the contact pressure calculated across the whole mat area. ICM is then defined by the pressure changes occurred between two consecutive time periods as follows:

$$\mathrm{ICM} = \left[\left(P_{(i)} - P_{(i-1)}\right) > \left(0.05 \cdot P_{avg}\right)\right] \tag{1}$$

where $P_{(i)}$ and $P_{(i-1)}$ are the values of mean contact pressure at (i)-th and (i-1)-th time instants respectively, and P_{avg} represents the average of mean contact pressure calculated for the whole trial duration.

(2) The second method, proposed by Le et al. [17], takes into account the peak pressure recorded on the mat during the trial. In this case the ICM is defined as a change in peak pressure which exceeds ± 6.4 kPa with respect to the steady state (i.e. the mean peak pressure calculated on the whole test duration). The 6.4 kPa threshold was arbitrarily selected by the authors "upon an average of values in relation to discomfort" [17] considering the findings of several previous similar studies [13].

(3) The method proposed by the authors, calculates the number of ICM starting from the COP coordinates. At first, the COP time-series are de-trended (to attenuate/remove any sensor's drift effect) and the number of ICM is calculated separately for antero-posterior (AP) and medio-lateral (ML) direction. In this case an ICM occurs when the COP coordinate (i.e. the displacement from the mean position) is above or below a predefined threshold that was chosen by means of an iterative method and optimized at a value of 21.2 mm. In this way, the number of calculated ICM results of the same order of magnitude when compared with the previous two methods.

All calculations were carried out using dedicated custom software developed under the Matlab® (The MathWorks, Inc, Natick, MA, USA) environment. For each approach, the number of ICMs was calculated within 15-min intervals, in order to avoid any specific trend within the time interval, as suggested by Fenety et al. [10].

To facilitate the comparison of the results obtained from the different approaches, for each trial, the 15-min ICMs series were normalized by dividing each value by the maximum.

2.4 Statistical Analysis

The existence of possible differences in ICM values during the 4-hour shift was assessed, for each tested approach, using one-way analysis of variance for repeated measures (ANOVA RM), performed using the IBM SPSS Statistics v.20 software (IBM, Armonk, NY, USA). The independent variable was the time and the dependent variable the number of ICMs. The significance level was set at $p = 0.05$. Multiple comparison tests vs. baseline value of ICM number (ICM relative to the first 15-min interval) were performed. Finally, the relationship between ICM and time was assessed by calculating the Pearson product moment correlation analysis. Also in this case the level of significance was set at $p = 0.05$.

3 Results

Due to acquisition issues, data of 2 operators were discarded, thus the results here presented refer to 14 participants. A typical map obtained from the experimental trials is reported in Fig. 1. ANOVA RM revealed a significant effect of time on normalized ICMs number for all considered methods: Na algorithm [$F(14,182) = 36.06$, $p < .001$, $\eta^2 = .735$], Le algorithm [$F(14,182) = 8.80$, $p < .001$, $\eta^2 = .404$], COP_{AP} [$F(14,182) = 4.77$, $p < .001$, $\eta^2 = .268$] and COP_{ML} [$F(14,182) = 7.65$, $p < .001$, $\eta^2 = .371$]. Results of multiple comparisons vs. baseline showed significant differences for all methods except for COP_{AP} (Fig. 5). The results of the correlation analysis revealed that ICM is significantly positively correlated with time ($r = .847$, $p < .001$). An example of changes in ICM with increasing time is shown in Fig. 2 (Figs. 3 and 4).

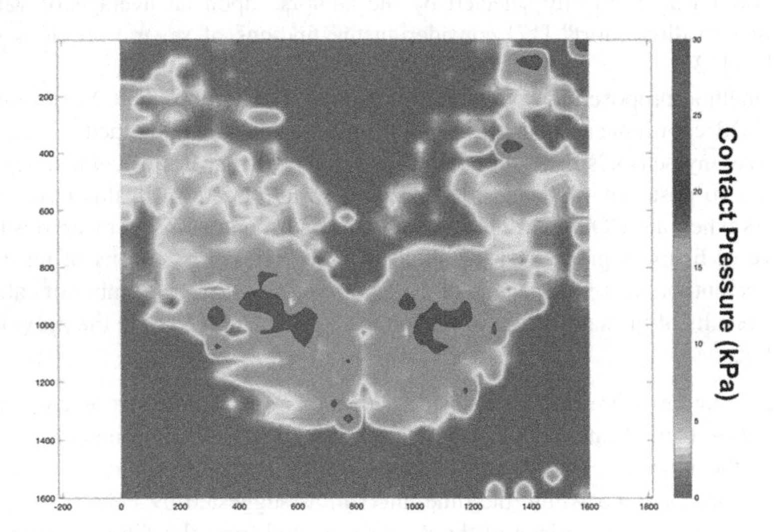

Fig. 1. Example of body-seat contact pressure distribution obtained during the trials (seat pan).

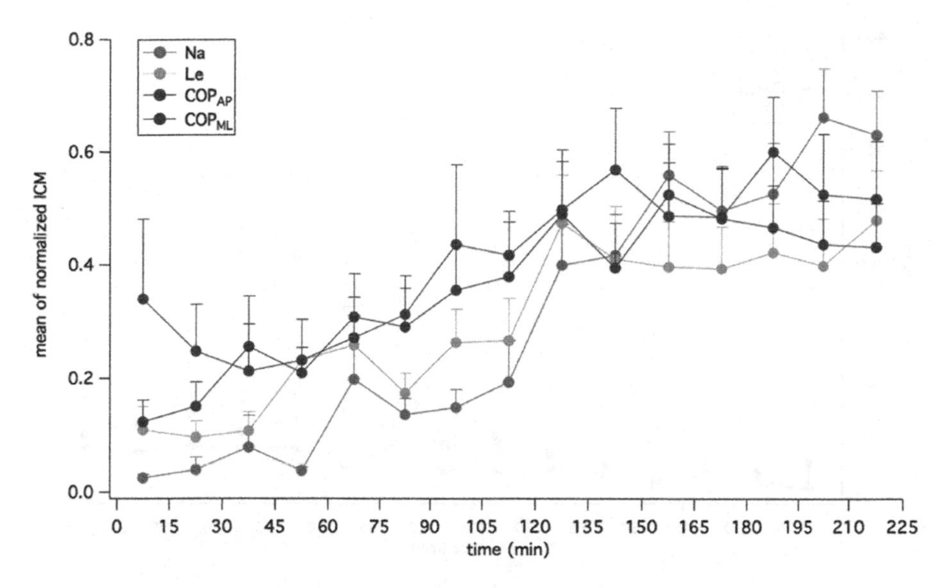

Fig. 2. Normalized ICM values. Bars indicate standard error.

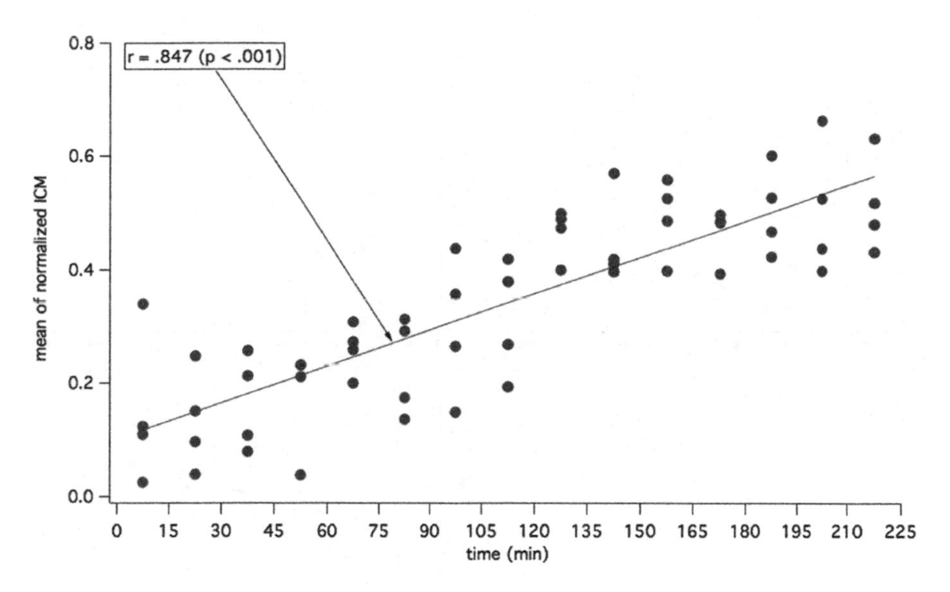

Fig. 3. Fitting of mean values of normalized ICM

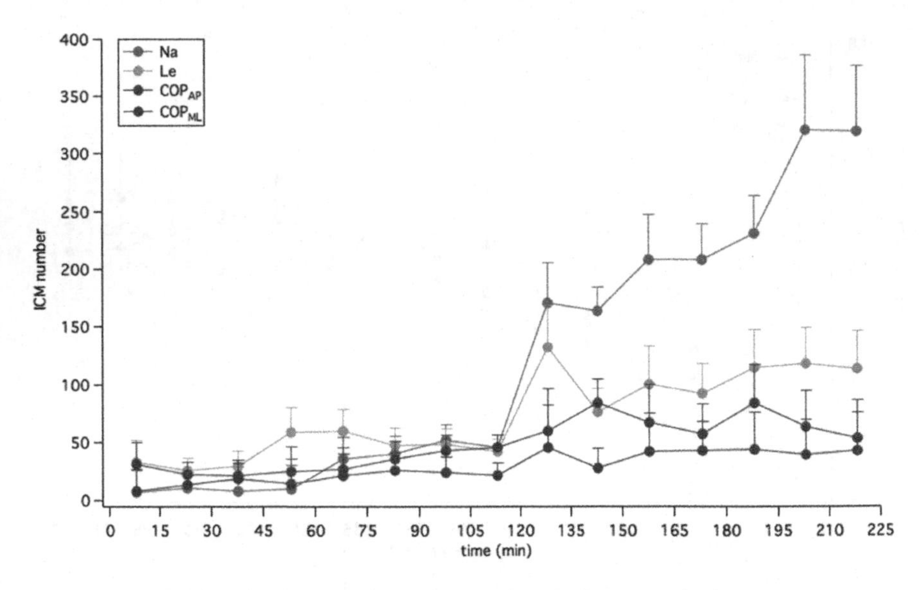

Fig. 4. Absolute ICM values. Error bars indicate standard error.

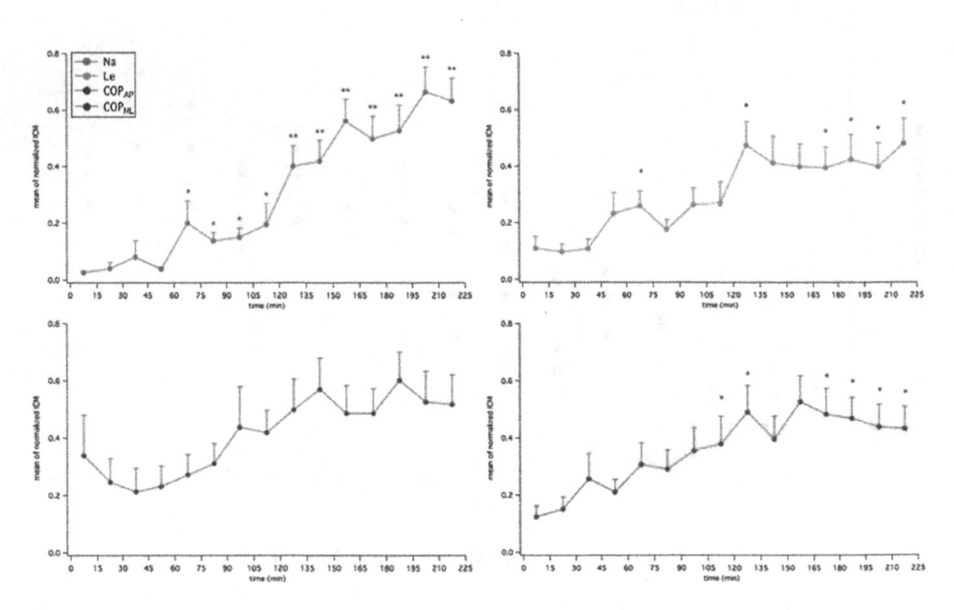

Fig. 5. Trends of normalized ICM during the simulated 4-h shift. Error bars indicate standard error. The symbols * and ** indicates a significant difference vs. baseline value ($p < 0.05$ and $p < 0.001$ respectively).

4 Discussion

The purpose of the present study was twofold: (1) to assess the existence of a possible trend for ICMs in a cohort of quay crane operators who perform a 4-h shift in a dedicated simulator and (2) to compare the results of different algorithms proposed to calculate the number of ICM, including a newly developed one based on COP position rather than on body-seat interface pressure. The employed methodology (i.e. monitoring of the body-seat contact pressure) can be considered suitable to achieve both goals, as several previous studies demonstrated that contact pressure is among the few quantitative parameters associated with perceived discomfort during constrained, seated posture [8]. Owing to this, the techniques which obtain the number of ICM on the basis of pressure changes may be considered reliable in predicting the overall discomfort level [17, 21].

Overall, our results suggest that, regardless the employed approach, as time progresses operators tend to perform more postural shifts and thus the number of ICM generally increases with time. Such findings are consistent with previous studies [3, 10, 12, 15, 20, 25] performed on drivers and office workers and this phenomenon is explained with the existence of a specific strategy to cope with the onset of discomfort. In particular, Na et al. [21] reported that seat-body pressure trends were associated with the driving period duration (i.e. contact pressure increased as the driving period progressed). Le et al. [17] hypothesized that individuals forced to spend long time in sitting posture continuously attempt to alleviate the discomfort through subtle movement. Fenety et al. [10], detected significant increases in ICM, calculated on the basis of the overall length of the COP trajectory during the trial.

Data obtained from the present study revealed that after 45–60 min from the beginning of the work shift, the number of ICM becomes significantly higher with respect to the baseline, regardless of the calculation technique employed. The only exception was represented by the COP_{AP} parameter, probably due to the fact that most postural shifts occur in the mediolateral direction, as the individual feel the need to relieve the high pressure concentrated in the ischial tuberosity regions.

However, even though we found a substantial agreement between the approaches previously proposed and our method based on COP trajectory analysis, we think that the latter is potentially more adequate to effectively detect an ICM for two reasons: firstly, the results provided by methods based on body-seat pressure are strictly dependent on the selected threshold for the contact pressure. This might represent a critical issue because pressure is influenced by anthropometry (i.e. individual body mass, fat distribution, etc.). Secondly, there are circumstances in which pressure changes do not necessarily reflect a macro-movement of the body over the seat for example as it happens when the control station underwent sudden random vibrations.

In contrast, the approach based on COP position, even though still dependent on the pressure distribution across the whole sensitive mat, allows a more straightforward identification of the ICM given the relation between postural shifts and COP. Moreover, COP data is able to provide information about the direction, the velocity and the magnitude of a postural shift while approaches based on pressure data only cannot.

Some limitations of the study should be acknowledged: the main critical issue is represented by the fact that we did not perform any assessment of either discomfort or fatigue as perceived by the operator, and thus we have no direct evidences of an increase of such parameters as the shift progresses. However, it is noticeable that previous studies reported that the number of ICM is correlated with fatigue in long-term tasks [9, 19, 23, 29, 30] and its estimation has been found reliable in predicting the discomfort onset. Furthermore, although the COP method allows to discriminate between displacements which occur either in ML or AP direction, this algorithm is not able to separate spontaneous movements from those required to perform the working task. This issue is crucial in the specific case of the quay crane operators here investigated, as they continuously move the trunk forward and backward to visually monitor the trajectory of the container from the ship to the shore and viceversa. Further studies are thus necessary to define specific strategies to isolate unnecessary postural changes to those associated with the performed task.

At last, the method here proposed is still dependent on the value of a predefined threshold. The definition of such value suffers from a certain subjectivity, that is strictly related with the lack of unambiguous definition of ICM.

5 Conclusions

The results of the present study confirm that the identification of ICM and the analysis of their variation with time may represent a useful tool to assess the existence of specific strategies adopted to reduce discomfort and fatigue. In the specific case of quay crane operators, monitored during a long-term shift performed in a dedicated simulation environment, the obtained data suggest that they attempt to mitigate the negative effects caused by the prolonged sitting posture by performing an increasing number of postural shifts, mainly in the ML direction.

Among the several methods available to calculate the number of ICM, starting from body-seat pressure data provided by pressure-sensitive mats, the one based on the analysis of COP trajectories seems the most suitable especially for the possibility to separate movements occurring in different directions. This feature might be of great help in discriminating voluntary (e.g. task-related) from unnecessary movements especially in all those conditions in which the expected trajectories of the trunk are known a priori.

Future developments of this study remain necessary, in particular to clarify the exact relationship between perceived fatigue and discomfort and number of ICM, in such a way to plan a possible use of this approach as non-invasive technique to monitor the physical workload of operators forced to adopt constrained sitting postures for a long time.

Acknowledgements. The authors wish to thank the Brieda Cabins di Rino Brieda e figlio S.r.l. (Porcia, Italy) for making available the control station installed in the quay crane simulator, and all the operators who participated in the study. The help of Ms. Chiara Deledda and Mr. Giacomo Fenza during the acquisition process was also greatly appreciated.

Conflicts of Interest. The authors report no conflicts of interest.

References

1. Andreoni G, Santambrogio GC, Rabuffetti M, Pedotti A (2002) Method for the analysis of posture and interface pressure of car drivers. Appl Ergon 33(6):511–522. https://doi.org/10.1016/s0003-6870(02)00069-8
2. As MA, Lukman SK, Ismail LH, Zakaria NA, Mahmood NH, Omar AH (2016) Automated prolonged sitting detection in office workplaces using kinect. ARPN J Eng Appl Sci 11(23):14025–14032
3. Bendix T, Winkel J, Jessen F (1985) Comparison of office chairs with fixed forwards or backwards inclining, or tiltable seats. Eur J Appl Physiol Occup Physiol 54(4):378–385. https://doi.org/10.1007/bf02337181
4. Bhatnager V, Drury CG, Schiro SG (1985) Posture, postural discomfort, and performance. Hum Factors 27(2):189–199. https://doi.org/10.1177/001872088502700206
5. Bruzzone A, Fadda P, Fancello G, D'Errico G, Bocca E, Massei M (2009) A vibration effect as fatigue source in a port crane simulator for training and research: spectra validation process. In: International conference on harbour, maritime and multimodal logistics modelling and simulation, vol 1, pp 77–86
6. Bruzzone A, Fadda P, Fancello G, Massei M, Bocca E, Tremori A, Tarone F, D'Errico G (2011) Logistics node simulator as an enabler for supply chain development: innovative portainer simulator as the assessment tool for human factors in port cranes. Simulation 87:857–874. https://doi.org/10.1177/0037549711418688
7. Corlett EN (1990) Evaluation of Industrial Seating. In: Wilson JR, Corlett EN (eds) Evaluation of human work. a practical ergonomics methodology. Taylor & Francis, London, pp 500–515
8. de Looze MP, Kuijt-Evers LF, van Dieën J (2003) Sitting comfort and discomfort and the relationships with objective measures. Ergonomics 46(10):985–997. https://doi.org/10.1080/0014013031000121977
9. Fenety A, Walker JM (2002) Short-term effects of workstation exercises on musculoskeletal discomfort and postural changes in seated video display unit workers. Phys Ther 82(6):578–589. https://doi.org/10.1093/ptj/82.6.578
10. Fenety PA, Putnam C, Walker JM (2000) In-chair movement: validity, reliability and implications for measuring sitting discomfort. Appl Ergon 31(4):383–393. https://doi.org/10.1016/s0003-6870(00)00003-x
11. Fleischer AG, Rademacher U, Windberg HJ (2007) Individual characteristics of sitting behaviour. Ergonomics 30(4):703–709. https://doi.org/10.1080/00140138708969762
12. Grandjean E, Jenni M, Rhiner A (1960) An indirect method for evaluation of the comfort feeling in sitting position. Int Z Angew Physiol 18:101–106
13. Gyi DE, Porter JM (1999) Interface pressure and the prediction of car seat discomfort. Appl Ergon 30(2):99–107. https://doi.org/10.1016/s0003-6870(98)00018-0
14. Jensen CV, Bendix T (1992) Spontaneous movements with various seated-workplace adjustments. Clin Biomech 7(2):87–90. https://doi.org/10.1016/0268-0033(92)90020-5 (Bristol, Avon)
15. Jurgens HW (1989) Body movement during observation tasks. Ann Physiol Anthropol 8 (1):33–35
16. Kolsch M, Beall AC, Turk M (2003) An Objective Measure for Postural Comfort. Proc Hum Factors Ergon Soc Ann Meet 47(4):725–728. https://doi.org/10.1177/154193120304700413
17. Le P, Rose J, Knapik G, Marras WS (2014) Objective classification of vehicle seat discomfort. Ergonomics 57(4):536–544. https://doi.org/10.1080/00140139.2014.887787

18. Leban B, Fancello G, Fadda P, Pau M (2017) Changes in trunk sway of quay crane operators during work shift: a possible marker for fatigue? Appl Ergon 65:105–111. https://doi.org/10.1016/j.apergo.2017.06.007
19. Liao MH, Drury CG (2000) Posture, discomfort and performance in a VDT task. Ergonomics 43(3):345–359. https://doi.org/10.1080/001401300184459
20. Michel DP, Helander MG (1994) Effects of two types of chairs on stature change and comfort for individuals with healthy and herniated discs. Ergonomics 37(7):1231–1244. https://doi.org/10.1080/00140139408964901
21. Na S, Lim S, Choi HS, Chung MK (2005) Evaluation of driver's discomfort and postural change using dynamic body pressure distribution. Int J Ind Ergon 35(12):1085–1096. https://doi.org/10.1016/j.ergon.2005.03.004
22. Pau M, Leban B, Fadda P, Fancello G, Nussbaum MA (2016) Effect of prolonged sitting on body-seat contact pressures among quay crane operators: a pilot study. Work 55(3):605–611. https://doi.org/10.3233/wor-162434
23. Pearson EJ (2009) Comfort and its measurement–a literature review. Disabil Rehabil Assist Technol 4(5):301–310. https://doi.org/10.1080/17483100902980950
24. Porter JM, Gyi DE (2002) The prevalence of musculoskeletal troubles among car drivers. Occup Med 52(1):4–12. https://doi.org/10.1093/occmed/52.1.4 (Lond)
25. Rieck VA (1969) Über die Messung des Sitzcomforts von Autositzen. Ergonomics 12(2):206–211 (in German)
26. Sammonds G, Fray M, Mansfield N (2017) Effect of long term driving on driver discomfort and its relationship with seat fidgets and movements (SFMs). Appl Ergon 58:119–127. https://doi.org/10.1016/j.apergo.2016.05.009
27. Sauter SL, Swanson NG, Waters TR, Hales TR, Dunkin-Chadwick R (2005) Musculoskeletal discomfort surveys used at NIOSH. In: Stanton N, Hedge A, Brookhuis K, Salas E, Hendrick H (eds) Handbook of human factors and ergonomics methods. CRC Press, Boca Raton/London/New York/Washington DC
28. Sjøgaard G, Jensen BR (2006) Low-level static exertions. In: Marras WS, Karwowski W (eds) Fundamentals and assessment tools for occupational ergonomics, 2 edn. Boca Raton, CRC Press LLC. Occupational ergonomics handbook, vol 1, pp 14.1–14.13
29. Søndergaard KH, Olesen CG, Søndergaard EK, de Zee M, Madeleine P (2010) The variability and complexity of sitting postural control are associated with discomfort. J Biomech 43(10):1997–2001. https://doi.org/10.1016/j.jbiomech.2010.03.009
30. Vergara M, Page A (2002) Relationship between comfort and back posture and mobility in sitting-posture. Appl Ergon 33(1):1–8. https://doi.org/10.1016/s0003-6870(01)00056-4

Classifying Normal and Suspicious Behaviours When Accessing Public Locations

Brendan Ryan[(✉)] and Aswin Vijayan

University of Nottingham, University Park, Nottingham NG72RD, UK
brendan.ryan@nottingham.ac.uk

Abstract. In order to identify and respond appropriately to suspicious behaviour in public locations, it can be important to understand the typical behaviours in a particular setting. An observation study has been carried out at two libraries to explore the entry behaviours as people approach card access barriers in these locations. Seven hours of video data have been collected over four observation periods. The recordings have been analysed by a researcher, using a behavioural framework that has been developed in previous research in a related study context. This framework was used as a prompt to look in more detail at a range of different aspects of behaviours that may be observable. The findings, including draft requirements for smart camera technologies to support security staff, were discussed in an interview with a security expert. The study identified common types of behaviour, with some differences evident when accessing the libraries alone or as part of a group. A number of interesting aspects of behaviour were observed on approach to, or at the point of interacting with, the access barrier for the buildings. These included observation of several behavioural cues that could be early indicators of anomalous behaviours that were in progress. The study has demonstrated the feasibility of adapting and using the behavioural framework to explore behaviours in new settings.

Keywords: Observing behaviour · Security · Surveillance technologies

1 Introduction

Security is becoming increasingly important, especially in light of several high profile incidents at transport locations or public buildings. Surveillance is one approach to improving security, with a goal of identifying suspicious behaviour and intervening to prevent an incident. This type of intervention to prevent incidents does happen in practice, though it is not always clear how people do this.

It is important to consider what is suspicious behaviour? A first step in understanding this is knowing what is normal behaviour? From this position it may be possible to identify when behaviour deviates from this [1]? Currently, it is hard to specify different varieties of behaviour and there is no clear characterisation of what is abnormal [2]. This can be behaviour that looks very distinct from what is typical [3], for example people running when everyone else is walking [4]. However, it can be difficult to determine whether what is being observed is different from normal behaviour (e.g. when does walking backwards and forwards in front of a bank become suspicious [5])?

S. Bagnara et al. (Eds.): IEA 2018, AISC 823, pp. 101–112, 2019.
https://doi.org/10.1007/978-3-319-96074-6_11

A second consideration is in relation to what can be observed, and furthermore, how this can be described and used as a basis for prevention. It is possible that technologies can assist in this area. Smart cameras are one example [6], where the system can complete some analysis on collected data using algorithms [7] and notify an operator [8] when unusual events have been identified in a particular setting [9, 10]. There is also the potential to anticipate events, providing an early warning about doubtful behaviours [11]. This can take some of the mental workload from security staff [6] and overcome problems with motivation during monitoring activities [3]. Such systems commonly rely on lists of abnormal behaviours [5], rules for what is classified as normal or abnormal [3] and statistical methods to identify uncommon behaviours [3]. In practical terms, this could include aspects of behaviour such as their body position, movements and gestures. Observations may be made more difficult in different lighting levels, with shadows and reflections [6]. There may also be technical difficulties in developing technologies to respond to the many different types of behaviour.

It is clear that there are many reasons why surveillance is difficult in practical contexts (i.e. knowing what to look for, in conjunction with the wide range of people accessing these locations and associated variation in what can be observed). This could be made easier by providing suitable prompts for the observer. Ryan described the development of a framework of behaviours that have been observed at railway stations before railway suicides [12]. Pre-existing data sources were analysed, producing a framework highlighting five aspects of observable behaviour: the display of emotion, appearance, posture/movements, activities and interactions. Potential future applications of the framework include improving the sensitivity of training for staff and using the contents to understand the range of visual cues that could be identified using new surveillance technologies. Understanding behaviour in this type of context is challenging and it is recognised that this initial framework needs further testing in a range of contexts, to add to the content and validate the existing classes and sub-classes of behaviour.

The current paper explains how the framework has been tested in a related context, studying behaviour at the point of access to public buildings. Behaviours of people entering two University libraries have been observed. The entrances to the libraries have card access barriers and require similar activities to those that are required at entry to railway stations with electronic gate lines or other public buildings with entry restrictions. The observations were carried out by a researcher who had limited experience of observations in research. The study therefore provides an opportunity to understand how the structure within the framework can be used by a relatively inexperienced observer in exploring how typical behaviours can be described, as well as giving initial indications of some deviant or suspicious behaviours in approaching and passing through the entry barriers. The findings can provide valuable input to the development of better surveillance methods, including what is needed for technologies such as smart camera systems.

2 Method

An observation study has been carried out on behaviours during access to two libraries at the University of Nottingham. The findings have been discussed with a security expert to validate conclusions from the observational study and clarify requirements for a technological support tool to assist security staff (i.e. a smart camera system). Observations were conducted at card access gates at the entrances to the two libraries. There were four periods of observation on separate days at the two library locations. Recordings of people were collected as they entered the libraries, using a high definition camera mounted on a tripod. This was placed in an unobtrusive position, but it was not hidden. The researcher remained in the location during the observation periods and recorded additional field notes. The study was approved by the Faculty of Engineering Ethics committee and senior staff at the libraries. It was not possible to obtain permission from all people accessing the libraries so a notice was displayed which explained that an observation study and video recording was being carried out at specified times in the area. Information on the project and details of an alternative access route were also listed.

A set of questions were prepared for the interview with the security expert. These questions included questions on the interviewee's role, what they look for in relation to normal and abnormal behaviours when observing CCTV and recollections of some common behaviours when people enter buildings. There were also several questions about use of smart cameras, whether they were familiar with these or if they had considered their use, how they decide where CCTV or smart cameras should be placed, and the types of interactions that they have with existing camera systems. The interview also used a scenario that was created from content from analysis of the camera data. An extract from this scenario is included later in Sect. 3. The interviewee was asked a number of questions about the scenario, including ranking of the level of importance of a number of requirements for a smart camera system that could support surveillance tasks in this type of location. These requirements are also explained later in Sect. 3.

Seven hours of recordings of access behaviours were collected. To analyse the content, an initial viewing was used to identify typical behaviours on entry to the library. A second viewing of the recorded material was used to focus in detail on atypical behaviours. The framework of behaviours [12] was used to prompt the search for a range of visual cues. The content of the framework is given in Table 1.

Findings have been summarised in descriptive text, supported by images that have been extracted from the recordings. Content from the observations was aggregated and used to create some common scenarios from the camera data. These scenarios were used in producing requirements for a smart camera system to monitor and react to typical and anomalous behaviours in this type of location. The recording of the interview with the security expert was transcribed, read and re-read, coded and analysed to identify and extract the interviewee's thoughts and conclusions on the questions that were posed during the interview.

Table 1. Behaviour framework, adapted from [12].

Description of behavioural types	Sub-classes
Emotional response	Non-specific visual indications
Appearance	Well-being Clothing Appearance in context
Posture/gestures/movements	Postures and movements, including those associated with emotional expression Movements around the location
Event/Activity/Actions (or inaction)	Waiting Searching Accessing unusual locations Interacting with possessions Normal activities at the location
Interactions/lack of interactions with people or the location	With the local environment In response to an outside intervention

3 Results

The observations focus on access behaviours immediately outside the libraries, on approach to the access barrier and on passing through the access barrier (or being inhibited by the access barrier in some cases). The analysis has considered people entering the library alone and in groups of two or more. There were 191 people entering alone during the study periods and 35 entering in groups.

For those entering alone, six types of behaviour were identified, as shown in Table 2. Five behaviour types were identified when entering in groups (Table 3). Classes from the behavioural framework [12] have been used in examining these examples of behaviour in more detail. Notable observations have been recorded in the Tables. Some of these observations are illustrated in photographs (Figs. 1a and b).

It was often possible to distinguish different emotional responses in facial expressions, when applying the framework in this context. The explanations of what was observed are not always clear in terms of description of exactly what was seen. These explanations tended to include the conclusions that the observer formed whilst observing the images (e.g. that people were happy, confused, frustrated, calm, embarrassed). An interesting aspect of the observations was that people could smile for different reasons, and at different points on their journey towards and through the access barrier. This included occasions when people were conversing with friends in groups, but also a situation where someone was going to do something that they should not be doing in this area (i.e. tailgating). Further investigation would be needed to see if it is possible to detect differences in intention from the way that people are smiling (or from similar facial expressions).

The category for appearance in the framework was generally less useful in prompting observations in this study. For example, the sub-category for visual indications of well-being wasn't commonly used as a prompt for observations, though did help draw attention to how staff may open the gates as people approached with walking

Table 2. Types of behaviour for those entering alone

Type of behaviour	Description of the behaviour	Categories and content from the behavioural framework [12]
Authorised entry with card	Card already in hand as they approach the entrance and on arrival scans card and enters library	*Emotional response* – "bland" facial expression *Posture/Gesture/Movement* – bending to scan the card when this is fixed to a lanyard *Interactions/lack of interactions* – Using a mobile phone
	Gets inside entrance, then waits and searches for card before entering the library	*Appearance* – common with those carrying many items, books, drinks etc. *Action* – search carried out without rushing, especially after glancing and checking that the area is not busy
Search for entry card	Search for card outside library for some time and if no card leaves the area	*Emotional response* – can be identifiable by facial expression on failing to find access card
Request for assistance	Person already knew they had no card or after searching finds no card, so raises hands at librarian to get their attention or walks towards the area where library staff are sitting	*Gesture/movement* – deviate from normal entry path towards the barriers and gesture for attention of staff or speak to staff member *Interactions* – speak with staff
Problems scanning entry card	Scans card but entrance gate does not open	*Emotional response* – facial expression indicating confusion *Action* – gaze at the card and scanner and re-scan, double checking. Look in direction of library staff *Interaction* – enter conversation with library staff
Entry with bulky items	University staff entering library with book trolley	*Interaction* – arrival noticed by library staff and the gate is opened for the staff member
Waiting	Staying outside the library for some time	*Emotional response* – often appear relaxed *Movements* – occasionally walking back and forth *Action* – occasional glances towards the building entrance *Interactions* – talking on the phone, limited interaction with other people

sticks or carrying bulky items. Similarly, observations of the type of clothing was not useful in understanding behaviour in this context. However, the observer did consider the prompt of appearance in context as a useful means of identifying some of the typical behaviours (e.g. standing near to the entrance gate or waiting outside) and

Table 3. Types of behaviour for those entering in groups

Type of behaviour	Description of the behaviour	Categories and content from the behavioural framework [12]
Authorised entry with card	Both people have card in hand as they approach library, ready for scanning	*Emotional response* – expression appears to be stress-free, happy, smiling *Activity* – waiting behind the first person where there is one access gate, selection of a separate access gate where there is more than one gate *Interactions* – conversations with other group members on approach to and during access through the gates
Request for assistance	One person with card ready in hand and another person waits to search for card to enter or informs staff they don't have card, either by going to them or by waving of hands	*Emotional response* – less likely to have a concerned facial expression when arriving as part of a group *Gesture* – to get the attention of the library staff *Movement* – walk towards library staff when they are located nearby *Action* – look in the direction of library staff *Interaction* – with others in the group
	Both people have no card so wait at entrance or raise hands to get attention from staff	*Emotional response* – both are relaxed *Gesture* - to draw attention of staff *Activity* – wait patiently near the gates, look towards the staff *Interaction* – casual conversation with friend whilst waiting, conversation with staff member
Problems scanning entry card	One person enters and second person's card not working	*Emotional response* – confused facial expression for the person whose card does not work, occasional smile, through embarrassment? *Activity* – person with successful entry waits after the gates and looks repeatedly at the staff. On some occasions the successful entrant does not notice the failure of entry for their friend and does not return to the entry gate *Interaction* - Eye contact between friends

(*continued*)

Table 3. (*continued*)

Type of behaviour	Description of the behaviour	Categories and content from the behavioural framework [12]
Deviant entry behaviours	Tailgating - One person with card and second person with no card but still enter library by following closely behind the first person	*Emotional response* – look calm or smile and exchange glances with friend on approach *Appearance* – closer to one another than in other entries of groups *Movement* – walking towards the same entry gate and walking quickly through the gate (Fig. 1a), may also include some use of the legs to restrict closure of the gate *Action* – glances around before entering library *Interactions* – constant conversation
	Jumping over entrance gate	*Emotional response* – change from a comfortable expression when the access card did not work. Smiling after access *Activity* – looks around the area and jumps over the barrier after access card did not function as expected (Fig. 1b) *Interaction* – limited prior to the activity of jumping, but expressive afterwards (laughing)
Waiting	Loitering outside entrance of library	*Emotional response* – difficult to observe at a distance, though some indications of laughing and having fun *Appearance* – common to see groups outside the library *Posture* – casual postures *Activity* – groups often do not enter the library after their conversation outside e.g. waiting for other friends to leave, people passing by stop to talk, place used as a meeting spot *Interaction* – conversation between friends, texting and speaking on phone

(a)

(b)

Fig. 1. Photographs of tailgating and jumping the barrier

abnormal behaviours (e.g. looking suspicious, though this is not described in detail) in this library setting.

The prompt of postures, gestures and movements drew the observer's attention to a wide range of behaviours, including speeds of approach to the gates in different circumstances, variations in the paths that people took towards the gates and returning back to the entrance gate to check on colleagues. This prompt also helped to identify raising hands to seek attention and various postures adopted during scanning for access and periods of waiting.

The event, activities, action prompt helped in drawing attention to periods of waiting (outside and inside the entrance), searching for access cards at different points of the approach to the access gate, activities associated with scanning for entry (e.g. putting down bags, dropping items, looking around the area, looking for staff for assistance), and some abnormal behaviours such as tailgating and jumping of the barriers.

The category for interactions or lack of interactions was another useful prompt for the observations, considering interactions with people (e.g. friends, library staff) or the location (e.g. how people approach and interacted with the scanning device and barrier and personal equipment (e.g. mobile phones).

The extract from a scenario in Fig. 2 was created from the analysis of the camera recordings.

It is a cloudy Monday morning David and John are having a conversation and smile as they approach the building. Before the entrance gate John realises that he has forgotten his access card and whispers about this in David's ear. On arrival, David and John walk side by side and soon John slows down to stay behind David. While John is staying behind, David quickly looks around the area and when he scans his card both have a smile and both quickly walk in to the library at the same time. As they walk off they smile again and chat. Next Rohit who is student of the university and his brother Rahul, who is visiting him, comes to the building. On their way to the library both are talking to each other and laughing. Rohit scans his card and gets in while Rahul is waiting near the entrance gate as he does not have an access card. Rahul raises his hands towards the place where the staff are sitting to get their attention. A member of staff comes near to him. Rahul and Rohit explain to staff about their visit and then the staff member allows Rahul to get in.

Fig. 2. Extract from a scenario created from the observations

The scenario and a selection of draft requirements for smart camera technologies to support security staff were used in the interview with the security expert. In response to a question on identifying behaviours, the expert mentioned that some identifiers of normal and abnormal behaviour were those that stand out as being different in a location (e.g. someone repeatedly visiting a location), being in an area that they should not be, or constantly looking or glancing around a location (such that it is evident that they are conscious of what is happening around them, looking to see if they are being observed or indicating that there is something wrong with the person). People in groups, especially if they were carrying things with them or wearing certain items of clothing (e.g. hoodies) may also rouse the suspicion of security staff, so that they would watch these more carefully. It was discussed how it might be difficult to identify intent associated with certain facial expressions (e.g. different types of smiles). However, it was elaborated that whilst there are many different types of smiles and circumstances where people are conscious about their smile, it may be possible to learn more about people's facial expressions (including smiles) by studying normal access behaviour.

When discussing current system details, existing CCTV is passive and does not alert security staff to problematic behaviour. Suspicious or abnormal behaviour would usually be identified by direct observation of security staff or notifications from burglar alarms. With regard to the potential for future use of smart cameras, it was explained how smart camera systems may be introduced in the near future for limited applications (e.g. car number plate recognition), though there appear to be no current plans for more ambitious use of this type of technology.

The set of requirements for smart camera technologies to support security staff was revised, taking account of feedback from the security expert and content within existing literature. Twenty two requirements were produced, some linked to what can be observed and some outlining functional requirements of the support system. Those relating to observable behaviours include:

- being able to identify and respond to different types of facial expressions
- gestures such as raising of hands for attention
- waiting near to the entrance in excess of a given time period
- anomalous behaviours such as jumping barriers or tailgating
- normal and alternative paths towards the entrance – including speed of movement and grouping or bunching of participants
- the types of interactions that could be precursors to deviant behaviours – such as exchanges of glances and certain facial expression,
- typical waiting behaviours.

Functional requirements included:

- ability to operate in different lighting levels
- ability to provide alarms to operators in certain circumstances
- ability to support wifi connectivity
- need to have sufficient battery life
- need for suitable size
- ability to retain or record relevant behaviours for a specified period of time, including ability to identify trends or common behaviours.

4 Discussion

Various types of behaviours associated with access to the study locations have been observed. The findings show that many aspects of the framework [12] have been useful in prompting observations of relevant behaviours, highlighting differences when accessing alone or in groups. This has produced some descriptive detail by a relatively inexperienced observational researcher on a range of behaviours. The level of description could be expanded in future studies, to clarify the limits of what can be observed and how this can be described.

This study has demonstrated the feasibility of broadening the application of the framework to new settings and the potential to use an adapted version of this to prompt the collection of descriptive content from closer inspection of various aspects of behaviour. This has given some indication of what is typical, as well as what can be suspicious, anomalous or unusual in this context. This type of knowledge, supported by outcomes from further observational studies, could be used more widely with potential application in railway stations, airports, banks, offices and shops [13, 14].

There are many interesting aspects of behaviour that have been observed so far. Different types of typical behaviours were classified, along with several anomalous behaviours that were not expected in this relatively short observational study. Several aspects of behaviour that were pre-cursors to later observable behaviours were of particular interest. These included observation of facial expressions (e.g. there may be different types of smiles – see also [15] re facial expressions), the interactions that forewarn of anomalous behaviours (e.g. the types of glances that may be exchanged between people colluding in deviant behaviour) and the different paths and movement speeds [3] that characterise how people approach the access barriers, either alone or in groups. There has been something about these aspects of behaviour that make these identifiable and potentially valuable for use in surveillance. These are of particular interest, because they can indicate that something is in progress or about to happen [1], enabling an earlier intervention. However, these are not always likely to be easily distinguishable from other behaviours, with various research and technical difficulties to overcome. Progress with understanding intent may be possible with further attention given to other simultaneous gestures and body postures (e.g. at the head and neck, [16, 17]. For some of the movements and activities that have been identified there is already technical capability within algorithms to support detection from video images. These include capability to detect deviations from a typical path [4, 18] and abnormal behaviours such as tailgating [8, 19] and jumping over fences [8]. The study has also revealed how others in the location may react in real time to visual prompts (e.g. when people are carrying unusual objects or trying to attract attention). There is potential to exploit these types of interactions and study the behaviours of people in response to these interactions with technology or people [20].

There are some limitations in the current study, with indications of things that need to be overcome in developing this work. Only one researcher analysed the video in this exploratory study. There are likely to be differences in what other observers may consider, with different viewing skills [21] and priorities for what should be examined in more detail (i.e. people may choose different parts of the seven hours of observational data for detailed analysis). It is also possible that patterns of behaviour may differ by time. This study was not carried out in term time at the University and greater crowding at other times of the year may influence what can be observed in this type of location. There was some limitation on what can be viewed on the camera, both in terms of the detail that is needed for some aspects of behaviour but also for clear views of the areas outside the libraries and on approach to the entrance gates. Further work will be needed to determine the technical ability to identify critical aspects of behaviour using passive and smart camera systems, through consultation with technical experts. Issues of privacy in the in-depth observation of behaviours will also need to be considered.

Overall, the study has demonstrated the feasibility of broadening the application of the behavioural framework, supporting the identification of typical and anomalous patterns of behaviour in new settings. Further studies are in progress, which will support the future development of the behavioural framework. These range from fundamental studies of the ability to observe selected aspects of behaviour from the framework, to use of the framework in field studies to observe and understand more about potential changes in a range of behaviours in response to design interventions.

References

1. Vats E, Chan CS (2016) Early detection of human actions—a hybrid approach. Appl Soft Comput 46:953–966
2. Chen Z, Tian Y, Zeng W, Huang T (2015) Detecting abnormal behaviors in surveillance videos based on fuzzy clustering and multiple auto-encoders. In: IEEE international conference on multimedia and expo (ICME), pp 1–6
3. Popoola OP, Wang K (2012) Video-based abnormal human behavior recognition—a review. IEEE Trans Syst Man Cybern Part C (Appl Rev) 42(6):865–878
4. Tehrani MA, Kleihorst R, Meijer P, Spaanenburg L (2009) Abnormal motion detection in a real-time smart camera system. In: Third ACM/IEEE international conference on distributed smart cameras, ICDSC. IEEE, pp 1–7
5. Shao Z, Cai J, Wang Z (2017) Smart monitoring cameras driven intelligent processing to big surveillance video data. IEEE Trans Big Data 99:1–13
6. Ozer B, Wolf M (2014) A train station surveillance system: challenges and solutions. In: Proceedings of the IEEE conference on computer vision and pattern recognition workshops, pp 638–643
7. Wolf W, Ozer B, Lv T (2002) Smart cameras as embedded systems. Computer 35(9):48–53
8. Kawamura A, Yoshimitsu Y, Kajitani K, Naito T, Fujimura K, Kamijo S (2011) Smart camera network system for use in railway stations. In: IEEE international conference on systems, man, and cybernetics (SMC). IEEE, pp 85–90
9. Rinner B, Wolf W (2008) An introduction to distributed smart cameras. Proc IEEE 96 (10):1565–1575
10. Rinner B, Wolf W (2008) A bright future for distributed smart cameras. Proc IEEE 96 (10):1562–1564
11. Hampapur A, Brown L, Connell J, Pankanti S, Senior A, Tian Y (2003) Smart surveillance: applications, technologies and implications. In: Proceedings of the joint fourth international conference on information, communications and signal processing, and fourth pacific rim conference on multimedia, vol 2. IEEE, pp 1133–1138

12. Ryan B (2018) Developing a framework of behaviours before suicides at railway locations. Ergonomics 61(5):605–626
13. Krahnstoever N, Tu P, Yu T, Patwardhan K, Hamilton D, Yu B, Greco C, Doretto G (2009) Intelligent video for protecting crowded sports venues. In: Sixth IEEE international conference on advanced video and signal based surveillance, AVSS 2009. IEEE, pp 116–121
14. Valera M, Velastin SA (2005) Intelligent distributed surveillance systems: a review. IEEE Proc Vis Image Signal Process 152(2):192–204
15. Ekman P, Friesen WV (1971) Constants across cultures in the face and emotion. J Pers Soc Psychol 17(2):124
16. Aviezer H, Trope Y, Todorov A (2010) 2) Body cues, not facial expressions, discriminate between intense positive and negative emotions. Science 338:1225–1229
17. Martinez L, Falvello VB, Aviezer H, Todorov A (2016) Contributions of facial expressions and body language to the rapid perception of dynamic emotions. Cogn Emotion 30(5): 939–952
18. Arroyo R, Yebes JJ, Bergasa LM, Daza IG, Almazán J (2015) Expert video-surveillance system for real-time detection of suspicious behaviors in shopping malls. Expert Syst Appl 42(21):7991–8005
19. Fujimura K, Yoshimitsu Y, Naito T, Kamijo S (2010) Behavior understanding at railway station by postures and the pseud-trellis analysis of trajectories. In: 13th international IEEE conference on intelligent transportation systems (ITSC). IEEE, pp 1116–1122
20. Vrij A, Granhag PA, Porter S (2010) Pitfalls and opportunities in nonverbal and verbal lie detection. Psychol Sci Public Interest 11(3):89–121
21. Schmidt RA (1982) Motor control and learning: a behavioural emphasis. Human Kinetics, Champaign

Data Mining Based Analysis of Hit-and-Run Crashes in Metropolitan City

Sathish Kumar Sivasankaran and Venkatesh Balasubramanian(✉)

RBG Lab, Department of Engineering Design,
IIT Madras, Chennai 600036, India
sathishkumarsmail@gmail.com, chanakya@iitm.ac.in

Abstract. Hit-and-run cases have also seen a steep escalation in the past few years. Recent published report by ministry of road transport and highways shows that 55,942 hit-and-run cases were reported which is 11.6% of the total accidents that occurred in the nation (MoRTH, 2016). In the current study, various factors associated with hit-and-run crashes were investigated for urban Chennai, a metropolitan city of India. A total of 65 variables under 20 factors were classified as driver, vehicle, crash and environmental factors. Data mining technique, classification and regression tree (CART) was used on the data related to 4818 hit-and-run crashes that occurred in Chennai urban between January 2015 and December 2016. The dataset was split in two as training and testing data with 50:50 ratios. The predictive accuracy of the model built with total of 65 variables was 92.29% for the training data and 92.19% for the testing data. The CART findings show that collision type is the most important variable associated with hit-and-run crashes. Other secondary variables associated were gender, driver age, vehicle type and light conditions. From the results of the present study, it can be concluded that CART algorithm can be a useful tool in determining and identifying potential causes of hit-and-run accidents.

Keywords: Road traffic accidents · Driver safety · Decision trees
Hit-and-run · Transportation human factors

1 Introduction

According to the official report by world health organization, traffic accidents alone cause around 3000 deaths per day and about two million death per year [1]. Nearly 90% of the accident occur in low and middle income countries. In India like other developing nations, road traffic accidents continues to cause a substantial number of death. In the year 2016, 4,80,652 road accidents caused injuries to 4,94,624 people and claimed around 1,50,785 lives. This data on an average resulted in 1317 accidents and 413 accidental deaths every day on Indian roads [2]. Hence Ministry of road transport and highways has committed to reduce the number of road accidents and fatalities by 50% by the year 2020 (NITI Aayog, 2016).

Considering different types of crashes which contribute to the loss of life, hit-and-run crashes is of special concern to traffic professionals and law enforcement officers A hit-and-run crash is generally one in which at least one of the drivers who are involved

S. Bagnara et al. (Eds.): IEA 2018, AISC 823, pp. 113–122, 2019.
https://doi.org/10.1007/978-3-319-96074-6_12

in the accident leaves the scene of accident without notifying the authorities. Such an act would increase the severity of the crash since around 85% of victims in the fatal crash would die within one or two hours [3–7]. Hence, hit-and-run crash is considered to be an serious offence in most of the countries including India. We also find that hit-and-run crashes increase the investigation and prosecution charges. The common reason which may influence drivers decision to flee from the spot of the accident or stay back depends upon the perception of the legal consequence which they may have to face subsequently.

The annual statistical report by Ministry of Road Transport and Highways (MoRTH), Government of India hit-and-run cases accounted for nearly 11.6% (55,942) of all accidents in India which is an increase from 10.9% (57,083) from the previous year. The number of persons killed due to hit-and-run was 22,962 which is 15.2% of total persons killed in accidents in the year 2016 (MoRTH, 2016) [2]. The detailed statistics of total accidents, no of persons injured and no of people killed due hit-and-run crashes for the past four year period is reported in the Fig. 1 below. Since hit-and-run crash is identified as the important contributor for traffic accidents, significant risk factors associated with hit-and-run crash is studied here.

In India, drivers are booked according to Section 304 of Indian penal code, where hit-and-run cases account for culpable homicide not amounting to murder due to careless driving behaviour of drivers. These cases results in maximum punishment of two years or compensation or both. Indian motor vehicle act, 1988 section 161 provides special provisions for victims in terms of compensation in case of hit-and-run. For fatal or grievous injury, compensation is summed up to Rs 25000, which is inadequate compared to other developed nations like Canada where drivers involved in hit-and-run who flee from site are penalized up to 10 years in prison or life

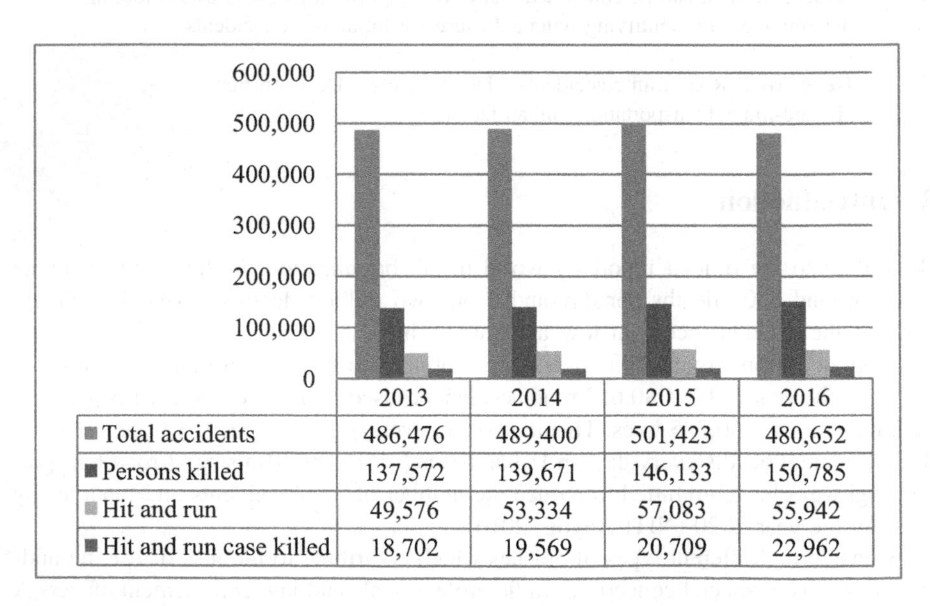

	2013	2014	2015	2016
Total accidents	486,476	489,400	501,423	480,652
Persons killed	137,572	139,671	146,133	150,785
Hit and run	49,576	53,334	57,083	55,942
Hit and run case killed	18,702	19,569	20,709	22,962

Fig. 1. Crashes reported in last four years for India. (Source: MoRTH reports 2012, 2013, 2014, 2015, 2016)

imprisonment. In China too, stringent laws prevail where drivers are penalized for a fixed term of over 3 years and less than 7 years whereas if accident resulted in death, imprisonment would be for fixed term imprisonment of not less than 7 years. The rate of hit-and-run crashes in India is higher compared to developed nations and states such as Singapore (1.83%), California (8.1%), Guangdong (7.7%) and shanghai (4.45%) in china. The current study is aimed at contributing to the existing literature by exploring the association of different factors with hit-and-run crashes.

The paper is further organized as follows: Sect. 2 deals with the methodology of the study along with brief description of the dataset used. Section 3 deals with results and discussions are done in Sect. 4. The final section provides the concluding remarks and the policy implications from the present study.

2 Methods and Materials

2.1 Data

The present study focuses on analysing hit-and-run crashes and attempts to identify significant risk factors associated with this crashes. For this purpose, we analyze the traffic accident data for the period between January 2015 and December 2016 from Chennai metropolitan city, India. The accident data are obtained from the Road Accident database management System (RADMS database), an official database maintained by the government of Tamilnadu. Each crash data is recorded by the traffic police which conduct on scene investigation. Together with other crashes, a total of 4818 samples was examined. Each sample crash data includes details such as demographic information, injury severity, vehicle characteristics, temporal and environmental conditions.

Chennai is the capital state of Tamilnadu located on the coromandel coast off the Bay of Bengal being one of the biggest cultural, social, economic and educational centre for south India. It is the fifth largest city in India covering an area of 175 km^2 with the population of 46,46,732 representing highest degree of urbanization and population density (District census Handbook, Chennai 2011). Due to Chennai rapid economic development and consequently vehicle growth, road traffic incidents are highest in Chennai among major cities in India. Taking this into account, this study aims to determine the risk factors associated with accidents relating to hit-and-run crashes in Chennai city. By assessing factors relating to hit-and-run in Chennai city, policy recommendations could be devised to reduce fatalities in Chennai city and can be extended to other cities nationwide.

2.2 CART

A decision tree is an predictive model which can be used for classification tasks. CART is an non-parametric tree model that assumes no relationship between the dependent variables and the independent variable. When the target variable is numerical type, regression tress are used whereas for the categorical classification tree is used. In the present study since the characteristics of accidents lead to Hit-and-run crashes or not classification tree is used in the present study.

Initially all the data is concentrated at the top node commonly referred as root node. It is further divided into child nodes based on the predictor variables which maximizes the homogeneity of the child variables. We find that data in the child node is more homogenous than the parent node. The process is continued until all the data have a greatest possible homogeneity. The final node which do not have branches is referred as the terminal node or leaf node.

$$\text{Gini(c)} = 1 - \sum\nolimits_j p^2(j|c) \tag{1}$$

$$p(j|c) = \frac{p(j,c)}{p(c)} \tag{2}$$

$$p(j,c) = \frac{\pi(\text{j})\text{N}_\text{j}(\text{C})}{N_j} \tag{3}$$

and

$$p(c) = \sum\nolimits_j p(j,c) \tag{4}$$

where,

$j =$ *number of target variable or classes*,
$\pi(\text{j}) =$ prior probability for class j,

$p(j/c) =$ *conditional probability of a case being in class j provided that is in node m*

$N_j(c) =$ *number of cases of class j of node*,
$N_j =$ *number of cases of class j in roof node*

Gini index (GI) is the measure of degree of purity of the node, when the value of GI is zero, all nodes are pure. In CART technique we aim to achieve the maximum purity also to avoid over fitting pruning is done.

The importance of each variable/predictor in the model is evaluated using the below equation

$$VIM(x_j) = \sum\nolimits_{t=1}^{\tilde{T}} \frac{n_t}{N} \Delta Gini(S(x_j,t)) \tag{5}$$

where $\frac{n_t}{N}$ = proportion of observation in the dataset that belong to node t; N- total number of observation in training set; $\Delta Gini(S(x_j,t))$ = reduction of Gini index on the basis of variable x_j. The value is calculated for all considered variables in which most important variablc has highest value.

To evaluate the model performance, ROC curve and classification accuracy is used to evaluate the model performance. If the area under the ROC curve is greater than 0.9, model has outstanding discrimination whereas if the area under the ROC curve is 0.8 the model is having excellent discrimination.

3 Results

For the present study, Chennai city crash data maintained by Government of Tamilnadu RADMS database have been used. Since the scope of the present study is to identify the factors which contribute to hit-and-run crashes, all crashes containing data related to hit-and-run crashes were extracted and a total of 4818 crashes were identified for the present study.

The crashes data was split into two subsets of 50:50 ratio for the training set and the test set respectively. Accuracy for the training and the testing sets were around 92% which shows that CART model is powerful in classifying the samples. The AUC for the testing and training sets were around 0.8 which shows that model has excellent discrimination.

The resultant decision tree obtained after applying the CART algorithm shows 8 nodes (5 terminal nodes). Figure 2 shows the resulting decision tree, ID number is displayed at each node, total number of crashes at each node and the classification of crashes whether it was a hit-and-run crash or not are presented at each node. The root

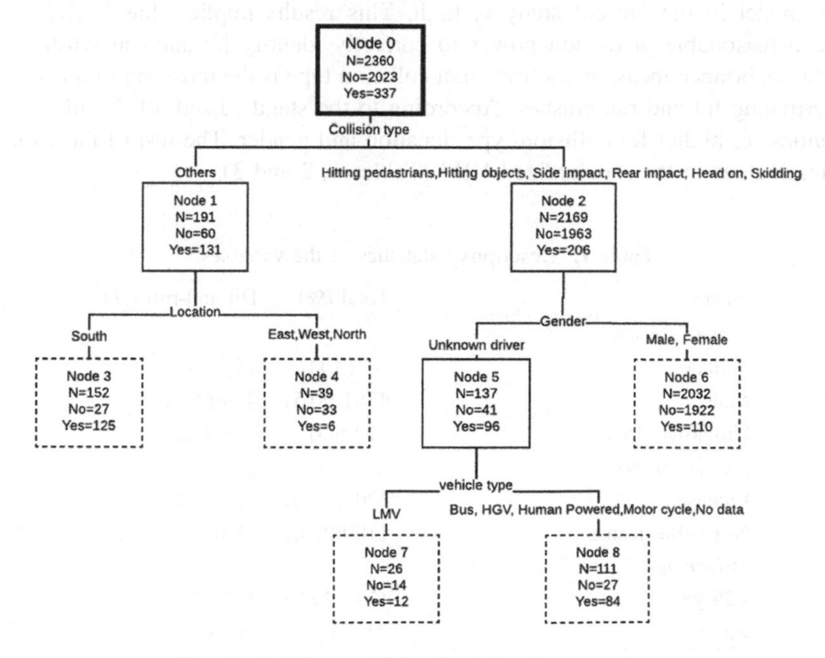

Fig. 2. CART model for the hit-and-run crashes

node was presented with darker box whereas the terminal nodes ware presented as a dashed box. From the decision tree it is clear that collision type, location, gender and vehicle type are the main splitters which implies that these are the important variables for decision making in drivers to flee from the accident spot. The percentage of hit-and-run crashes in the terminal nodes varied from 5% to 82%.

4 Discussions

On interpreting the decision tree we find that terminal nodes 3 and 8 have sizable observations and have potential of causing hit-and-run crashes and the tree was thoroughly examined. The tree was spilt by the collision type, followed by the gender and location of accident spot. To the left side of the tree, at node 1 we find that classification tree shows higher percentage for the hit-and-run crashes due to other type of collision such as overturning, overturning without collision and simple driver errors. Node 1 is further spilt into terminal nodes 3 and 4 based on the variable location. The results show that hit-and-run crashes are more frequent in south Chennai region. Turning to the right side of the tree, node 2 is split into child nodes based on the variable gender. At node 5 it is identified that unknown drivers significantly contribute to hit-and-run crashes and they flee from the spot of the accident. This node is further split into terminal nodes 7 and 8 based on the variable vehicle type. The findings show that compared to LMV vehicles drivers, drivers who rode buses, heavy goods vehicle, powered motorcycles and human powered vehicles are more likely to flee from the spot.

The classification tree model for the hit-and-run crashes showed that around 92% overall prediction accuracy for the training and testing test. The predictive power of the CART model in the present study is high. This results implies that CART model produced reasonable prediction power to correctly identify hit-and-run crashes. The variable importance measure confirms that collision type is the most important variable in determining hit-and-run crashes. According to the standardized VIM table we find that values are higher for collision type, location and gender. The rest of the variables have less values in the standardized VIM (Tables 1, 2 and 3).

Table 1. Descriptive statistics of the variables.

Factors	Total (%)	Hit-and-run (%)
Driver's gender		
Female	147(3.1)	15(2.2)
Male	4361(90.5)	439(65.5)
Unknown driver	310(6.4)	216(32.2)
License status		
License	3709(77.0)	624(93.1)
Not valid license	1109(23.0)	46(6.9)
Driver age		
<25 yrs	1343(27.9)	132(19.7)
>65	74(1.5)	9(1.3)

(*continued*)

Table 1. *(continued)*

Factors	Total (%)	Hit-and-run (%)
25–44	2284(47.4)	243(36.3)
45–65	818(17.0)	75(11.2)
No data	299(6.2)	211(31.5)
Illumination level		
Day	2819(58.5)	374(55.8)
Night	1999(41.5)	296(44.2)
Peak hours		
Evening peak hours	809(16.8)	112(16.7)
Morning peak hours	685(14.2)	90(13.4)
Non peak hours	3324(69.0)	468(69.9)
Day of the week		
Week end	1420(29.5)	221(33.0)
Working day	3398(70.5)	449(67)
Location		
East	491(10.2)	61(9.1)
North	795(16.5)	47(7.0)
South	2403(49.9)	515(76.9)
West	1129(23.4)	47(7.0)
Light condition		
Darkness - with street lights on	1340(27.8)	181(27.0)
Darkness without street light	693(14.4)	105(15.7)
Daylight	2722(56.5)	338(50.4)
No data	63(1.3)	46(6.9)
Season		
Autumn	1628(33.8)	156(23.3)
Spring	946(19.6)	177(26.4)
Summer	1075(22.3)	229(34.2)
Winter	1169(24.3)	108(16.1)
Severity		
Fatal/Grievous injury	3070(63.7)	593(88.5)
Simple injury/Vehicle dam	1748(36.3)	77(11.5)
Collision type		
Head on	1422(29.5)	80(11.9)
Hit objects	123(2.6)	15(2.2)
Others	383(7.9)	271(40.4)
Pedestrian	1702(35.3)	221(33.0)
Rear	597(12.4)	39(5.8)
Side	366(7.6)	25(3.7)
Skidding	225(4.7)	19(2.8)

(continued)

Table 1. (*continued*)

Factors	Total (%)	Hit-and-run (%)
Vehicle type		
Bus	215(4.5)	27(4.0)
HGV	235(4.9)	19(2.8)
Human powered	236(4.9)	38(5.7)
LMV	1387(28.8)	170(25.4)
Motor cycle	2517(52.2)	258(38.5)
No data	228(4.7)	158(23.6)
Road conditions		
Good	4809(99.8)	669(99.9)
Not good	9(0.2)	1(0.1)
Central divider		
No	2635(100.0)	535(79.9)
Yes	2183(45.3)	135(20.1)
Traffic control		
Control	112(2.3)	14(2.1)
No control	4706(97.7)	656(97.9)
Road category		
Highway	4371(90.7)	631(94.2)
Not a highway	447(9.3)	39(5.8)
No of lanes		
More than one lane	101(2.1)	13(1.9)
single	4717(97.9)	657(98.1)
Traffic movement		
One-way	99(2.1)	7(1.0)
Two-way	4719(97.9)	663(99.0)
Landmark		
Not a public place	448(9.3)	45(6.7)
Public place	4370(90.7)	625(93.3)
Primary cause		
Dangerous overtaking	97(2.0)	14(2.1)
Inappropriate speed	4276(88.8)	592(88.4)
Inattentive lane change and turning	237(4.9)	33(4.9)
No data	50(1.0)	20(3.0)
Others	72(1.5)	8(1.2)
Violating rules	86(1.8)	3(0.4)

Table 2. Performance of the model

Sample	Observed injury	Predicted injury			AUC
		No	Yes	Accuracy	
Training	No	1969	54	92.29%	0.819
	Yes	128	209		
Testing	No	2050	75	92.19%	0.833
	Yes	117	216		

Table 3. Variable importance

Independent variable	Importance
Collision type	0.23
Location	0.20
Gender	0.16
Driver age	0.07
License	0.07
Severity	0.07
Central divider	0.07
Primary cause	0.07
Vehicle type	0.04

5 Conclusions

Many previous studies have investigated the factors associated with crashes. However on the other hand, few studies have specifically focussed on factors associated with hit-and-run crashes. Since hit-and-run crashes contribute significantly to crash severity and saw a steep escalation in the recent past particularly in developing economies like India, It is necessary to identify the main factors influencing the mortality caused due to hit-and-run crashes. In the present study, CART model was used which has several advantages over other data mining methods. In CART method, we do not require to set variables in advance. So we can identify most significant variables and remove the insignificant variables. For example, in this study we find that collision type and location are the most significant variables. These variables have to be focussed while developing countermeasures to mitigate accidents. We find that hit-and-run crashes are more likely to happen in South Chennai region. Hence area dominance by the police officials can reduce the accidents and prevent driver from flee from the spot of the accidents.

The findings of the current study highlight the fact that drivers decision to stay or leave the spot of accident is decided by specific situations of crash occurrence. Hence measures should be taken to prevent the drivers from escaping the accident spots. For examples, installation of more surveillance cameras in the identified public spots, compulsory insurance for vehicles to reduce the damage due to loss, strict licensing activity for drivers involved in hit-and-run, increased punishments and penalties for offending drivers by making major amendments to the existing motor vehicle legislature (Motor Vehicle Act, 1988). Further, promotion of public awareness through legal education, awareness camps, advertisements, social media and display boards either by traffic police or nongovernmental organizations to reduce accident severity due to hit-and-run crashes.

References

1. WHO (2008) World report on road traffic injury prevention. World Health Organization
2. Road accidents in India 2012–2016. Ministry of road transport and highways (MoRTH), Transport Research Wing, Government of India
3. Roess P, Prassas E, McShane W (2004) Traffic engineering, 3rd edn. Prentice Hall, Englewood Cliffs
4. Tay R, Kattan L, Sun H (2010) Logistic model of hit-and-run crashes in Calgary. Can J Transp 4(1)
5. Tay R (2008) Marginal effects of increasing ageing drivers on injury crashes. Accid Anal Prev 40(6):2065–2068
6. Tay R, Barua U, Kattan L (2009) Factors contributing to hit-and-run in fatal crashes. Accid Anal Prev 41(2):227–233
7. Solnick S, Hemenway D (1995) The hit-and-run in fatal pedestrian accidents: victims, circumstances and drivers. Accid Anal Prev 27(5):643–649

Analysis of Driver Injury Severity in Metropolitan Roads of India Through Classification Tree

Sathish Kumar Sivasankaran and Venkatesh Balasubramanian(✉)

RBG Lab, Department of Engineering Design, IIT Madras, Chennai 600036, India
sathishkumarsmail@gmail.com, chanakya@iitm.ac.in

Abstract. Reducing the injury severity from traffic accidents is most important step in mitigating accidents occurring in developing economies like India where two way roads are more common in cities. The number of deaths due to accidents has rose from 83,491 in 2005 to 1,36,071 in 2016 as per the latest reports of ministry of road transport and highways, government of India (MoRTH, 2016). To explore the factors contributing to injury severity in such roads, non parametric classification tree is used since it does not assume any underlying assumption between target variable and the predictors. CART (Classification and Regression tree), a classification tree establishes empirical relation between injury severity outcomes and variables including driver, vehicle, crash and environmental factors. The present study analyzed traffic crash data of single lane two way roads of Chennai city pertaining to period from January 2015 to December 2016. The final dataset included a total of 5271 crash information after excluding incomplete and missing data. This finalized dataset was split into two subsets, training and testing data and the classification models reported an accuracies of 63.4% and 61.5% for the training and testing data. The results indicated that collision type and vehicle type were the two important variables affecting the severity of injury. The findings of this study will help in determining influential factors so that countermeasures to reduce the severity of injury in urban cities can be developed.

Keywords: Road traffic accidents · Decision tree · Injury severity
Data mining · Transportation safety

1 Introduction

According to the official report by world health organization, traffic accidents alone cause around 3000 deaths per day and about two million death per year [1]. Nearly 90% of the accident occur in low and middle income countries. In India like other developing nations, road traffic accidents continues to cause a substantial number of death. In the year 2016, 4,80,652 road accidents caused injuries to 4,94,624 people and claimed around 1,50,785 lives. This data on an average resulted in 1317 accidents and 413 accidental deaths every day on Indian roads [2]. Hence Ministry of road transport

S. Bagnara et al. (Eds.): IEA 2018, AISC 823, pp. 123–131, 2019.
https://doi.org/10.1007/978-3-319-96074-6_13

and highways has committed to reduce the number of road accidents and fatalities by 50% by the year 2020 (NITI Aayog, 2016).

The annual statistical report by Ministry of Road Transport and Highways (MoRTH), Government of India, We could find a steep escalation in number of fatal crashes from 83,491 in 2005 to 1,36,071 in 2016. The total share of fatal and grievous injuries accounted for 28.3% and 25.1% of the total crashes that occurred in India as per latest reports of MoRTH. The situation is alarming to note that fatal and grievous injury severity has increased by 63% and 17.6% during the period from 2005 to 2016 [2]. The detailed statistics of total accidents, no of persons injured and no of people killed for past 10 year period is reported in the Fig. 1 below. Since injury severity is identified as the important contributor for traffic accidents, significant risk factors associated with hit-and-run crash is studied here.

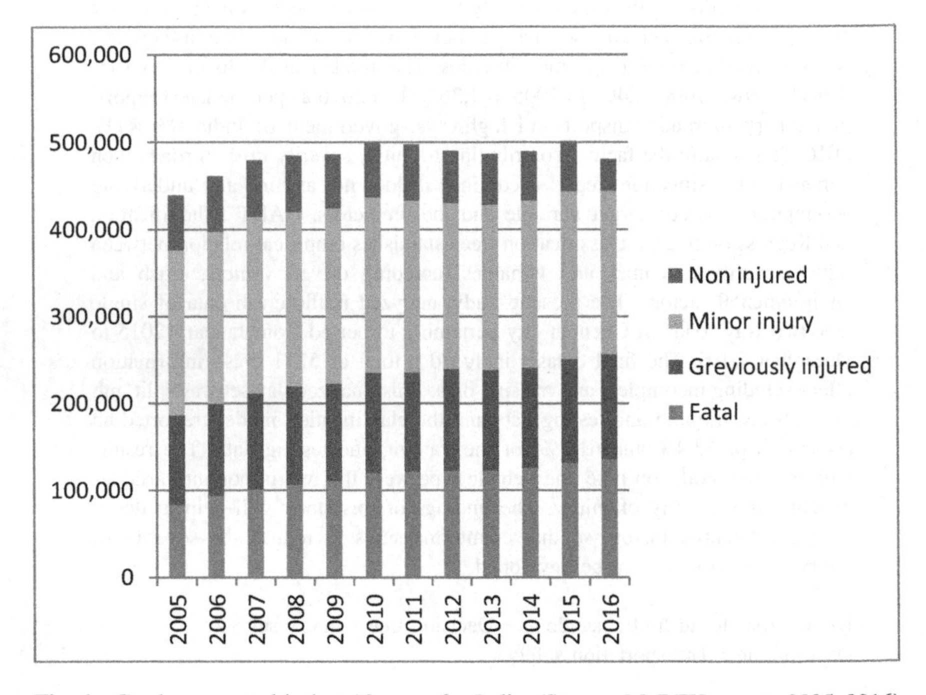

Fig. 1. Crashes reported in last 10 years for India. (Source: MoRTH reports 2005–2016)

Many researchers in the field of traffic safety have started focussing on factors that contribute to crash severity. Identifying factors relating to accident severity not only help in redesigning infrastructural factors, provide inputs for vehicle manufacturers, help in incorporate human factors in design considerations. It also reduce the mortality rate in developing nations like India. Generally studies on traffic injury severity use generalized linear models to fit their data [3–8]. The major drawback in such linear models is pre-assumption of linear relationship between the independent variable (injury severity outcome) and dependent variable (factors causing accidents). When no such pre-assumptions are assumed, data mining approaches would be an alternative preferred approach. Also such approaches have following advantages (1) no problem in case of outliers present (2) no prior assumption is made between variables and apriori probabilistic knowledge is not required (3) has ability to handle both discrete variable as wells as categorical variables with more categories at a time and (4) large amount of information could be extracted from the data.

The aim of the current study is to identify the contributing factors associated with accident severity of Road Traffic Accidents (RTA). In case of huge amount of accident data, data mining approaches prove to be efficient method to identify meaningful patterns and models (Han and Camber, 2006). Data from police crash database for urban Chennai city, was available and hence data mining approach was implied for our study. To the best of knowledge of the authors, the present study is the first of its kind to be conducted in Indian City using data from reliable data source.

The paper is further organized as follows: Sect. 2 deals with the methodology of the study along with brief description of the dataset used. Section 3 deals with results and discussions are done in Sect. 4. The final section provides the concluding remarks and the policy implications from the present study.

2 Methods and Materials

2.1 Data

The present study focuses on analysing injury severity of the crashes and attempts to identify significant risk factors associated with this crashes. For this purpose, we analyze the traffic accident data for the period between January 2015 and December 2016 from Chennai metropolitan city, India. The accident data are obtained from the Road Accident database management System (RADMS database), an official database maintained by the government of Tamilnadu. Each crash data is recorded by the traffic police which conduct on scene investigation. Together with other crashes, a total of 4818 samples was examined. Each sample crash data includes details such as demographic information, injury severity, vehicle characteristics, temporal and environmental conditions.

Chennai is the capital state of Tamilnadu located on the coromandel coast off the Bay of Bengal being one of the biggest cultural, social, economic and educational centre for south India. It is the fifth largest city in India covering an area of 175 sq. km with the population of 46,46,732 representing highest degree of urbanization and population density (District census Handbook, Chennai 2011). Due to Chennai rapid economic development and consequently vehicle growth, road traffic incidents are highest in Chennai among major cities in India. Taking this into account, this study aims to determine the risk factors associated with accidents relating to hit-and-run crashes in Chennai city. By assessing factors relating to injury severity in Chennai city, policy recommendations could be devised to reduce fatalities in Chennai city and can be extended to other cities nationwide.

2.2 CART

A decision tree is an predictive model which can be used for classification tasks. CART is an non-parametric tree model that assumes no relationship between the dependent variables and the independent variable. When the target variable is numerical type, regression tress are used whereas for the categorical classification tree is used. In the present study since the characteristics of accidents lead to Hit-and-run crashes or not classification tree is used in the present study.

Initially all the data is concentrated at the top node commonly referred as root node. It is further divided into child nodes based on the predictor variables which maximizes the homogeneity of the child variables. We find that data in the child node is more homogenous than the parent node. The process is continued until all the data have a greatest possible homogeneity. The final node which do not have branches is referred as the terminal node or leaf node.

$$\text{Gini(c)} = 1 - \sum_j p^2(j|c) \tag{1}$$

$$p(j|c) = \frac{p(j,c)}{p(c)} \tag{2}$$

$$p(j,c) = \frac{\pi(\text{j})\text{N}_\text{j}(\text{C})}{N_j} \tag{3}$$

and

$$p(c) = \sum_j p(j,c) \tag{4}$$

where, $j = number\ of\ target\ variable\ or\ classes$, $\pi(\text{j}) =$ prior probability for class j,

$$p(j/c) = conditional\ probability\ of\ a\ case\ being\ in\ class\ j\ provided\ that\ is\ in\ node\ m$$

$$N_j(c) = number\ of\ cases\ of\ class\ j\ of\ node,$$

$$N_j = number\ of\ cases\ of\ class\ j\ in\ roof\ node$$

Gini index (GI) is the measure of degree of purity of the node, when the value of GI is zero, all nodes are pure. In CART technique we aim to achieve the maximum purity also to avoid over fitting pruning is done.

The importance of each variable/predictor in the model is evaluated using the below equation

$$VIM(x_j) = \sum_{t=1}^{\tilde{T}} \frac{n_t}{N} \Delta Gini(S(x_j,t)) \tag{5}$$

where $\frac{n_t}{N}$ = proportion of observation in the dataset that belong to node t; N- total number of observation in training set; $\Delta Gini(S(x_j,t))$ = reduction of Gini index on the basis of variable x_j. The value is calculated for all considered variables in which most important variable has highest value.

To evaluate the model performance, ROC curve and classification accuracy is used to evaluate the model performance. If the area under the ROC curve is greater than 0.9, model has outstanding discrimination whereas if the area under the ROC curve is 0.8 the model is having excellent discrimination.

3 Results

For the present study, Chennai city crash data maintained by Government of Tamilnadu RADMS database have been used. Since the scope of the present study is to identify the factors which contribute to injury severity, The original extracted dataset consists of 7952 crashes. After cleaning the data finally, 5271 crashes were considered for analysis in which 3254 records were fatal/grievously injuries and the remaining 2017 records were related to simple injury/vehicle damage only. The dependent variable considered in the present study was the injury severity with two outcomes: KSI (Killed/seriously injured) and SI (simple injury) the independent variables were driver, environmental, crash, vehicle and road characteristics. Table 1 provides the complete information regarding the injury severity for all the independent variables included for analysis in the present study along with their frequency distribution (Table 2).

Table 1. Descriptive statistics of the variables.

Variables	Values	Total count (%)	Severity	
			KSI (%)	SI (%)
Driver's gender	(1) Female	174(3.3)	101(1.9)	73(1.4)
	(2) Male	5097(96.7)	3199(60.7)	1898(36.0)
License status	(1) Valid license	4011(76.1)	2543(48.2)	1468(27.9)
	(2) Not valid license	1260(23.9)	757(14.4)	503(9.5)
Driver's age	(1) <25 Yrs	1572(29.8)	994(18.9)	578(11.0)
	(2) 25–44 Yrs	2659(50.4)	1643(31.2)	1016(19.3)
	(3) 44–65 Yrs	941(17.9)	597(11.3)	344(6.5)
	(4) >65 Yrs	99(1.9)	66(1.3)	33(0.6)
Weather conditions	(1) Cloud	64(1.2)	32(0.6)	32(0.6)
	(2) Fine	5193(98.5)	3258(61.8)	1935(36.7)
	(3) Rainy	14(0.3)	10(0.2)	4(0.1)
Light conditions	(1) Darkness- with street light on	1507(28.6)	961(18.2)	546(10.4)
	(2) Darkness-without street light	723(13.7)	474(9.0)	249(4.7)
	(3) Day light	3041(57.7)	1865(35.4)	1176(22.3)
Season	(1) Autumn	1843(35.0)	1254(23.8)	589(11.2)
	(2) Spring	988(18.7)	563(10.7)	425(8.1)
	(3) Summer	1188(22.5)	728(13.8)	460(8.7)
	(4) Winter	1252(23.8)	755(14.3)	497(9.4)

(continued)

Table 1. (*continued*)

Variables	Values	Total count (%)	Severity	
			KSI (%)	SI (%)
Visibility level	(1) Not Good	2174(41.2)	1390(26.4)	784(14.9)
	(2) Good	3097(58.8)	1910(36.2)	1187(22.5)
Peak hours	(1) Non peak hours	3627(68.8)	2266(43.0)	1361(25.8)
	(2) Evening Peak hours	892(16.9)	570(10.8)	322(6.1)
	(3) Morning peak hours	752(14.3)	464(8.8)	288(5.5)
Day of week	(1) Week end	1566(29.7)	1008(19.1)	558(10.6)
	(2) Working day	3705(70.3)	2292(43.5)	1413(26.8)
Central divider	(1) No	2610(49.5)	1707(32.4)	903(17.1)
	(2) Yes	2661(49.5)	1593(30.2)	1068(20.3)
Traffic control	(1) Controlled	154(2.9)	92(1.7)	62(1.2)
	(2) No control	5117(97.1)	3208(60.9)	1909(36.2)
Land mark	(1) Public place	4805(91.2)	3042(57.7)	1763(33.4)
	(2) Not a public place	466(8.8)	258(4.9)	208(3.9)
Collision type	(1) Head on	1690(32.1)	960(18.2)	730(13.8)
	(2) Hit object	147(2.8)	80(1.5)	67(1.3)
	(3) Others	411(7.8)	350(6.6)	61(1.2)
	(4) Pedestrian	1455(27.6)	926(17.6)	529(10.0)
	(5) Rear impact	744(14.1)	440(8.3)	304(5.8)
	(6) Side impact	474(14.1)	296(5.6)	178(3.4)
	(7) Skidding	350(6.6)	248(4.7)	102(1.9)
Vehicle type	(1) Bus	306(5.8)	223(4.2)	83(1.6)
	(2) HGV	271(5.1)	179(3.4)	92(1.7)
	(3) Human powered	282(5.4)	161(3.1)	121(2.3)
	(4) LMV	1550(29.4)	922(17.5)	628(11.9)
	(5) Motorcycle	2862(54.3)	1815(34.4)	1047(19.9)
Primary cause	(1) Inappropriate speed	4692(89.0)	2951(56.0)	1741(33.0)
	(2) Dangerous overtaking	107(2.0)	63(1.2)	44(0.8)
	(3) Inattentive lane changing and turning	268(5.1)	158(3.0)	110(2.1)
	(4) Others	86(1.6)	56(1.1)	30(0.6)
	(5) Violating rules	118(2.2)	72(1.4)	46(0.9)

Table 2. Performance of the model

Sample	Observed injury	Predicted injury		
		KSI	SI	Accuracy
Training	KSI	1490	160	63.4%
	SI	825	215	
Testing	KSI	1424	226	61.5%
	SI	768	163	

Table 3. Variable importance

Independent variable		Importance	Normalized importance
Collision type	ACT	0.015	100.0%
Vehicle type	VEH	0.005	31.9%
Land mark	LNM	0.004	25.4%
Light conditions	LTC	0.002	15.8%
Seasons	SEA	0.002	13.3%
Central divider	CEN	0.002	10.1%
Primary cause	CAU	0.001	9.3%
Visibility level	SGC	0.001	6.1%
Peak hours	PKH	0.000	2.9%
Drivers age	AGE	0.000	2.2%
Junction control	JUN	0.000	1.0%
License status	LIC	0.000	1.0%
Weather conditions	WTC	0.000	0.9%
Driver's gender	SEX	0.000	0.1%

4 Discussions

Classification tree contains a total of 24 nodes in which around 13 nodes are identified as terminal nodes (node no's 3, 4, 11, 19, 20, 21, 22, 14, 15, 16, 23, 24, 18). The node number, total number of crashes, classification based on the two categories are indicated at each node. The tree generated 7 splitters and 9 terminal nodes (Nodes 3, 4, 11,

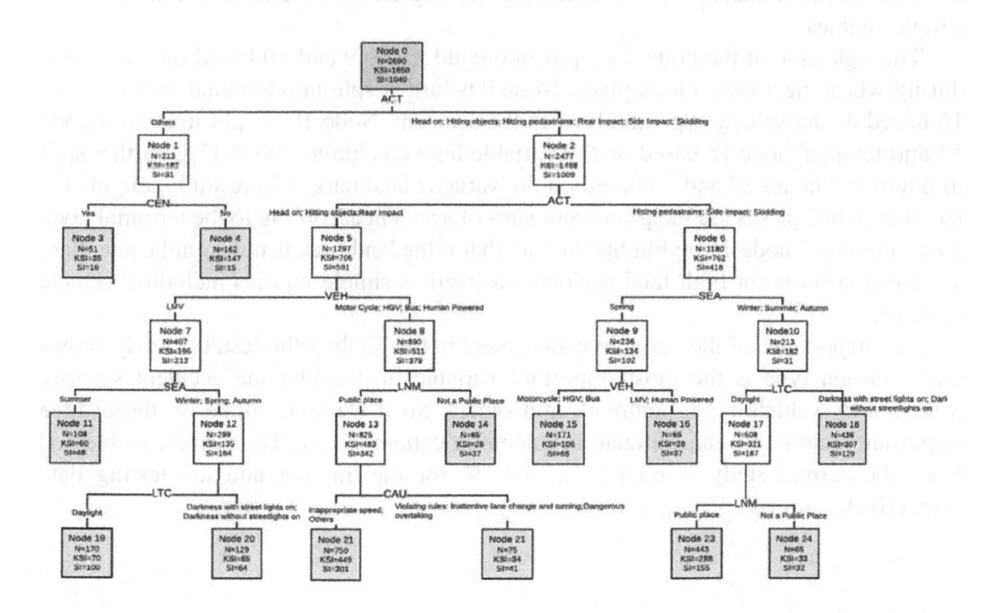

Fig. 2. CART tree for injury severity in urban Chennai

20, 21, 15, 23, 24,18) has higher potential of causing fatal/serious injury whereas the remaining terminal nodes (nodes 19, 22, 14, 16) results in simple injury and vehicle damage (Fig. 2).

The root node, Node 0 is divided into two child nodes 1 and 2 based on the variable collision type which shows that collision type is the most important variable in deciding the injury severity on a single lane two way roads of Chennai urban city. The next best variable for the splitting purpose is vehicle type. Considering the left side of the tree we find that node 1 is divided into child nodes 3 and 4 based on the presence of central divider. Both node 3 and 4 are the terminal nodes. The results highlight the fact presence of central divider, the severity of the crashes among the drivers is less percentage compared to absence of the central divider.

Further proceeding, considering the node 2, it is further divided into child nodes 5 and 6 based on the collision type. We find that collision type frequently acts as the splitter in tree growth shows the importance of the variable in determining the accident severity. Further proceeding down the node 5, the tree gets further split into child nodes 7 and 8 based on the type of the vehicle. It can be decoded form the tree that during the summer season, when the vehicle involved is LMV and the collision is of type head on, rear impact, hitting fixed objects, the severity of the crash would be fatal according to the Terminal node 11. For other type of vehicles, in the public place the severity would be just simple injury according to the terminal node 14. Proceeding the node 12, it splits the node into terminal nodes 19 and 20 based on the light conditions. In presence of the daylight, the severity would be fatal and for darkness with and without street lights the severity would be equal chances for fatal and simple injuries. Considering the node 13 and proceeding downward we find when the primary cause for accident is due to exceeding lawful speed, the severity of the crashes are fatal and for other causes such as violating rules, inattentive lane changing, turning etc the severity of the crash is just simple injuries.

The right side of the node 2 is split into child nodes 9 and 10 based on the season during which the accident took place. Node 9 is further split into terminal nodes 15 and 16 based on the vehicle type involved in the accident. Node 10 is split into child node 17 and terminal node 18 based on the variable light conditions. Node 17 is further split into terminal nodes 23 and 24 based on the variable landmark. The result highlights the fact that public places are the prominent sites of accident according to the terminal node 23 and terminal node 24 highlights the fact that if the landmark is not a public site there are equal chances for both fatal accidents as well as simple injuries including vehicle damage.

The importance of the variable is discussed in the Table 3 the results clearly shows that collision type is the most important variable in deciding the accident severity followed by vehicle type, landmark and others. So if we concentrate on these three important factors, we can mitigate accidents to certain extent. The accuracy obtained from the current study is 63.4% and 65.1% for the training and the testing data respectively.

5 Conclusions

In this study we find that collision type is most important determinant for the injury severity. Among this collision type, head on crashes and pedestrian hit crashes contribute significantly to fatal injuries in Chennai city accounting for nearly 50% of the crashes occurring in the city which is higher compared to national statistics. Hence countermeasures for these collision type can significantly mitigate accident severity in Chennai city. To get a deeper understanding of the collisions, modelling individual crashes can provide mechanism of crashes. Though crashes occur due to unforeseen circumstances, strategies such as slowing down the vehicle, flashing of headlights, blasting of horn through electronic means could add advantage well ahead before occurrence of crash. All other related cause of collisions can be reduced through public education, strict adherence to traffic laws and stringent punishments to law breakers. Pedestrian crossing zones and barriers in important roads especially in public places could mitigate pedestrian related accidents. Further, it was found that certain vehicle and environmental related factors such as vehicle type, landmark, light conditions and seasons are identified to have effect on the injury severity.

References

1. WHO (2008) World report on road traffic injury prevention. World Health Organization
2. Road accidents in India 2012–2016, Ministry of Road Transport and highways (MoRTH), Transport Research Wing, Government of India
3. Al-Ghamdi AS (2002) Using logistic regression to estimate the influence of accident factors on accident severity. Accid Anal Prev 34(6):729–741
4. Breault JL, Goodall CR, Fos PJ (2002) Data mining a diabetic data warehouse. Artif Intell Med 26(1):37–54
5. Kashani AT, Mohaymany AS (2011) Analysis of the traffic injury severity on two-lane, two-way rural roads based on classification tree models. Saf Sci 49(10):1314–1320
6. Kashani AT, Rabieyan R, Besharati MM (2014) A data mining approach to investigate the factors influencing the crash severity of motorcycle pillion passengers. J Saf Res 51:93–98
7. de Ona J, López G, Mujalli R, Calvo FJ (2013) Analysis of traffic crashes on rural highways using Latent Class Clustering and Bayesian Networks. Accid Anal Prev 51:1–10
8. Montella A, Aria M, D'Ambrosio A, Mauriello F (2012) Analysis of powered two-wheeler crashes in Italy by classification trees and rules discovery. Accid Anal Prev 49:58–72

A Psycho-Ergonomic Approach of the Street-Crossing Decision-Making: Toward Pedestrians' Interactions with Automated Vehicles

Stéphanie Cœugnet[1(✉)], Béatrice Cahour[2], and Sami Kraïem[1]

[1] Institut VEDECOM, 77, rue des Chantiers, 78000 Versailles, France
{stephanie.coeugnet-chevrier, sami.kraiem}@vedecom.fr
[2] CNRS i3 Telecom ParisTech, 46 Rue Barrault, 75013 Paris, France
beatrice.cahour@telecom.paristech.fr

Abstract. A confident interaction between pedestrians and automated vehicles will be a milestone in the acceptability of the use of these new vehicles. The automated vehicle should be able to offer the pedestrian a safe and non-uncertain street-crossing situation. The aim of the study was to perform an analysis of the activity of street-crossing, focusing on the decision-making in a natural environment.

The study was carried out with a sample of 20 participants A triangulation of Ergonomics methods (explicitation interview, video recordings and questionnaires) was conducted. Immediately after each last crossing, participants were invited to participate to an elicitation interview. In total, 73 street-crossings were described. Cognitive, perceptive activities and emotional feelings were identified, classified and analyzed.

The results highlight some different patterns of perceptive and cognitive activities between risky and non-risky decision-making. They show the complexity of the decision-making that cannot be reduced to a specific moment but constitutes a continuously updated process containing perception, action, cognition and social aspects. Moreover, a strong link with the emotional feelings is highlighted. In the context of the future interaction between the pedestrian and the automated vehicle, this study is a first step toward AVs which will have to anticipate and react appropriately to the street-crossing behavior of the pedestrians.

Keywords: Street-crossing · Decision-making · Pedestrian
Automated vehicles

1 Introduction

1.1 The Future Interactions Between Pedestrians and Automated Vehicles

At a time when the automated vehicle upsets the codes of the automotive industry, its biggest challenge is not only to detect other vehicles or to avoid an obstacle on the road but also to manage with the human behavior of other road users, and especially,

S. Bagnara et al. (Eds.): IEA 2018, AISC 823, pp. 132–141, 2019.
https://doi.org/10.1007/978-3-319-96074-6_14

pedestrians. Indeed, crossing the street is a complex activity in a dynamic environment involving cognitive, social, emotional and physiological processes. This activity can be dangerous because of the presence of vehicles on the streets and the resulting risk of being hit but also because of dangerous pedestrian behavior, whether voluntary or because of an incorrect estimation of the street-crossing situation. It requires the pedestrian (Arman et al. 2015) to seek an adapted location in which to cross the street, (Boroujerdian and Nemati 2016) to identify both the correct location and the correct moment to analyze the traffic, combining information provided from different directions, (Cahour et al. 2016) to select an adapted gap, and (Charron et al. 2012) to adapt that gap to the crossing time and then coordinating the motor activity with the continuous perception of road traffic. The ultimate goal of the future research is that the automated vehicle can facilitate this activity by a preventive strategy, i.e., by adapting its behavior to avoid any critical or uncertain situations (Rothenbucher et al. 2016). Thus, using game theory to analyze the interactions between pedestrians and autonomous vehicles, Millard-Ball (2016) concludes that pedestrians would be able to act with impunity, without worrying about the traffic. A recent study (Kraiem et al. 2018) has effectively shown in an experimental setting that pedestrians cross the street more often when they are in interaction with an automated vehicles (shuttles or taxi-robots) than with a conventional vehicle, especially when they have a great degree of acceptability of the automated vehicles. To ensure a safety context, it becomes imperative to better understand how the street-crossing decision-making of the pedestrian is made.

1.2 Literature Review About Street-Crossing Decision-Making and Limitations for the Automated Vehicles

The literature review demonstrates the interest in the study of street-crossing with a various elements to be taken into account in the decision-making process. The mostly studied notion is the gap acceptance that is defined as the available time between two vehicles that the pedestrian perceives and estimates sufficient to cross the street (Boroujerdian and Nemati 2016; Koh et al. 2014; Pawar and Patil 2015). However, many of these studies do not define the method of this gap calculation and what it really corresponds to. In addition, the moment of the triggering of the gap diverges according to the studies, being either the moment when the pedestrian stops at the curb, or the moment when the pedestrian puts a foot on the road. Other concepts are also used in this literature, such as arrival time, collision time and margin of safety (Lobjois and Cavallo 2009; Pawar and Patil 2015). If these measures are helpful, they have to be conducted from the point of view of the automated vehicle to a moving pedestrian.

Secondly, the major conclusion of research about the influence of the social context, the cognition and the individual characteristics on the street-crossing is that the individual does not cross the street in the same way if s/he is alone or in a group (Faria et al. 2010; Yagil 2000; Zhou et al. 2009). Pedestrians would cross the street more than twice if another pedestrian crosses before him/her. With respect to red pedestrian crossings, the results are more contradictory, with some studies showing an increase or a decrease in risk taking in a group rather than alone. Moreover, other elements have been studied such as pedestrian violations, waiting time, attention and anticipation in dual task conditions (see for example, Arman et al. 2015; Neider et al. 2010). Nevertheless, these dimensions have been tested separately or out of a natural context.

Thirdly, non-verbal communication has been shown as an important factor in decision-making, with some actions or gestures helping pedestrians to know whether the driver is giving them priority or not (Zhuang and Wu 2014). However, all of these studies reveal a lot of inaccuracies and shortcomings regarding the interactions between the different elements, to be bridged for a safe interaction between the pedestrian and the automated vehicle. In particular, the temporality of decision-making is lacking but in a dynamic environment, decision-making elements are continually updated (Endsley 1995; Hoc 1996). For example, regarding gap acceptance, one of the non-asked research questions would be how the attention and perception are involved both in this task of street-crossing but also in other contextual elements such as the presence of other pedestrians or obstacles. Indeed, a gap can be considered acceptable by the pedestrian who could decide not to cross because of the behavior of other pedestrians. The laboratory studies do not allow to treat these aspects since the simulated traffic never behaves as the real traffic (e.g. adaptation of the speed, of the trajectory to the behaviors of the pedestrian) and whatever the methodologies used, the measurements of the gap testify only of an extrinsic evaluation, not taking into account the subjective experience of the pedestrian.

For the autonomous vehicle, it seems essential that the gap and the related evaluation are considered before the pedestrian puts his/her foot on the road because at that moment, the pedestrian is considered as an obstacle and therefore the autonomous vehicle stops automatically. In other words, the current research must focus on a pre-crossing context leading to the crossing decision in a dynamic environment and not only on an observation of the accepted gap. The position of the gaze would also be another important indicator of the decision-making process in a given context (Geruschat et al. 2003) and it could be identified by the autonomous vehicle (e.g., context of the crossing, pedestrian behavior, social influence).

1.3 Aim of the Study

The aim of the study was to perform an analysis of the activity of street-crossing, focusing on the decision-making in a natural environment. For this purpose, a methodology allowing us to access to the reflexive and pre-reflexive activities of the action selected. The explicitation interview is a method developed by Vermersch (2009; 1994) for gathering precise elements of human phenomenological experience. The goal of the explicitation interview is for the interviewee to describe the experience of a specific moment, i.e., the lived experience, being defined as "what people have experienced subjectively during their activities, which includes the entire stream of actions, thoughts, emotions and perceptions that occur at a given moment while performing an activity, of which the actors are either aware at the time or can be subsequently made aware" (Cahour et al. 2016, p. 11). This methodology was associated with the analysis of two types of video recordings: one from the point of view of the participant and one from a global view of the environment. We assumed that risky decision-making would be influenced by social elements, vehicles behaviors and communication with the drivers. The risk taking would be linked to a different emotional experience.

2 Method

2.1 Participants

A sample of twenty participants, 10 of whom were women, were included in the study. On the basis of the inclusion criteria, they were between the ages of 25 and 45 (M = 36.9; SD = 5.16), had normal to corrected vision and no motor disabilities.

Participants were treated in accordance with the Helsinki Declaration, and all gave their written informed consent one week prior to the date of the experiment.

2.2 Experimental Setting

Urban Routes

The experiment was conducted in Versailles (France). Two urban routes of approximately one kilometer in length were selected for their diversity in terms of urban environment (containing high and residential traffic and different types of pedestrian crossings). The first route comprised three pedestrian crossings: two with pedestrian lights and one without traffic lights and with the possibility of crossing in two stages. The second route also contained three pedestrian crossings without traffic lights including one two-stage crossing. Each course was represented on a paper map showing a view of the entire area from the point of departure to the point of arrival. To ensure that all participants took the same sidewalks and crossed the streets in the same places, checkpoints were represented on the map where participants were to gather information (e.g., opening hours of a shop, available services of car repairer). This also allowed participants to be given a travel objective to be closer to real conditions.

Explication Interview

Immediately following the last street crossing of each route, the participant was invited to participate in an explicitation interview. Our interviews were focused on the street-crossing processes before, during and after the effective crossing of the street. The experimenter brought the participant through the evocation when s/he had the intention to cross the street and on the entire process of decision-making that followed until the actual crossing, following the temporal development of the activity: the elements s/he perceived in the environment; what s/he was attentive to; what s/he mentally assumed, evaluated, decided, and imagined; and what s/he felt at the different moments of the street-crossing. The data analysis focused on a total of 73 explicitated street-crossing decision-makings.

Video Recording

Throughout the routes, the participant was equipped with camera glasses to record the sight directions, for example, in terms of left sight, right sight or sight in front of him/her, and the perceived environmental elements. The video recording were coded in terms of number of gazes. Moreover, they validated the elicited street-crossing context (see Fig. 1).

Fig. 1. A street-crossing from the point of view of the pedestrian (at the left with camera glasses worn by the participant) and from the point of view of the experimenter walking behind the pedestrian (at the right with a camera)

2.3 Procedure

Participants received the instructions explaining the two routes they had to negotiate and the information regarding the necessity to move as they generally do. Then they were equipped with the camera glasses. The map of the first course was given to them. The participants negotiated the first route; then the experimenter joined them to begin the first explicitation interview in a car parked near the last street crossing (beyond the visibility of the pedestrian crossing). The EI was recorded. The second map was then given to the participants. Immediately after the second route was completed, the second interview was conducted in our office.

3 Results

3.1 Decisional Factors

In the sample of the 73 street-crossings, 4 decision-making factors triggering a non-risky street crossing were identified: (1) the presence of elements of reassurance, (2) a favorable evaluation of the coupling of distance and speed, (3) a sign from the driver and (4) the waiting time (see Fig. 2).

Moreover, 5 decision-making factors triggered risky street crossing: (1) an assessment bias of the situation, (2) social influence, (3) emotional rumination, (4) a habit of breaking the rules and (5) time pressure (see Fig. 3).

3.2 Information Seeking

The results indicated a significant difference in the number of gazes per decision according to whether the participant did or did not make a risky decision ($U = 368$; $p < .03$; without risky decision-making: $M = 4.34$; $SD = 3.42$; with risky decision-making: $M = 2.81$; $SD = 2.31$). When the participant made a risky decision, fewer gazes were observed. Additionally, participants who took risks reported significantly fewer perceptive activities involved in the decision-making ($\chi 2$ [1; N = 73] = 7.08; $p < .01$) and fewer anticipation processes ($\chi 2$ [1; N = 73] = 7.39; $p < .01$).

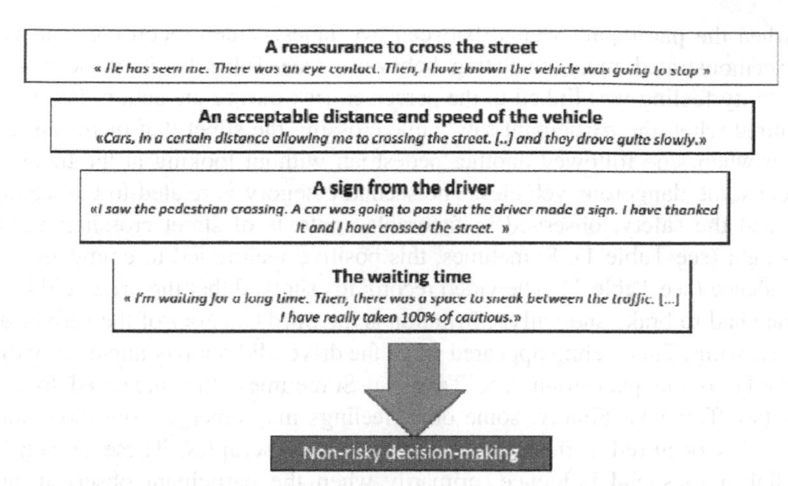

Fig. 2. Decision-making factors triggering a non-risky street crossing and examples of verbatim.

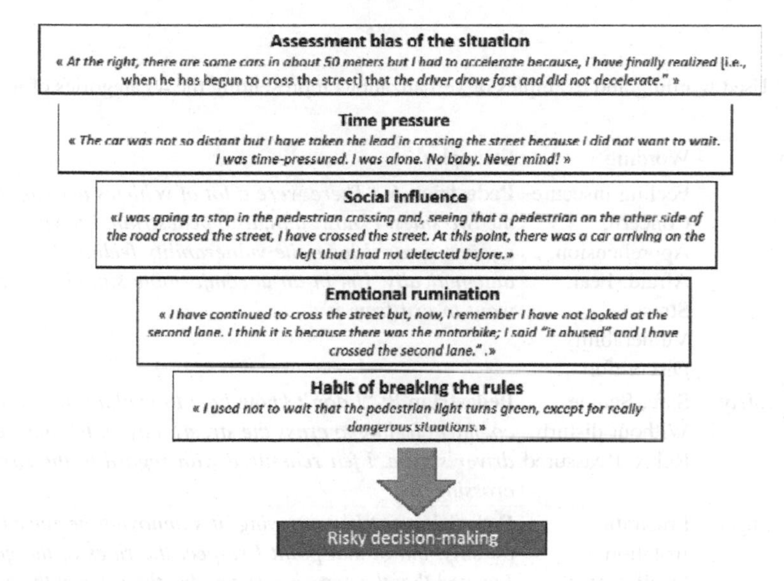

Fig. 3. Decision-making factors triggering a risky street crossing and examples of verbatim.

3.3 Reported Feelings

We precisely categorized the verbalizations; we particularly highlighted 4 categories in the area of worry/insecurity, irritation/anger, serenity/safety and surprise/other feelings. Table 1 presents examples of verbalizations representing these 4 categories and the wording used by the participants to describe their feelings.

The first category concerns the realm of worry and insecurity (see Table 1). These feelings were primarily reported in two global contexts: when the traffic was dense and/or when the participant waited between two lanes. Video recordings and verbalizations demonstrated more hesitating behaviors (see Table 1). In some cases, the fear/insecurity feeling was linked to the perceived risk during the action. For example, this occurred when the participant saw while crossing the street that there was a close vehicle or when s/he followed another pedestrian without looking at the traffic when there were some dangerous vehicles. The second category is related to the feelings of serenity and the safety, observed in favorable contexts of street crossing: when the traffic is light (see Table 1). Sometimes, this positive feeling led to confidence, if not overconfidence (see Table 1). The video recordings showed that the driver did not plan to stop and had to brake suddenly.). Irritation is the third category of the verbalizations of street crossing. This feeling appeared when the driver did not pay attention or did not give priority to the pedestrian (see Table 1). Sometimes, the anger led to a risky decision (see Table 1). Finally, some other feelings may emerge from more singular contexts. This occurred with surprise, amazement or scruples. These feelings were always linked to social influence, primarily when the participant observed another pedestrian crossing the street while s/he waited for the pedestrian light to turn green or s/he did not want to influence a more vulnerable pedestrian.

Table 1. Used wording and examples of verbalizations representing the 4 categories of reported feelings

Category	Wording	Examples
Worry insecurity	Feeling insecure Concern, Apprehension, Afraid, Fear, Stress Vulnerability Discomfort	**Pedestrian 6**: "*There were a lot of vehicles arriving from the two sides. I know, that, as a pedestrian, I'm vulnerable against a car. I have the vulnerability feeling. So automatically, I'm in an uncomfortable situation and I want to exit from it*"
Serenity safety	Safe, Serene Without disturb Relax, Reassured	**Pedestrian 8**: "*I don't know how to explain it [i.e., a confident feeling to cross the street] maybe. It's maybe the driver's gaze. I felt reassured with regard to the street-crossing*"
Irritation anger	Frustration Irritation Disillusioned Anger Annoyed	**Pedestrian 6**: "*It's annoying, it's annoying because it's a bit silly, but at that point I respect the rules of the game. I waited that the person respect also the rules of the game. (...) The driver is dry in the car, I'm in the rain, so, uh, at that moment I wanted to say a lot of insults*"
Surprise and other feelings	Surprise Amazed Absence of scruple	**Pedestrian 7**: "*I was surprised she was crossing the street because I had been waiting for a while*"

4 Discussion

The primary aim of this study was to better understand street-crossing decision-makings by considering both the complexity and the dynamics of decision process. Collected data emphasized non-exhaustive elements of street-crossing by showing different decisional factors triggering either a non-risky decision or a risky decision.

A risky decision was associated with fewer gazes before the street-crossing, leading to a reduction of the amount of information absorbed and a limited anticipation scope in the near space. A risky decision was also associated with more negative emotions related to anger and irritation. This result is consistent with the related literature since emotions have been described as a factor of risky decision-making, producing a selective or accelerated process of information treatment (Mathews and MacLeod 1985; Yiend 2010). In the context of street-crossing decision making, emotion may be induced by an anterior state (e.g., a negative event in a previous activity or a stressful state), independent of the context of the street-crossing activity. Conversely, emotion may emerge directly from the street-crossing situation, for example, when the pedestrian feels worried about crossing the street in a dense traffic context or because no driver has stopped although the pedestrian had the right of way. Whatever the trigger of the emotion, it influences the street-crossing decision-making process ant the related risk level. Because of the found links between risk, emotion, and information seeking, further studies will aim at investigating the engaged process between emotion and information treatment during the street-crossing. The occurrence of negative valence of emotion may be analyzed within the theory of appraisal (Lazarus and Folkman 1984), the pedestrian assessing the situation as inconsistent with the achievement of his/her goal (i.e., to cross the street to do something).

Concerning the future studies to conduct, in the present study we restricted the sample age (i.e., to 25 to 45 years old); however, some elements of the literature show that street crossing, as well as the related evaluation mechanisms, differ according to life stages (e.g., in children, Charron et al. 2012; in elders, Dommes et al. 2015). Thus, it will be interesting to extend our protocol with younger and older participants.

To conclude, in the respect of automated vehicles development, studies aiming to better understand how natural decision-making takes place in a variety of contexts may lead to the construct of mathematical models that can be used for the pedestrian detection algorithms to give a probability that s/he crosses the street in a given context. Since our result show also the importance of a non-verbal communication, the pedestrian detection can also be accompanied by a communicating solution on the vehicle. It would substitute the non-verbal component between the pedestrian and the driver and it would reduce both the feeling of uncertainty of pedestrians and the negative outcomes towards the new technology (Kraiem et al. 2018). This point is a strong issue for the multidisciplinary community in Ergonomics, working about the automated vehicles.

References

Arman MA, Rafe A, Kretz T (2015) Pedestrian gap acceptance behavior, a case study: Tehran. In: Transportation research board, 15–2217. https://trid.trb.org/view/1337505

Boroujerdian AM, Nemati M (2016) Pedestrian gap acceptance logit model in unsignalized crosswalks conflict zone. Int J Transp Eng 4(2):87–96

Cahour B, Salembier P, Zouinar M (2016) Analyzing lived experience of activity. Le Travail Humain 79(3):259. https://doi.org/10.3917/th.793.0259

Charron C, Festoc A, Guéguen N (2012) Do child pedestrians deliberately take risks when they are in a hurry? An experimental study on a simulator. Transp Res Part F: Traffic Psychol Behav 15(6):635–643. https://doi.org/10.1016/J.TRF.2012.07.001

Dommes A, Le Lay T, Vienne F, Dang N-T, Beaudoin AP, Do MC (2015) Towards an explanation of age-related difficulties in crossing a two-way street. Acc Anal Prev 85:229–238. https://doi.org/10.1016/j.aap.2015.09.022

Endsley M (1995) Toward a theory of situation awareness in dynamic systems. Hum Factors 37:32–64

Faria JJ, Krause S, Krause J (2010) Collective behavior in road crossing pedestrians: the role of social information. Behav Ecol 21(6):1236–1242. https://doi.org/10.1093/beheco/arq141

Geruschat DR, Hassan SE, Turano KA (2003) Gaze behavior while crossing complex intersections. Optom Vis Sci: Official Pub Am Acad Optomet 80(7): 515–28. http://www.ncbi.nlm.nih.gov/pubmed/12858087

Hoc JM (1996) Supervision et contrôle de processus - La cognition en situation dynamique. Presses Universitaires de Grenoble, Grenoble

Koh PP, Wong YD, Chandrasekar P (2014) Safety evaluation of pedestrian behaviour and violations at signalised pedestrian crossings. Saf Sci 70:143–152. https://doi.org/10.1016/j.ssci.2014.05.010

Kraiem S, Bel M, Coeugnet S (2018) Prediction of the road users' behaviors beyond an autonomous vehicle. In: 29th international congress of applied psychology, Montréal

Lazarus RS, Folkman S (1984) Coping and adaptation. In: Gentry WD (ed) The handbook of behavioral medicine. Guilford Press, New York, pp 282–325

Lobjois R, Cavallo V (2009) The effects of aging on street-crossing behavior: from estimation to actual crossing. Acc Anal Prev 41(2):259–267. https://doi.org/10.1016/J.AAP.2008.12.001

Mathews A, MacLeod C (1985) Selective processing of threat cues in anxiety states. Behav Res Ther 23(5):563–569. https://doi.org/10.1016/0005-7967(85)90104-4

Millard-Ball A (2016) Pedestrians, autonomous vehicles, and cities. J Plan Educ Res 38:1–7. https://doi.org/10.1177/0739456X16675674

Neider MB, McCarley JS, Crowell JA, Kaczmarski H, Kramer AF (2010) Pedestrians, vehicles, and cell phones. Acc Anal Prev 42(2):589–594. https://doi.org/10.1016/J.AAP.2009.10.004

Pawar DS, Patil GR (2015) Pedestrian temporal and spatial gap acceptance at mid-block street crossing in developing world. J Saf Res 52:39–46. https://doi.org/10.1016/j.jsr.2014.12.006

Rothenbucher D, Li J, Sirkin D, Mok B, Ju W (2016) Ghost driver: a field study investigating the interaction between pedestrians and driverless vehicles. In: 2016 25th IEEE international symposium on robot and human interactive communication (RO-MAN), pp 795–802. IEEE. https://doi.org/10.1109/ROMAN.2016.7745210

Vermersch P (1994) L'entretien d'explicitation (ESF), Paris

Vermersch P (2009) Describing the practice of introspection. J Conscious Stud 16:20–57

Yagil D (2000) Beliefs, motives and situational factors related to pedestrians' self-reported behavior at signal-controlled crossings. Transp Res Part F: Traffic Psychol Behav 3(1):1–13. https://doi.org/10.1016/S1369-8478(00)00004-8

Yiend J (2010) The effects of emotion on attention: a review of attentional processing of emotional information. Cogn Emotion 24(1):3–47. https://doi.org/10.1080/02699930903205698

Zhou R, Horrey WJ, Yu R (2009) The effect of conformity tendency on pedestrians' road-crossing intentions in China: an application of the theory of planned behavior. Acc Anal Prev 41(3):491–497. https://doi.org/10.1016/J.AAP.2009.01.007

Zhuang X, Wu C (2014) Pedestrian gestures increase driver yielding at uncontrolled mid-block road crossings. Acc Anal Prev 70:235–244

How Do People Move to Get into and Out of a European Cabin-Over-Engine Truck?

Jennifer Hebe(✉) and Klaus Bengler

Institute of Ergonomics, Technical University of Munich,
Boltzmannstraße 15, 85747 Garching, Germany
jennifer.hebe@tum.de

Abstract. The purpose of this study was the identification of motion strategies for truck ingress and egress and the analysis of influencing factors. To shorten the development process, digital human models (DHM) are being used to simulate and assess human-machine-interfaces. Before being able to simulate a human motion with DHM, this movement needs to be well analyzed. For example, we analyzed which influencing factors lead to a change in motion. Therefore, this study considered the ingress and egress motion of 36 truck drivers into and out of two real vehicles, one with two steps, representing distribution transport, and one with three steps, representing long haul transport. The body height affected the choice of starting foot for the egress motion. This was the only impact factor that this study confirmed. Although no clear motion strategies, constantly used by every driver, could be determined, we identified frequently occurring general sequences of behavioral motion patterns for the ingress and egress motion.

Keywords: Truck ingress · Truck egress · Motion strategy

1 Introduction

In this day and age, more and more computer-aided procedures are being used in the development process to shorten the development time, and thus to reduce the development costs. On behalf of ergonomics, digital human models are being used to consider the different users' requirements. In the automotive industry, RAMSIS is the most widespread digital human model [1]. On the basis of its large anthropometric database, RAMSIS can predict and analyze probability-based driving postures. Further strengths of RAMSIS are static analyses for reachability and visibility [1]. For an ergonomically optimized product, it is important to consider human motions like car ingress and egress in the vehicle design as well [2]. To simulate a motion, the details of the motion process need to be well known. For example, we need to find out what kind of strategy occurs, and which factors influence motion sequences or the choice of motion patterns. For car ingress and egress, different motion patterns and impact factors have been analyzed in several studies [3–6]. For trucks, there exists much less knowledge about the driver's movements, even though there is a higher number of motions in the everyday life of a truck driver than in the everyday life of a car driver [2].

S. Bagnara et al. (Eds.): IEA 2018, AISC 823, pp. 142–151, 2019.
https://doi.org/10.1007/978-3-319-96074-6_15

The existing knowledge of motion strategies for truck ingress and egress is limited to two works. Shorti [7] considered egress motion patterns for US trucks, while Chateauroux, Wang and Roybin [8] focused on the egress motion for COE (Cabin-Over-Engine) trucks which are common in Europe. As reported by Shorti [7] egress motions can be divided into two major groups: *facing the cab* and *facing outward*. Additionally, he classified the foot behavior by the leading foot at each step. Shorti [7] distinguished a step-by-step behavior, where each footstep was made with the same leading foot, and a step-over-step behavior, where the leading foot alternated from step to step. For the truck egress on US trucks, most participants (59.5%) used the step-over-step behavior beginning with the left foot [7]. The hand behavior was defined according which was the leading hand during the initial transition phase. The majority of the subjects led with the left hand (64.4%) regardless of the facing direction or the hand preferences [7]. Since the layout of US and EU trucks is vastly different, the results from Shorti [7] cannot be transferred to European trucks [8, 9]. For the egress motion in COE trucks, subjects only used the tactic *facing the cab* [8]. However, Chateauroux et al. [8] observed two kinds of egress strategies: *Right Foot First* (RFF) and *Left Foot First* (LFF) depending on the leading foot from the cabin floor to the uppermost step. Additionally, they found four different hand strategies, based on the time when a hand was moved and the foot location during that movement [8].

This study focused on the ingress and egress motion of COE trucks. To record natural movements, the study was conducted on real vehicles and a valid truck driver's license was the prerequisite to participate. The hypothesis of this study is that experienced truck drivers developed a motion strategy for entering and exiting the truck over time. Since most truck drivers are long haul drivers or distribution drivers, they are used to a certain kind of vehicle and optimized their motion strategy in relation to this vehicle. The main difference of COE trucks is the number of steps. Hence, a two-step and a three-step truck were chosen, assuming that subjects would adapt their strategy to the changing number of steps (even/uneven). Additionally, possible influencing factors, like age and body height, were requested in a demographic survey.

2 Methods

2.1 Study Design

In the following section, the subjects are compared considering anthropometric characteristics to the participants from Bothe [10], who conducted a study with 152 truck drivers (*Reference Truck*), and to the SizeGermany database [11] (*Reference SizeGermany*) of the German male population, if values are available. Thirty-six subjects joined in this study. Two of them had to be excluded since they did not fulfil the prerequisite of having a truck driver's license. The remaining 34 subjects ranged between the age of 23 and 64 years ($M = 42.85$, $SD = 12.08$; $M_{ReferenceTruck} = 44$). On average the subjects had their driver's license for 19.15 years ($SD = 12.63$; $M_{ReferenceTruck} = 20$). Eleven subjects are familiar with two-step trucks, while 17 subjects are used to three-step trucks. Six subjects marked the option "others", which combines all participants who drive a truck with one step, four steps, or who frequently change their trucks regarding the number of steps. This distinction between the number of steps the drivers were used to was called *step experience*. The thirty-two male and two female subjects had a body height between 1.52 m to 1.92 m (M = 1.80, SD = 7.65; $M_{ReferenceTruck}$ = 1.76; $M_{ReferenceSizeGermany}$ = 1.80). This study used a within-subject design.

Observation of the ingress and egress motion took place outside in a parking spot using a real truck. Two real vehicles, instead of a mock-up, were taken as test vehicles. In order to monitor changes in motion strategies dependent on the number of steps, we chose a two-step truck used in distribution transport, and a three-step truck representing long haul transport. The motions of the participants were tracked by the motion capture system CAPTIV based on initial measurement units. The motion capture system was calibrated again after the fifth ingress and egress motion to eliminate sensor drifts. Additionally, two GoPro Hero3 devices, one from a side view and one from a top view, recorded the participants' motions into and out of the trucks for qualitative analyses.

To reduce the Hawthorne effect, the subjects were invited to participate a study about the seating comfort of the driver's seat and thus, did not know the purpose of the study. After an introduction to the seating comfort concept, including a definition and an explanation on how to rate seating comfort, a demographic survey and an anthropometric measurement of the subjects followed. The anthropometric measurement considered the measurement of the segment lengths (forearm, upper arm, sitting height, thigh, shank) and the range of motion (shoulder, elbow, wrist, hip, knee, ankle). The subjects conducted the ingress and egress motion six times. We only analyzed the sixth repetition [12]. After the final sixth trial, the participant went on to the next truck and repeated the process. The starting order of the two trucks was randomized.

2.2 Data Analysis

Following the study of Rigel [6], we analyzed the video data in a qualitative way. In order to determine motion strategies, motion phases were first established. Motion phases had to occur sequentially [6]. Therefore, the leading body part specified the motion phases [12]. Two key frames, one at the beginning and one at the end, framed one motion phase. Each contact of the leading foot with a structured component with which it had not been in contact before counted as a key frame. For example, when a subject used the step-by-step strategy [7], only the first contact including a step would count as a key frame. The trailing foot would be neglected in the definition of key frames. For the ingress motion into a long haul truck, seven key frames and therefore six motion phases were set. For the egress motion, six key frames framed five motion phases (see Fig. 1). The missing step of a distribution truck in comparison to a long haul vehicle led to five motion phases for the ingress and four motion phases for the egress motion (see Fig. 1). The initial position for the ingress motion was a standing posture facing the truck, with both feet on the ground and arms hanging next to the body. The ingress ended with the driving posture: sitting on the driver's seat with both hands at the steering wheel and both feet on the cabin floor. For the egress motion, the initial position and the final position were switched.

We presented the results of the behavioral motion patterns in dendrograms. To analyze existing factors which may influence the choice of motion strategy, a stepwise

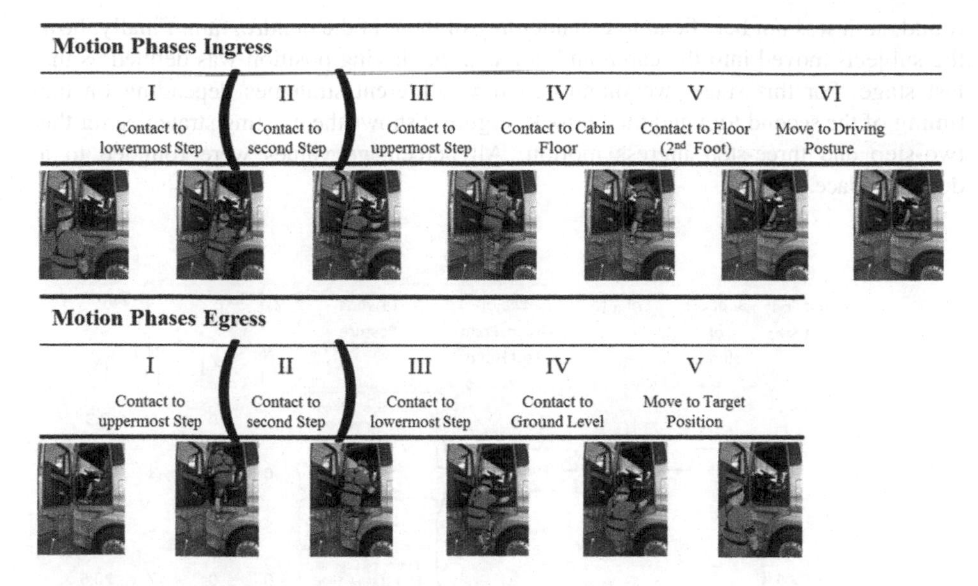

Fig. 1. Motion phases for the ingress and egress motion of a three-step truck.

logistic regression considered the sixth motion repetition of each calculation. As predictors, the factors *step experience*, *foot preference*, *age*, *BMI* and *body height* were considered. All logistic regressions were conducted twice in order to consider each truck. For the ingress motion, the model used the first foot on the lowermost step (left/right) and the first foot on the cabin floor (left/right) as output variables, while the first foot on the uppermost step (left/right) worked as a criterion for the egress.

3 Results

3.1 Dendrogram of Truck Ingress and Egress Motion Strategies

According to Lu, Tada, Endo and Mochimaru [13], a dendrogram was used to cluster the existing behavioral motion patterns. For the ingress motion, the dendrogram consisted of five stages. The first stage was set up in relation to the moving foot: *Which is the first foot on the lowermost step?* The next stage differentiated between *Pull* and *Push* strategies. Therefore, the timing of the grasping position of the hands was significant. If both hands were grasping a grasping point at the time the leading foot arrived on the first step, it was defined as a *Pull*-Strategy, since the hands supported the ascent movement. If only one hand grasped a grasping point, or if there was no contact between hands and any grasping point, a *Push*-Strategy was identified, since mainly the leg muscles performed the ascent movement. Following the distinction between a step-by-step and step-over-step foot strategy by Shorti [7], the third stage was established as a *Climb* stage. Thereafter, identifying the major grasping points for the left hand marked the next stage. For the right hand, a great variety of grasping options was found, so it was not beneficial to consider any of them in the dendrogram. Finally, how the subjects moved into the cabin and attained the driving position was defined as the last stage. For this stage, we monitored three different strategies depending on the timing of the second foot and the buttock. Figure 2 shows the existing strategies for the two-step and three-step ingress motion. All percentage values were rounded to a decimal place.

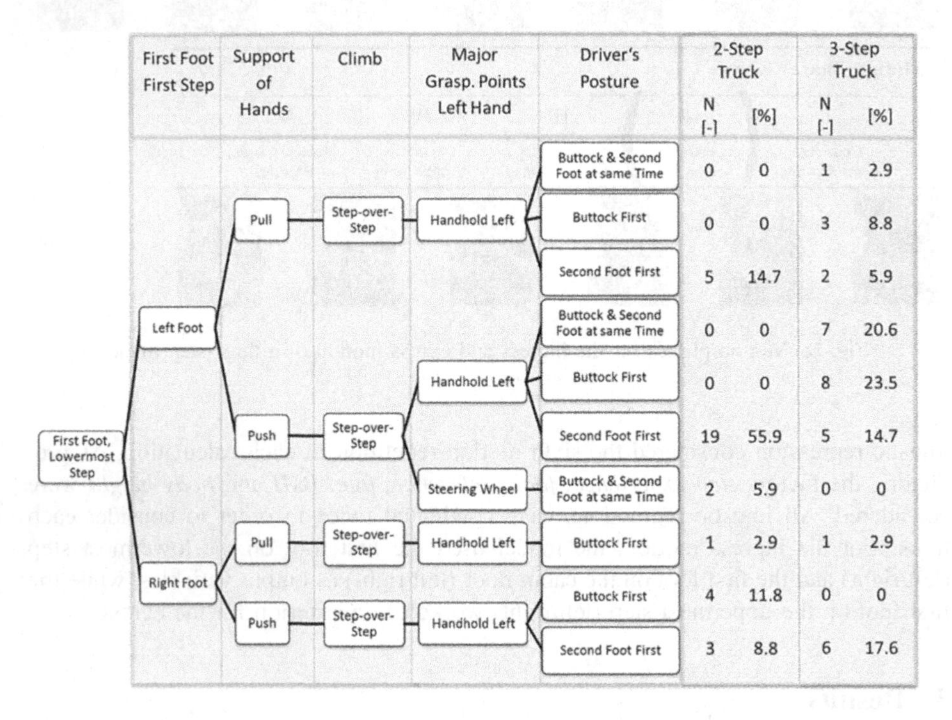

Fig. 2. Dendrogram of the ingress motion of a two-step and three-step truck.

For the two-step truck, less variation occurred in the dendrogram than for the three-step truck. The most frequently used ingress strategy for the two-step truck started with a leading left foot and a *Push*-strategy, followed by a step-over-step climb, with the left handhold as the major grasping option and finally both feet were placed on the cabin floor before the subject sat down on the driver's seat. Nineteen subjects used this ingress pattern. Five subjects used this strategy with a *Pull*-variation (see Fig. 2).

For the three-step ingress, twenty participants started with the left foot as the leading body part, pushed themselves up, climbed the steps step-over-step and used the left handhold as the major grasping point. In the last stage, these twenty drivers chose different strategies ranging from *Buttock First* (23.5%) to *Buttock and Second Foot at the same Time* (20.6%) and *Second Foot First* (14.7%). Most subjects, starting with the right foot, pushed themselves up and climbed step over step with the left hand on the left handhold most of the time and entered the cabin with the second foot first (17.6%).

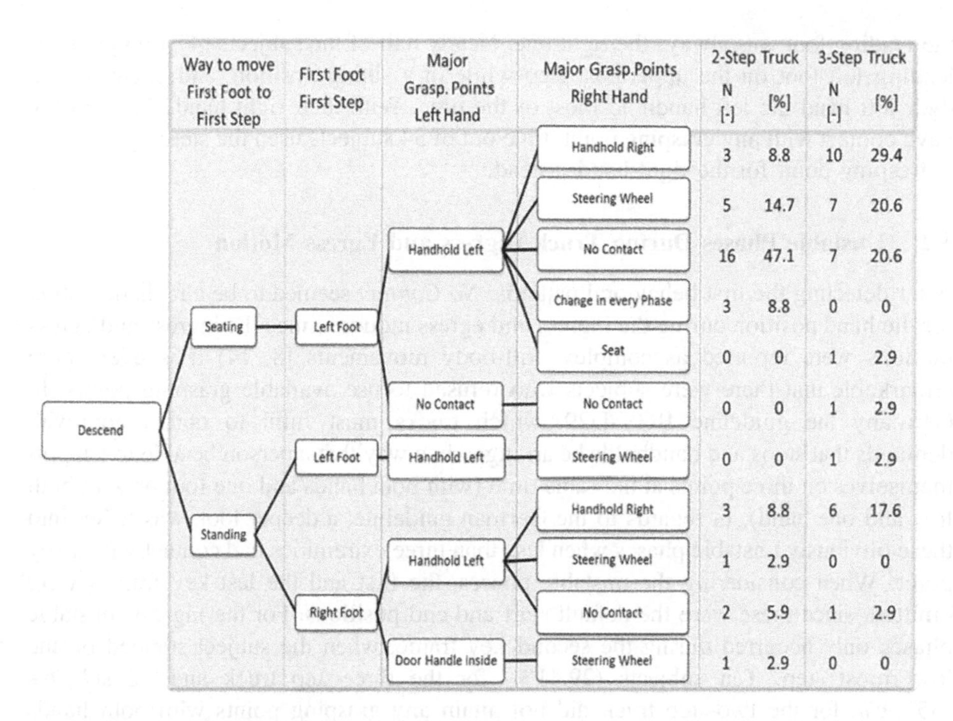

Fig. 3. Dendrogram of the egress motion of a two-step and three-step truck.

Only 33 subjects appeared in the dendrogram of the three-step ingress, since one skipped the lowermost step during ingress motion.

The egress motion was divided into four separate and distinct stages including shorter motion phases. The first stage described the posture in which the subject put the first foot to the uppermost step. In a *Seating* position, subjects glided sideways onto their seat and stood with the foot towards the step while the pelvis remained on the seat. A *Standing* position was reached when the subject stood up to the cabin sill before putting the first foot on the uppermost step. This position is comparable to the first stage of the foot strategies from Chateauroux et al. [8]. The next stage distinguished which was the leading foot on the uppermost step. The last two stages focused on the main grasping option for the left and right hand.

For the three-step egress, twenty-four subjects put their left foot to the uppermost step while being in a sitting position and grasping the left handhold mostly with the left hand. These twenty-four drivers varied grasping, with the right hand, between *Handhold Right* (29.4%), *Steering Wheel* (20.6%) and *No Contact* (20.6%). When participants stood up before they started descending, they usually put the right foot to the uppermost step (20.6%). One subject used nearly no grasping points during the three-step egress, because he jumped from the second step down to the ground level.

For the two-step egress motion, significantly fewer variations appeared in the first three stages. When subjects started descending from a sitting position, their leading foot was always the left one. When they started descending from a standing position, the leading foot was always the right one. Nearly half of the subjects (47.1%) put their leading left foot on the uppermost step while in a sitting position, and grasped with their left hand the left handhold most of the time. With their right hand, they did not have contact with any grasping point. Five out of 34 subjects used the steering wheel as a grasping point for the right hand instead.

3.2 Unstable Phases During Truck Ingress and Egress Motion

After detecting the first behavioral patterns, *No Contact* seemed to be a realistic option for the hand position during the ingress and egress motion. Since the ingress and egress motions were reported as complex, full-body movements [8, 14] it is even more remarkable that there were subjects who refused to use available grasping points. In Germany the guideline BGV-D29, which trucks must fulfil to obtain approval, demands that steps and handholds be arranged in a way that a person be able to support themselves on three points at the same time (with both hands and one foot or with both feet and one hand). In regards to the German guideline, a deeper look was taken into these obviously unstable phases when less than three extremities had contact with a key point. When considering the unstable phases, the first and the last key frames were omitted, since these were the default start and end positions. For the ingress, unstable phases only occurred during the second key frame, when the subject stepped on the lowermost step. Ten subjects (29.41%) for the three-step truck and 12 subjects (35.29%) for the two-step truck did not attain any grasping points with both hands during this phase. These seemed to be the subjects who pushed themselves up the first step only using their leg muscles. Small subjects especially needed to use this *Push* tactic to climb into the three-step truck, since they could hardly reach both handholds from the ground level. These circumstances are aggravated by the peculiarity of a COE truck. Due to the fact that the stairs are located in front of the front axis, they are off-center to the door opening, and therefore to the handholds (see Fig. 4).

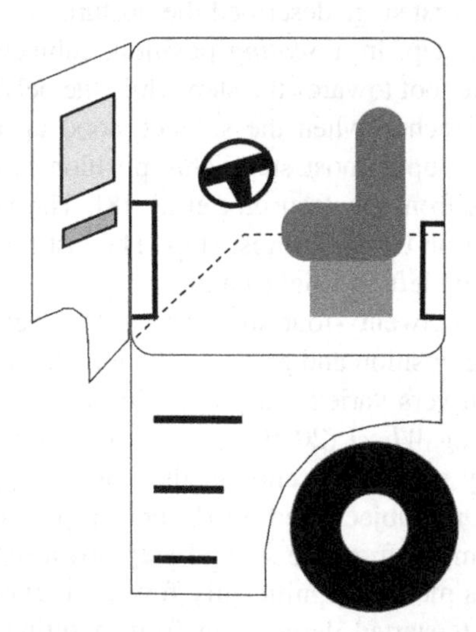

Fig. 4. Ingress geometry of COE truck; the steps are located off-center to the handholds.

A few subjects (two for the three-step truck, one for the two-step truck) created an unstable situation in the penultimate motion phase of the ingress motion, when the second foot arrived on the cabin floor. Here again, the hands caused the unstable phase. During the climbing phases, no unstable situations occurred. For the egress motion, unstable situations arose towards the end of the motion. The participants stood with one foot on the middle step, while the other one was on the lowermost step and the hands were at their individual grasping points. In the next sequence the foot on the middle step moved to the ground level. Before this foot arrived, the subjects released the hand grasping points and let themselves "fall down" to the ground. In this phase, they had only one contact point: the foot on the lowermost step. Twenty-four subjects used this tactic for the three-step truck and 30 subjects used it for the two-step truck.

3.3 Impact Factors on Motion Strategies for Truck Ingress and Egress

From literature, we know some impact factors on motion strategies. On the operator side, the body height and BMI [9, 15] as well as age and agility [16] influence ingress and egress motion strategies. On the vehicle side, the ingress configuration does have an impact on the motion pattern [14], especially as different door angles lead to a change in motion strategy [8]. The impact of the step configuration and the number of steps has not yet been proven [8, 9, 15].

The logistic regression showed that none of the predictors considered had a significant impact on the choice of foot strategy for the ingress motion, not even the step experience. Assuming that each truck driver used the same ingress strategies each time, a change in the foot strategy was expected between the ingress motion into the two-step truck and the three-step truck, depending on the number of steps the driver was used to. Neither the starting foot on the first step nor the first foot on the cabin floor seemed to be decisive for the choice of the ingress motion strategy. These results reflected the circumstances that no consistent motion strategy for ingress or egress motion could be defined. Random checks of the video data showed that the subjects even changed their motion strategies within the six repetitions. For the egress motion, the logistic regression indicated the predictor *body height* as a significant impact factor on the choice of foot leading to the uppermost step (see Table 1). The taller a subject is, the higher the probability that the foot leading to the uppermost step will be the left one. ($p_{two\text{-}step\ truck} < .05$; $p_{three\text{-}step\ truck} < .05$). The G*Power test [17] confirmed that the sample size was adequate for the logistic regression. Even a larger sample size would not have any effect on the results.

Table 1. Logistic regression for impact factors on egress motion strategy.

	95% CI for Odds Ratio			
	b	Lower	Odds	Upper
Included				
Constant	1.752			
Body Height	−1.845 *	0.032	0.158	0.772

R^2 = .229 (Hosmer & Lemeshow), .208 (Sox & Snell), .326 (Nagelkerke)
Modell $\chi^2(1) = 7.919$, * p = .023

4 Discussion

Despite preliminary considerations, like using real vehicles and inviting only people with a valid truck driver's license, the fact that it was an experiment influenced the behavior of the subjects. The data of the motion capture system are neglected for this work, but the motion sensors attached by straps to the participants had an impact on the subjects' motion behavior. During the trials, we did not create any time pressure. The subjects got an acoustic signal to start the ingress and egress motions. Besides that, the subjects could decide on their own how fast they ascended and descended. In reality, truck drivers usually work under time pressure and are not focused on the ingress and egress motion [8].

5 Conclusion

Finally, how do people move to get into and out of a COE truck? To answer this question, we conducted a study with 36 truck drivers entering and exiting two real test vehicles, one with three steps, representing the long haul sector, and one with two steps, representing distribution transport. Certain reoccurring behavioral motion patterns could be identified (see Figs. 2 and 3). However, it must be noted that it was difficult to define a consistent motion strategy, since random checks showed that the drivers used different motion strategies within the six repetitions.

In summary, most of the subjects started the ingress motion with their left foot (76.47%) both for the three-step and two-step truck. In this study, none of the participants used the step-by-step pattern [7], neither for the ingress nor the egress. For the ingress motion, the participants used either the *Pull* or the *Push* strategy, depending whether they used hand grasping points to pull themselves up the steps, or they pushed themselves up the lowermost step using only their leg muscles and reached the hand grasping points during the next step. The major grasping points were the handholds, but there are many variations on the chronological sequence used for the grasping points. Especially for the right hand, nearly every participant used his own chronological sequence for grasping. For the egress, most subjects (79.41%) started with their left foot on the uppermost step. In particular, large subjects put their left foot on the uppermost step. Therefore, they slid sideways on the driver's seat and stepped onto the uppermost step in a sitting position. This movement was supported by the left hand grabbing the left handhold, pulling the body sideways on the seat. These results indicate that the truck cabins under investigation do not support any specific motion strategy. The subjects changed their motion patterns even despite six, temporal proximity repetitions. Also risky and consciously unsafe behavior could be applied successfully. Changes in the cabin layout could result in more consistent and safer motion patterns.

This work strongly focuses the vehicle part, since the study was conducted on real vehicles with a variation in number of steps. Further studies should concentrate on the anthropometric part with subject groups differing in parameters like body height, BMI, and age.

References

1. Van der Meulen P, Seidl A (2007) Ramsis - the leading Cad tool for ergonomic analysis of vehicles. In: Duffy, V.G. (ed.) Digital Human Modeling, HCII 2007, vol 4561, pp 1008–1017. Springer, Heidelberg
2. Daimer AG - Press Information. http://media.daimler.com/marsMediaSite/en/instance/ko.xhtml?oid=9913400. Accessed 18 Feb 2018
3. Chateauroux E, Wang X, Tasbot J (2007) A database of ingress/egress motions of elderly people. SAE Technical Paper 2007-01-2493
4. Reed M, Huang S (2008) Modeling vehicle ingress and egress using the human motion simulation framework. SAE Technical Paper 2008-01-1896
5. Andreoni G, Rabuffetti M, Pedotti A (1997) New approaches to car ergonomics evaluation oriented to virtual prototyping. EURO-BME course on methods & technologies for the study of human activity & behaviour, pp 19–20
6. Rigel S (2005) Entwicklung und Validierung einer Methode zur quantitativen Untersuchung der Ein- und Ausstiegsbewegung in einen Pkw. Dissertation. Technical University of Munich, Germany
7. Shorti R (2016) Step negotiation biomechanics during truck cab egress and the effects of anthropometrics and cab design on driver fall biomechanics etiology. Dissertation. University of Utah, USA
8. Chateauroux E, Wang X, Roybin C (2012) Analysis of truck cabin egress motion. Int J Hum Factors Model Simul 3(2):169–186
9. Reed MP, Ebert SM, Hoffman SG (2010) Modeling foot trajetories for heavy truck ingress simulation. In: 9th international conference on applied human factors and ergonomics, AHFE 2010, Miami, USA
10. Bothe A (2015) Analyse dynamischer Sichtsituationen zur ergonomischen Auslegung von Kamera-Monitor-Systemen (KMS) in schweren Nutzfahrzeugen. Dissertation. Technical University of Darmstadt, Germany
11. Human Solutions GmbH - SizeGERMANY. Die deutsche Reihenmessung, https://portal.sizegermany.de/SizeGermany/pages/home.seam. Accessed 25 Jan 2018
12. Cherednichenko A, Assmann E, Bubb H (2006) Computational approach for entry simulation. SAE Technical Paper 2006-01-2358
13. Lu J-M, Tada M, Enda Y, Mochimaru M (2016) Ingress and egress motion strategies of elderly and young passengers for the rear seat of minivans with sliding doors. Appl Ergon 53 (Part A):228–240
14. Bubb H, Bengler K, Grünen RE, Vollrath M (2015) Automobilergonomie, 1st edn. Springer Vieweg, Wiebaden
15. Reed MP, Hoffman SG, Ebert-Hamilton SM (2011) The influence of heavy truck egress tactics on ground reaction force. In: NIOSH (National Institute for Occupational Safety and Health) (ed.) International Conference on Fall Prevention and Protection 2010, pp 192–195. DHHS(NIOSH) Publication No.2012-103
16. Brenner M (2013) Altersspezifische Ergonomie im Fahrzeug. Dissertation. Technical University of Munich, Germany
17. Heinrich Heine Universität Düsseldorf, Allgemeine Psychologie und Arbeitspsychologie, G*Power: Statistical Power Analyses for Windows and Mac. http://www.gpower.hhu.de/. Accessed 28 Feb 2018

Corporate Robot Motion Identity

Jakob Reinhardt[(✉)], Jonas Schmidtler, and Klaus Bengler

Chair of Ergonomics, Technical University of Munich,
85747 Garching, Germany
Jakob.reinhardt@tum.de

Abstract. Mobile robotic systems are increasingly merging into human dominated areas and therefore will interact and coordinate with pedestrians in private and public spaces. To ease intuitive coordination in human-robot interaction, robots should be able to express intent via motion. This will enable an observer to quickly and confidently infer the robot's goal to establish productive encounters. For long-term interaction, trajectories in straight drive or curvature have been optimized for this purpose. In addition, short-term movement cues are perceivable changes in motion parameters and direction of movement that can be utilized to express intent in a non-verbal manner. For example, yielding priority to a person via a short back-off movement cue, as opposed to merely a stop, provides the possibility of legible and agreeable robot navigation. In the service design domain, the front line personnel's behavior is a crucial quality factor of how an organization is perceived by customers and society. Recent developments show that mobile robotic systems are increasingly supplementing a service company's front line personnel. Companies such as Starship Technologies or Deutsche Post apply service robots for transportation purposes. Integrating robot motion into an organization's visual identity to communicate the visual cues of what the organization wants to express could contribute to the customer experience. In order to provide movement cues that are not only legible, but convey an inherent personality of the robot carrying out the task and therefore reflect on the organization's public image, we discuss aforementioned factors for consideration when developing a corporate robot motion identity. We integrate service quality domains and affected human roles for application in the creative practice of designing motion. Thus, recognizable movement cues which are designed to express intent to coexisting and cooperating pedestrians in an everyday context can be tailored to what an organization wants to express to its environment, customers or other stakeholders.

Keywords: Service robots · Corporate identity · Service design
Human-robot interaction · Motion planning · Legibility · Movement cues

1 Mobile Robots Are a New Type of Front Line Personnel

Technical achievements allow an increasing application of mobile robots to assists or replace human service personnel. Especially the quality of service operations is based on the quality and character that is expressed by the interaction between the human user and the service agent – may it be human or technical.

S. Bagnara et al. (Eds.): IEA 2018, AISC 823, pp. 152–164, 2019.
https://doi.org/10.1007/978-3-319-96074-6_16

Human service operators use different communication channels to interact. These include nonverbal as well as verbal signals and cues covered by terms like mimics, hand or body gestures, prosodic cues and natural language. In addition, and correlated to these, motion of the agent is a relevant means of communication. Especially mobile systems will benefit from a sophisticated design and application of this very natural way to communicate. These aspects shall be discussed in their relevance and characteristics in this survey.

1.1 What Front Line Personnel Does, Influences a Company's Image

The behavior of its front line personnel influences the public image of an organization. For example, flight attendants have recently made it to newspaper headlines by insulting customers or using violence [1–3]. These incidents lead to malicious publicity for some aviation companies. Usually, an organization's staff is advised to behave in a certain way to show a specific degree of politeness. This human behavior is an umbrella term for activities in response to external or internal stimuli [4].

As strategic decision, an organization aims to create a recognizable corporate identity (CI) and image through the behavior and appearance of its staff to communicate the visual cues of what the organization wants to express [5]. Front line personnel such as flight attendants show identity through their common look (e.g. corporate outfit) and movements (e.g. standardized safety instructions) and can therefore easily be distinguished from the staff of other organizations and recognized as associated to a brand. Incidents in connection with bad behavior of front line personnel cause severe trouble and dealing with this is a task for the domain of service design.

Service design has been defined as the activity of planning and organizing people, infrastructure, communication, and material components of a service in order to improve its quality and the interaction between service provider and customers [6]. This includes front line personnel behavior which is especially linked to immediate customer experience [7, 8]. It is a crucial quality factor of how an organization is perceived by customers and society [7]. To measure this, service quality has been operationalized through five dimensions: reliability, assurance, tangibles, empathy, and responsiveness [9]. In Sect. 3.1 we will discuss how these dimensions are affected by CI and associated motion planning.

1.2 Mobile Service Robots Are on the Rise

Imagine a typical service encounter with front line personnel: Would you accept a rude delivery person? A person can show inappropriate spatial behavior like stepping close to you neglecting social distances [10] or entering your door without permission. You might consider another delivery service next time you order.

Would you accept if your next delivery is carried out by a robot? Mobile robotic systems are increasingly supplementing an organization's front line personnel. Companies such as Starship Technologies or Deutsche Post apply robots for transportation purposes for their services already. Based on the advance of this technology, statistics show that humans will increasingly mingle with autonomous robots [11]. The total number of service robots sold in 2016 rose by 24% compared with the previous year [12].

The International Federation of Robotics defines a service robot as a robot that performs useful tasks for humans or equipment excluding industrial automation application [13, 14]. They are categorized into personal service robots and professional service robots. A personal service robot is a robot used for a non-commercial task, usually by lay persons. Examples are domestic cleaning robots or servants. A professional service robot is a robot used for a commercial task, usually operated by a properly trained operator. These include cleaning robots for public places, delivery robots and surgery robots in hospitals [13, 14]. Although a distinction into personal and professional robots seems adequate for purposes of owner statistics or liability issues, a different perspective could be reasonable considering concerns in human-robot interaction. In some situations, a professional service robot used for delivery will interact with humans just like a personal service robot for domestic cleaning does. When robots perform tasks that were previously executed by human service personnel, the question is, how to establish general rules for such robots. Therefore, a review to structure the parameters to consider in HRI is needed.

1.3 Human-Robot Spatial Interaction

The umbrella term human-robot interaction (HRI) describes the domain of research that deals with the investigation, understanding and design of mutual acting of human and robot [15]. Besides motion, this includes all auditory, haptic or visual communication via displays. Designing motion of these systems is a rather urgent issue. Most of the coordination between humans in public spaces is accomplished by observing the other's movements [16]. Integrating robots into this context means that at first, their movements have to fit in. In the next years, robots will interact with humans who lack previous experience with robots. The first encounters will determine a great degree of future acceptance of robots.

For human-robot spatial interaction (HRSI), motion behavior plays a predominant role [17]. A robot, just like human service personnel, should behave properly. This means, it has to keep acceptable distances [18] and make its intentions clear [19, 20]. Different spatial limits and movement strategies may be considered depending on whether the interaction is focused or unfocused [21]. A typical scenario for focused interaction is when a robot approaches a customer, possibly to deliver a parcel. A typical scenario for unfocused interaction is passing a pedestrian. It is comprehensible that robots need a different set of applicable movement strategies to solve these very different situations in a suitable way.

HRI can also be classified into human-robot coexistence, human-robot cooperation, and human-robot collaboration [22]. If there is a common workspace for both entities and they act at the same time, the HRI can be labeled as a human-robot coexistence. In addition, humans and robots are working on the same goal in a human-robot cooperation. Should direct contact occur, the interaction can be labeled as a human-robot collaboration.

To classify the human tasks in such varying scenarios, Scholtz [23] have proposed five different roles of humans in HRI, namely, supervisor, operator, mechanic, bystander and peer. In Sect. 3.2 we will discuss how these roles are affected by CI and associated motion planning.

1.4 What Can Go Wrong?

Acceptance and perceived sociability of mobile robots are the first criteria that are affected when robots show inadequate behavior [21]. This can also reduce the reputation of an organization that sends robots to its customers. When developers program robots to show a certain scripted or adaptive movement, there is a strong need to find the adequate strategy. Hancock [24] suggests morality principles for robots to become inherent characteristics of the robot operating system. Accordingly, there is a need for programmed morality in autonomy. The human should thereby always remain the upper hand on the design of moral behavior of autonomous machines. Thereby the specific coding and effect of applied principles is a challenge ergonomists face.

If inadequate robot behavior is widely applied in the infrastructure, humans might adapt to this behavior and a reversion of learned behavior cannot easily happen [24]. An example to illustrate this is right-hand and left-hand traffic. Even though it causes inefficiency to manufacturers developing cars for right-hand and left-hand steering, and task-load for drivers, traveling between countries, a unification of such deeply implemented rules cannot take place because whole societies have adapted to them.

Movements that are not designed properly or communicate unwanted intentions to customers can have malicious impact. Robots that perform a rather dominant behavior over pedestrians might meet efficiency goals, though the respective organization might be perceived as impolite. A malicious effect on an organization's reputation cannot easily be reversed. In contrast, a company that aims to deliver parcels which applies robots that always give priority to humans, may face challenges in the efficiency that their robots will achieve. The pedestrians could learn such behavior and may take advantage and hinder an always-submissive robot. This suggests, the behavior has to be properly calibrated.

2 How to Do It Right

2.1 Movement Cues Are Social Cues of Mobile Robots

In the following we take an attitude consistent with previous research to base human-robot interaction on human-human interaction [17]. Robots should reflect human social norms and show a consistent set of behaviors [25] that have common sense [21, 26]. Developing robots according to what is common sense among humans implies to consider human social behavior to determine the necessary components of robot social behavior.

Motivated by ethological research, Mehu and Scherer [27] prepared the terms social signals and social cues for the use in human-human interaction. Individuals have to organize social information in a way that helps them take adaptive decisions [27]. The management and assessment of the social environment are achieved through the production and perception of social signals and cognitive processing of social cues [27]. Social signals are acts or structures that influence the behavior or internal state of other individuals. In ethology, a social signal is displayed by means of a structural component that is solely designed to communicate the social signal [27]. An example from ethology shall demonstrate this idea. A peacock's tail is made up of elongated upper tail feathers and nearly all of these feathers end with an elaborate eye-spot. As a visual ornament, these eyes, if feathers are put up to form a wheel, are used to scare enemies off. Here, as well as between humans, social signals can be labeled as explicit forms of communication [27]. Social cues describe a set of mostly nonverbal messages to notify others about the own internal state and intention [21]. These cues can be facial expressions, body posture, proximity, neuromuscular and physiological activities [27]. Social cues are entities that have not evolved as communicative units but still can serve as method to derive information [27]. The peacock's feathers, whose size and visuals are primarily necessary to scare enemies, are also spotted by peahen. To them, they communicate an indication of healthy viable offspring. Social cues can be labeled as implicit forms of communication [27].

We can use the ethological metaphor to classify modalities for communication of robots. A signal is displayed by means of a device that is solely designed to communicate the signal. This could be a display, attached to the robot, that shows a request or a stop sign. Cues make use of technical equipment that primarily exist for other pragmatic functions. A mobile robot is comprised of technical equipment (e.g., engines, undercarriage, wheels) that enables it to drive. This equipment's function primarily is, to make the robot mobile. If motion of a robot is additionally used to communicate intent, status or a CI in short-term movement variations, it can be labeled as a movement cue.

When human and robot share the same space, rules to establish a social order must be identified [21]. For example, this includes social space conventions where similar values as observed between humans may be appropriate [21]. In coordinative assembly tasks, carried out by two persons, also humans show movement cues to avoid collisions or express the intent of giving priority to the co-working person via a short back-off movement with their hand [28]. Such examples suggest robots' behavior may be designed precisely according to human behavior. However, robots do not comprise the same embodiment as humans and therefore their behavior has to be adapted to their simplified appearance. A challenge is also to design social cues so they are linked with the right social signals [21]. Attention, empathy or politeness are examples of information that can be detected by observation of multiple social cues [21]. Vinciarelli et al. [29] provide a list of social cues linked to social signals. However, these authors' description of cues is too generic to be transferable to applications in robotics. E.g., "walking" or "posture" as descriptions of cues do not determine any spatial representation of cues in terms of technical parameters like "speed" or "acceleration". To target this issue, we have previously documented a robotic back-off movement cue with the parameters "path" and "speed" to be reproduced in other robotic appliances [30].

2.2 Legible Movement Cues Can Represent the Language for Robots

Motion as a language is a necessary and sufficient way for robots to communicate with pedestrians. Currently, nonverbal communication represents more than sixty percent of the communication between two people or between a speaker and a group of listeners [31]. This share is likely to be even higher in areas where verbal communication diminishes (e.g., everyday traffic scenarios, pedestrian precinct and sidewalks). Robots will populate these crowded and noisy public spaces where they have to coordinate with many pedestrians in a short period of time. Due to the disturbance by noise and the limited reach, auditory feedback or explicit signals appear to be ineffective and inefficient in these situations. Instead, we promote motion as a modality, in order to elicit elegant and esthetic ways for communication, visible by many observers.

Humans have to build internal mental models of the robot to resolve spatial conflicts effectively [17]. Inspired by human motor control, which relies on minimizing cost [32], developers have commonly applied cost-functions to robot control [33]. Cost-functions allow for long-term planning and collision avoidance, however, the robot appears to move rigidly in coordinative situations with humans. It is then difficult for pedestrians to infer a status or intentions of the robot and thus building an internal mental model is hindered.

To ease intuitive coordination, robots should be able to express intent via motion [20]. Such legible motion enables an observer to quickly and confidently infer the robot's goal [19]. Therefore, it has to happen within a person's information process space [34]. For long-term cooperation, trajectories in straight drive [35] or curvature [19, 36] have been optimized for this purpose. One design approach to robot behavior could take the human status into account and adapt robot movements to changes in the human behavior in real-time. Overly adaptive robot motion however, has led to more confusion about the robot's intentions and goals [33, 37]. On the other hand, predefined and consciously designed movement strategies [36] might be better understood, remembered and even associated to a brand. Hence, scripted movements might be prioritized over efficiency based cost-function strategies in short-term coordination. Scripted movement cues are planned changes in motion parameters and direction of movement that can be utilized to express intent in a non-verbal manner [30]. For example, a short back-off movement cue as opposed to merely a stop when a robot aims to yield priority to a person, has the potential to promote legibility and has increased human trust in the system in a previous experiment [30]. Other examples of applied movement cues, communicating respective intentions are, "boost" by shortly accelerating towards a person [21, 38], and "wiggle" by turning from side to side [39]. These studies broach the issue that such movement cues may enable ways of communicating effectively and efficiently, enabling humans to better infer a robot's intention. Organizations that aim to integrate robots in their services could benefit from this research. They could communicate the right intentions through expressive robot movements.

3 Using Movement Cues to Express Corporate Identity

Integrating robot motion into a company's visual identity to communicate the visual cues of what the company wants to express [5] could contribute to the customer experience [8]. In the area of human-computer interaction, many companies understand motion CI, as a part of visual identity. For instance, the movements triggered and observed by the user when opening and closing windows or browsing through lists in popular operating systems (Apple iOS – Rubber Band) and websites (Google Material Design – Floating Action Button), are neatly and consciously designed. Corporations such as Google and Apple provide human interface guidelines to make apps associable to the brand [40, 41]. Developers are specifically advised here, that "motion provides meaning. Motion respects and reinforces the user as the prime mover. [...] Motion is meaningful and appropriate, serving to focus attention and maintain continuity" [40]. "Fluid motion and a crisp, beautiful interface help people understand and interact with content while never competing with it" [41].

We claim that organizations can also benefit from a coherent and legible motion identity of their automatic and autonomously navigating robots. They will be recognized by many stakeholders and ultimately human-centered design of motion behavior will create an advantage over companies that base their robots' motion merely on optimizing technical efficiency. If this is systematically developed at an organization, we call this a corporate robot motion identity (CRMI). Thus, recognizable movement cues which are designed to express intent can be tailored to what an organization wants to express to its environment, customers or other stakeholders. An organization can distinguish its robots from the ones supervised by other companies through the CRMI.

3.1 Affected Service Quality Domains

Service quality can be assessed through five dimensions: reliability, assurance, tangibles, empathy, and responsiveness [9]. Here, we will elaborate how these dimensions should be understood when implementing a CRMI in an effective way.

Reliability. This dimension describes the ability to perform the service dependably and accurately [9]. For a robot's movement cues this means, they should be designed to communicate intentions and a CI reliably. Only if a cue and its intended effect works for a broad variety of people, it can have a positive effect on an organization. Additionally, pedestrians will be confused if a movement cue is executed differently each time it is observed. Design elements such as logos of companies do not change regularly either. In the same way movement cues should be reliable. The observer can then rely on the robots' affiliation and be confident about the robots' intention.

Assurance. Here, the knowledge and courtesy of employees and their ability to inspire trust and confidence are assessed [9]. Also movement cues must not lower the human trust in the system. Trusting the robot is the basis for trusting an organization. In a positive way, expressive movement cues can be used to actively increase trust. A robot yielding priority to humans and acknowledging personal space shows courtesy.

Tangibles. This is the category to integrate all activities regarding the physical appearance of a service company [9]. Traditionally, this would imply the design of logos, packaging, advertisements, and the looks of service personnel. In the context of robotic personnel, it is possible to assign the esthetics of movement cues to the domain of tangibles. Abrupt stopping, unnatural accelerations or turns might be perceived unaesthetic and unattractive by observers, thereby effecting a corporate image. The impression of movements can also communicate a marketing objective of a company to the observers. A movement cue that has rather harsh accelerations or sharp turns may be linked to an athletic culture of a company.

Empathy. This comprises caring and individualized attention, an organization provides its customers [9]. Empathy is the category of service quality where issues such as politeness are allocated. This is closely related to the way a robot's intentions are recognized. The customers or pedestrians will create an opinion about robot's politeness or intelligence. An expressive movement cue such as the back-off movement cue can represent an organization's politeness. However, a company that wants to communicate an image of strength, should consider applying back-offs rather sparsely.

Responsiveness. In order to be considered responsive, willingness to help customers and prompt service should be provided [9]. A movement cue can give this immediate feedback as opposed to rigid cost-function based guidance. As stated in Sect. 2.2, scripted movement cues can have a responsive appearance. Preprogrammed execution of a movement has a smaller time delay compared to the computational effort of cost-based control algorithms. Since also humans show immediate unconscious movement reactions as a direct response to external stimuli, an immediate human-like feedback via movement cue may support the perceived responsiveness of robots.

3.2 Affected Human Roles

In the following we discuss five different roles of humans in the context of implementing and being exposed to CRMI. The human role can be categorized into supervisor, operator, mechanic, bystander and peer [23].

Supervisor. The supervisor is responsible for the robots control mechanism, managing the overall situation of a robot. The supervisor can step in on the robot control and modify plans in long-term [23]. In the case of representing a CRMI, supervisors have to follow corporate guidelines to modify robot behavior. The supervisor may be in charge of designing and implementing corporate movement cues.

Operator. The operator intervenes in case of flawed robot control and takes over control manually or changes software in short-term [23]. This can be done from close proximity with a remote control or from an operator workstation at a remote desktop. An operator has to be familiar with the CRMI so that also a manual control expresses visual identity via consciously executed movements.

Mechanic. The mechanic changes or develops physical capabilities of a robot and hardware (e.g., in case of malfunction) [23]. In doing so, a mechanic needs to verify if the robot's mechanical parts enable the execution of CRMI (e.g., after development, modification or repair). The physical capabilities of the robot must enable the movement cues in terms of parameters like accelerations or speed. A heavy robot might have too much inertia to show a specifically designed acceleration that is part of a movement cue. Thus, movement cues that are designed according to CRMI guidelines, might be executed in the wrong way.

Peer. While Scholtz [23] referred to peers as teammates of robots (person working with the robot), we refer to peers as customers or persons receiving the respective service. We assume this is the closest interaction between robots and stakeholders. Here, often human-robot cooperation takes place [22]. These might be situations when the service robot provokes human interaction (e.g., receiving an item). Peers are exposed to the effect of movement cues. A peer is the one who will judge a robot or organization for impolite behavior. Oppositely, these peers can be surprised by a well-designed CRMI that they can repeatedly expect when dealing with a robot sent from the respective company.

Bystander. A bystander mostly coexists with a robot, yet also needs an internal mental model of robot behavior to understand the consequences of the robot's actions [23]. A robot needs movement cues in order to act legibly for example when passing bystanders or yielding priority to them in front of narrow sections. Bystanders are not the end customers (peers), however they can affect and can be affected by the organization. Affecting the organization could mean stepping in front of a robot and hindering it from its task. Consciously designed movement cues however, can also work as an advertisement to bystanders who have previously not known the respective organization.

3.3 Who Defines Requirements, Limits, and Best Practices?

The question is no longer, if robots will operate within our daily life, but rather how they should behave in our daily life. Simultaneously it is not yet clear who defines requirements, limits, and best practices for HRSI. Right now, we see several possible HRSI design and behavior drivers, who are able to shape and provide guidelines.

Leading Corporations. Corporations (developing robotic hard- and/or software) will have the possibility to use their current market leadership or use this chance to get to market leadership. Apple, for instance, shaped the way we are nowadays interacting with our mobile devices such as smartphones and tablets. This innovation and marketing driven disruption paved the way, not only for fast follower companies (e.g., Samsung, LG, etc.), but also introduced a new and, by customers perceived and marketing conveyed as "intuitive", way to interact with electronic devices.

Driver 1: Market leadership provides power to early shape interactions and therefore determines the behavior of not only their own robots.
Driver 2: Marketing innovative interaction before anyone ever interacted with the system can enable perceived intuitive interaction.

Corporations Together. As more and more infotainment systems (e.g., navigation, telecommunication, portable devices) entered the automotive domain, driver distraction (diverted attention away from the primary driving task to secondary or tertiary tasks) became a major issue to solve and consider in future developments. Automobile manufacturers decided to collaboratively design guidelines to address this challenge. The European Statement of Principles (ESoP), the Japan Automobile Manufacturers Association (JAMA), or the Alliance of Automobile Manufacturers (AAM) are examples for different corporation driven consolidations of design guidelines. In this context cultural, market, and attitude differences are visible in form of partially diverging requirements.

Driver 3: Corporation driven consolidation of robot behavior guidelines to provide generalizability and transferability.

Standardization and Regulation Bodies. International standardization organizations (e.g., ISO), government bodies (e.g., German StVO, where road traffic regulations are summarized), and universities, can provide rules and recommendations to enable legible behavior of robots.

Driver 4: Government driven consolidation of robot behavior guidelines to provide generalizability and transferability.

Users Themselves. This group will be able to provide valuable insights that have to be considered. Artificially intelligent systems will adapt to different users and pedestrians in certain scenarios and environments. Evoked via either active and explicit interaction (e.g., waving or yelling at a robot) or rather unconscious, passive, and implicit interaction (e.g., initializing a person's presence and preferences via movement and body language or even via her smartphone signals).

Driver 5: Adaptive robots perceiving, learning, and adapting via changing and fixed human behavior.

4 Conclusion

We opened the scope of this paper by talking about recent incidents with bad behavior of service personnel. With the increase of robotic systems in customer interactions, we discussed that service robots also fall into this category regarding an organization's service issues. If it is not designed consciously, inadequate behavior of robots can lead to negative reputation for an organization. Motivated by the theory of social cues we suggest movement cues as a language for mobile service robots. This includes to provide legibility of those cues. We call the whole process of systematically designing robotic motion of an organization: The development of a Corporate Robot Motion Identity. We included the service quality domains from service design theory that are affected by this approach. Additionally, we discussed the human roles within an organization and external stakeholders, that are affected by the implementation of corporate robot motion identity. The companies' staff is affected with designing the behavior or reacting appropriately when intervention is needed. The stakeholders are pedestrians and customers that coexist and cooperate with the robots. They are exposed to the strategy that a company wants to achieve with the designed robot motion. We discussed several drivers that could lead to the standards of how we will interact with robots in the future. The standards for behavior could be driven independently by corporations that are market leaders, by conglomerates of companies or by governmental institutions. Past events (e.g. the advance of smartphones) have shown that it is possible that the market leading companies will define the widely applied and accepted robot behavior. A transfer of guidelines from and to other domains such as the automotive domain seems plausible also. Manufacturers of automated cars are already embracing the task of designing the movements of their cars so that the driver and passengers are satisfied. Research highlights the necessity of motion as a way to communicate externally with pedestrians [42]. However, semantics of the body language vary with different cultures [21]. When humans utilize social cues, sometimes they differ between cultures, context and individually. The laboratory research of robotic movement cues can only give a guideline on what communicative value motion can be utilized for. Field studies are necessary to evaluate movement cues in the context of a whole service, cultural aspects, repeated interaction between customers and robots, and a longer-term exposure of customers to a service company.

References

1. Fortin J (2017) Delta Passenger Restrained After Trying to Open Exit Door, Charges Say. New York Times, 08 July 2017
2. Haag M (2017) Video Shows Airport Attendant Punching Passenger Holding a Child. New York Times, 31 July 2017
3. Victor D, Stevens, M (2017) United Airlines Passenger is Dragged From an Overbooked Flight. New York Times, 10 April 2017
4. VandenBos G (2007) APA Dictionary of Psychology, 2nd edn.
5. Balmer JMT (2001) Corporate identity, corporate branding and corporate marketing - seeing through the fog. Eur J Mark 35(3/4):248–291
6. Mager B (2009) Touchpoint. J Serv Des 1(1):20–29
7. Andreassen TW, Kristensson P, Lervik-Olsen L, Parasuraman A, McColl-Kennedy JR, Edvardsson B, Colurcio M (2016) Linking service design to value creation and service research. J Serv Manag 27(1):21–29
8. Zomerdijk LG, Voss CA (2010) Service design for experience-centric services. J Serv Res 44:1–41
9. Parasuraman LA, Zeithaml VA, Berry LL (1988) SERVQUAL: a multi-item scale for measuring consumer perceptions of the service quality. J Retail 64(1):12–40
10. Hall ET (1969) The hidden dimension. Doubleday & Company, Inc. Anchor Books, Garden City, New York
11. Hancock PA, Billings DR, Schaefer KE (2011) Can you trust your robot? Ergon Des 19(3):24–29
12. International Federation of Robotics (2017) Executive summary world robotics 2017 service robots. In: World robotic report - executive summary, pp 12–19
13. International Federation of Robotics (2016) World Robotics Report 2016 Service Robots, VDMA, 2016, pp 9–12
14. DIN EN ISO 8373:2012 - Robots and robotic devices - Vocabulary, no. November 2010 (2011)
15. Goodrich MA, Schultz AC (2007) Human-robot interaction: a survey. Found Trends Hum Comput Interact 1(3):203–275
16. Ju W (2015) The design of implicit interactions. Synth Lect Hum Centered Inf 8(2):1–93
17. Schubö A, Vesper C, Wiesbeck M, Stork S (2007) Movement coordination in applied human-human and human-robot interaction. In: HCI and usability for medicine and health care, pp 143–154
18. Lauckner M, Kobiela F, Manzey D (2014) 'Hey robot, please step back!'- exploration of a spatial threshold of comfort for human-mechanoid spatial interaction in a hallway scenario. In: IEEE RO-MAN, pp 780–787
19. Dragan AD, Lee KCT, Srinivasa SS (2013) Legibility and predictability of robot motion. In: IEEE/ACM conference on human-robot interaction

20. Lichtenthäler C, Kirsch A (2016) Legibility of robot behavior: a literature review. HAL-archives <hal-01306977>
21. Rios-Martinez J, Spalanzani A, Laugier C (2015) From proxemics theory to socially-aware navigation: a survey. Int J Soc Robot 7(2):137–153
22. Schmidtler J, Knott V, Hölzel C, Bengler K (2015) Human centered assistance applications for the working environment of the future. Occup Ergon
23. Scholtz J (2003) Theory and evaluation of human robot interactions. In: Proceedings of the 36th annual hawaii international conference on system sciences 2003, vol 3, p 10
24. Hancock PA (2017) Imposing limits on autonomous systems. Ergonomics 60(2):284–291
25. Bartneck C, Forlizzi J (2004) A design-centred framework for social human-robot interaction. RO-MAN 2004:591–594
26. Barraquand R, De Europe A, Ismier S, Crowley JL (2008) Learning polite behavior with situation models. In: ACM/IEEE international conference on human-robot interact, pp 209–216
27. Mehu M, Scherer KR (2012) A psycho-ethological approach to social signal processing. Cogn Process 13(2):397–414
28. Moon A, Parker CAC, Croft EA, Van Der Loos HFM (2011) Did you see it hesitate empirically grounded design of hesitation trajectories for collaborative robots. IEEE/RSJ Intell Robots Syst (IROS) 2011:3–8
29. Vinciarelli A, Pantic M, Bourlard H, Pentland A (2008) Social signals, their function, and automatic analysis: a survey. In: Proceedings of the 10th international conference on multimodal interfaces - IMCI 2008, pp 61–68
30. Reinhardt J, Pereira A, Beckert D, Bengler K (2017) Dominance and movement cues of robot motion: a user study on trust and predictability. In: IEEE International Conference on Systems, Man, and Cybernetics 2017, pp 1493–1498
31. Hogan K, Stubbs R (2003) Can't get through: eight barriers to communication. Pelican Publishing, Gretna
32. Rosenbaum D (2009) Human motor control. Academic Press, New York
33. Kruse T, Kirsch A, Khambhaita H, Alami R (2014) Evaluating directional cost models in navigation. In: Proceedings of the 2014 ACM/IEEE international conference on human-robot interaction - HRI 2014, pp 350–357
34. Kitazawa K, Fujiyama T (2010) Pedestrian vision and collision avoidance behaviour: investigation of the information process space of pedestrians using an eye tracker. Pedestr Evacuation Dyn 2008:95–108
35. Knight H, Thielstrom R, Simmons R (2016) Expressive path shape (swagger): simple features that illustrate a robot's attitude towards its goal in real time. In: IEEE/RSJ International Conference on Intelligent Robots and Systems (IROS), pp 1475–1482
36. Reinhardt J, Schmidtler J, Körber M, Bengler K (2016) Follow me! wie roboter menschen führen sollen. Z Arbeitswiss 70(4):203–210
37. Garrell A, Sanfeliu A (2010) Model validation: robot behavior in people guidance mission using dtm model and estimation of human motion behavior. In: IEEE RSJ International Conference on Intelligent Robots and Systems, pp 18–22
38. Müller J, Stachniss C, Arras KO, Burgard W (2008) Socially inspired motion planning for mobile robots in populated environments. In: International Conference on Cognitive Systems (CogSys), 2008 January, pp 85–90
39. Fink J, Lemaignan S, Dillenbourg P, Rétornaz P, Vaussard FC, Berthoud A, Mondada F, Wille F, Franinovic K (2014) Which robot behavior can motivate children to tidy up their toys? design and evaluation of 'Ranger'. In: ACM/IEEE international conference on human-robot interaction 2014, pp 439–446
40. Google (2018) Google material design. https://material.io/guidelines/#introduction-principles

41. Apple (2018) Human Interface Guidelines. https://developer.apple.com/ios/human-interface-guidelines/overview/themes/
42. Fuest T, Sorokin L, Bellem H, Bengler K (2018) Taxonomy of traffic situations for the interaction between automated vehicles and human road users. In: Advances in human aspects of transportation AHFE 2017

Identifying the Effects of Visual Searching by Railway Drivers upon the Recognition of Extraordinary Events

Daisuke Suzuki[1](✉), Kana Yamauchi[1], and Satoru Matsuura[2]

[1] Railway Technical Research Institute, 2-8-38 Hikari-cho, Kokubunji-shi, Tokyo, Japan
suzuki.daisuke.55@rtri.or.jp

[2] Hokkaido Railway Company, 1-1 Kita11, Nishi15, Chuo-ku, Sapporo, Japan

Abstract. The purpose of this study is to investigate effective visual-searching behaviours for recognising extraordinary events based on the eye movements of railway drivers. 121 railway-company drivers participated in our study using a driving simulator. An eye tracker equipped with the simulator measured the drivers' eye movements. The given driving scenario was a multi-task scenario in which the main task was to stop the simulated train before a ground-device malfunction. The important sub-task was to recognise an extraordinary event, in this case, the subsidence of a railway track to their right. Participants who braked before passing the subsidence were identified as part of the recognising group; those who did not brake until after passing the subsidence were identified as part of the non-recognising group. Logistic-regression analysis was conducted, with the driver's group as the objective variable. The explanatory variables were the means and standard deviations of gaze duration and horizontal and vertical visual angles, as well as the driver's age and duration of driving experience. The variables used to improve the possibility of recognising subsidence were the standard deviation of the gaze duration, the means of the horizontal and vertical visual angles and the driver's age. The standard deviation of the gaze duration had the largest influence among these four variables.

Keywords: Visual searching · Railway driver · Recognition
Extraordinary event

1 Introduction

Railway drivers are required to cope with various extraordinary events such as vehicle problems, signal troubles and ground-device malfunctions. Railway companies often conduct vocational-training sessions using a driving simulator to improve drivers' skills in coping with extraordinary events [1]. When focusing on the manner in which drivers recognise such extraordinary events, visual searching is crucial.

There have been several studies about the visual-searching practices of railway drivers. Groeger et al. [2] investigated the visual behaviour of 10 train drivers and showed that approximately 50% of the time spent approaching signals was used to scan the visual scene. The remaining time was spent fixating on railway signage and

S. Bagnara et al. (Eds.): IEA 2018, AISC 823, pp. 165–173, 2019.
https://doi.org/10.1007/978-3-319-96074-6_17

infrastructure, locations beside the track and signals. Luke et al. [3] analysed the visual behaviour of 86 drivers whilst operating in-service trains, revealing that the signal aspect, which is colour of the signal, preceding aspect, which is colour of the next signal, signal type and signal complexity are important factors affecting this behaviour. Naweed et al. [4] examined the tasks and activities of urban passenger-train drivers during daily railway driving in order to understand the nature of the visual demand in their task activities. Their study showed that railway driving under the urban environment requires a mastery of key visual and technical driving skills. Although there have been studies on visual searching during ordinary railway-driving situations, few have examined the relationship between visual searching and recognition of extraordinary events.

In the field of automobile research, there have been several studies attempting to correlate visual searching and driver skill levels. Mourant et al. [5] investigated differences in visual searching by novice and experienced drivers. According to their results, novice drivers frequently sampled the curb in order to verify or estimate vehicle-lane alignment, whereas experienced drivers looked farther than novices during neighbourhood driving. Underwood et al. [6] identified the fixation-sequence modes of drivers with different levels of experience to answer the question of whether their different accident liabilities can be associated not only with their distribution of attention but also with the subject of their attention. Differences in sequences of fixations were found between novice and experienced drivers on the three types of roads (rural, suburban and dual-carriageway), with experienced drivers showing greater sensitivity overall and with some stereotypical transition in the visual attention of the novices. Konstantopoulos et al. [7] focused upon experiential differences in visual attention. The results showed that driving instructors had an increased sampling rate, shorter processing time and broader scanning of the road than did learner drivers. This broader scanning of the road may be explained by the mirror-inspection pattern, which revealed that driving instructors fixated more on side mirrors than did learners.

In this way, these studies indicated that experiential differences led the differences of drivers' visual-searching behaviour. It is important to examine the relationship between such behaviour and the recognition of extraordinary events. Therefore, the purpose of this study is to investigate effective visual-searching behaviour in terms of recognising extraordinary events based on the eye movements of railway drivers.

2 Method

2.1 Apparatus

We analysed simulator-training data for actual drivers. Figure 1 shows a railway-driving simulator for vocational training made by Mitsubishi Precision Company, Limited. An eye tracker equipped with the simulator measured the drivers' eye movement. Calibration was conducted before the measurement. The sampling rate was 30 Hz. The front view appeared on a 42-in. display whose width was 930 mm and height was 520 mm. The distance between the display and the driver's eye point was 900 mm. The visual angle of the display was 27.32° and 16.11°.

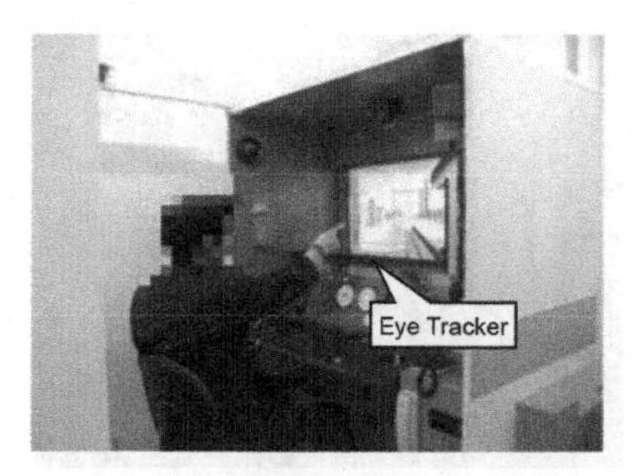

Fig. 1. A railway-driving simulator for vocational training

2.2 Participants

Overall, 121 drivers of a railway company participated in our study; the ages of these participants ranged from 23 to 59 years (mean was 41 years and standard deviation was 11). The driving experience of the participants ranged from 1 to 33 years (mean was 16 and standard deviation was 8). They participated in our study as periodical training once every six months.

2.3 Driving Scenario

A multi-task driving scenario was given, with the main task being to stop the simulated train before a ground-device malfunction (Fig. 2). The participants were provided with an operating statement of ground-device malfunction before station B at station A. In this task, participants were required to focus on distance posts on their left to identify the exact location at which to stop. The important sub-task was to recognise an extraordinary event, namely the subsidence of the railway track on their right (Fig. 3). Participants were required to brake if they recognised this subsidence. In addition, the participants continued driving up to the location of ground-device malfunction.

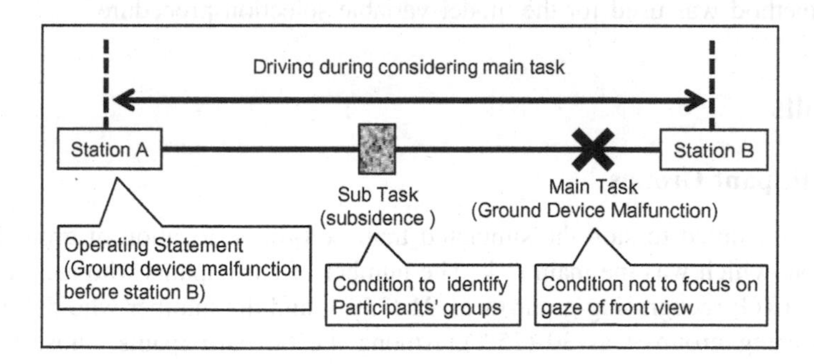

Fig. 2. Outline of the driving scenario

Fig. 3. Appearance of subsidence

Participants who braked before passing the subsidence were identified as part of the recognising group (RG), and those who did not brake until after passing the subsidence were identified as part of the non-recognising group (NG).

2.4 Analysis Item

Data related to eye movements were recorded using a non-contact eye tracker incorporated into our driving simulator. Data were classified as a 'gaze' if four or more subsequent frames (0.133 s) were observed with the same point of regard. The gaze was extracted using the Sight Tracker Editor software made by Emovis Corporation.

Gaze analysis began approximately 30 s before the subsidence, where the participants had not yet recognised it. The gazes of 66 participants whose eye movements were detected accurately by the eye tracker were analysed.

Next, we conducted logistic-regression analysis to identify the effects of visual searching in the driving simulator upon the recognition of extraordinary events. The objective variable was the participant group, and the explanatory variables were the means and the standard deviations of the gaze duration and the horizontal and vertical visual angles, as well as the driver's age and duration of driving experience. The stepwise method was used for the model variable selection procedure.

3 Results

3.1 Participant Groups

No participant failed to stop the simulated train before the location of ground-device malfunction, which was the main task. The number of participants who recognised the subsidence (RG: recognising group) was 91 (75%), and the number who did not (NG: non-recognising group) was 30 (25%). Among the 66 participants whose gaze was analysed, the number in the RG was 47 (71%), and that in the NG was 19 (29%).

3.2 Logistic-Regression Analysis

Table 1 shows the list of objective and explanatory variables used in the logistic-regression analysis.

Table 1. List of objective and explanatory variables.

No	Objective variable	Explanatory variables							
	Participant group	Mean of the gaze duration (0.1 s)	Standard deviation of the gaze duration (0.1 s)	Mean of the horizontal visual angle (deg)	Standard deviation of the horizontal visual angle (deg)	Mean of the vertical visual angle (deg)	Standard deviation of the vertical visual angle (deg)	Age (years old)	Driving experience (years)
1	RG	7.55	6.35	0.73	6.35	−4.48	3.02	58	23
2	RG	3.28	4.73	2.05	2.45	−5.85	7.92	58	33
3	NG	5.33	4.35	−3.64	4.04	−4.53	2.19	59	31
4	NG	12.70	13.53	−3.08	2.16	−2.61	1.13	57	25
5	RG	11.22	10.94	−0.53	3.71	−5.31	1.29	57	23
6	NG	4.87	3.59	−2.71	2.63	−5.74	1.85	57	32
7	NG	3.83	3.01	−4.56	2.56	−1.90	1.75	57	25
8	RG	5.27	5.32	−1.47	4.61	−3.64	2.07	57	23
9	RG	11.34	15.08	−5.75	2.69	−2.57	3.00	56	31
10	RG	5.98	8.44	2.10	2.86	−6.76	2.47	55	30
11	RG	6.53	7.15	−1.92	3.44	−3.70	1.01	56	23
12	NG	4.10	3.61	−4.37	3.94	−4.37	1.74	55	23
13	RG	5.71	8.09	0.87	3.71	−6.33	1.85	54	30
14	RG	55.59	92.19	−2.29	1.47	−3.73	1.69	53	21
15	RG	7.14	8.34	0.94	1.97	−1.05	1.32	52	23
16	NG	8.30	9.78	−3.23	7.56	−3.44	3.82	57	30
17	NG	3.66	2.75	−5.12	2.47	1.47	3.12	54	16
18	NG	4.46	2.97	−7.51	4.01	−1.61	1.42	53	25
19	RG	6.57	6.24	−0.51	2.85	−1.98	2.31	53	24
20	NG	5.11	4.42	−0.97	3.05	−5.39	2.83	52	20
21	RG	10.21	19.27	−2.01	3.63	−7.48	4.42	54	24
22	RG	6.43	8.15	−2.23	2.97	−3.73	3.19	52	23
23	RG	11.38	15.07	0.24	3.17	−3.71	1.18	42	20
24	RG	4.77	3.32	−0.92	2.14	−3.49	1.90	44	17
25	NG	9.65	9.94	−2.74	3.67	−3.64	1.62	41	17
26	RG	6.77	5.72	−8.00	1.38	−1.35	0.98	41	19
27	RG	10.18	9.03	0.30	1.02	−2.37	0.94	41	19
28	RG	7.03	7.88	−1.13	3.26	−4.27	2.06	39	18
29	RG	6.12	5.38	−5.47	2.78	0.03	1.82	39	17
30	RG	7.24	7.01	2.59	2.12	−3.49	1.45	38	15
31	RG	8.54	10.07	−2.47	5.58	−2.70	2.09	38	16
32	NG	6.38	7.99	−0.92	3.61	−5.50	2.49	37	16
33	RG	13.68	15.50	−1.32	1.71	−3.72	0.72	37	5
34	RG	8.50	7.36	−2.85	4.04	−4.50	0.86	37	15
35	NG	5.81	3.71	−3.15	3.99	−2.33	1.43	39	14
36	RG	6.58	7.37	−5.09	3.42	−3.26	3.74	38	16

(*continued*)

Table 1. (*continued*)

No	Objective variable	Explanatory variables							
	Participant group	Mean of the gaze duration (0.1 s)	Standard deviation of the gaze duration (0.1 s)	Mean of the horizontal visual angle (deg)	Standard deviation of the horizontal visual angle (deg)	Mean of the vertical visual angle (deg)	Standard deviation of the vertical visual angle (deg)	Age (years old)	Driving experience (years)
37	RG	10.97	10.48	−5.47	2.40	−1.57	1.07	36	11
38	RG	17.68	27.65	−3.59	6.00	1.32	1.97	38	16
39	RG	14.46	22.59	−4.18	4.79	−1.48	1.26	37	15
40	NG	8.58	4.45	−2.09	2.03	−3.89	1.95	37	15
41	NG	6.27	8.87	−0.99	2.50	−4.90	1.55	36	12
42	NG	6.27	4.82	−1.98	3.33	−6.04	2.26	35	12
43	RG	5.69	4.11	−3.73	5.09	−4.31	1.87	35	14
44	RG	9.85	13.40	0.01	3.81	−3.71	1.60	35	8
45	RG	5.67	6.35	−1.12	2.24	−4.89	3.37	35	12
46	RG	13.03	16.98	−3.07	3.13	−3.97	0.76	41	15
47	RG	9.96	11.68	−1.43	4.24	−2.62	2.83	37	15
48	NG	6.16	4.80	−9.22	6.56	−3.30	3.05	36	15
49	NG	6.42	4.34	−8.39	3.21	0.04	2.01	34	6
50	RG	8.50	8.49	−2.92	4.79	−2.60	1.80	33	9
51	NG	9.06	7.31	−3.37	2.88	−4.97	1.38	34	10
52	RG	18.17	20.63	−0.60	1.28	−3.32	1.19	33	8
53	RG	21.12	53.08	−3.01	1.23	−4.44	1.79	33	11
54	RG	7.26	9.06	−0.67	2.90	1.42	0.90	32	11
55	RG	7.12	6.40	−0.79	3.56	−6.25	1.34	32	6
56	RG	6.00	5.24	−1.49	4.57	−0.76	1.97	30	5
57	RG	4.58	3.02	−7.67	3.87	−0.30	1.97	30	7
58	RG	20.51	32.00	−3.07	3.47	−1.54	1.60	33	6
59	NG	6.25	4.65	−1.07	3.47	−5.08	1.93	36	9
60	RG	7.89	8.03	0.22	2.40	−2.89	1.27	34	9
61	RG	8.64	7.68	−3.39	4.90	−1.35	2.07	27	2
62	RG	11.23	11.46	−1.78	4.19	−2.70	1.88	29	7
63	RG	8.45	7.66	−4.89	4.38	−1.83	1.80	24	1
64	RG	7.80	7.73	−3.87	3.75	−3.46	1.65	23	1
65	RG	3.90	2.82	−3.49	3.55	−2.41	3.33	23	1
66	RG	10.39	12.92	−2.38	2.70	−7.50	3.47	26	2

Variance-information factors (VIFs) were calculated to evaluate multicollinearity. It is said that there is a possibility of multicollinearity if the maximum VIF is above 10 or the average VIF is considerably more than 1 [8]. The maximum VIF among the explanatory variables of this study was 1.27. We judged that there is no problem of multicollinearity.

Table 2 shows the results of the logistic-regression analysis. The odds ratio revealed that recognition of subsidence was significantly associated with the standard deviation of the gaze duration, the mean horizontal and vertical visual angles and the driver's age. The following findings were obtained from the odds rate.

Table 2. Results of logistic-regression analysis.

Explanatory variables	Partial regression coefficient	Standardized partial regression coefficient	Odds ratio	95% confidence interval of the odds ratio		*p* value	*: *p* < 0.05
				Lower limit	Upper limit		
Mean of the gaze duration (0.1 s)	−0.53	−3.64	0.59	0.31	1.14	0.11	
Standard deviation of the gaze duration (0.1 s)	0.60	7.74	1.83	1.12	3.00	0.02	*
Mean of the horizontal usual angle (deg)	0.51	1.24	1.66	1.09	2.51	0.02	*
Mean of the vertical visual angle (deg)	0.57	1.13	1.77	1.05	2.98	0.03	*
Driver age (years old)	−0.09	−0.92	0.92	0.85	0.99	0.02	*
Constant	7.41		1659.81	6.41	429850.30	0.01	*

- With all other variables being constant, the possibility of recognising subsidence increases by 1.83 times when the standard deviation of the gaze duration increases by 0.1 s.
- When all other variables are constant, the possibility of recognising the subsidence increases by 1.66 times when the mean horizontal visual angle increases by 1°.
- When all other variables are constant, the possibility of recognising the subsidence increases by 1.77 times when the mean vertical visual angle increases by 1°.
- When all other variables are constant, the possibility of recognising subsidence changes by 0.92 times when driver's age increases by 1 year.

The standardised partial-regression coefficient revealed that the standard deviation of the gaze duration has the largest influence among the four variables.

3.3 Relationship Between Horizontal Visual Angle and Gaze Duration

Figure 4 shows the relationship between horizontal visual angle and gaze duration for each participant from both groups in order to understand the actual gaze behaviour. The horizontal axis of the scatter diagram represents the visual angle, and the vertical axis represents the gaze duration. Each plot represents gaze points. Both groups looked in front of them for a long time and looked to their right/left for a short time. With regards

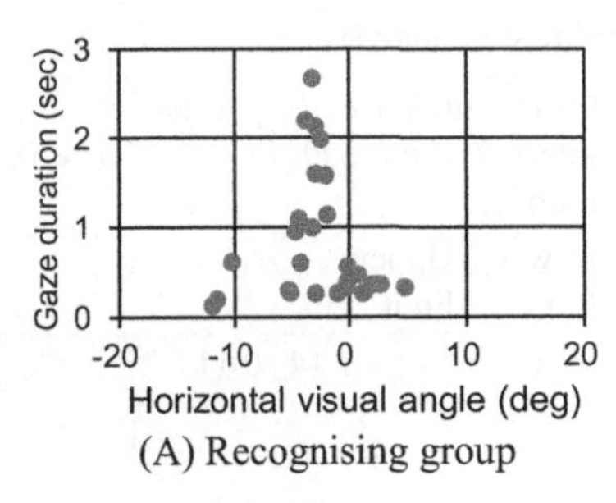

(A) Recognising group

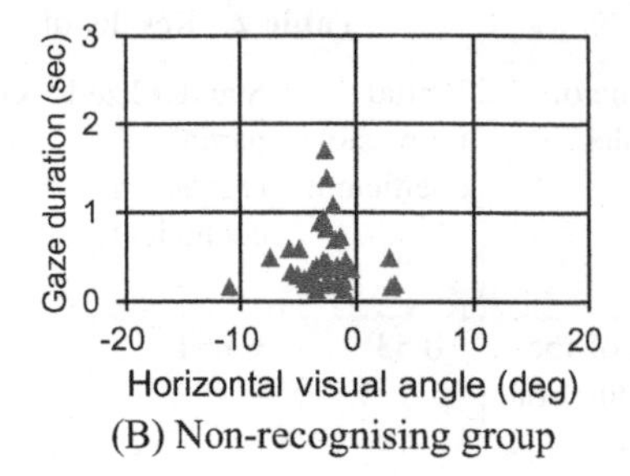

(B) Non-recognising group

Fig. 4. Relationship between horizontal visual angle and gaze duration

to the characteristics of the RG, there was a tendency to mix long and short gazes. On the other hand, there were few gazes by the NG that lasted more than 2 s.

4 Discussion

The results of the logistic-regression analysis revealed that the standard deviation of the gaze duration, mean horizontal visual angle and mean vertical visual angle affected the recognition of the subsidence. With regards to gaze duration, drivers with larger dispersion in the individual were usually able to recognise the subsidence. For gaze points (Fig. 4), there was a tendency for long gazes to exist towards the centre and short gazes to exist on the right/left sides. It is thought that drivers gazed for a long time to search for extraordinary events in front of them and for a short time on their right/left to confirm distance posts, railway signage and traffic signals. From the findings, it is possible that mixing long and short gazes enables drivers to simultaneously recognise the subsidence far away and confirm the distance posts nearby.

With regard to gaze position, drivers with a tendency to look to their lower left were usually unable to recognise the subsidence. More specifically, overlooking the subsidence could be attributed to excessively concentrating on the distance posts (which existed on the left) because the subsidence existed on the right in the driving scenario of this study. The drivers needed to look at the distance posts in order to cope with the ground-device malfunction.

From the above findings, it is thought that excessively concentrating on an existing extraordinary event can cause drivers to overlook a potential secondary extraordinary event. Effective visual-searching behaviour for recognising extraordinary events is to mix long and short gazes to search in front and confirm features to the right and left.

Finally, older drivers were usually unable to recognise the subsidence probably because of gradual weakening of their visual performance. Further study is required as this study did not quantitatively measure visual-performance metrics such as peripheral-vision area.

5 Conclusion

In this study, logistic-regression analysis was conducted to identify the effects of visual searching by railway drivers upon the recognition of extraordinary events. Drivers were characterised according to whether they recognised or did not recognise rail subsidence, with this categorisation constituting an objective variable. The explanatory variables were the means and standard deviations of gaze duration and horizontal and vertical visual angles, as well as the driver's age and duration of driving experience. The following results were revealed.

- The variables that improved the possibility of extraordinary event recognition were the standard deviation of the gaze duration, the mean of the horizontal visual angle, the mean of the vertical visual angle and the driver's age. The standard deviation of the gaze duration had the largest influence among these four variables.
- Drivers with larger individual gaze dispersion were usually able to recognise the subsidence. From the findings, it is possible that mixing long and short gazes enables drivers to simultaneously recognise subsidence far away and to confirm distance posts nearby.
- Overlooking the subsidence could be attributed to excessively concentrating on the distance posts, which existed to the driver's left, whereas the subsidence occurred to their right.

References

1. Endoh H, Omino K (2013) Practical system for implementing vocational training program for improving train driver skills for coping with abnormal situations. Q Rep RTRI 54:237–242
2. Groeger JA, Bradshaw MF, Everatt J, Merat N, Field D (2003) Pilot study of train drivers' eye-movements. University of Survey Technical Report for Rail Safety and Standards Board, London
3. Luke T, Brook-Carter N, Parkes AM, Grimes E, Mills A (2006) An investigation of train driver visual strategies. Cogn Technol Work 8:15–29
4. Naweed A, Balakrishnan G (2014) Understanding the visual skills and strategies of train drivers in the urban rail environment. Work 47:339–352
5. Mourant RR, Rockwell TH (1972) Strategies of visual search by novice and experienced drivers. Hum Factors 14:325–335
6. Underwood G, Chapman P, Brocklehurst N, Underwood J, Crundall D (2003) Visual attention while driving: sequences of eye fixations made by experienced and novice drivers. Ergonomics 46:629–646
7. Konstantopoulos P, Chapman P, Crundall D (2010) Driver's visual attention as a function of driving experience and visibility. Accid Anal Prev 42:827–834
8. Netter J, Wasserman W, Kutner MH (1983) Multicollinearity, influential observations, and other topics in regression analysis-II. In: Applied linear regression models. Richard D Irwin Inc., Homewood, pp 377–416

Rally Driver's Eye Movements When Driving the Corner on Gravel Road – Differences Between World Rally Championship and National Championship Drivers

Tomoyasu Hariyama(✉) and Takaaki Kato

Keio University, Fujisawa, Kanagawa, Japan
htomo@sfc.keio.ac.jp
http://hpl.sfc.keio.ac.jp

Abstract. We focus on rally drivers' eye movements and behaviors when they are driving around a corner on a gravel road in this research. Recent studies of visual strategy regarding visual control in driving have been conducted in a driving simulator and on a racing course or paved roads. However, this study was conducted in the real environment of gravel roads and with real vehicles. In this study, the difference between the accomplished and intermediate rally driver's eye movements and fixation points were defined. The study's purpose was to understand how expert drivers move their eyes while driving around a corner on a gravel road. Twelve subjects, who were rally drivers in Japan, participated in the experiment; the accomplished driver had won the World Rally Championship (WRC) and was the fast rally driver in Japan. Each subject's eye movement was measured with mobile eye tracker (Tobii Pro Glasses 2, Tobii, Sweden) and operated at 50 Hz. The eye tracker recordings were analyzed using Frame-By-Frame analysis. This study suggests that the eye movements of accomplished drivers allow them to control the car safely and be conscious of oncoming corners. Furthermore, by looking and thinking ahead like accomplished drivers, the cognitive load associated with driving is likely to decrease; furthermore, it may serve to activate their memories of previous rallies.

Keywords: Gravel · Eye movement · Driving · Rally · Gaze behavior

1 Introduction

Many researches on automobile drivers have been conducted so far. Driver's perception in driving is said to be visual information over 90% [1]. When driving a curve, it is said that there is a close relationship between the movement of the line of sight, the head and the movement of the steering operation [2].

The skilled driver gazes at the top of the curve about 2 s before driving the curve. Then, when driving the curve, keep a close eye on the TP, and watch the OP after passing the apex of the curve. In this way, it is said that the skilled driver tends to see far away to acquire the previous information. On the other hand, non-proficient drivers tended to pay close attention to the car's present (NP) and found that they are driving

S. Bagnara et al. (Eds.): IEA 2018, AISC 823, pp. 174–178, 2019.
https://doi.org/10.1007/978-3-319-96074-6_18

from the car with relatively close information. In addition, there are few cases where we conducted gaze research under motor sports conditions [3, 4, 6, 7].

In the circuit driver, after moving the line of sight to the TP, it turned out that the steering wheel and the head were moving in a form following it [5].

This research searched that how rally drivers search the curve visually and how they behave under unpaved road environment. It is very difficult to consider safely in unpaved environments and to carry out without accidents. Also, in unpaved roads (gravel and snowy roads), many accidents have occurred in recent years, driving ordinary drivers is dangerous in conducting experiments. Therefore, in consideration of accidents and safety, subjects of subjects participated in twelve top Japanese drivers and occupied unpaved road circuit.

2 Method

2.1 Participants and Experimental-Method

The number of all subjects was twelve. Everyone has experiences of gravel rally and has participated in the domestic championship. The subjects were limited to men. One subject has played in WRC, and he has won the championship and became a series champion in the past.

The gravel test road was about 3.2 km with no lane markings. The course was announced on that day and asked everyone to practice before the experiment for few times. The experiment was done only once for each driver. After completing the experiment, we interviewed each subject for few questions. Each subject's gaze was measured with mobile eye tracker (Tobii Pro Glasses 2, Tobii, Sweden) and operated at 50 Hz. The eye tracker recordings were analyzed using Frame-By-Frame analysis. The gaze ratios in each range were calculated and compared.

2.2 Data Analyses

The analysis range was roughly divided into two methods (see Fig. 1).

Fig. 1. The analysis section was set to after apex from the turn-in section of the curve (Fig. 1. left). And we separate the road in 7 areas [Red area: path edge (outside), Yellow area: road (outside), Green area: road, Blue area: road (inside), Purple area: path edge (inside), Orange area: OP, Black area: other] (Fig. 1. right). (Color figure online)

3 Result

3.1 Frame-by-Frame Analysis

The frame by frame analysis differences between the WRC driver and national championship drivers are as bellow (see Fig. 2).

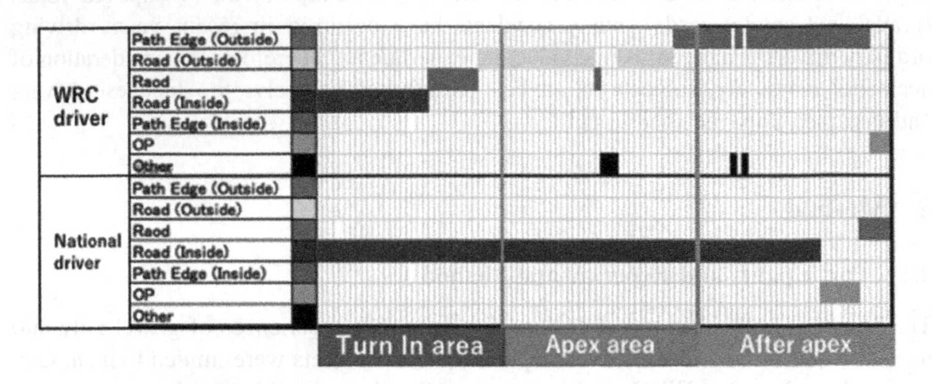

Fig. 2. From the turn-in section to the after apex, the national driver gazed at the road (inside) and after the second half of the after apex tended to watch the exit of the curve.

On the other hand, the WRC driver was gazing at various ranges. In the turn-in section we are closely watching the inside of the curve, but as the car slides, he was watching in the order of the center of the road, outside of the road, path edge (outside).

3.2 Gaze Proportion of 7 Areas

We summarized the gaze movement distribution range of national championship drivers and WRC drivers during curve running (see Fig. 3).

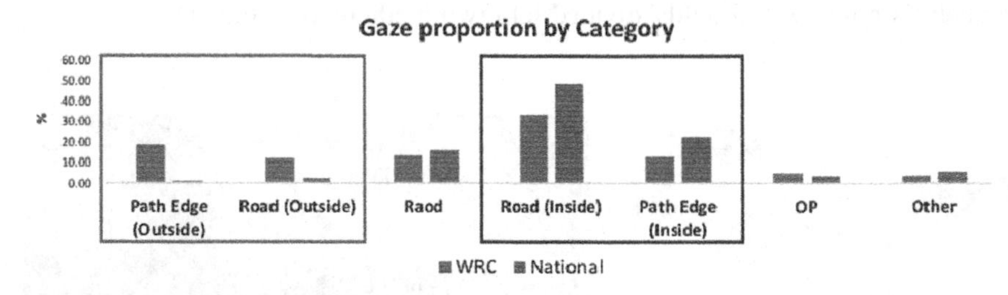

Fig. 3. We found that the WRC driver is gazing at the path edge (outside) and road (outside). On the other hand, national championship driver gazes that the road (inside) and path edge (inside).

3.3 Mean and Standard Deviation of the Horizontal Viewing Angle Velocity

We summarized the horizontal line-of-sight travel speed average and standard deviation of national championship driver and WRC driver (see Fig. 4).

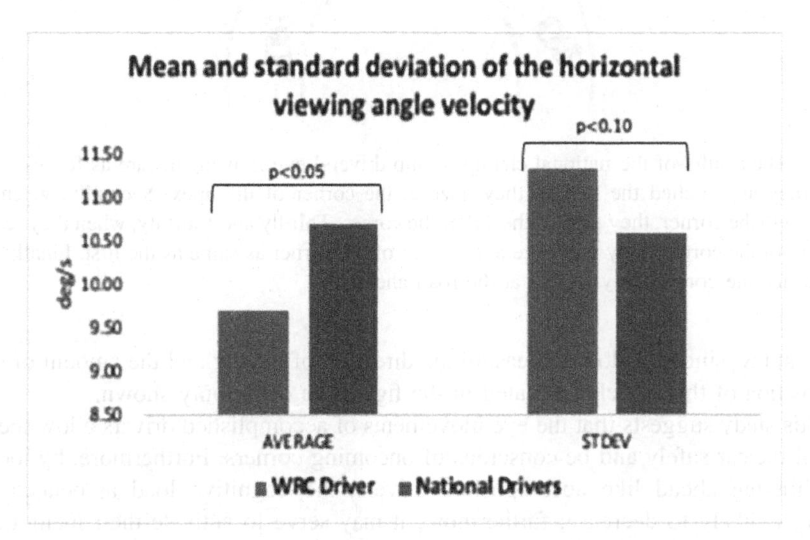

Fig. 4. There is a difference in the average value of the line of sight data on the horizontal axis (X axis), Indicating that there is a significant difference. To investigate whether there is a difference in the standard deviation of the horizontal axis (X axis) of the line of sight movement.

4 Conclusion

These results suggest that the way of use of gaze is different between WRC driver and national championship driver in unpaved road. The gaze strategy of the WRC driver is summarized in Fig. 5, and the gaze strategy of national championship driver is Fig. 6.

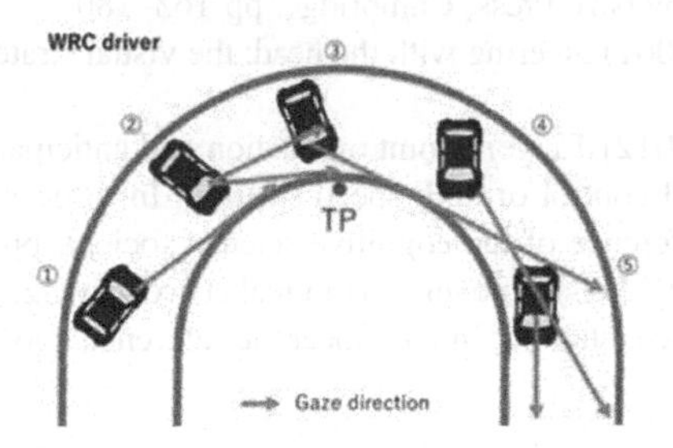

Fig. 5. The results of the WRC driver gaze behavior are as follows. First, when he approached a corner, he set his line of sight at the apex of the corner. Second, when he turned the corner, he moved his eyes in two ways. One way is he gaze at the road and checked whether there were any fallen rocks or obstacles on the road. The other way is he gazes at the apex of the corner. Third, when he reached the apex of the corner, he moved his eyes outside of the road. Fourthly, after turning the corner, he set his line of sight at the next corner or road ahead. Finally, after they turned the corner, they looked at the road ahead.

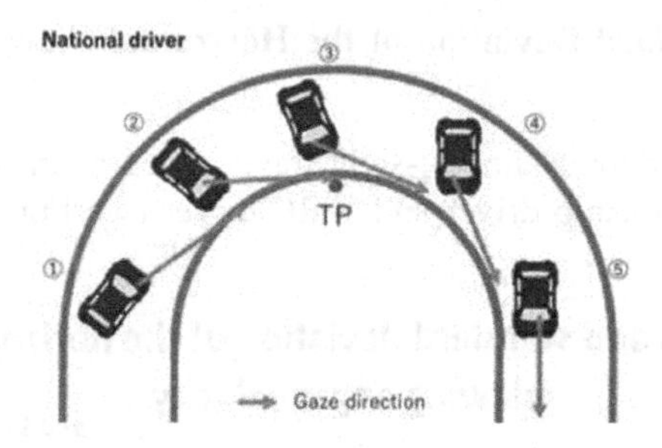

Fig. 6. The results of the national championship drivers' eye movements are as follows: First, when they approached the corner, they gaze at the corner of the apex. Secondly, when they turned in to the corner, they gaze at the TP of the corner. Thirdly and fourthly, when they reached the apex of the corner, they will gaze at the apex of the corner as same as the first. Finally, after they turned the corner, they looked at the road ahead.

In this study, since we do not measure the direction of the car and the amount of slide, the position of the vehicle indicated in the figure are commonly shown.

This study suggests that the eye movements of accomplished drivers allow them to control the car safely and be conscious of oncoming corners. Furthermore, by looking and thinking ahead like accomplished drivers, the cognitive load associated with driving is likely to decrease; furthermore, it may serve to activate their memories of previous rallies.

References

1. Hartman E (1970) Driver vision requirements, SAE Technical Paper Series, Hillsdale, 629–630
2. Land MF (1992) Predictable eye-head coordination during driving. Nature 359:318–320
3. Land MF, Lee DN (1994) Where we look when we steer. Nature 369(30):742–744
4. Land MF (1998) The visual control of steering. In: Harris LR, Jenkin M (eds) Vision and Action. Cambridge University Press, Cambridge, pp 163–180
5. Land MF, Tatler BW (2001) Steering with the head: the visual strategy of a racingdriver. Curr Biol 11(15):1215–1220
6. Lappi O, Lehtonen E (2012) Tangent point orientation and anticipated trajectory curvature - a field study on the visual control of high speed steering. In: Proceedings of CogSci2012, the thirtyfourth annual conference of the cognitive science society, pp 1846–1851
7. Lappi O, Lehtonen E (2013) Eye-movements in real curve driving: pursuit-like optokinesis in vehicle frame of reference, stability in an allocentric reference coordinate system. J Eye Mov Res 6(1):1–13 (4)

Automated Driving: The Potential of Non-driving-Related Tasks to Manage Driver Drowsiness

Veronika Weinbeer[1,2], Tobias Muhr[1(✉)], and Klaus Bengler[2(✉)]

[1] AUDI AG, Ingolstadt, Germany
{veronika.weinbeer,tobias.muhr}@audi.de
[2] Technical University of Munich, Munich, Germany
bengler@tum.de

Abstract. This study investigated the reactivation potential of non-driving-related tasks during a simulated automated drive. In total, 71 participants took part in this experiment. After a relaxation phase, the sample was divided into three groups that were given different non-driving-related tasks (a dictation, a sport activity and a relaxation task). In this study, a rating greater than 7 on the Karolinska-Sleepiness Scale (KSS) was considered the system limit "Driver Drowsiness". It was found that targeted use of non-driving-related tasks has potential as a suitable option for managing driver drowsiness. As no participant of the Dictation or Sport activity group exceeded level 7 on the KSS after the reactivation phase. Even after the effectiveness phase, there was still a major difference between the number of participants exceeding level 7 between the Dictation and Sport activity group compared to the Relaxation group. Results of this study should be considered for the design of human-vehicle interactions for Automated Driving Systems.

Keywords: Automated Driving · Drowsiness · Sleepiness
Non-driving-related tasks · Reactivation potential · Effectiveness
Drowsiness management

1 Introduction

1.1 Driver Drowsiness and Automated Driving

This article defines the term "drowsiness" as "a transitional state between wakefulness and sleep" [1]. According to Johns [2], in this state periods occur that are accompanied by "a lack of awareness". Further, drowsiness can be distinguished from fatigue by its "fluctuating nature". On the Karolinska-Sleepiness Scale (KSS), which ranges from 1 ("extremely alert") to 9 ("extremely sleepy - fighting sleep") [3], KSS levels greater than 7 may be allocated to the state "drowsiness" [4]. Also, one study found an exponential connection between drowsiness and deterioration of performance [2]. The occurrence of omission errors and of the incident and accident risk increased slightly at KSS 7 and strongly when KSS ratings were greater than 7 [2, 5]. However, it was also observed that some participants performed well even when their KSS rating was greater than 7 [2].

S. Bagnara et al. (Eds.): IEA 2018, AISC 823, pp. 179–188, 2019.
https://doi.org/10.1007/978-3-319-96074-6_19

For safe implementation of Automated Driving Systems, it is relevant to know whether drivers are able to take over control safely in case of a request to intervene. So far, some driving simulator studies exist in which sleepiness or automation duration did not negatively influence take-over time or quality [6–8]. In addition, drowsiness did not significantly influence the hands-on and driver intervention times in a Wizard-of-Oz study. However, in this experiment some startle reactions were observed in the event of a request to intervene [9]. Further, a negative influence of sleepiness on take-over performance (lateral acceleration) [10] and on the time until regaining situational awareness [11] was found in driving simulator studies.

Hence, in this article it is also assumed, that driver drowsiness can represent another limit of a Human Machine System including automation, and thus needs to be managed during an automated drive [12].

1.2 Automated Driving and Non-driving-Related Tasks

In the "Driver Availability" concept, the current driver state is determined by a driver's arousal level, by the type of actually conducted non-driving-related tasks, and by motivational aspects. If a driver needs to take over control from an Automated Driving System, a target driver state must be achieved within a given time budget [13]. Different studies evaluated the influence of standardized and of naturalistic non-driving-related tasks (for an overview see [14]). However, the type of non-driving-related tasks had none (e.g. [15]) or only a minor impact on take-over performance [16]. Hence, non-driving-related tasks might be a suitable approach to avoid or at least to postpone the system limit "Driver Drowsiness" during an automated drive.

So far, few studies exist that have investigated the influence of non-driving-related tasks on drowsiness development or fatigue. One study found, that the use of non-driving-related task can reduce driver fatigue [17]. Further, other studies showed that participants' drowsiness developed more slowly when they executed a non-driving-related task compared to being inactive [6] or when participants performed a motivational compared to a tiring non-driving-related task [7]. Also it is known, that measures against driver drowsiness were intensively studied for manual driving (for an overview see [18]). However, the negative influence of distraction, for example due to the cell-phone use, has been clearly demonstrated in the context of manual driving (e.g. [19]).

In contrast to manual driving, drivers might be able to perform very different types of non-driving-related tasks during an automated drive, even for a longer period. Hence, based on the present state of research, this study aims to investigate the reactivation potential and effectiveness of different non-driving-related tasks to avoid or to postpone the system limit "Driver Drowsiness".

2 Method

2.1 Participants

Seventy-one employees of the AUDI AG took part in this experiment. The sample consisted of 24 women and 47 men. On average participants were 31.90 (*SD* = 8.08) years old and held their driver's licenses for 14.03 years (SD = 7.48). Participants were asked to register for this study only if they are usually able to read as a passenger without feeling sick. Further, participants were asked to abstain from all caffeinated beverages for one hour prior to the experiment.

2.2 Test Vehicle and Test Track

A right-hand-drive vehicle (A4 sedan) was used to simulate an Automated Driving System in a real driving environment (see Fig. 1). The test vehicle was equipped with additional driving school mirrors so that the participants were also able to observe the surrounding traffic. Further, the vehicle was equipped with a 6-in. tablet, which was attached in front of the passenger's seat. This tablet showed the pilot status (*pilot active*), speed in km/h and indicators. Further, a 12-in. tablet was integrated into the center console. On this tablet, the applications of the various non-driving-related tasks were presented. These applications guided the participants through the entire experiment. As long as an Automated Driving System was simulated, a curtain was attached between the participant and the investigator. The investigator simulated the system behavior of a possible future highway pilot. Thus, the maximum speed was 130 km/h and lane changes were performed conservatively. The assistant systems adaptive cruise control and lane keeping assist were not used in this study, as this would represent a state-of-the-art system rather than a future highway pilot. The study was conducted on the A9 autobahn in Germany. The experiment started at the highway service station Koeschinger-Forst.

Fig. 1. Test vehicle

2.3 Experimental Design and Measures

The examination consisted of three parts. Part A used a within-subject design to assess the effectiveness of drowsiness generation. Part B consisted of a reactivation and an effectiveness phase, because these two phases allow investigation of the reactivation potential and effectiveness (even after the actual reactivation) of non-driving-related tasks. In Part C a follow-up survey regarding the experience when executing the non-driving-related tasks of Part B was conducted.

As a dependent variable, the Karolinska-Sleepiness Scale (KSS) was used [3]. Figure 2 presents the experimental design and the timing of questioning (KSS_1, KSS_2, KSS_3, KSS_4, and KSS_5). Further, to check whether the tasks were perceived differently, participants assessed the "In-game GEQ" at the end of this study. Based on 14 items, seven components can be calculated [20]. These components are Competence, Sensory and Imaginative Immersion (abbreviated as Immersion in this article), Flow, Tension, Challenge, Negative affect, and Positive affect.

The German version of the GEQ was used in this study (see [21]). However, the items "I was interested in the game's story" and "I had to put a lot of effort into it" were slightly adjusted for the purpose of this study to "I was interested in the content of the task" and "I had to put a lot of effort into the task".

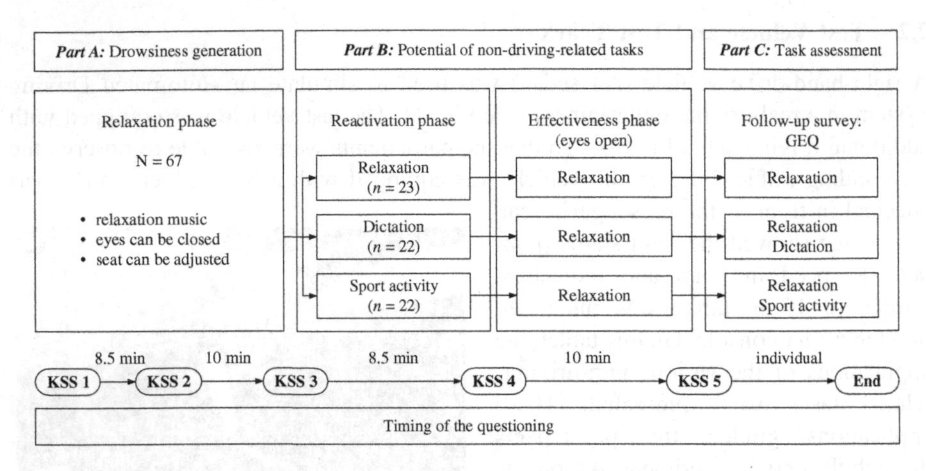

Fig. 2. Experimental design and timing of the drowsiness assessment

2.4 Procedure and Non-driving-Related Tasks

Participants were randomly assigned to one of the three different groups. Four experiments were conducted per day. These started at 8:00 a.m., 10:15 a.m., 1:00 p.m. and 3:15 p.m.. These start times were permuted among groups to distribute the effect of circadian rhythm. Each trial was scheduled to last for a maximum of 2 h.

Before the examination, participants were informed that a highway pilot will be simulated and that the investigator will drive the vehicle all the time.

During *Part A*, which was identical for all groups, relaxation music was played. Further, participants were informed that they should adjust the volume and/or seat position in the relaxation phases in such a way that they were able to relax as well as possible. During *Part A,* participants were also allowed to close their eyes. However, they were also informed that, if possible, they should avoid falling asleep during the entire test drive.

Three applications were developed to provide the different non-driving-related tasks and to ask participants (by an audio request) to rate their current drowsiness level at certain points. In *Part B* participants first experienced their group-specific task. These are presented in the following.

- Relaxation: In this case, the relaxation group can be considered as a control group. Participants of this group were asked to continue relaxing. However, from that moment, they were asked to keep their eyes open.

- Dictation: Different studies showed that a large number of users would use the driving time to conduct tasks, such as "texting" (e.g. [22]). Hence, we decided to use a dictation as a non-driving-related task, as this requires typing different words for a limited period.
- Sport activity: Further, we decided to test a sport activity (Handytrim fitness device) in this experiment for two reasons. First, it is a task that can definitely not be done during a manual drive. Second, using the travel time to improve physical fitness could increase users and the societal benefit generated by Automated Driving Systems.

In the subsequent effectiveness phase, all participants were asked to relax while keeping their eyes open. The aim of this phase was to assess the effectiveness of the reactivation phase.

3 Results

The significance level of the statistical analysis was .05. The data of four participants were not recorded due to a system crash during the experiment. Hence, in total data of 67 participants were analyzed.

3.1 Part A: Drowsiness Generation

The KSS ratings increased significantly within the three measurement times, as assessed by a Friedman test ($\chi^2 = 49.22$, $p < .001$). At the beginning of the test drive (at KSS_1) the mean KSS rating was 4.48 ($SD = 1.59$), and further increased to 5.54 ($SD = 1.41$) at KSS_2 and reached 6.15 ($SD = 1.63$) at the end of *Part A* (KSS_3).

3.2 Part B: Distribution Functions of the KSS Ratings After the Reactivation and Effectiveness Phase

KSS scores greater than 7 can be allocated to the driver state "Drowsiness" [4]. Based on this, for the following analysis it is considered that KSS ratings greater than level 7 would lead to the system limit "Driver drowsiness". Thus, the number of participants exceeding a KSS level of 7 is of great importance, as this would represent the number of users who would not be able to use the automated system any longer. The reactivation potential is considered high, when the number of participants exceeding a critical drowsiness state, in this case a KSS rating greater than 7, is small. A reactivation is considered to be effective, if the number of participants exceeding a critical drowsiness state, remains small even after the actual reactivation (in this case after 10 further minutes of relaxation — which represents a worst-case consideration). Hence, the distribution functions of the KSS ratings at KSS_4 and KSS_5 were determined.

$$KSS_{rating} = \begin{cases} x \leq 7 \; drowsiness\ is\ not\ considered\ a\ system\ limit \\ x > 7 \; drowsiness\ is\ considered\ a\ system\ limit \end{cases} \tag{1}$$

Participants whose KSS rating was greater than 7 at KSS_3 were excluded from the calculation of distribution functions (n = 10), as their drowsiness state would already represent the system limit "Driver drowsiness" (see formula 1). In total, 57 data sets could be analyzed (Relaxation: n = 18, Dictation: n = 19, and Sport Activity: n = 20). The distribution functions of these groups were calculated at KSS_4 and at KSS_5 (see Fig. 3).

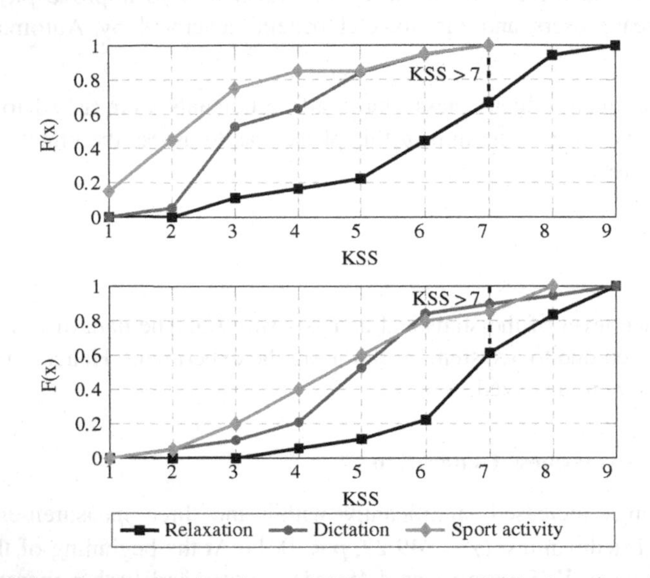

Fig. 3. Cumulative distribution functions of KSS ratings at KSS_4 (top figure) and at KSS_5 (bottom figure)

After the reactivation phase, no participant of the Dictation and Sport activity group exceeded KSS level 7. In contrast to this, the number of participants who reached level 8 or level 9 on the KSS was 33.34% in the Relaxation group. At KSS_5 (after the effectiveness phase) 38.89% of the Relaxation group exceeded level 7 on the KSS. The number of KSS ratings greater than 7 increased to 10.52% in the Dictation group and to 15% in the Sport activity group, during the effectiveness phase.

3.3 Part C: Assessment of the Different Non-driving-Related Tasks

The KSS distribution functions clearly show that there is a difference between the Sport activity and the Dictation group compared to the Relaxation group at KSS_4 and KSS_5. However, the differences between the KSS distributions of the Dictation and Sport activity group seem rather small. This raises the question whether the Dictation and Sport activity tasks were perceived differently. To check this, participants rated 7 categories according to the In-game GEQ (see Sect. 2.3). Figure 3 provides an overview of the results.

For this analysis, the significance level was adjusted to p = .007, due to the multiple comparisons (Bonferroni correction). Overall the assessment of the Dictation and

Sport activity tasks did not differ significantly: Competence (U = 158.5, z = −0.910, p = .363), Immersion (U = 107.5, z = −2.383, p = .017), Flow (U = 153.0, z = −1.070, p = .285), Tension (U = 144.5, z = −1.357, p = .175), Challenge (U = 122.0, z = −2.222, p = .026), Negative affect (U = 99.0, z = −2.597, p = .009), and Positive affect (U = 134.0, z = −1.670, p = .120).

In addition, it was assessed in which components the tasks of the reactivation and effectiveness phase were perceived differently (Dictation compared to the Relaxation task and Sport activity compared to the Relaxation task). The significance level was again adjusted to p = .007. The assessment of the Dictation task compared to the Relaxation task differed significantly in the following categories: Flow (z = −3.185, p = .001), and Challenge (z = −3.517, p < .001). The Sport activity task differed significantly compared to the Relaxation task in the following categories: Competence (z = −2.975, p = .003), Immersion (z = −3.280, p = .001), and Challenge (z = −3.517, p < .001), as assessed by a Wilcoxon test.

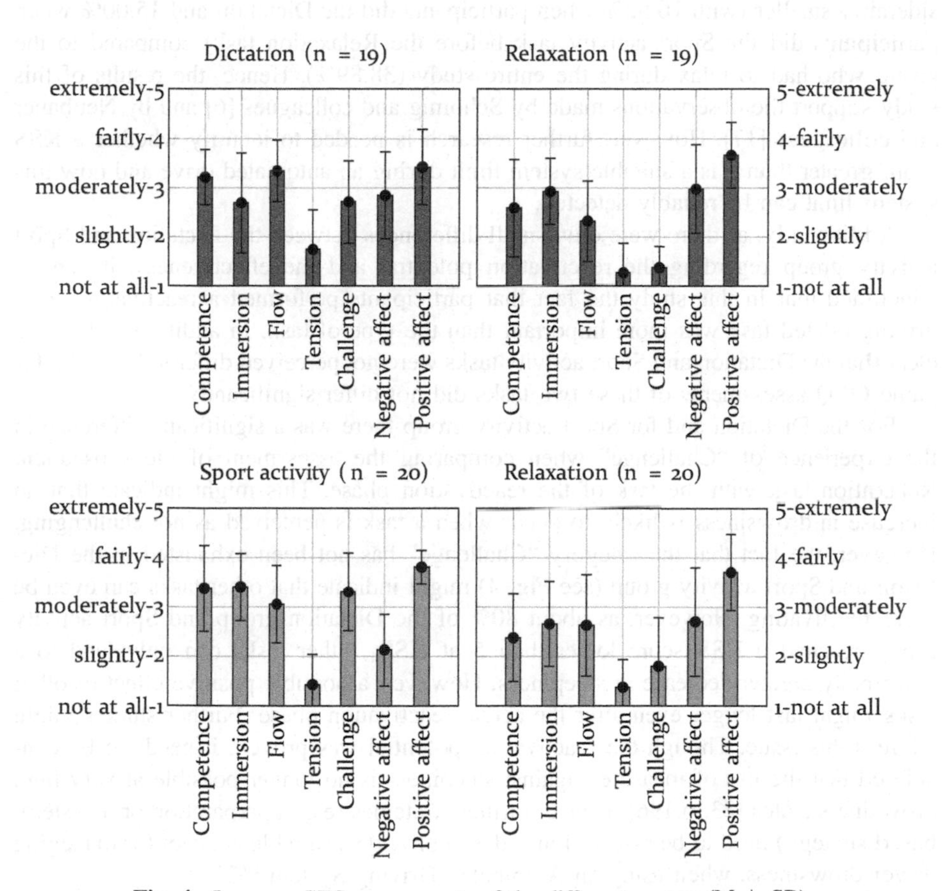

Fig. 4. In-game GEQ assessments of the different groups (M ± *SD*)

4 Discussion

The drowsiness generation by relaxing music within *Part A* of the study can be considered successful. Ten participants already reached a KSS score greater than 7 at KSS_3 (i.e. in about 20 min). Hence, the drowsiness generation phase should not be longer for this study purpose. This is because KSS scores greater than 7 are considered a system limit in the further analysis (see formula 1). Thus, the data of participants who already exceeded this level cannot be used for calculation of the KSS distribution functions.

Besides this, it became clear that the driver state drowsiness can be reached rather quickly. This finding is in line with other studies, which also showed that drowsiness occurs quickly during an automated drive [11, 12].

Further, this study proved the reactivation potential of non-driving-related tasks. This became clear, as no participant of the Dictation and Sport activity group exceeded KSS level 7 at the end of the reactivation phase. In addition, after the subsequent effectiveness phase the number of participants exceeding a KSS level of 7 was considerably smaller (with 10.52% when participants did the Dictation and 15.00% when participants did the Sport activity task before the Relaxation task) compared to the group who had to relax during the entire study (38.89%). Hence, the results of this study support the observations made by Schömig and colleagues [6] and by Neubauer and colleagues [17]. However, further research is needed to identify whether a KSS score greater than 7 is a suitable system limit during an automated drive and how this system limit can be reliably detected.

Additionally, as there were only small differences between the Dictation and Sport activity group regarding the reactivation potential and the effectiveness, it can be concluded that in this study the fact that participants performed a reactivating non-driving-related task was more important than the type of task. In addition, it became clear that the Dictation and Sport activity tasks were not perceived differently, as the In-game GEQ assessments of these two tasks did not differ significantly.

For the Dictation and for Sport activity group there was a significant difference in the experience of "Challenge" when comparing the assessment of the subsequent Relaxation task with the task of the reactivation phase. This might indicate that an increase in drowsiness is likely to occur when a task is perceived as not challenging. However, the fact that the category "Challenge" has not been exhausted in the Dictation and Sport activity group (see Fig. 4) might indicate that other tasks can even be more reactivating. However, as about 80% of the Dictation group and Sport activity group reached a KSS score lower than 5 at KSS_4, other tasks can only lead to a marginally greater decrease in sleepiness. However, a possible positive effect of other tasks might last longer even after the actual reactivation phase. Further studies might address this issue. Though the reactivation potential was proven, it needs to be considered that the use of measures against sleepiness is no longer possible at very high drowsiness states [23, p.196]. Hence, further strategies (e.g. a preparation or a system-based strategy) need to be assessed in order to provide a suitable concept for managing driver drowsiness, when using an Automated Driving System [12].

5 Conclusion

In this study, it was found that non-driving-related tasks have the potential to be a suitable option for managing driver drowsiness, as no participant of the Dictation or Sport activity group exceeded level 7 on the KSS after the reactivation phase. Even after the effectiveness phase, there was still a major difference between the number of participants exceeding level 7 between the Dictation and Sport activity group compared to the Relaxation group. Future studies might also evaluate the potential of other naturalistic non-driving-related tasks and especially their link to the effectiveness of the reactivation. The results of this study should be considered when designing human-vehicle interactions of Automated Driving Systems.

Acknowledgment. This work results from the joint project Ko-HAF – Cooperative Highly Automated Driving and has been funded by the Federal Ministry for Economic Affairs and Energy based on a resolution of the German Bundestag.

References

1. Johns M (1998) Rethinking the assessment of sleepiness. Sleep Med Rev 2(1):3–15
2. Johns MW (2007) Drowsy Driving and the Law: A Submission to the Tasmania Law Reform Institute, vol. Issues Paper No 12
3. Åkerstedt T, Gillberg M (1990) Subjective and objective sleepiness in the active individual. Int J Neurosci 52(1–2):29–37
4. Johns MW (2009) What is excessive daytime sleepiness. In: Fulke P, Vaughan S (eds) Neuroscience research progress series, Sleep deprivation: causes, effects and treatment. Nova Science Publishers, New York, pp 1–37
5. Ingre M et al (2006) Subjective sleepiness and accident risk avoiding the ecological fallacy. J Sleep Res 15(2):142–148
6. Schömig N, Hargutt V, Neukum A, Petermann-Stock I, Othersen I (2015) The interaction between highly automated driving and the development of drowsiness. Procedia Manufact 3:6652–6659
7. Jarosch O, Kuhnt M, Paradies S, Bengler K (2017) It's out of our hands now! Effects of non-driving related tasks during highly automated driving on drivers' fatigue. In: International driving symposium on human factors in driver assessment, training and vehicle design
8. Feldhütter A, Gold C, Schneider S, Bengler K (2017) How the duration of automated driving influences take-over performance and gaze behavior. In: Schlick CM et al (eds) Advances in ergonomic design of systems, products and processes. Springer, Heidelberg, pp 309–318
9. Weinbeer V et al (2017) Highly automated driving: How to get the driver drowsy and how does drowsiness influence various take-over aspects? 8. Tagung Fahrerassistenz
10. Goncalves J, Happee R, Bengler K (2016) Drowsiness in conditional automation: proneness, diagnosis and driving performance effects. In: 2016 IEEE 19th international conference on intelligent transportation systems (ITSC), Rio de Janeiro, Brazil, pp 873–878
11. Vogelpohl T, Kühn M, Hummel T, Vollrath M (2018) Asleep at the automated wheel-Sleepiness and fatigue during highly automated driving. Accid Anal Prev
12. Weinbeer V, Bill J-S, Baur C, Bengler K (2018) Automated driving: subjective assessment of different strategies to manage drowsiness. In: de Waard D et al (eds) Proceedings of the human factors and ergonomics society europe chapter 2017

13. Marberger C et al (2018) Understanding and applying the concept of "driver availability" in automated driving. In: Stanton N (ed) Advances in human aspects of transportation, vol. 597. Springer, Cham, pp 595–605
14. Naujoks F, Befelein D, Wiedemann K, Neukum A (2018) A review of non-driving-related tasks used in studies on automated driving. In: Stanton N (ed) Advances in human aspects of transportation, vol. 597. Springer, Cham, pp 525–537
15. Radlmayr J, Gold C, Lorenz L, Farid M, Bengler K (2014) How traffic situations and non-driving related tasks affect the take-over quality in highly automated driving. Proc Hum Factors Ergon Soc Annu Meet 58(1):2063–2067
16. Gold C, Happee R, Bengler K (2017) Modeling take-over performance in level 3 conditionally automated vehicles. Accid Anal Prev
17. Neubauer C, Matthews G, Saxby D (2014) Fatigue in the automated vehicle. Proc Hum Factors Ergon Soc Annu Meet 58(1):2053–2057
18. Hashemi Nazari SS, Moradi A, Rahmani K (2017) A systematic review of the effect of various interventions on reducing fatigue and sleepiness while driving. Chin J Traumatol 20 (5):249–258 Zhonghua chuang shang za zhi
19. Strayer DL, Drews FA, Crouch DJ (2006) A comparison of the cell phone driver and the drunk driver. Hum Factors 48(2):381–391
20. IJsselsteijn WA, de Kort YAW, Poels K (2013) The game experience questionnaire. https://pure.tue.nl/ws/portalfiles/portal/21666907
21. Engl S Mobile gaming: Eine empirische Studie zum Spielverhalten und Nutzungserlebnis in mobilen Kontexten. Magisterarbeit, Information und Medien, Sprache und Kultur, Universität Regensburg, Regensburg
22. Pfleging B, Rang M, Broy N (2016) Investigating user needs for non-driving-related activities during automated driving. In: The 15th international conference, Rovaniemi, Finland, pp 91–99
23. Hargutt V (2002) Das Lidschlussverhalten als Indikator für Aufmerksamkeits- und Müdigkeitsprozesse bei Arbeitshandlungen. Dissertation, Philosophische Fakultät III, Julius-Maximilian-Universität Würzburg, Würzburg

Effects of Verbal Communication with a Driving Automation System on Driver Situation Awareness

Taiki Uchida(✉), Toshiaki Hirano, and Makoto Itoh

University of Tsukuba, Tsukuba, Ibaraki 305-8573, Japan
t_uchida@css.risk.tsukuba.ac.jp

Abstract. At level 2 of SAE's driving automation, it is difficult for drivers to continue to monitor a driving automation system and the environment. An important issue to be addressed involves maintaining driver situation awareness to keep the driver in the control loop. In the study, we investigate verbal communication between the driver and system. We hypothesize that the driver can cognitively participate in vehicle operation even if he/she is not physically in the control loop. We use a driving simulator to examine how verbal communication affects driver situation awareness. We compare the following two conditions: (1) talking with the system and (2) not talking with the system during automated driving. Under the condition of talking with the system, the system asks the driver about the peripheral situation and/or vehicle control. The driver is required to respond to the system. In the experiment, two events occur during which the driver is expected to intervene during cruising. We measure the event response time, number of collisions, how the driver maneuvers the vehicle, and subjective usability by administering a questionnaire. The results indicate that the number of collisions are significantly higher under the condition of conversation than under the condition of no conversation. The event response time is significantly longer under the condition of conversation than the condition of no conversation. The aformentioned results indicate that the verbal communication does not improve driver situation awareness. There is no difference in the questionnaire score, and thus the verbal communication does not improve the usability of the driving automation system. The results indicate that drivers can potentially overestimate the extent to which they obtain information about driving situation only through conversation. The results provide important insights for designing systems to support driver situation awareness.

Keywords: Driving automation · Situation awareness · Verbal communication

1 Introduction

1.1 Background

In order to reduce car accidents, realization of driving automation systems is a national strategy in Japan [1]. However, several problems persist with respect to driver situation awareness while using driving automation system [2, 3]. At the SAE's driving automation level 2, a system executes longitudinal and lateral control [4]. Subsequently,

S. Bagnara et al. (Eds.): IEA 2018, AISC 823, pp. 189–198, 2019.
https://doi.org/10.1007/978-3-319-96074-6_20

a driver must monitor the system and exercise control when necessary to ensure vehicle safety. Nevertheless, it can be difficult for drivers to continue monitoring for a long period since the driving environment tends to be boring in monotonous roads such as highways [5]. Furthermore, it is suggested that a driver fails to recognize situation and exercise vehicle control in emergencies because drivers are distracted or absorbed in tasks other than driving [6, 7]. Thus, an important issue involves maintaining driver situation awareness during automated driving.

We distinguish the following four types of interaction between a driver and a driving automation system when the system operates.

(1) **Driver is out of the loop:** The driver does not interact with the system at all in neither in a physical nor a cognitive manner.
(2) **Driver is cognitively involved:** The driver is not physically in the loop and neither holds the steering wheel nor touches the pedals although the driver interacts with the system to manage the operation through visual and/or auditory information exchange.
(3) **Driver is physically although not cognitively involved:** The driver is physically in the loop as he/she holds the steering wheel or touches the pedals although there is no communication with the system to manage the operation neither through visual nor auditory information exchange.
(4) **Driver is both physically and cognitively involved:** In addition to (2), the driver holds the steering wheel or touches the pedals.

In condition (4), the driver understands the system behavior most easily among the four conditions. Conversely, the driver's workload may increase for a long period. Condition (3) is the typical design of current level 2 systems in the market although we believe this is not the optimal method for drivers to maintain situation awareness. Condition (1) is not allowed for level 2 driving automation systems. If the situation awareness improves under condition (2) than under condition (1) through only cognitive interaction, then condition (2) potentially constitutes a practical solution.

The use of voice communication for requests to intervene with the driver is proven as effective [8]. Therefore, the study also uses voice to implement cognitive interaction. Additionally, we consider that verbal communication between the driver and system is preferable to unilateral announcements from the system.

1.2 Purpose of the Study

We investigate the effects of verbal communication between a driver and a driving automation system on driver situation awareness. It is hypothesized that the verbal communication makes the driver naturally attain situation awareness on the driving environment. Furthermore, the system can confirm that the driver ensures situation awareness by driver's response to the system.

2 Experiment

2.1 Apparatus

Figure 1 shows the experimental setup for our experiment. It consists of a gaming controller, three 80-in. screens, two speakers, and a laptop computer for controlling conversations.

We used the DS-nano made by AST-J to simulate vehicle operation.

Fig. 1. Devices used in the experiment

2.2 Participants

There were 40 participants who satisfied the following conditions.

- They possess a valid driving license.
- They drive a vehicle on a daily basis.
- They are healthy.

We paid 820 Japanese yen per hour to the participants as compensation. When a participant arrived at the laboratory, the purpose and details of the experiment were explained and especially with respect to the protection of participant safety and privacy. The participants signed on an informed consent form after they received the explanation.

We divided the 40 participants into two groups, namely the conversation group and the no conversation group.

2.3 Driving Automation System

The features of the driving automation system used in the experiment are shown below.

Longitudinal Control

- The system maintains its own vehicle velocity at 100 km/h when there is no other vehicle ahead in the current lane
- The system maintains its own vehicle velocity as identical to that of the leading vehicle

- The system can recognize the vehicle in front although it cannot recognize stopped objects on the road

Lateral Control

- The system cruises along the lane and prevents deviation
- The system changes the lane after 3 s if right/left signal is turned on. The signal level can only be operated by the driver. Thus, the system cannot perform autonomous lane changes.

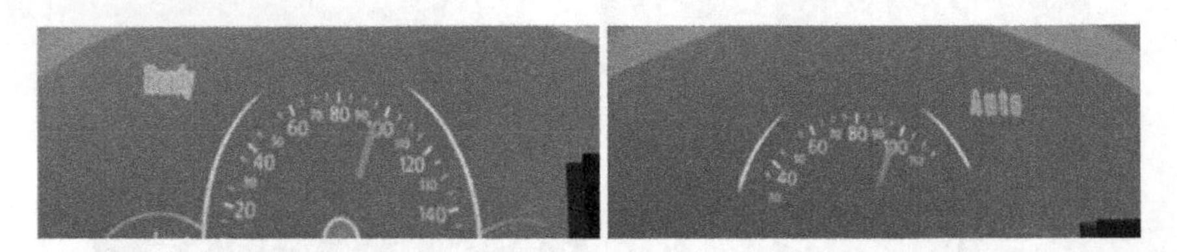

Fig. 2. State of the system

The operation state of the system is shown on the instrument panel. The panel shows "Ready" when the situation is suitable for automated control such that the driver can commence driving automation (as shown in the left side of Fig. 2). The panel shows "Auto" when the system commences operation (as shown in the right side of Fig. 2). The system ends the operation, and "Auto" disappears when the driver pushes the control button, pushes the brake pedal, or turns a steering wheel.

Table 1. Contents of verbal communication

No.	Question (System)	Response (Driver)	Response (System)
1	"I detect a vehicle ahead. May I follow this vehicle?"	"Yes"	"OK"
		"No"	to No. 3
2	"Speed is falling"	–	to No. 3
3	"May I overtake front vehicle?"	"Yes"	"OK. Confirm the safety and turn on the winker"
		"No"	"OK. I continue following"
4	"This is overtaking lane. May I go back to the left lane?"	"Yes"	"OK. Confirm the safety and turn on the winker"
		"No"	"OK. I continue cruising straight"
5	"An exit ramp is 1 km away. May I switch to manual driving?"	"Yes"	"OK. I switch to manual driving after 5 s"
		"No"	"OK. I continue cruising straight"

2.4 Verbal Communication

All conversations are initiated as a question from the system to the driver. The system informs the driver as to how the system recognizes the situation. It also proposes an action. The driver is expected to answer in terms of either a "yes" or "no" to the proposed action. Table 1 shows the contents of question and response.

The system voices are prepared in WAV files, and they are played with the "say" command outputs on Mac OS X. The voice begins to flow automatically based on driving situation. An experimenter sitting near the participant executes the system response by inputting Yes/No key on keyboard based on the answer from the participant.

2.5 Scenarios

The driving course in the experiment involves a 30 km-long expressway in a straight line with three merging points. There are a few other vehicles to the extent that there is no jamming and they operate at 90–100 km/h. It takes approximately 20 min for one trial.

There are two events (events A and B) in the experiment. The events are risky such that the system is unable to deal with the same, and thus the driver must resolve the event. There are four types of event occurrences as follows: no event, event A only, event B only, and both events A and B. The place where the event occurs is randomized to prevent drivers from predicting when an event occurs. The contents of the event are described below.

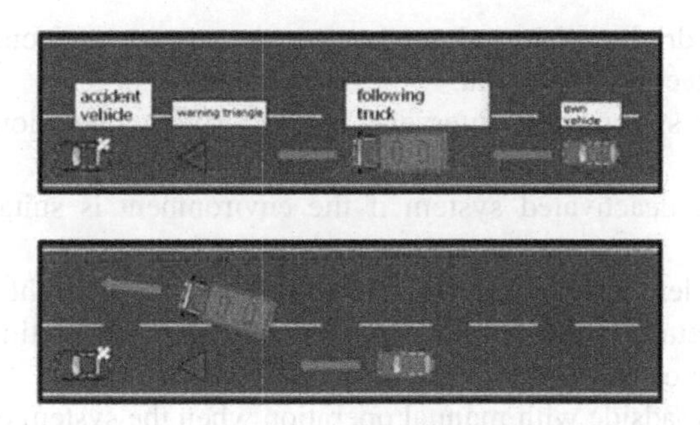

Fig. 3. Event A

Event A – accident vehicle (Fig. 3): An accident vehicle is stopped ahead on the road, and a warning triangle is put on the left lane. Drivers are unable to perceive the same until the host vehicle approaches extremely near the accident vehicle because a large truck leading the way hides the front view. When the distance from the own vehicle to the accident vehicle corresponds to 100 m, the large truck suddenly changes lanes to the right. The participant must avoid a collision himself/herself while operating the vehicle or heavily pushing the brake pedal since the system is unable to recognize the stopping object.

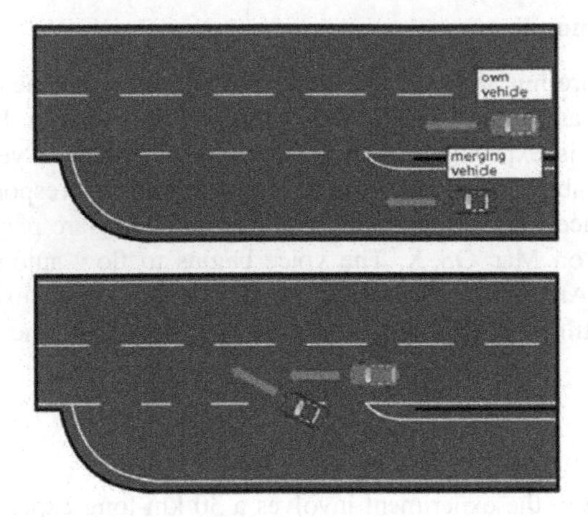

Fig. 4. Event B

Event B – merging vehicle (Fig. 4): A merging vehicle suddenly appears in front of the own vehicle. The system is unable avoid a collision due to the extremely short distance between vehicles. The driver must avoid the collision while performing the steering maneuver or pushing the brake.

2.6 Instruction and Subtask

The contents of instruction for participants are described below.

- To sit on the driving seat and to monitor the system and environment without touching the steering and pedal.
- To cancel the system and intervene in the driving operation for safety when necessary.
- To resume the deactivated system if the environment is suitable for automated driving.
- To keep to the left lane. If the vehicle changes lanes to the right to pass the leading vehicle, then return to the left lane immediately after completing the passing.
- To answer Yes or No to the question given from the system.
- To stop at the roadside with manual operation when the system ends at the terminal of the driving course.

Additionally, a few vehicles travel at low speeds corresponding to 80 km/h in the conversation group, and thus participants are instructed to overtake them.

Furthermore, participants are assigned a subtask during experimental driving. The subtask is the SuRT [9]. Figure 5 shows an example of the subtask display. The display is divided into four areas. It is considered as "correct" if a participant touches the area that includes a larger circle, and it is considered as "incorrect" if he/she touches another area. If the display is touched, the next problem immediately appears. Participants continue to play the game to the maximum possible extent while assigning priority to

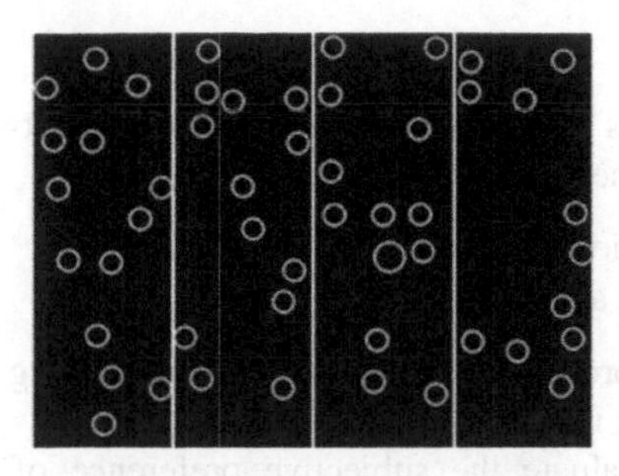

Fig. 5. Example of subtask display

vehicle safety. A laptop computer with touch display is used for the subtask and mounted on the left side of the steering (Fig. 1). It is reachable from the driver's seat.

2.7 Procedure

Table 2 shows the schedule of the experiment. Participants execute training driving after they receive an explanation on the experiment. Training driving is executed thrice: to learn how to operate the simulator, to learn how to operate the driving automation system, and to learn how to operate the subtask. Additionally, the conversation group is trained for the verbal communication with the system. Experimental driving is executed four times as follows: no event, only event A, only event B, and both events A and B. A 5-min break is set up between the experiments.

Table 2. Schedule of the experiment

Guidance	10 min
Training Driving 1–3	15 min
(Training Driving 4)	5 min
Break Time	5 min
Experimental Driving (No Event)	20 min
Break Time	5 min
Experimental Driving (Event A or B)	20 min
Break Time	5 min
Experimental Driving (Event B or A)	20 min
Break Time	5 min
Experimental Driving (Event A and B)	20 min
Questionnaire	5 min

2.8 Measurements

Comparison of the values of all variables is performed by a two-sided t-test. The main dependent variables in the experiment are as follows:

- The number of collisions in events
- The response time to an event

Furthermore, we record participant behavior while driving to evaluate the extent of smoothness of a driver's reactive maneuver.

Additionally, we evaluate the subjective preference of the system and verbal communication through a questionnaire with a Likert scale including 13 levels. The questionnaire items are as follows:

- Do you trust the system?
- Do you think the system is safety?
- Did the system help your driving?
- Could you understand what the system said?
- Do you think verbal communication is troublesome?
- Did verbal communication help your driving?

3 Results

3.1 The Number of Collisions

The left of Fig. 6 shows the mean number of collisions. In event A, the number of the conversation group exceeds that of the no conversation group although a significant difference between the two groups is absent ($t(18) = 1.14$, $p = 0.268$). In event B, the number of crashes in the conversation group significantly exceeds that in the no conversation group ($t(18) = 3.13$, $p < 0.01$). Our interpretation of the result is that the drivers in the conversation group consider verbal communication as the only information source for situation awareness and only monitor the situation when the system asks the driver. However, drivers are unable to perceive events since the system does not ask when events occur.

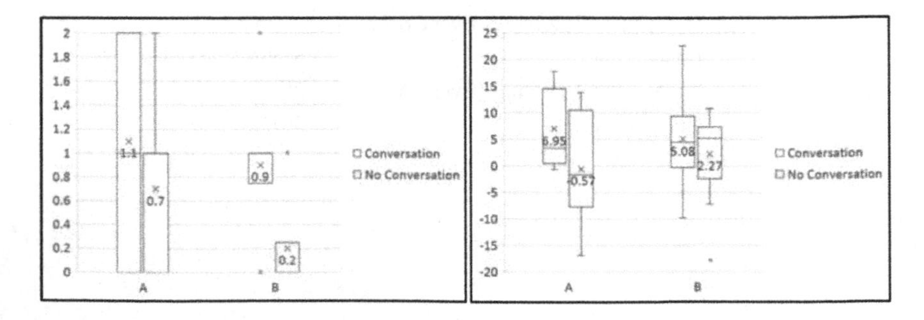

Fig. 6. Mean number of collisions (left) and the mean response time (right)

3.2 Response Time

The right side of Fig. 6 shows the mean response time. However, this is excluded in the case in which a participant does not perceive the event. In event A, response time of the conversation group significantly exceeds that of the no conversation group ($t(29) = 2.39$, $p < 0.05$). In event B, the time taken by the conversation group is also longer than that of the no conversation group although the difference is not significant ($t(29) = 0.954$, $p = 0.348$). The conversation group frequently does not perceive the event more than the no conversation group. Given the number of collisions, the drivers in the conversation group tend to not perceive the events.

3.3 Questionnaire

The left side of Fig. 7 shows the mean score of questionnaire on the driving automation system. The larger score indicates an increase in degree of agreement with respect to the question, and the lower score indicates a decrease in degree of agreement with respect to the question. With respect to each question, a significant difference between the two groups is absent ($t(18) = 0.357$, $p = 0.725$; $t(18) = 1.44$, $p = 0.168$; $t(18) = 0.388$ $p = 0.703$). It appears that participants do not think the system is safe based on experience wherein the system is unable to deal with emergency events. However, they do not assign a low value for system reliability since the content of the question from the system is not incorrect.

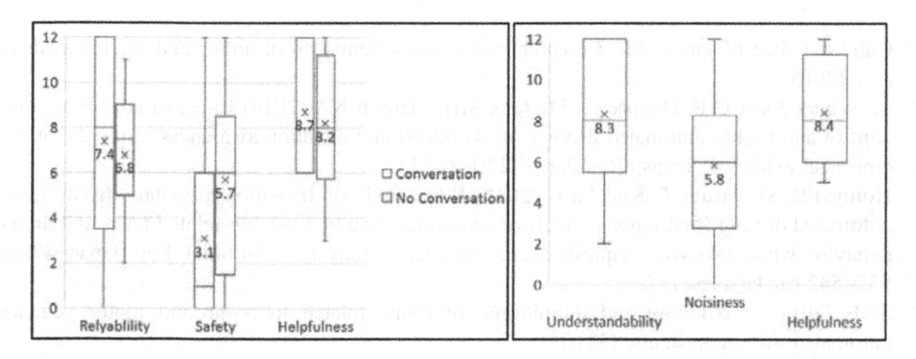

Fig. 7. Mean score of questionnaire on the system (left) and verbal communication (right)

The right side of Fig. 7 shows the mean score of the questionnaire on verbal communication. As shown in the figure, participants consider that verbal communication is not extremely troublesome. Verbal communication with the system does not constitute a heavy load for driving.

4 Conclusions

The results of the experiment indicate that drivers in the conversation group were not easily aware of the event more than those in the no conversation group since the number of collisions in the former group exceeds that in the latter and the response time is longer. Therefore, verbal communication with the system did not improve driver situation awareness in the experiment.

Our interpretation of the results is that the drivers in the conversation group think that they should only perceive the driving environment when the system asks the driver, i.e., they place excessive trust in the system. Furthermore, the main content of verbal communication involved following or overtaking the front vehicle, and drivers in conversation group did not think about it independently. Conversely, drivers in the no conversation group must form a decision independently, they frequently interact with the driving environment, and this resulted in increased attention.

Driver situation awareness may be improved by changing the content of verbal communication. Future studies should consider increasing sources for situation awareness by adding sounds that do not require driver response or that are not directly linked to the driving operation.

Acknowledgements. The study was supported by JSPS KAKENHI 15H05716.

References

1. Cabinet Office of Japan: SIP Research and development plan of automated driving system. p. 1 (2015)
2. de Winter Joost C F, Happee R, Martens MH, Stanton NA (2014) Effect of adaptive cruise control and highly automated driving on workload and situation awareness: a review of the empirical evidence. Transp Res Part F 27:196–217
3. Homma R, Wakasugi T, Kodaka K (2016) Basic study on transition to manual from highly automated driving (first report)-effects of drowsiness and non-driving-related-tasks and driver behavior when take-over-requests were provided. Trans Soc Automot Eng Japan 47(2): 537–542 (in Japanese)
4. SAE (2014): Taxonomy and definitions for terms related to on-ground motor vehicles automated driving systems. J3016
5. Inagaki T (2012) Design of human-machine symbiotic – search for "human centered automation". Morikita Publishing, Tokyo (in Japanese)
6. Zeeb K, Buchner A, Schrauf M (2016) Is take-over time all that matter? the impact of visual-cognitive load driver take-over quality after conditionally automated driving. Accid Anal Prev 92:230–239
7. Merat N, Jamson AH, Lai FCH, Carsten OMJ (2012) Highly automated driving, secondary task performance and driver state. Hum Factors 54:762–771
8. Politis I, Brewster S (2015) Language-based multimodal displays for the handover of control in autonomous car. In: Proceedings of the automotive user interfaces and interactive vehicular applications, pp 3–10
9. ISO (2012): Road vehicles — Ergonomic aspects of transport information and control systems — Calibration Tasks for Methods which Assess Demand Due to the Use of In-Vehicle Systems. ISO/TS 14198

The Root Causes of a Train Accident: Lac-Mégantic Rail Disaster

İbrahim Öztürk[1,2(✉)], Gizem Güner[3], and Ece Tümer[3]

[1] Safety Research Unit, Department of Psychology, Middle East Technical University, 06800 Ankara, Turkey
ibrahmoztrk@gmail.com

[2] Department of Psychology, Çanakkale Onsekiz Mart University, 17100 Çanakkale, Turkey

[3] Department of City Planning, Middle East Technical University, 06800 Ankara, Turkey
g.guner1246@gmail.com, ecetumer93@hotmail.com

Abstract. Oil transportation with trains is one of the most important and dangerous areas of transportation. Many agents play different roles throughout the route, indicating different responsibilities in case of an accident. The analysis of transportation accidents is important to determine the roles of these agents and their strengths and weaknesses. The root causes analysis (RCA) is one of the methods used to understand the story of the accident and improve safety concerns. In the present study, the Lac-Mégantic Rail Disaster was analyzed by using different RCA methods and considering organizational safety culture. The accident happened in the 6th July 2013 and resulted in 47 deaths and destruction of 40 buildings and 53 vehicles. By using the detailed reports about the accidents and using different methods of RCA, the contributory factors of the accidents were identified. In general, it has been found that human factors, organizational safety culture, and engineering related factors were active throughout the different stages of the accident. The chronology and timeline of the accident were used to determine possible barriers and underlying causes. Lastly, fishbone diagram was constructed to represent the active factors throughout the accident. In addition to the RCA, the organizational safety culture of the company was also interpreted based on the accident information and possible root causes. Overall, the interpretation of the root causes of these accidents is important for the development of safe transportation.

Keywords: Root-Cause Analysis · Organizational safety culture
Transportation accident investigation · Lac-Mégantic Rail Disaster

1 Introduction

Accident analysis might be used for different purposes such as identification of contributory factors and developing effective countermeasures. The causes of an accident could be classified as root cause, direct cause, and contributing causes. Root causes are defined as core causes representing set of deficiencies. Interventions to root causes have the ability to solve possible future problems [1]. The RCA aims to identify latent or

S. Bagnara et al. (Eds.): IEA 2018, AISC 823, pp. 199–208, 2019.
https://doi.org/10.1007/978-3-319-96074-6_21

root cause factors by applying series of analysis processes. Moreover, considering a system-based approach by identifying both factors directly related to the accident and latent factors associated with the organization and other components provide better understanding of the accident [2]. Analysis based on different dimensions of accidents also results in more effective and comprehensive interventions [3]. In the present paper, the Lac-Mégantic Rail Accident is investigated by using different methods related to root causes and organizational safety culture.

1.1 Information Regarding the Accident

The train, MMA-002 – Montreal, Maine & Atlantic Railway (MMA), began its travel from Farnham at 13:55 to Saint John, New Brunswick, Canada on 5th July 2013. The train included 72 tank cars with 7.7 million liters of petroleum crude oil, a buffer box car, and a locomotive consist. The locomotive consist included 5 locomotives and one special-purpose caboose, a VB car [4]. The train was operated by the locomotive engineer (LE) who is familiar with the environment and transportation route. The locomotive engineer was alone and controlling train in the lead locomotive, MMA5017. The LE reported several mechanical problems with the lead locomotive that result in problems controlling the speed of the train throughout the transportation. After reaching to Nantes where a stop took place at 22:50, the LE used automatic air brakes on a descending grade. He also used independent air brakes and hand brakes on the locomotive consist, the control car and the buffer car [4, 5]. After releasing the automatic air brakes, the effectiveness of handbrake was tested without releasing the locomotive independent air brakes. Finally, the LE reported to MMA's yard office that the train was secured. Then, the LE called the rail traffic controller (RTC) in Bangor, Maine about the mechanical problems during the travel and excessive black and white smoke that was coming while the train was parked. Both, the LE and the rail traffic controller agreed the mechanical problems will be repaired in the morning [4, 5].

At 23:30, a local taxi was called to take the LE to a hotel. The taxi driver realized the smoke and oil that leap to the windshield of the car and then talked about the smoke with the LE. The LE stated that he talked with Montreal, Maine & Atlantic Railway and agreed to leave it in that situation [4].

At 23:40, the Nantes Fire Department received a call reporting a fire on the train. The Sûreté du Quebec (SQ) informed the Farnham RTC about the fire. MMA sent a track foreman to the fire department at Nantes after not reaching to the LE. When the track foreman reached to the field, the firefighters used emergency fuel cut-off switch; thus, the fuel source was removed from the fire and the fire was put out. After that, the Farnham Rail Traffic Controller was informed about the condition of the train by both the firefighters and the track foreman. After a discussion about the condition of the train with rail traffic controller, everyone left the area [4]. Before 01:00 on 6th July, the air in the train's brake system was leaked and that result in the decrease in the force that was holding the train from moving. During that time, there was no locomotive working and, after that, around 01:00, the train began to move to Lac-Mégantic that is 11.5 km away from the stop location of the train. The derailment of the train took place near the center of the town after approximately 15 min. The derailment resulted in the spillage of the 6 million liters of crude oil. After the derailment, multiple explosions and a large fire took place [4, 5].

Meanwhile, the locomotive consist did not derail but separated into two and continued to move. The two sections stopped 145 m away from each other (see Picture 1). At 03:30, the MMA officials secured the locomotive consist by using hand brakes [4].

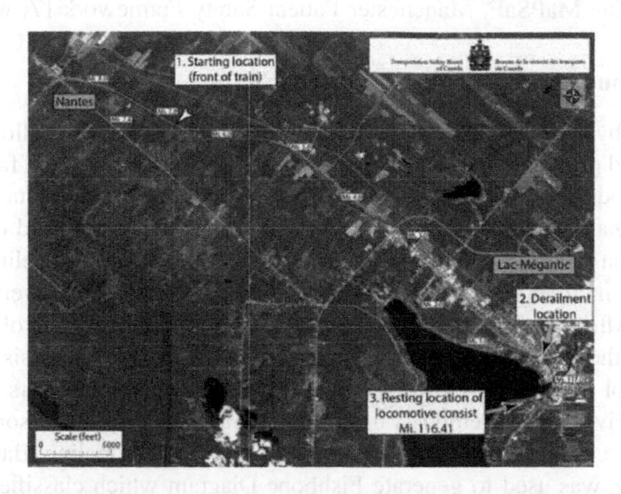

Picture 1. Three Locations of the Accident [4]

The temperature was between 20.5 °C and 21.7 °C before 01:00 and 21.2 °C after 01:00. Moreover, the speed of the wind was 5 km/h before 01:00. After 01:00, there was no wind. The locomotive was a single-person train and the LE got enough sleep before the travel. The LE had also completed a total number of 60 trips with MMA-002 and 20 of them were single-person trains. The train included both automatic and independent brakes. Moreover, the black box of the train, locomotive event recorder also recorded much information regarding the accident [4].

Overall, the accident resulted in 47 deaths and destruction of 40 buildings and 53 cars (see Picture 2). Around 6 million liters of crude oil contaminated 31 hectares area. 100 000 L of crude oil mixed in with lakes and sewer systems. Only around 740 000 L of crude oil were saved from the derailed tank cars out of 6.7 million liters. The clean-up operations began only 2 days after the derailment [4].

Picture 2. The accident site [4]

2 Method

In the present study, RCA Toolkit developed by National Patient Safety Agency [6] of England and The MaPSaF, Manchester Patient Safety Framework [7] were used.

2.1 Design and Procedure of the Present Study

As suggested by Cerniglia-Lowensen [8], the RCA took place as following different stages, case and problem identification, data gathering, identification of factors by using different methods and finally developing future suggestions. The data related to the accident was gathered from the report of Transportation Safety Board of Canada [4]. The information taken from the report was used to structure the timeline of the accident. Timeline method was used to draw the chronological flow of events involved in the accident. After forming the timeline of the accident, possible control measures that failed during the accident were determined by using barrier analysis method. The possible control measures were used to construct the five main factors of Five Whys method. The Five Whys method is used to determine underlying reasons of the accident [9]. All the information gathered from the reports and Timeline, Barrier Analysis, and Five Whys was used to generate Fishbone Diagram which classifies the determinants of the accident by visualizing the major causes [10].

In addition to the RCA, The MaPSaF, Manchester Patient Safety Framework, is developed to understand the perceived importance of safety in the organization. The framework was originally designed for health care organizations. It includes nine dimensions with five levels of safety culture [6]. The levels of safety culture are pathological (A), reactive (B), bureaucratic (C), proactive (D) and generative (E). In pathological level, safety issues are evaluated as a waste of time and money. In reactive safety culture, safety practices are concerned only when there is an accident. Bureaucratic safety culture generates different systems for the management of safety issues that are not used systematically. Proactive safety culture systematically works on safety issues. Finally, generative safety culture approaches safety issues as an internal part of the whole system within the organization [6, 7]. The related factors of the accident were used to evaluate safety culture of the organization.

3 Results and Discussion

3.1 Timeline

The Timeline method was used to highlight important points regarding the chain of events. The highlighted events were used to evaluate the accident by analyzing its components. In the timeline, the times of the events and elapsed time between these events might be used to determine information gaps or questionable areas that need to be clarified. Since the timeline method provides prior information and general picture regarding the accident, the Fig. 1 forms a basis for the rest of the methods.

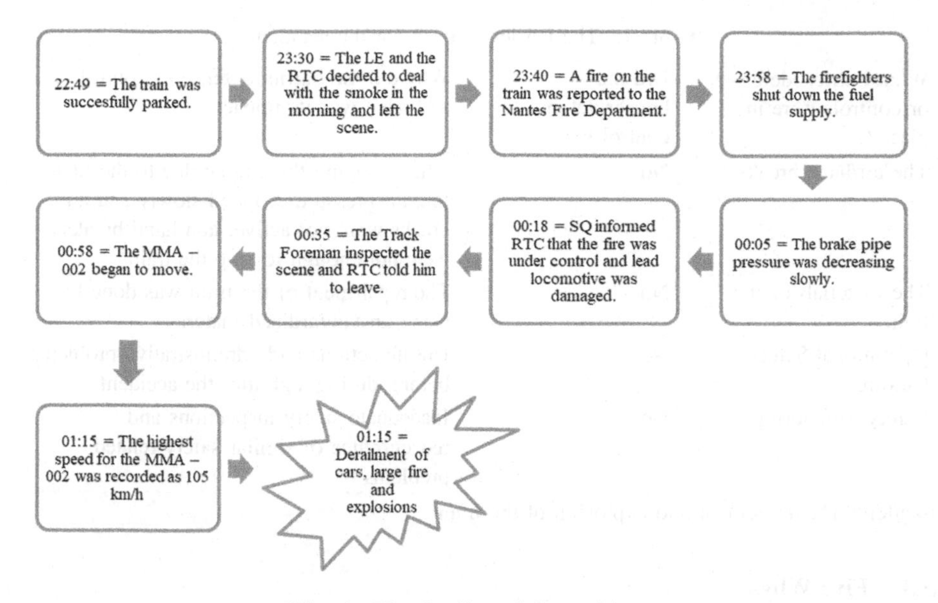

Fig. 1. The timeline of the accident.

3.2 Barrier Analysis

For the main accident which is the second fire and the explosions, barrier analysis was conducted by using the information gathered from the timeline (Table 1). The main mechanical and procedural failures were identified as air/hand breaks, materials of the train, operational safety culture, and safety monitoring. The barriers were constructed in a way that moves from initial factors to more general, comprehensive, and behind the scenes factors such as safety monitoring. The first barrier was the air/hand breaks which could prevent the train from moving and stop the derailment. The insufficient hand breaks and not active air breaks resulted in the movement of the train indicating the failure of the first barrier. Inadequate defenses, such as break system failures were the final line before the occurrence of an accident [11]. The second barrier was the material of the train used for repairment. Using unstandardized materials resulted in fire and the failure of this barrier. The third and more comprehensive barrier was the operational safety culture. The decisions taken before, during and after the accident caused different events and leading to failure of the third barrier. The final barrier was determined as safety monitoring. The inadequate safety inspections and reoccurrence of similar safety related problems led to failure of the final barrier.

Table 1. The barrier analysis for the accident

What barriers/defenses or controls were in place?	Did the barrier/defense or control work?	Why did the barrier/defense or control fail and what was its impact?
The air/hand breaks	No	After stopping the engine due to the first fire, air pressure dropped slowly and air breaks were not active, and hand breaks were not enough to stop the train
The materials of the train	No	The repairment of the train was done by using unstandardized materials
Operational Safety Culture	No	Unsafe actions and administrative problems before, during and after the accident
Safety Monitoring	No	Inadequate safety inspections and reoccurrence of similar safety related problems

Incident: The derailment and explosion of the train

3.3 Five Whys

Beginning with the Lac-Mégantic train derailment followed by multiple explosions and from the contributory factors to the accident and outcomes of the accident, five whys were determined. Each why included the main factors that resulted in several sub-factors affecting the accident process and the severity of the accident (see Fig. 2). The first why was related to the last event before the derailment identified as the movement of the train. The second why was about the one of the chains resulting the movement of the train and occurring because of unstandardized materials. The answer of the second why formed the third why. The answer of third why is the reason behind the use of unstandardized materials. The third why led to fourth and fifth whys as the application unsafe practices and insufficient inspections of safety issues.

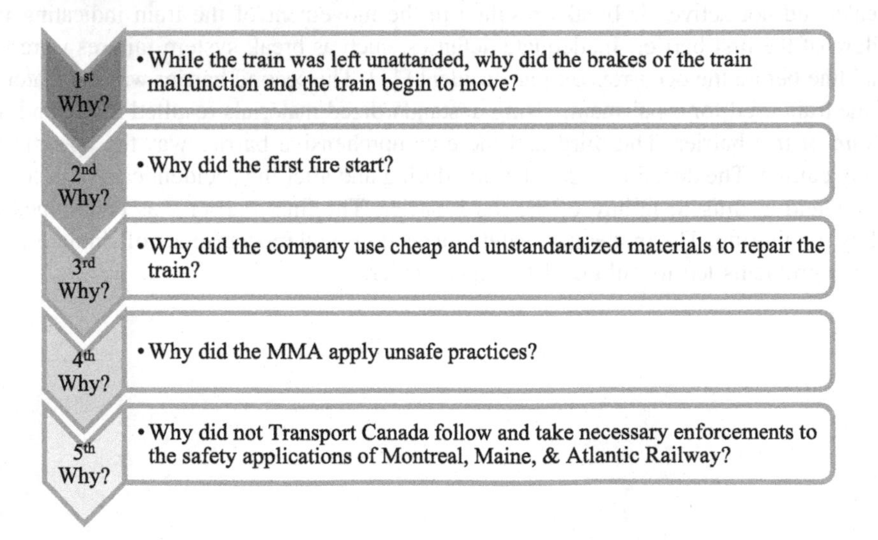

Fig. 2. Five whys analysis of the accident

3.4 Fishbone Diagram

By using the Timeline, Barrier Analysis and Five Whys, the contributory factors were determined and used to construct the fishbone diagram (Fig. 3). The Fishbone Diagram helps to understand how the accident occurred and what the roles of these factors were during the accident. When all these factors were considered, different sub-factors were determined for all factors except individual factors. For example, for the task factors, the improper hand brake test conducted by the locomotive engineer was evaluated as one of the determinants of the accidents and outcomes. In addition to this, some of the factors included more than one sub-factor, which can be interpreted as these factors were active in multiple phases of the accident and they had dynamic and various effects throughout the accident process. For instance, equipment and resource factors involved unstandardized locomotive materials and maintenance, and break and air pressure problems sub-factors.

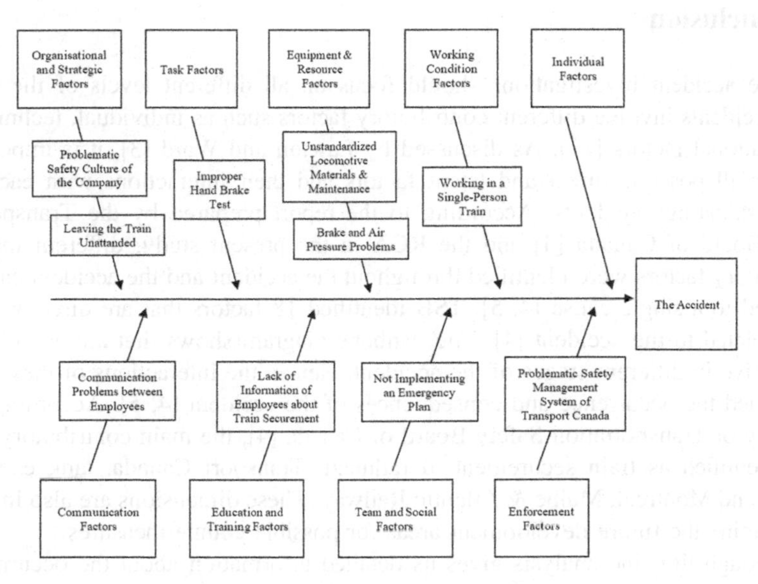

Fig. 3. The Fishbone Diagram of the accident causes

3.5 Organizational Safety Culture

As discussed by Lacoursière, Dastous and Lacoursière [5] effective safety culture in railway organizations could result in decreased number of accidents. According to the general definitions of organizational safety culture, some of the dimensions which are available for the evaluation in the reports were reviewed for the accident. In general, the organizational safety culture of the MMA was evaluated as being reactive because of all problems with safety issues and lack of systematic safety culture procedures [4, 5]. Moreover, there have been many indicatives of problems in organizational safety culture such as leaving the train unattended [5]. It has been also found that the employees were not getting effective training in terms of train securement [4]. Education and experience of stuff is an important factor for railroad industry [12]. In addition to the problematic areas on Montreal, Maine & Atlantic Railway, Transport Canada performed enforcement operations that might be evaluated as insufficient in terms of safety [4]. Insufficient safety management systems for railways performed by Transport Canada might have increased the risk in railways [5].

As seen by the reports [4] and evaluations according to the MaPSaF [6], the organizational safety culture of the company should be moved from being reactive to bureaucratic and proactive which might be succeeded by developing new countermeasures. According to the report of Transportation Safety Board of Canada [4], different countermeasures were applied after the accident. In the future, some interviews might be applied to understand changes in the applications and general organization safety culture of the MMA.

4 Conclusion

Effective accident investigations should focus on all different levels of the system. Train accidents involve different contributory factors such as individual, technical and organizational factors [12]. As discussed by Lawton and Ward [3], it is important to consider all possible direct and latent factors and their interactions with each other while evaluating accidents. According to the report prepared by the Transportation Safety Board of Canada [4] and the RCA in the present study, different main and contributory factors were identified throughout the accident and the accident cannot be attributed to a single cause [4, 5]. TSB identified 18 factors that are directly or indirectly related to the accident [4]. The fishbone diagram shows that almost all factors were active in different stages of the accident. Hence, the interactions of these factors determined the occurrence and consequences of the accident [4, 5]. According to the summary of Transportation Safety Board of Canada [4], the main contributory factors were identified as train securement, derailment, Transport Canada, tank cars, locomotive, and Montreal, Maine & Atlantic Railway. These dimensions are also important to determine the future development areas for possible countermeasures.

Although timeline analysis gives us detailed information about the occurrence of the accident and events after that, the five whys show that there are other contributory factors affecting the occurrence of the accidents months and years ago. The decisions made just before the accident and years earlier could affect the occurrence of an accident [12].

The RCA shows that there are many factors from different general areas such as human factors, engineering, education, and enforcement. As indicated by Groeger [13], the Es of road safety might be used to cover all the problematic areas by including safety culture factors and suggesting different future countermeasures. The organization was evaluated as showing ineffective organizational safety culture in terms of management of the risk. Canadian Centre for Policy Alternatives [14] and Transportation Safety Board of Canada [4] also expressed that the accident resulted in the development of some regulation awareness.

Finally, while constructing the factors, different reports were evaluated and findings were presented by using certain methods. However, it should be noted that other methods such as the Swiss Cheese Model and the Bow-Tie Model might also be used to present comprehensive findings [1, 11]. It might be suggested that reports based on accidents should include detailed information, which can be used as a source for future studies. As stated by Cerniglia-Lowensen [8], the first step of an effective RCA is the data collection. Since the present study is prepared based on different reports, future analysis should also consider this limitation.

Acknowledgements. The authors wish to thank Prof. Dr. Türker Özkan and Yeşim Üzümcüoğlu Zihni for reviewing a draft of this manuscript and providing valuable feedback.

References

1. Matsika E, Ricci S, Mortimer P, Georgiev N, O'Neill C (2013) Rail vehicles, environment, safety and security. Res Transp Econ 41:43–58. https://doi.org/10.1016/j.retrec.2012.11.011
2. Reason J (1990) Human error. Cambridge University Press, New York
3. Lawton R, Ward NJ (2005) A systems analysis of the Ladbroke Grove rail crash. Accid Anal Prev 37:235–244. https://doi.org/10.1016/j.aap.2004.08.001
4. Transport Safety Board of Canada (2014) Railway Investigation Report R13D0054, Runaway and Main Track Derailment, Montreal Maine and Atlantic Railway Freight Train MMA-02 Mile 0.23, Sherbrooke Subdivision, Lac-Mègantic, Quebec, 06 July, 2013, Minister of Public Works and Government Services Canada, Transportation Safety Board, Ottawa, Cat no TU3–6/13–0054E, ISBN 978-1-100–24860-8
5. Lacoursière JP, Dastous PA, Lacoursière S (2015) Lac-Mègantic accident: What we learned. Process Saf Prog 34(1):2–15. https://doi.org/10.1002/prs.11737
6. National Patient Safety Agency (2004) Root cause analysis toolkit. http://www.nrls.npsa.nhs.uk/. Accessed 27 Aug 2010
7. Parker D (2009) Managing risk in healthcare: understanding your safety culture using the Manchester Patient Safety Framework (MaPSaF). J Nurs Manag 17(2):218–222. https://doi.org/10.1111/j.1365-2834.2009.00993.x
8. Cerniglia-Lowensen J (2015) Learning from mistakes and near mistakes: using root cause analysis as a risk management tool. J Radiol Nurs 34(1):4–7. https://doi.org/10.1016/j.jradnu.2014.11.004
9. Ohno T, Bodek N (1988) Toyota Production System: Beyond Large-Scale Production. Produc-tivity Press, New York
10. Jayswal A, Li X, Zanwar A, Lou HH, Huang Y (2011) A sustainability root cause analysis methodology and its application. Comput Chem Eng 35:2786–2798. https://doi.org/10.1016/j.compchemeng.2011.05.004
11. Suryoputro MR, Sari AD, Kurnia RD (2015) Preliminary study for modeling train accident in Indonesia using Swiss Cheese Model. Procedia Manuf 3:3100–3106. https://doi.org/10.1016/j.promfg.2015.07.857

12. Reinach S, Viale A (2006) Application of a human error framework to conduct train accident/incident investigations. Accid Anal Prev 38:396–406. https://doi.org/10.1016/j.aap.2005.10.013
13. Groeger JA (2011) How many e's in road safety? In: Porter BE (ed) Handbook of Traffic Psychology. Elsevier Inc, USA, pp 3–12
14. Canadian Centre for Policy Alternatives: Lac-Mègantic, Loose Ends and Unanswered Questions. https://www.policyalternatives.ca/sites/default/files/uploads/publications/National%20Office/2015/01/Lac_Megantic_Loose_Ends_and_Unanswered_Questions.pdf. Accessed 28 Apr 2018

Study on the Analysis of Traffic Accidents Using Driving Simulation Scenario

Ya-Lin Chen[1,2], Min-Yuan Ma[2], Po-Ying Tseng[1], Yung-Ching Liu[3], and Yang-Kun Ou[1(✉)]

[1] Department of Creative Product Design, Southern Taiwan University of Science and Technology, Tainan, Taiwan
ouyk@stust.edu.tw

[2] Department of Industrial Design, National Cheng Kung University, Tainan, Taiwan

[3] Department of Industrial Engineering and Management, National Yunlin University of Science and Technology, Yunlin, Taiwan

Abstract. This study investigated the visual search ability and number of collisions of subjects in different age group when encountering dangerous situations in a traffic environment. A total of 32 subjects were recruited in this study, which included the 12 young group (18–25 years old) and 12 elderly group (65–75 years old) respectively. Subjects wore the eye tracker to drive inside the driving simulator. There were 5 possible real road risky driving scenarios designed for the experiment. The result demonstrated that elderly group had a longer fixation duration than young group, and more fixation counts as well. However, the result found that the elderly group had no significantly higher than young group in the number of collisions. The results of the study were expected to assist in understanding the visual search behavior of elderly group in the road risk situation.

Keywords: Elderly · Risk awareness · Eye tracking · Driving simulation

1 Introduction

Transportation has become necessities of life. However, there are also many traffic accidents in life. According to government statistics, there were 215,703 traffic accidents in Taiwan in 2017. The deaths number was up to 1,126 of the accidents. According to the U.S. Transportation Authority, human factors have accounted for 92.6% of incidents in traffic accidents, and mostly due to improper attention, inattention, excessive speed, and distraction of drivers; Environmental factors have accounted for 33.8%, and mostly were because of the affect of signs, obstacles or buildings on the road (Treat et al. 1977). The road environment and thoughtless driving behavior tend to decrease the driving safety (Zhao and Rong 2013; Engström et al. 2013). According to the statistical survey of National Highway Police Bureau, National Police Agency Ministry of the Interior, most traffic accidents were caused by improper driving behavior (NPA 2016). Besides, the driver's inattention and improper attention were also one of the main reason of road hazards.

S. Bagnara et al. (Eds.): IEA 2018, AISC 823, pp. 209–216, 2019.
https://doi.org/10.1007/978-3-319-96074-6_22

Horswill (2008) and others have indicated that elderly drivers have visual impairments in driving due to degeneration of sensory function. Attebo et al. (1996) have also found that the driver's ability of visual attention decreases with the increase of age. As a result, the incidence of driving accidents has increased. Therefore, it is necessary to be capable of noticing relevant information and ignoring irrelevant information which the information may be involved in complex traffic scenes in the driving environment when driving.

Driving is a complexity operation that requires great concentration. People have to use multiple sensory when driving a vehicle, such as: sense of sight, hearing, or touch, etc., in which the visual sense is especially matters. There are studies have indicated that people's decision-making while driving affects by situation awareness. Good situation awareness can help people construct proper mental status and make appropriate decisions. The higher the situation awareness is, the more accurate it is to predict what may happen in the future and to reduce the chances of danger as well (Hartel et al. 1991). With the increase of age, the visual quality, sensory ability, and mental action ability have all gradually declines, which causes the perception ability to the surrounding information declined (Bolstad 2001). Drivers had to quickly define information in traffic environment in the driving activities, such as: traffic signs, warning signs, speed, etc. At the same time to understand what will happen during the driving process, such as pedestrians, lane departures, etc., in order to determine and perform the behavioral decisions to assure the driving safety, and to avoid traffic accidents as well. Elderly also have to predict or to detect where and when the danger may occur in the surrounding area and adjust their behavior (Scialfa et al. 2011). Therefore, effectively detects dangers can assure enough reaction time for elderly drivers to avoid traffic accidents. According to studies, visual attention is extremely important in driving behavior. However, the driver's the perception ability to the surrounding information declines with the increase in age. Therefore, the purpose of this study is to investigate whether the elderly drivers are able to react appropriately when facing complex driving situation and the upcoming dangerous event to avoid traffic accidents.

The studies have indicated that the driver's situation awareness and the attention ability declines with the increases in age. Therefore, the intension of study is to learn more about the reaction of drivers in different age groups when they encounter dangerous events in a complex road situation. The purpose of this study was as follows: (1) To investigate whether the drivers in different age group can effectively detect and react to dangerous events? (2) Whether the eye attention behavior results is significant difference between elderly drivers and young drivers from the eye tracker when the dangerous situation occurs? (3) To investigate the detection and reaction method of elderly drivers who may potentially affect traffic safety incidents by using driving simulator experiment.

2 Research Method

This study used the risk situation test to investigate the visual search patterns of the elderly group in the traffic environment. As well as which areas were the visual focus point. In order to understand the fixation position of the elderly when they encounter dangerous events.

2.1 Experience Subjects

A total of 32 subjects were recruited in this study, 16 subjects were from 65–75 years old and 16 subjects were from 18–25 years old. Each subject's visual acuity must achieve (or achieved after correction) 0.8 or more, no color blindness (through Ishihara Color Test), and had road driving experience. Subjects understood and agreed to participate in this experiment. The consent form must be signed before the experience, and subjects received a $10 for traveling expenses allowance after completed the experiment.

2.2 Experience Equipment

Eye Tracker

The Eye tracker (Mobile eye XG) uses two cameras to record the individual location of eyeballs and environment. One of the cameras was used to measure eyeball movement and degree of pupil dilation; Another camera was used to photograph the environment which seen by the subjects. Then, integrated the eye fixation position with the environment position after correction. The eye tracker's appearance was like a pair of glasses and unlimited to indoor or outdoor environments while operating. Users who wore the eye tracker was capable of free movement comprehensively and therefore can be integrate into daily life. Besides, there was also a lens frame specially developed for who wears the glasses, and there was no necessary to give up the glasses that have been properly adapted (Fig. 1). The eye tracker calibrates the rotation angle of eyeball to correspond the subject's fixation position through correction procedure. The eye tracker sampling frequency was 25/30 Hz, and the error was within 0.5°.

Fig. 1. Eye tracker

Driving Simulator

The STISIM Model 100WS driving simulator developed by STI (System technology, Inc.) was used in scenario construction in this experiment (see Fig. 2). The driving environment was constructed by SDL (Scenario Definition Language V8.1) to create scripts that simulate the road environment. The script was designed as a two-way and two-lane road section, and the maximum speed limit was 60 km/hr. A total of five dangerous traffic environment scripts were designed in the experiment, which including: motorcycle driving against the traffic, right turn without play the direction lights, linked car make a right turn and etc. Each traffic script was about 20s–30s long. The subject was asked to press a button immediately when realized a sudden event occurred when watching the video in order to collect the reaction time, the near miss event, and the correct rate. The total experiment was about 1 h.

Fig. 2. Driving simulator

Road Driving Script

The simulation script in this study was wrote by SDL (Scenario Definition Language V8.1). The road environment design was mainly based on urban roads. The criteria for the road design were: road type, turns, density of opposite directions, crossroads, and density of roadside houses. The urban road script design was based on the study of Avinoam et al. (2009), which were a two-way single lane, multi-turns, more crossroads, and more frequent coming vehicles on the opposite side. A total of five dangerous driving scenarios (S1–S5) were designed, and all of the five driving environments were in accordance with real events.

S1. There was a crossroad in front of a straight road; a vehicle truck (T) was turning to the left on the opposite lane; a pedestrian (P) was crossing the road simultaneously on the opposite lane (Fig. 3a); the truck in the front turned left, and the pedestrians on the opposite lane crossed the street.

S2. There was a crossroad in front of a straight road; The driver (D) would drive straight forward with a truck (T) in front; the truck (T) would turn left at the intersection and block the driver's sight that there was a vehicle (V) turning left on the opposite lane (Fig. 3b); the front vehicle turned left; the vehicle on the opposite lane turned left.

S3. There was a crossroad in front of a straight road; the driver (D) would drive straight forward and there was a truck (T) parked next to the road and blocked the driver's sight; a pedestrian (P) was at the zebra crossing and was going to cross the road (Fig. 3c); pedestrians crossed the street.

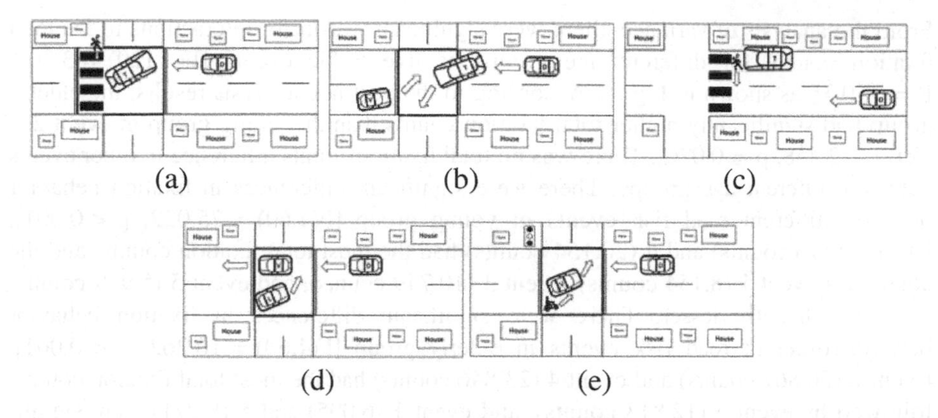

Fig. 3. Driving scenario settings

S4. There was a crossroad in front of a straight road; the driver (D) drove straight forward with the car in front (L); the car (L) accelerated past the road at the crossroad, meanwhile there was a vehicle (V) turning left on the opposite lane (Fig. 3d); the vehicle on the opposite lane turned left.

S5. At a crossroad, the traffic signal was green and the driver (D) drove straight forward; there was a car (V) turned left on the opposite lane, and followed by an illegal left-turned motorcycle (S) (Fig. 3e); The motorcycle made an illegal left-turn.

2.3 Experience Design

This study was a 2-factor experiment with 2 (different age groups: young vs. elderly) × 5 groups of audio driving videos. The dependent variables of this study were divided into two parts. The first part collected the numbers and durations of visual attention area of the subjects when distinguishing the dangerous traffic environment; The second part collected and distinguished the number of collisions in dangerous traffic hazards. The factors in each group were designed with counterbalance method to avoid learning or sequential effects.

3 Results

3.1 Eye Movement Behavior

Total Fixation Counts of Risk Event

The results of ANOVA analysis indicated that there were significant differences in the total fixation count of risk events between different age groups and five traffic events [$F(1,30) = 4.526$, $p < 0.042$; $F(4,120) = 32.751$, $p < 0.001$]. The elderly group had more total fixation counts than young group (elderly: 15.54 counts vs. Young: 12.19 counts). In the five different traffic event, the event 2 (21.43 counts) and Event 4 (22 counts) had the most total fixation counts, followed by event 3 (11 counts), and event 1 (7.125 counts) and event 5 (6.990 counts) had the least total fixation counts.

From the analysis of variance, there were significant two-factor interactions in the total fixation counts for different age groups × five traffic events [$F(4,120) = 3.302$, $P = 0.013$], as shown in Fig. 4. According to the post hoc analysis results, the elderly group had significantly higher total fixation counts than the young group in event 2 [$t(30) = -2.458$, $p = 0.020$]. There was no total fixation counts difference in other events between different age groups. There were significant differences in fixation behavior between different road risk events in young group [$F(4,60) = 25.022$, $p < 0.001$]. Events 2 (16 counts) and 4 (20.154 counts) had the most total fixation counts, and the counts for event 1 (8.154 counts), event 3 (10.749 counts), and event 5 (5.909 counts) were significantly lesser. There were significant differences in fixation behavior between different road risk events in elderly group [$F(4,60) = 16.262$, $p < 0.001$]. Event 2 (26.867 counts) and event 4 (23.846 counts) had the most total fixation counts, followed by event 3 (12.813 counts), and event 1 (6.095) and 5 (8.071) were significantly least counts.

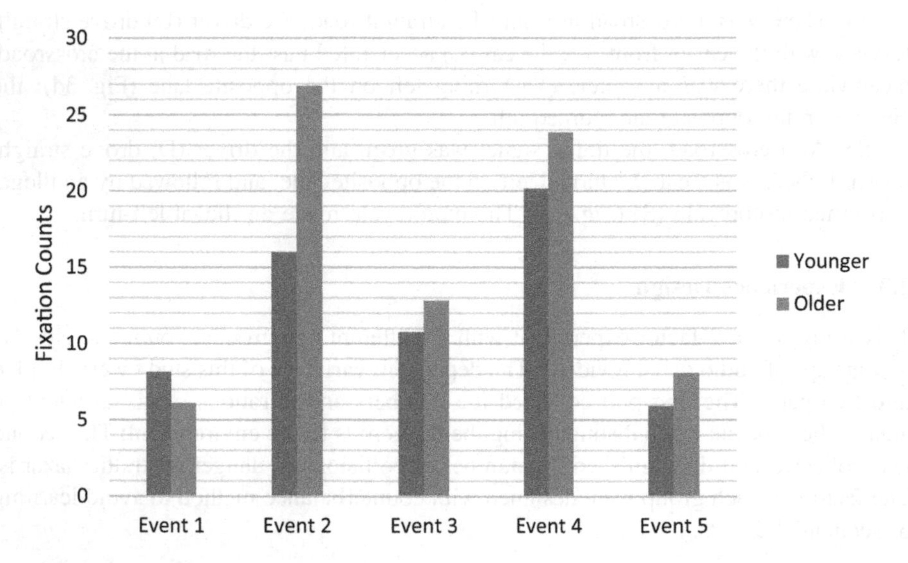

Fig. 4. Total fixation counts of different risk events for elderly group and young group

Total Fixation Duration of Risk Event

The results of the ANOVA analysis demonstrated that there were significant differences between different groups and the five traffic events in total fixation duration of risk events [$F(1,30) = 4.494$, $p < 0.042$; $F(4,120) = 28.604$, $p < 0.001$]. Elderly group had longer total fixation duration than young group (Elderly: 4.833s vs. Young: 3.609s). In the five different traffic incidents, event 2 (7.530s) and event 4 (6.107s) had the longest total fixation duration, followed by event 3 (3.621s), and event 1 (2.083s) and event 5 (1.764s) were the shortest.

3.2 Numbers of Collision

The results of the chi-squared test demonstrated there was no significant difference in the numbers of collision between the elderly group and young group under different traffic situations. Which showed that there was no higher risk of collision for the elderly in this five traffic scenarios.

4 Discussion

The result shows that among the subjects in elderly group demands a much more total fixation counts and longer duration than the subjects in young group. Past studies indicated that the increased time in total fixation duration was related to the complexity of the task which the subject performed. The elderly group had a longer total fixation duration than young group when the two groups are operating the same task. It showed that elderly group who had suffered a larger burden on the event. Therefore, the total fixation duration and counts increased (Bhise 1986).

The result demonstrates that the total fixation duration of elderly group is 1.24s longer than the young group. Even though it would affect the driving behavior, it was still within acceptable driving safety range when the driver's attention left from the road for 1–2s. However, there was immediate and significant impact on traffic safety when the driver's attention left the road for more than two seconds (Zwahlen et al. 1988; French 1990). Therefore, in this situation, the increase of total fixation duration of the elderly group may affect road safety and increase driving risk.

The event 1, event 3, and event 5 have shorter total fixation duration and lower total fixation counts than the others in the road risk scenario. The reason might be that these 3 road risk scenarios are the violations of motorcycle and pedestrians. The driver might pay more attention about the possible collision risk of the car ahead because the car collision could possibly result in a greater injure to the driver. Therefore, the driver ignores the more fragile objects and groups. Thus, the motorcycle and pedestrians play a riskier role in the road environment.

There are also limitations in this study. This study copies the real world possible road risk situation to a simulated environment experience in order to reduce the research risk. However, there might possibly certain differences between driving in the simulator and driving on the real road. The physiological variables and driving behavior variables of the subjects do not present in this study. It is suggested that both of the variable data can be adopted and collected in order to better understand drivers' physiological and psychological reactions when they are confronting with the risk situation in the future studies.

References

National Police Agency Ministry of the Interior, R.O.C - accident statistics (2017). http://stat.motc.gov.tw/mocdb/stmain.jsp?sys=100&funid=b3303

Avinoam B, David S, Oron-Gilad T (2009) Age, skill, and hazard perception in driving. Accid Anal Prev 42(2010):1240–1249

Bhise VD, Forbes LM, Farber EI (1986) Driver behavioural data and considerations in evaluating in-vehicle controls and displays. Presented at the transportation review board 65th annual meeting, Washington, DC

Bolstad CA (2001) Situation awareness: does it change with age? Proceedings of the human factors and ergonomics society annual meeting, vol 45, no 4, pp 272–276

Engström J, Monk CA, Hanowski RJ, Horrey WJ, Lee JD, McGehee DV, Regan M, Stevens A, Traube E, Tuukkanen M, Victor T, Yang CYD (2013) A conceptual framework and taxonomy for understanding and categorizing driver inattention. European Commission, Brussels

French RL (1990) In-vehicle navigation-status and safety impacts. Technical papers from ITE's 1990, 1989, and 1988 conference, pp 226–235

Hartel CE, Smith K, Prince C (1991) Defining aircrew coordination: searching mishaps for meaning. Paper presented at the sixth international symposium on aviation psychology, Columbus

Scialfa CT, Deschênes MC, Ference J, Boone J, Horswill MS, Wetton M (2011) A hazard perception test for novice drivers. Accid Anal Prev 43:204–208

Treat NS, Tumbas ST, McDonald D, Shinar RD, Hume RD, Mayer RL, Stansifer, Castallen NJ (1977) Tri-level study of the cause of traffic accidents: final report

Zhao X, Rong J (2013) The relationship between driver fatigue and monotonous road environment. In: Computational intelligence for traffic and mobility

Zwahlen HT, Adams CC, Debald DP (1988) Safety aspects of CRT touch panel controls in automobiles. In: Gale AG, Freeman MH, Haslegrave CM, Smith P, Taylor SP (eds) Vision in vehicles II. Elsevier, Amsterdam, pp 335–344

Studying a New Embarking and Disembarking Process for Future Hyperloop Passengers

Danxue Li[1], Wilhelm Frederik van der Vegte[1], Mars Geuze[2], Marinus van der Meijs[2], and Suzanne Hiemstra-van Mastrigt[1(✉)]

[1] Faculty of Industrial Design Engineering, Delft University of Technology, Landbergstraat 15, 2628 CE Delft, The Netherlands
S.Hiemstra-vanMastrigt@tudelft.nl

[2] Hardt Hyperloop, Paardenmarkt 1, 2611 PA Delft, The Netherlands

Abstract. This paper presents an embarking and disembarking process for the hyperloop, a future high-speed transportation of passengers and goods in tubes. A concept of the (dis)embarking process has been designed and tested with two experiments. The first experiment was performed to compare the new concept to one that is more similar to the current embarking setup of trains on the aspects of efficiency and experience. Participants were asked to (dis)embark in the test settings that simulate the new concept and the conventional situation with luggage. As a result, new passenger flow saves 40% of the time for vehicles to stay on the platform. Follow-up questionnaires and interviews with the participants show that the proposed passenger flow gives a better experience in terms of efficiency, seamlessness and friendliness. The new solution increases the number of doors, which increases the manufacturing complexity and the chance of failure. Narrowing the door size minimizes this effect. Subsequently, a second experiment has been carried out to study the influence of door width on (dis) embarking efficiency and passenger experience following a similar method. It turns out that narrowing the door width does not noticeably influence the embarking time, but the disembarking time does increase. Interviews show that half of the participants sense a negative experience with narrower doors, while the other half do not notice a difference.

Keywords: Boarding · Passenger flow · Luggage solution

1 Introduction

1.1 Background

For years, people have been travelling on road, rail, water and by air. Each mode of transportation has its unique (dis)embarking process and they have not changed much over the years. New transportations, such as hyperloop, present the opportunity to design the (dis)embarking process from the passenger perspective without being restrained from legacy requirements.

Hyperloop is a network of tubes with a low pressure environment that reduce air resistance, allowing a high cruising speed with low energy use (Musk 2013) [1]. In this

S. Bagnara et al. (Eds.): IEA 2018, AISC 823, pp. 217–229, 2019.
https://doi.org/10.1007/978-3-319-96074-6_23

network, transfers and intermediate stops are reduced by having vehicles travel to their destination directly, making it comparable to air travel in terms of passenger process.

There will be a station where passengers get on and off the hyperloop vehicle. Because of the high throughput and departure frequency of the hyperloop system, preparing and planning in advance, and thus luggage pre-collecting of any kind before the boarding process will also not be necessary. Passengers will be guided to one of the platforms with their luggage and get on a vehicle that is almost ready to depart. This system demands an efficient (dis)embarking process that includes both passengers and their luggage, if any, right in front of the vehicle.

Research by Mas et al. [2] shows that in three specific aircraft models, the most efficient way of boarding is not assigning seats and let passengers board and choose a seat they like. Other studies state that, in theory, the Steffen method (where passengers board separately in a given order, illustrated in Fig. 1) is the fastest, because this maximizes the utilization of the aisle, especially for when placing hand luggage (Jaehn and Neumann 2014) [3]. However, the Steffen method can hardly be used in practice, since asking groups of travellers to board separately may decrease customer satisfaction (Steffen 2008) [4].

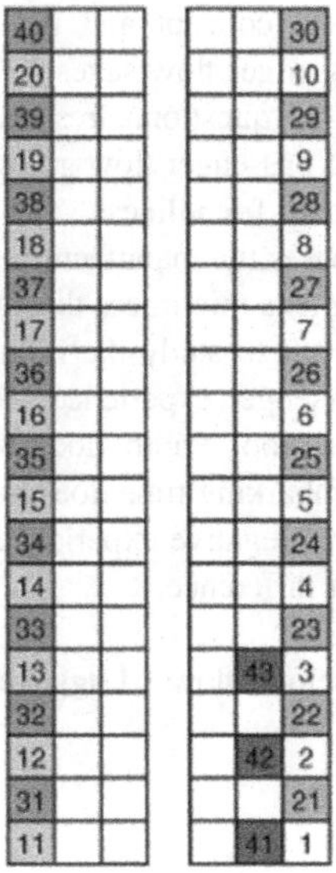

Fig. 1. Steffen method for boarding aircrafts (Steffen 2008).

Self-organizing phenomena in pedestrian crowds have been studied by Helbing et al. [5, 6], revealing dynamics in different ingress and egress environments. Helbing et al. [6] discovered that high interaction frequencies and high numbers of braking or avoidance manoeuvres slow down the average velocity in the desired direction of motion. Bottlenecks form another factor that reduces the pedestrian flow, and the longer the bottleneck, the slower the flow (Helbing et al. 2002) [5]. Besides the literature study, an interview on crowd dynamics with J. Li, an expert on crowd behaviour at Delft University of Technology, has also lead to valuable insights (Li 2018) [7]. She stated that any kind of hesitation in the crowd will cause a chain of reaction delays and slow down the flow or even create conflicts and stampedes in the crowd. Li also mentioned that, in an overcrowded situation, getting into a bottleneck is dangerous because the unawareness of what is happening in the bottleneck causes pressure and anxiety among the crowd. Therefore, eliminating bottlenecks and avoiding passenger hesitations will be beneficial to the (dis)embarking process.

1.2 Proposed Design Solution

Based on the results from the literature research, a design solution for the hyperloop boarding process is proposed [8]. Our design proposition separates the passenger compartments from luggage space (underneath passenger compartments) and influences the (dis)embarking flow by the arrangement of doors and seats. Multiple doors on both sides of the vehicle allow passengers to embark and drop the hold luggage on the one side and disembark and pick up the luggage on the other side (Fig. 2).

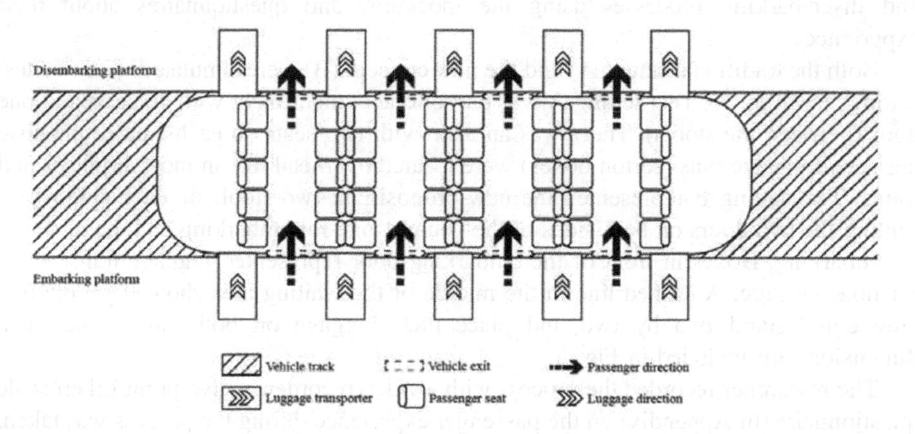

Fig. 2. Concept sketch of the new (dis)embarking solution (top view of hyperloop vehicle on platform).

1.3 Objectives

The aim of the study reported in this paper is to assess the efficiency and the experience of this new way of (dis)embarking by answering two research questions:

1. *Does the proposed design solution achieve higher efficiency and better user experience than the traditional (dis)embarking model?*

And, since the size of the doors affects the complexity of the system and narrow doors are preferred, the second research question:

2. *Does the width of the doors influence the efficiency and passenger experience while (dis)embarking?*

The two research questions were addressed separately in two experiments. After the description of the experiments, the results are discussed and a further improved design solution is presented based on the results.

2 Experiment 1: A Comparison Between the New Design and Traditional Embarking and Disembarking Model

The first experiment was conducted to determine if the new design achieves a higher efficiency and better user experience than the traditional (dis)embarking model.

2.1 Method of Experiment 1

Tests with mockups were performed in a laboratory environment. Ten people (9 males and 1 female) between the age of 24 and 53 years old volunteered to participate in the study. They were recruited from the university and from companies with different backgrounds. The study included observations of participants during their embarking and disembarking processes using the mockups, and questionnaires about their experiences.

Both the traditional setup (A) and the new concept (B) were simulated with two test settings (see Fig. 3). Test setting A was part of one compartment with 10 seats and one door (between the stools). There was an aisle with two seats on each side of the aisle per row. Luggage bins (carton boxes) were located overhead like in most airplanes and trains. Test setting B represented the new proposition: two stools on each side representing the two doors on both sides of the vehicle, one for embarking and the other for disembarking. Boxes in front of the embarking door represented luggage transporters for hold luggage. A dashed line in the middle of the waiting area showed people that they could stand two by two and place their luggage on both sides. The main dimensions are included in Fig. 3.

The researcher recorded the process with a video recorder. A five-point Likert scale questionnaire (in Appendix) on the passenger experience during the process was taken, including questions regarding whether it was clear what to do, and what level of convenience and comfort were experienced.

After the experiment, the videos of the test were analyzed and the time it took for each process was noted separately: (i) total embarking time: all embarking passengers get from the departure platform to their seats with their luggage well placed; (ii) total disembarking time: all disembarking passengers get from their seats to the arrival platform; (iii) total vehicle-on-platform time: the sum of the previous two; (iv) average embarking time per passenger: from in front of the door to being seated with their belongings well-placed; (v) average disembarking time per passenger: from standing up from their seats to getting out of the door with their belongings; and (vi) the average time of the (dis)embarking process per passenger: the sum of the previous two.

The average embarking and disembarking times per passenger were recorded separately. The sum of these two average times is the average time each passenger spends on (dis)embarking. Average scores for the experience of the new concept (B) and the traditional model (A) were collected using the questionnaires.

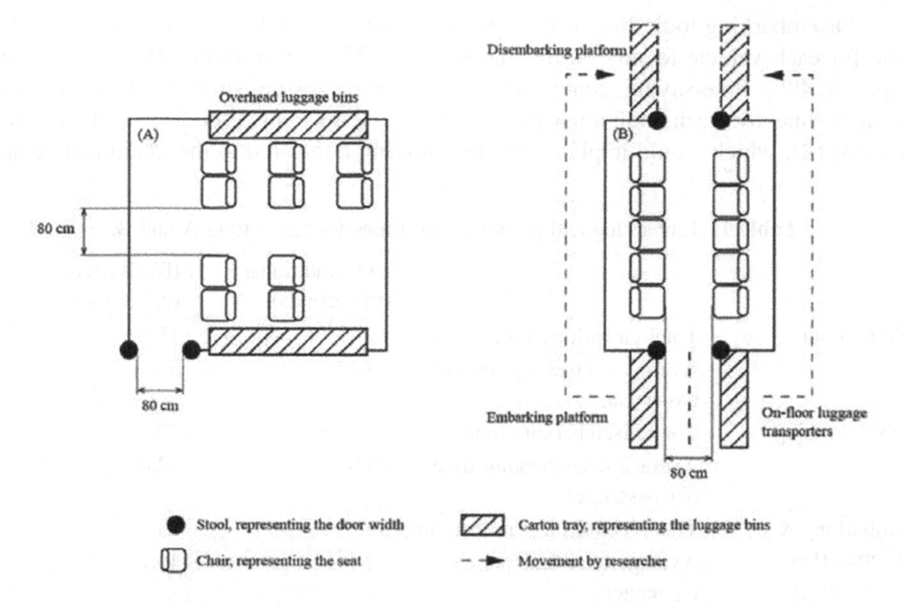

Fig. 3. Test setting A: traditional (dis)embarking experience (left). Test setting B: new (dis) embarking experience (right).

The ten participants were first introduced to test setting A and participants were each given an empty suitcase or a backpack as their personal belongings. They entered the vehicle and looked for a seat when the door was indicated to be open; Once all participants were seated with their belongings placed in the carton boxes, they were told that the door was closed and the vehicle would depart. After 30 s, the researcher indicated that the vehicle arrived at the destination and the door was opened. Participants collected their belongings and walked out through the door.

The same 10 passengers were then introduced to test setting B and participants used the same suitcases and backpacks from the previous test. Participants were told to wait on the embarking side. When they were told the vehicle arrived and the door opened, they placed their belongings in the cardboard boxes on both sides of the entrance door, walked into the aisle and sat down. Then the researcher moved their luggage behind the seats and told them the luggage was automatically moved into the luggage compartment under the passenger floor. They were then told that the door was closed and the vehicle departed. After 30 s, they were told the vehicle arrived at the destination and the door on the disembarking side was open. At this moment, the researcher moved the luggage boxes to the disembarking side on both side of the exit door. Participants walked out of the door on the right and picked up their belongings as they walked out. After the experiment, participants were asked to fill in the questionnaires.

2.2 Results of Experiment 1

Table 1 shows the comparison between test setting A and B in terms of (dis)embarking efficiency. Embarking in the traditional model (A) took 29 s and in concept (B) it took

15 s. Disembarking took 26 s in the case of A and 18 s in the case of B. In total, the time for each vehicle to stay on the platform was 33 s for concept (B), which would imply a 40% time-saving compared to traditional (dis)embarking. Moreover, the average time for each individual passenger to embark and disembark is 15.7 s for concept (B), which would imply a 50% time-saving compared to the traditional setup.

Table 1. Embarking and disembarking times for test setting A and B.

		(A) Traditional (in seconds)	(B) Concept (in seconds)
Embarking	Total embarking time	29	15
	Average embarking time per passenger	13.1	3.3
Disembarking	Total disembarking time	26	18
	Average disembarking time per passenger	18.9	12.4
Embarking & disembarking	Total vehicle on-platform time	55	33
	Average total time per passenger	32	15.7

Based on the questionnaire responses, the average user-experience scores for both setups were determined (Fig. 4). On 13 out of a total of 15 investigated aspects, the concept achieves a better user experience than the traditional setup. On the embarking platform, the new concept scored lower on making clear what to do, but it was experienced as more comfortable and pleasant compared to traditional model. During both

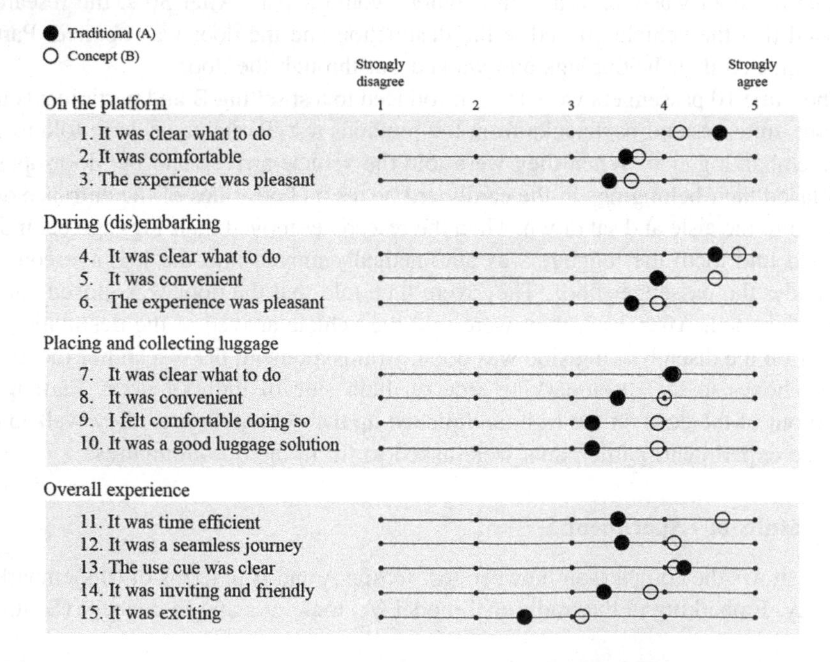

Fig. 4. Average experience score of traditional model (A) and concept (B). (n = 10)

(dis)embarking and placing/collecting luggage, the concept scored better on clarity, convenience, comfort and pleasantness. The overall experience of the concept was more time efficient, more seamless, less clear, more inviting, friendlier and more exciting than the traditional model. On one hand, the concept was considered considerably more time efficient, seamless, inviting and exciting; on the other hand, the procedure on the platform appeared to be less self-explaining. Overall, on 16 out of 18 investigated aspects, the concept achieves a better user experience than the traditional setup.

3 Experiment 2: Influence of Door Width on Passenger Embarking and Disembarking

In the proposed design concept for hyperloop, there are multiple doors on each side of the vehicle for embarking and disembarking. Since the vehicles will be operating in a depressurized environment, each door added to the chassis will need reinforcement around it to maintain the desired strength and stiffness. Therefore, the smaller the size is for each door, the better it can be realized from an engineering and economics point of view. The second experiment was conducted to investigate whether the width of the doors influences the efficiency and passenger experience.

3.1 Method of Experiment 2

In this follow-up experiment, test setting C was realized by reducing the door width of test setting B from 80 cm to 40 cm. It was intended to have passengers (dis)embark one by one. Therefore, unlike test setting B, C has the luggage transporters on only one side (see Fig. 5), while the other dimensions stayed the same as B. The instructions given and the procedure of the test remained the same as with test setting B. The experiment was video recorded and the same measurements as in experiment 1 were made by analyzing the video recording. The researcher also looked for behaviour differences between B and C during (dis)embarking, when passengers were looking for a seat or placing and collecting luggage, to explain possible different results in measurements. After the test, a semi-structured interview was conducted to compare the experience of B and C as well as to identify the reasons for the differences in experience.

3.2 Results Experiment 2

Passenger Behaviour
In the experiment, all participants first placed their luggage on the luggage transporter and stood in a line to wait for the vehicle. When the door opened, they entered one by one (spontaneously) and picked random seats. Passengers chose to sit in the seat that was neither next to anyone nor opposite to anyone in zigzag pattern (Fig. 6). This was different from test setting B, where participants lined up in two rows before embarking and 8 out of 10 of them sat on the same side as the line they had been in. When the vehicle arrived, participants walked out of the vehicle and picked up their luggage. Comparing to setting B, people in the back had to wait longer for the first ones to pick up their luggage.

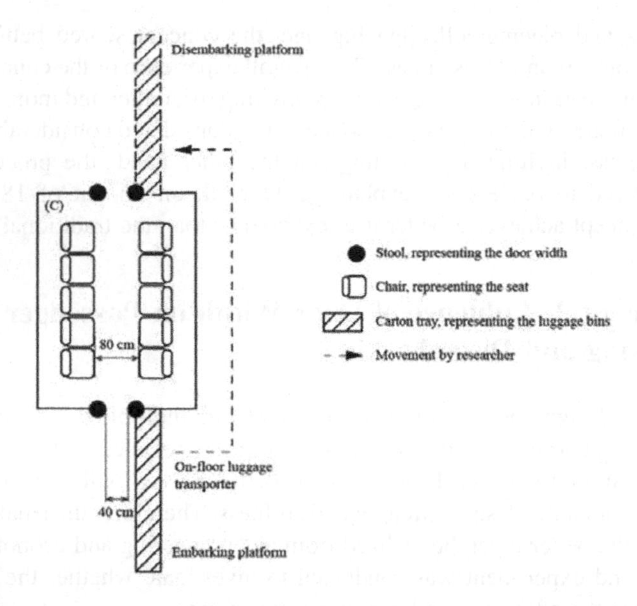

Fig. 5. Test setting C: the (dis)embarking concept with narrow doors (40 cm).

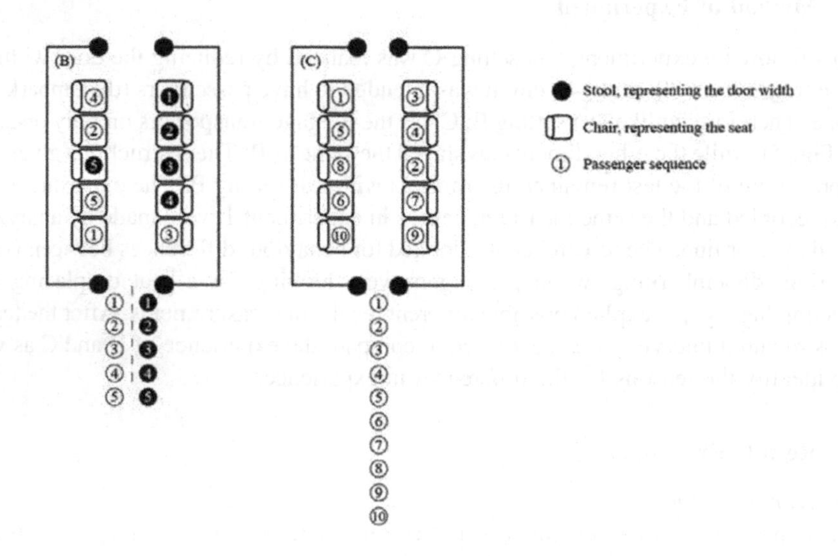

Fig. 6. Participants embark sequence in test setting B (left) and C (right).

(Dis)embarking Times

Analyzing the video recording of the experiment, narrow doors had little influence on the total embarking time (Table 2). However, the total disembarking time increased 39% and the disembarking time per passenger increased 27% on average. The video showed that a jam occurred during disembarking the narrow door setup. Some participants had to wait for others to collect their luggage first in order to be able to exit the door and collect their luggage. In the wide door setting (B), no jam was observed. In total, the time for the vehicle to stay on platform was 40 s, compared to 33 s for wide doors and the embarking and disembarking time per passenger was 19.3 s for narrow doors, compared to 15.7 s for wide doors.

Table 2. (Dis)embarking time comparison between setup B and C.

		(B) Wide (in seconds)	(C) Narrow (in seconds)
Embarking	Total embarking time	15	15
	Average embarking time per passenger	3.3	3.6
Disembarking	Total disembarking time	18	25
	Average disembarking time per passenger	12.4	15.7
Embarking & disembarking	Total vehicle on-platform time	33	40
	Average total time per passenger	15.7	19.3

Passenger Experience
The interview revealed that 5 out of 10 participants did not feel an obvious difference regarding the overall experience of wide and narrow doors. Among the other 5 participants, 4 of which mentioned that there was a longer queue for embarking through a narrow door and that it felt slower and more cramped; two of them felt that participants before them completely blocked the disembarking flow when picking up their luggage at the door in the setup with narrow doors.

4 Discussion

For the envisaged hyperloop system, a vehicle consists of 5 compartments, each with 10 passenger seats, carrying 50 passengers in total. This means that 5 compartments of 10 passengers (dis)embark at the same time. Our assumption has been that the time it takes for all passengers to (dis)embark will be the same as in the test settings experiments.

Compared to the traditional (dis)embarking process, our small-scale experiment to evaluate the proposed concept presented in the introduction suggests a 50% time-saving on average for each passenger and 40% for the overall (dis)embarking process. It also suggested that a better passenger experience could be achieved in terms of comfort, convenience and the perceived efficiency. In the new concept, narrowing the width of the doors seemed to reduce efficiency and led to a negative passenger experience.

Eliminating bottlenecks can be one of the reasons that increases efficiency. The new concept increases passenger flow when the width of the doors and the width of the aisle are the same (in test setting B), which corresponds to Helbing et al.'s [5] findings regarding bottlenecks. Another aspect of the design is that the path for passengers is designed to reduce hesitations. In the traditional (dis)embarking model, passengers in the main aisle are unconsciously making decisions at every line of seats. With the new design, fewer decisions are necessary and decisions are easier to make, as there is only one aisle with all seats facing the aisle. Passengers immediately get a glance of the environment and, therefore, less hesitations are involved in the process, which according to Helbing et al. [6], increases the velocity of flow. Furthermore, without turnings during (dis)embarking and interactions while dealing with luggage, less conflicts occur in the passenger flow. Less braking and avoidance during (dis)embarking also makes the flow faster as described in Helbing et al.'s [6] research on pedestrian flow.

Reducing the width of the doors (in test setting C) seems to cause a bottleneck in the process of (dis)embarking and in theory that can be the reason for smaller flow velocity (Helbing et al. 2002) [5]. However, the embarking time of the narrow doors is the same as the wide doors, which means that in this setup, door width does not directly reduce efficiency. For disembarking, the test with narrow doors takes more time than the one with wide doors, which may be due to the luggage pick up process. In the narrow door setup, people could only place and pick up their luggage from the luggage transporter on one side while in the other setup, they have a wider on platform space to pick up luggage on both sides. The narrow-door setup could be improved by moving the luggage collection area few meters away from the vehicle exit so that people collecting their luggage can get out of the way. In terms of passenger experience, the longer queue at the embarking platform could perhaps be reduced by placing luggage transporters on both sides of the queue. Passengers arriving later can make use of the luggage system when the previous luggage is loaded.

Since it was a test with only 10 participants for one time each experiment, the exact time for embarking and disembarking mentioned in this chapter can only serve as a first indication and repetition of the experiments is recommended. Other limitations of the experiments are listed below:

- Out of the 10 participants, 6 were acquaintances of the researcher, which may have made the interview results more positive than it would be with non-acquaintances;
- All participants experienced the different test settings in the same sequence (A, B, C). Later tests might have been more efficient because they were familiar with the environment and procedure; in order to prevent a learning effect, future tests should counterbalance this effect by mixing the order;
- Participants did not (dis)embark in the same order and their personal behaviour might have an influence on the result. This effect can be minimized by repeating the tests with other participants;
- When comparing test setting B and C, the variables are the door width and the arrangement of the luggage system (two lines in B and one line in C). For further research, it is suggested to control the luggage transporter and only change the width of the doors;

- Participants were given empty suitcases in the test, while heavier luggage could require more time and effort and would make the test more realistic;
- The perception of door width was tested with two stools instead of solid walls and this might have made the experiments more efficient since they could see the interior before embarking as well as the luggage collection area before disembarking.

It is recommended to take the aforementioned aspects into account when repeating the experiments.

5 Conclusion

In this paper, two experiments have been carried out to evaluate a new concept of passenger (dis)embarking for a future hyperloop environment. The new concept separates the vehicle into compartments of 10 seats in two rows facing an aisle in the direction perpendicular to the travel direction. Passengers embark from one side, walk into the seats and disembark from the other. The luggage compartment is located underneath the passenger floor and the luggage is dropped during passenger embarking and collected right after disembarking. Results of these experiments indicate that the new design can reduce (dis)embarking time and improve passenger experience compared to traditional model of boarding (train/aircraft). The width of the door does influence the disembarking time but no noticeable influence was found on the embarking time. In this test setting, a narrow door design did not only increase disembarking time but also had a negative influence on passenger experience. However, suggestions have been made to adjust the narrow-door setup to minimize the influence on time efficiency and passenger experience for further research.

Acknowledgements. The authors are grateful to Prof. P. Vink for the latest information in the field of aviation passenger experience. They appreciate that J. Li pointed out the interesting study by Helbing et al. (2002 and 2005). Special thanks to Buccaneer Delft for providing the testing environment and connections to the participants.

Appendix: Questionnaire for Experiment 2

PART 2: Questionnaire

Test A

	Strongly Disagree 1	2	3	4	Strongly Agree 5
When I was on the platform:					
It was clear what to do	☐	☐	☐	☐	☐
It was comfortable when waiting for the vehicle	☐	☐	☐	☐	☐
The experience was pleasant	☐	☐	☐	☐	☐
During boarding and disembarking:					
It was clear what to do	☐	☐	☐	☐	☐
It was convenient to get in and get seated and the other way around	☐	☐	☐	☐	☐
The experience was pleasant	☐	☐	☐	☐	☐
Regarding the way to place and collect my large luggage:					
It was clear what to do	☐	☐	☐	☐	☐
It was convenient	☐	☐	☐	☐	☐
I felt comfortable doing so	☐	☐	☐	☐	☐
It was a good luggage solution	☐	☐	☐	☐	☐
Overall experience:					
It was time efficient	☐	☐	☐	☐	☐
It was a seamless journey	☐	☐	☐	☐	☐
The process was clear to me	☐	☐	☐	☐	☐
It was an inviting and friendly (dis)embarking experience	☐	☐	☐	☐	☐
It was an exciting (dis)embarking experience	☐	☐	☐	☐	☐

Test B

	Strongly Disagree 1	2	3	4	Strongly Agree 5
When I was on the platform:					
It was clear what to do	☐	☐	☐	☐	☐
It was comfortable when waiting for the vehicle	☐	☐	☐	☐	☐
The experience was pleasant	☐	☐	☐	☐	☐
During boarding and disembarking:					
It was clear what to do	☐	☐	☐	☐	☐
It was convenient to get in and get seated and the other way around	☐	☐	☐	☐	☐
The experience was pleasant	☐	☐	☐	☐	☐
Regarding the way to place and collect my large luggage:					
It was clear what to do	☐	☐	☐	☐	☐
It was convenient	☐	☐	☐	☐	☐
I felt comfortable doing so	☐	☐	☐	☐	☐
It was a good luggage solution	☐	☐	☐	☐	☐
Overall experience:					
It was time efficient	☐	☐	☐	☐	☐
It was a seamless journey	☐	☐	☐	☐	☐
The process was clear to me	☐	☐	☐	☐	☐
It was an inviting and friendly (dis)embarking experience	☐	☐	☐	☐	☐
It was an exciting (dis)embarking experience	☐	☐	☐	☐	☐

References

1. Musk E (2013) Hyperloop Alpha. SpaceX. http://www.spacex.com/sites/spacex/files/hyperloop_alpha-20130812.pdf. Accessed 3 Mar 2017
2. Mas S, Juan AA, Arias P, Fonseca P (2013) A simulation study regarding different aircraft boarding strategies. Springer, Heidelberg
3. Jaehn F, Neumann S (2014) Airplane boarding. Eur J Oper Res 244(2015):339–359
4. Steffen JH (2008) Optimal boarding method for airline passengers. J Air Transp Manag 14 (3):146–150
5. Helbing D, Farkas IJ, Molnar P, Vicsek T (2002) Simulation of pedestrian crowds in normal and evacuation situations. Pedestr Evacuation Dyn 21(2):21–58
6. Helbing D, Buzna L, Johansson A, Werner T (2005) Self-organized pedestrian crowd dynamics: experiments, simulations, and design solutions. Transp Sci 39(1):1–24
7. Li J (2018). Crowds inside out: understanding crowds from the perspective of individual crowd members' experience. PhD thesis (submitted, to be published in October 2018). Delft University of Technology
8. Li D (2017) (Dis)embarking Hyperloop: design of the process and infrastructure for passengers. Master thesis (under embargo until August 2018). Delft University of Technology

Field Observations of Interactions Among Drivers at Unsignalized Urban Intersections

Evangelia Portouli[1(✉)], Dimtiris Nathanael[2], Kostas Gkikas[2], Angelos Amditis[1], and Loizos Psarakis[2]

[1] Institute of Communication and Computer Systems, 9 Iroon Polytechniou Street, 15773 Zografou, Athens, Greece
v.portouli@iccs.gr

[2] School of Mechanical Engineering, National Technical University of Athens, 9 Iroon Polytechniou Street, 15780 Zografou, Athens, Greece

Abstract. Interactions among drivers are an essential part of the driving task and need to be considered in the design of interaction strategies of automated vehicles. Interactions between drivers relevant to left and right turns in unsignalized urban intersections were recorded via an eye glass mounted gaze sensor. Participants were asked to retrospectively comment aloud on the process of their decision making for each case of interaction. The typical sequences of actions that were observed relevant to left and right turns may be used as a basis for designing turning strategies for automated vehicles. Establishing eye contact was considered as a good means to convince the other driver to yield, while avoiding eye contact was interpreted as unwillingness to do so. Vehicle edging was intentionally used by participants so that the other coming drivers could better see them. Automated vehicles should consider using edging and directed communication to other drivers in their interaction strategies.

Keywords: Drivers' interactions · Communication · Automated vehicles

1 Introduction

Interactions among road users are an essential component of driving activity. It is empirically known that, on several occasions, drivers somehow reach a common agreement about their future motion plan before starting a manoeuvre. Drivers may also decide to enforce their motion plan by starting a manoeuvre that will oblige other drivers to react. Such interactions involve explicit communication, such as eye contact, gestures or vehicle signals, and multiple means of implicit cues, such as approach speed, which are used by drivers to communicate and anticipate intent and to influence each other's behaviour. Although the exact means can differ across different regions and cultures, such interactions allow the effective coordination of future motion plans between different road users.

Such capabilities are currently missing in automated vehicles, although they are expected to enter the traffic at several levels of automation. Their interaction with other road users is typically dominated by the principle of collision avoidance. But practice shows that this results in non-human-like (robotised) behaviour of the vehicle, whose

S. Bagnara et al. (Eds.): IEA 2018, AISC 823, pp. 230–237, 2019.
https://doi.org/10.1007/978-3-319-96074-6_24

actions can actually be quite frustrating. Anecdotal evidence from evaluation activities in the framework of several research projects shows the frustration of other drivers encountering an automated vehicle in real conditions. Typically, human drivers cannot predict the automated vehicle's actions as it does not behave according to their expectations. To safely and efficiently integrate automated vehicles in traffic, it must be ensured that such vehicles can interact with other road users in an intuitive and expectation-conforming manner [1]. This will allow both the on-board users of the vehicle and the surrounding road users to coordinate their planned actions accordingly.

Although studies of drivers' interactions with pedestrians are available [2–4], studies of drivers' interactions with other drivers are rare. One study that specifically studied interactions between drivers relevant to lane changes in an urban ring-road shows that drivers based their predictions about the evolution of the traffic scene mostly on implicit cues, such as disturbances of the expected smooth motion of a vehicle that could not be attributed to road geometry or to obstacles on the road [5]. The findings suggest that more in-depth studies of interactions are needed, as the explicit vehicle communication signals are not consistently used by drivers [6] and even more, the same signal may be interpreted differently according to context. For example, flashing headlights may be interpreted either as "stop, I will pass first" or as "pass first, I will stop".

The objective of the present work was to observe interactions between drivers relevant to turns in unsignalized urban intersections in order to develop guidelines for the design of appropriate interaction strategies for automated vehicles.

2 Method

An on-road, video-assisted observational study with retrospective commentary by drivers was designed and conducted so as to collect empirical evidence relevant to drivers' interactions.

Experienced drivers were asked to drive their own passenger car in a predefined urban course, while wearing an eye glass mounted gaze sensor. The course consisted of a circular route of 0.75 km which was driven 5 times by each driver (participant). The total course length was 3.75 km and the mean driving duration was 18 min. The course included left turning from a two-way street and right turning from a smaller to a two-way street. Turns were not regulated by a traffic light and given the traffic density it was expected that there would be a lot of interactions between drivers relevant to the left and right turns. Example traffic scenes are shown in Fig. 1.

After the end of the driving session, each participant was asked to watch selected parts of the eye gaze video recording and to comment aloud on the process of his/her decision making for each case of interaction with another driver. Verbal protocols offer a way to record the human thought process [7] and have been used in driving studies [5].

Afterwards, an analyst watched the participant's eye gaze and scene video as well as his/her retrospective commentary, and labelled the interactions between the participant and another driver. An interaction start was defined as the time point when (i) the participant had to wait for a gap in the oncoming traffic before turning or (ii) the

Fig. 1. Examples of eye gaze video recording relevant to left turn from two-way street with oncoming traffic (left) and right turn to two-way street (right)

participant started turning knowing that the oncoming driver would have to modify his/her vehicle motion. For each interaction, the analyst labelled the type of the interacting vehicle and whether the other driver reacted. The signals or cues by the participant and his/her vehicle and by the other driver and his/her vehicle and their sequence were labelled for each interaction.

3 Results

Twenty-one experienced drivers, 10 males and 11 females, participated in the observations. Their mean age was 39.1 years (median 38 years, standard deviation 11.7 years) and their mean driving experience was 18.5 years.

The observed interactions, the interacting vehicle type and the observed reactions by other drivers per manoeuvre are shown in Table 1. In 146 of the 188 observed left turns and in 126 of the 179 observed right turns, an interaction was started by the participants. In 62 and 60 cases respectively, the other driver reacted. In 23 of 25 interactions with drivers of large vehicles, the other driver reacted to the interaction started by the participant. Only 7 out of 58 motorcycle riders reacted to the interaction started by the participant.

Table 1. Interaction starts and other drivers' reactions per manoeuvre

Manoeuvre type	Number of turns	Number of interactions started by the participants	Number of interactions where the other driver reacted
Left turn from 2-way street	188	146 (64 passenger cars, 36 taxis, 16 large vehicles, 30 motorcycles)	62 (26 passenger cars, 18 taxis, 14 large vehicles, 4 motorcycles)
Right turn to 2-way street	179	126 (63 passenger cars, 26 taxis, 9 large vehicles, 28 motorcycles)	60 (33 passenger cars, 15 taxis, 9 large vehicles, 3 motorcycles)

The signals or cues by the participants are shown in Table 2. The row "Nothing observed" refers to interactions where the participant started turning knowing that the other driver would have to react and slow down. A relevant accompanying comment by the participants was *"I am sure that he/she has seen me, so I can turn, because I know that he/she can and will yield"*. Vehicle edging, flashing headlights and gesture/nodding was followed by a reaction by the other driver in most of the interactions when they were used. The turn indicator alone was not so effective, especially for right turns when the other driver, coming from the left of the participant's vehicle, could not perceive the right turn indicator. One participant specifically mentioned his/her intense gazing towards the other drivers as a means to enforce his/her priority on them.

Table 2. Signals or cues by the participants

Observed signal/cue by the participant	Left turn from 2-way street		Right turn to 2-way street	
	Number of started interactions (N = 146)	Number of interactions with another driver's reaction (N = 62)	Number of started interactions (N = 126)	Number of interactions with another driver's reaction (N = 60)
Turn indicator	119	40	66	21
Turn indicator + Edging	17	17	10	10
Turn indicator + Edging + Headlights	2	2		
Turn indicator + Gesture/Nodding	1	1		
Turn indicator + Gesture/Nodding + Edging	1	1		
Edging	1		18	12
Gesture/Nodding			3	2
Nothing observed	5	1	29	15

The signals or cues by the other drivers are shown in Table 3. The other driver's deceleration or stopping was always followed by the participant turning in front of the other vehicle. The same holds true when the other driver made a gesture/nodded or when the other driver used the turn indicator. The latter because it indicated a change in the other vehicle's trajectory in a way that a conflict with the participant's vehicle was not to be expected. The headlights by the other driver did not always result in the participant turning in front of the other vehicle, so the interpretation of this signal is rather done complementary to other signals and cues. Acceleration and use of horn by the other driver was not followed by the participant turning; therefore, these signals were rather interpreted as other's intention to not yield.

The sequences of labelled signals/cues for 61 interactions relevant to left turns where the other driver reacted are shown in Fig. 2, while the sequences for 60 interactions relevant to right turns to two-way street where the other driver reacted are depicted in Fig. 3. On the left, the signals and cues by the participants are drawn, along with the number of observations for each signal or cue (encircled). Similarly, the right boxes represent the signals or cues by the other drivers. The arrows depict the sequence of actions.

Table 3. Signals or cues by the other drivers

Observed signal/cue by the participant	Left turn from 2-way street		Right turn to 2-way street	
	Number of started interactions (N = 146)	Number of interactions with another driver's reaction (N = 62)	Number of started interactions (N = 126)	Number of interactions with another driver's reaction (N= 60)
Gesture/Nodding			1	1
Headlights	7	4	2	1
Horn	1			
Accelerate			2	
Decelerate	22	22	22	22
Decelerate + Gesture	3	3	2	2
Decelerate + Headlights	1	1		
Decelerate + Headlights + Gesture			1	1
Stop	25	25	24	24
Stop + Gesture	4	4	1	1
Stop + Headlights	1	1		
Stop + Horn	1	1		
Turn indicator			4	3
Opportunity due to another event			3	2
Nothing observed	81	1	64	2

According to Fig. 2, a typical action sequence of for an interaction relevant to a left turn in the specific location evolved as follows: the participant turned the indicator on and decelerated; if the oncoming driver reacted and stopped the vehicle while the participant decelerated, the participant turned; else the participant came to a full stop and waited. In these circumstances, the participant frequently edged, namely moved the vehicle a bit forward, possibly trying to make an oncoming driver to yield. Sometimes, the participant flashed headlights to the oncoming driver, made a gesture, nodded or tried to achieve eye contact with the oncoming driver. When an oncoming driver decided to yield, he/she decelerated. Sometimes, the other driver flashed headlights or made a gesture/nods towards the participant. Then, the participant turned.

According to Fig. 3, a typical actions sequence during an interaction relevant to a right turn in the specific location is the following: the participant decelerated, turned on the indicator and searched the environment for oncoming traffic; frequently the participant came to a full stop; sometimes the participant edged the vehicle a bit forward; more rarely, the participant gazed towards the oncoming drivers, trying to achieve eye contact, and made gestures or nods; when the other driver decided to respond, he/she normally decelerated or stopped the vehicle; then the participant turned.

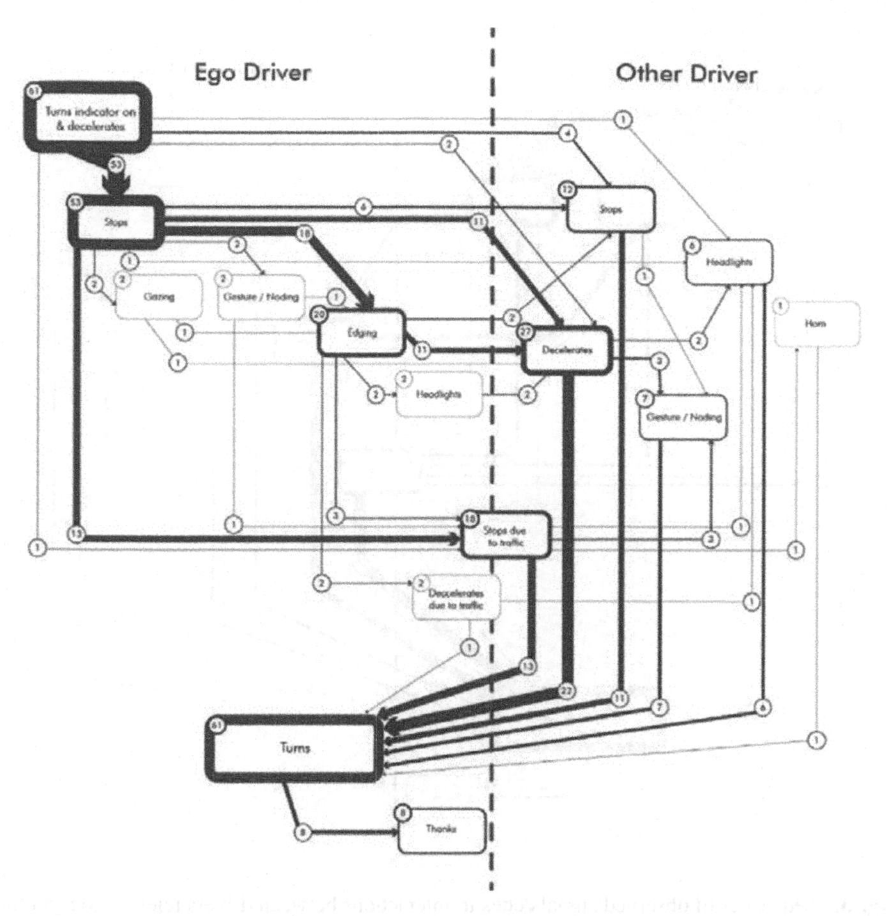

Fig. 2. Sequences of observed signals/cues in interactions between drivers relevant to left turns

4 Discussion

The objective of this work was to observe and codify interactions among drivers relevant to left and right turns in urban environment, in order to derive guidelines for designing appropriate interaction strategies for automated vehicles. Specifically, the typical sequences of actions that were observed may be used for embedding "human-like" turning strategies and interaction patterns to automated vehicles, i.e. interaction patterns compatible to the norms and expectations of human road users, so that such vehicles are more efficiently integrated in traffic. The signals and cues observed in the present study along with some findings from the drivers' video-assisted retrospective commentaries are discussed below.

The observations indicate that drivers were trying to monitor the other driver's gaze orientation, to check whether the other driver had perceived them or if he/she was distracted, and planned their future motion accordingly. For example, if the other driver was using his/her phone, the participants were more conservative as regards the time

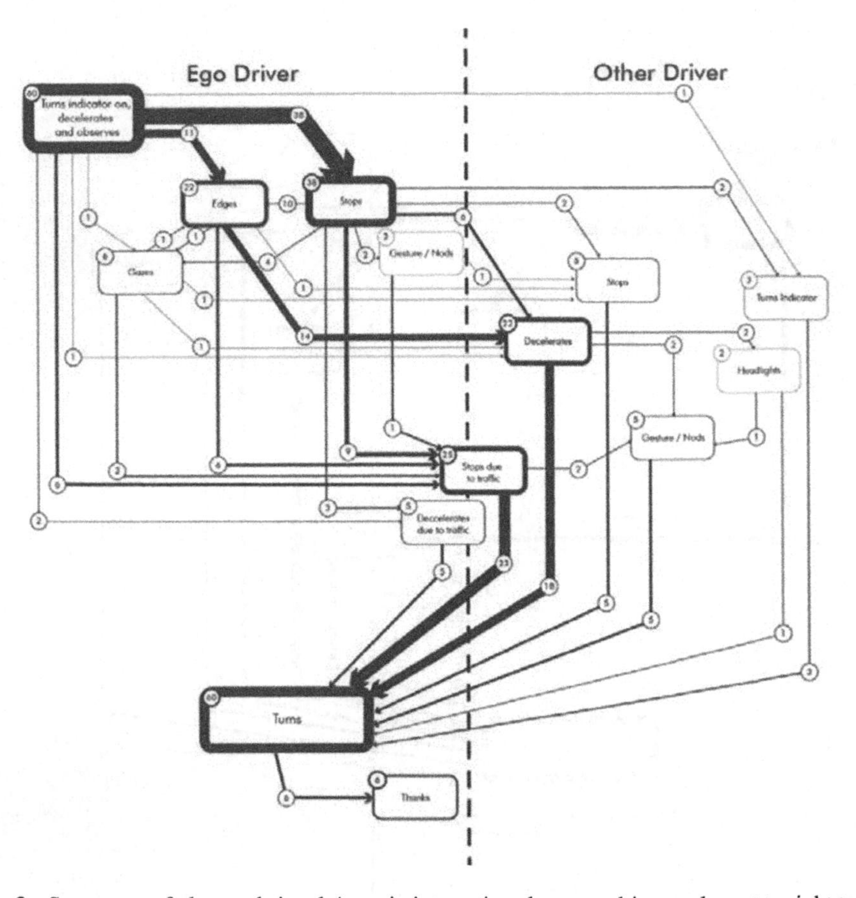

Fig. 3. Sequences of observed signals/cues in interactions between drivers relevant to right turns

gap when they would start turning. Additionally, participants seemed to be affected by their estimated waiting time before turning. If they expected that they would not wait long, for example when only one vehicle was oncoming, they would wait for it to pass before turning. If, on the other hand, multiple vehicles were coming in a row and they would have to wait for long, they tended to start turning at shorter time gaps.

According to the drivers' commentaries, achieving eye contact and intense gazing were good means for "convincing" the other driver to yield. On the contrary, when the other driver intentionally avoided looking towards the participant, although he/she normally should, this was interpreted as "he/she will not yield priority". Drivers also reported flashing their headlights in left turns as a means to attract the other driver's attention. Edging was also reported as being done intentionally to attract another drivers' attention.

Some additional cues were also reported by drivers to anticipate the evolution of the situation. For example, a motorcyclist's foot being lowered to the ground was interpreted as intending to stop. The presence of people at the bus stop created the expectation that the bus would stop. Also, drivers took advantage of opportunities due to side events. For example, a pedestrian crossing the street was "a green light" for the participants to start turning, since they expected that the oncoming driver would slow down to avoid hitting the pedestrian.

In conclusion, automated vehicles should consider progressive edging in their interaction strategies for low speeds manoeuvres and directed communication to other drivers, simulating the social conventions revealed in the present work. Although safety must be always safeguarded, designers of automated vehicles may consider that human drivers sometimes start turning without any observable reaction by the other driver, when they feel confident that the other driver has seen them and has time/distance to slow down and yield. Finally, an explicit signal by an automated vehicle to inform the other drivers that it will yield may be beneficial for the traffic flow and efficiency.

Acknowledgement. This work is a part of the interACT project. interACT has received funding from the European Union's Horizon 2020 research & innovation programme under grant agreement no 723395. Content reflects only the authors' view and European Commission is not responsible for any use that may be made of the information it contains.

Compliance with Ethical Standards. Recruitment of participants and data collection was conducted in accordance with National Technical University of Athens ethics procedures concerning research involving human participants.

References

1. Parkin J, Clark B, Clayton W, Ricci M, Parkhurst G (2016) Understanding interactions between autonomous vehicles and other road users: a literature review. Project Report. University of the West of England, Bristol. http://eprints.uwe.ac.uk/29153
2. Schneemann F, Gohl I (2016) Analyzing driver-pedestrian interaction at crosswalks: a contribution to autonomous driving in urban environments. In: Intelligent vehicles symposium (IV). IEEE, pp 38–43
3. Rasouli A, Kotseruba I, Tsotsos JK (2017) Agreeing to Cross: How Drivers and Pedestrians Communicate. https://arxiv.org/abs/1702.03555
4. Gueguen N, Eyssartier C, Meineri S (2016) A pedestrian's smile and drivers' behavior: when a smile increases careful driving. J Saf Res 56:83–88
5. Portouli E, Nathanael D, Marmaras N (2014) Drivers' communicative interactions: on-road observations and modelling for integration in future automation systems. Ergonomics 57 (12):1795–1805
6. Ponziani R (2012) Turn signal usage rate results: a comprehensive field study of 12,000 observed turning vehicles. SAE Technical Paper 2012-01-0261
7. Ericsson KA, Simon HA (1993) Protocol analysis: verbal reports as data. MIT Press, Cambridge

Can Simple Anthropomorphism Change People's Perception of Self-driving Vehicle Accidents?

Qianru Guo and Peng Liu(✉)

Tianjin University, Tianjin 300072, China
pengliu@tju.edu.cn

Abstract. Although self-driving vehicles (SDVs) promise to reduce traffic crashes largely, they cannot eliminate all crashes and they might create new crashes. Efforts to understand how the public responds traffic accidents caused by SDVs or involved with SDVs are very limited. Our research aims to understand whether simple anthropomorphic features (giving a SDV a human name and a male photo) change participants' perceived severity of a traffic accident involving with a SDV or caused by the SDV and whether certain post affective responses (affect evoked by the accident information, negative affect by SDV, trust in SDV) account for the change in perceived severity. In a 2 (SDV: normal vs. anthropomorphic) * 2 (accident cause: self-caused vs. other-caused) between-subject design (N = 260), we found that the simple anthropomorphic design and the type of accident cause did not affect perceived severity and these three affective responses.

Keywords: Self-driving vehicles · Anthropomorphism · Affect Trust

1 Introduction

More than 1.2 million road accident fatalities occur annually throughout the world [1]. More than 70% traffic crashes were directly or indirectly the results of human errors [2]. As an innovation in transportation, self-driving vehicles (SDVs) promise to largely reduce traffic crashes and increase road safety. At the same time, SDVs could bring other social benefits, including reducing air pollution, increasing fuel efficiency, and increasing the mobility of people who are unable to drive [3–5].

However, SDVs cannot eliminate all traffic crashes [4, 6]. SDVs might perform worse than human drivers in complex weather and environment conditions. Current road tests did not confirm SDVs' safety advantage over human drivers [7, 8]. SDVs might also create new serious risks (e.g., accidents caused by cyber attacks), cause serious injuries and fatalities similar to human drivers. In addition, eliminating human control from driving may reduce people's safety perceived during the driving. Designing SDVs to be safe, trustworthy, and reliable is critically important.

We are interested in how people respond to accidents associated with SDVs and whether people's perceptions are influenced by accident-related and vehicle-related

S. Bagnara et al. (Eds.): IEA 2018, AISC 823, pp. 238–244, 2019.
https://doi.org/10.1007/978-3-319-96074-6_25

characteristics. Its underlying implication is that we may relieve people's concerns about SDVs when they are involved in accidents and reduce their resistance to SDVs. It might be important for SDVs at its initial stages.

1.1 Anthropomorphism

Anthropomorphism plays a role in affecting people's perception on external actors (e.g., the automated system driving the vehicle for people). Anthropomorphism is defined as a process of inductive inference whereby people attribute to nonhumans human characteristics [9]. As stated by Waytz et al. [9], "[a]nthropomorphizing a nonhuman does not simply involve attributing superficial human characteristics (e.g., a humanlike face or body) to it, but rather attributing essential human characteristics to the agent (namely a humanlike mind, capable of thinking and feeling)" (p. 113).

Several empirical studies were conducted to understand anthropomorphic features affect people' affective and behavioral responses of autonomous vehicles. Waytz et al. [9] found that attaching anthropomorphic features (gender and voice in their study) to an autonomous vehicle lead participants to blame this vehicle less for an accident caused by another driver and lead participants to be more relaxed in this accident. A humanlike machine might receive more trust from users. Similar results were found by Lee et al. [10], who investigated how anthropomorphic cues (humanlike appearance and high autonomy in their study) of an unmanned driving system affect people's perception of this agent. They found that humanlike appearance lead participants to report higher social presence, perceived safety, perceived intelligence, affective trust (evaluated by "likeable," "enjoyable," and "positive"); and high autonomy increased all measures including these four measures and cognitive trust. Lee et al. [10] also reported that anthropomorphism might affect intelligence and trust through the mediating effect of social presence. Humanizing an agent can make this agent to be more accepted [11]. To sum up, these studies suggest that anthropomorphic features can change people's affective and behavioral responses.

1.2 Research Aim and Hypotheses

We aim to understand how simple anthropomorphic cues can affect people's perceptions of traffic accidents involving SDVs. The simplest way to anthropomorphize a nonhuman agent is to give it human physical characteristics such as humanlike facial expressions and bodily forms. In our study, the anthropomorphic SDV was given a human name ("Anda") and a male photo.

We first assume that people might *less* negatively evaluate a traffic accident involving an anthropomorphic SDV than that involving a normal SDV, although the two accidents have the same undesirable outcome. The reasoning for this assumption is that anthropomorphic of an autonomous agent can mitigate blame for the agent's involvement in a bad outcome [9]. Waytz et al. [9] found that participants blamed the autonomous vehicle significantly *less* in the anthropomorphic condition than in the normal condition. They suggested a clear relationship between anthropomorphism and responsibility attribution. Another indirect evidence for this assumption is that social psychology research [12] found that the same negative outcome was evaluated different when caused by humans and other agents.

If the above assumption is confirmed to be true, we suspect that two types of affective mediators can explain it. The first one is trust. Research [9, 10] argues that a humanlike agent can increase people's willingness to trust the agent. de Visser et at. [13] found that anthropomorphic agents were related to greater trust resilience, a higher resistance to breakdowns in trust. Thus, we assume that after the accident people trust higher in the anthropomorphic SDV than in the normal SDV.

Another type of affective mediators is affects evoked by accident scenario and by SDV. The affect heuristic by Slovic and his colleagues [14] suggests that people might rely on affect to judge risks and benefits of specific hazards. According to Siegrist and Sütterlin [12], people may rely on the affect evoked by a technology or by a piece of information to evaluate the severity of the technology or of that implied in the information. More negative affect experience, people more negatively evaluate the technology. Anthropomorphic features have been associated with higher positive affective responses (e.g., liking [9] and affective trust [10]). Thus, we assume people experience *less* negative affect evoked by the accident scenario involving an anthropomorphic SDV and also *less* negative affect evoked by the anthropomorphic SDV.

In addition, we consider whether the above hypotheses hold in different conditions. The cause of accident is manipulated two types: caused by other driver (other-caused) and caused by the SDV (self-caused). To sum up, we will examine whether people perceive the severity of a traffic accident involving a SDV with simple anthropomorphic features *less* than that of the same accident involving a normal SDV and whether post trust, affect evoked by the accident information, and affect evoked by SDB can account for the difference in perceived severity. Our study expects to increase our understanding about how people make decisions and judgments related to traffic accidents associated with autonomous vehicles.

2 Methodology

2.1 Participants

Overall, 260 college students (134 female, 126 male) participated in the online survey.

2.2 Material and Procedure

It was 2 (SDV: normal vs. anthropomorphic) * 2 (accident cause: self-caused vs. other-caused) between-subject design. First, 137 participants completed the survey assuming SDV caused the accident and they were randomly assigned to one of two conditions: normal ($n = 75$) and anthropomorphic ($n = 62$). Participants were given a short description of SDV: "*The automated driving system takes over speed and steering control completely, on all roads and in all situations. The driver or passenger sets a destination via a touchscreen. The driver or passenger cannot drive manually and perform interventions, because the vehicle does not have a steering wheel. Self-driving enables the driver (i.e., passengers) to perform more non-driving activities, such as* *reading a book, watching a film, surfing the Internet, playing their phones, dealing with their working affairs, sleeping, and so on and so forth.*" The SDV was assumed to have the same safety performance as the average performance of all human drivers. In the anthropomorphic condition, the SDV was given anthropomorphic features, named "Anda", and given a gender (male) and a male photo. Then, participants in the normal condition received the following information: "*On an urban road, a passenger was taking this SDV to the destination. However, due to the SDV's fault, a traffic accident occurred and injured the passenger.*" In the anthropomorphic condition, the term "SDV" was replaced by the SDV's human name ("Anda"). Participants were asked to answer the following question about the scenario: "*What feelings did you experience due to this information?*" This question measured affect evoked by scenario, with 10 levels (*very negative* = 1, *very positive* = 10). This question including the following questions were adapted from [12]. The next questions asked participants to rate their fear, dread, and trust if they were required to ride the SDV/"Anda" (*very low* = 1; *very high* = 10). Fear and dread were used to measure negative affect evoked by SDV"Anda" (Cronbach's α = .89). The final question was: "*How do you judge the severity of this accident?*" (*very low* = 1; *very high* = 10).

Another 123 participants completed the survey assuming SDV caused other vehicles and they were randomly assigned to one of two conditions: normal (n = 61) and anthropomorphic (n = 62). Similar questions were asked, with the only one difference in the accident scenario which indicated that the accident was other vehicle's fault.

3 Results

Table 1 shows the basic statistics (mean and standard deviation) of the constructs and their zero-order correlations (r). All correlations were significant ($ps < .01$). According to a rule of thumb (Evans 1996) for determining the strength of correlations (< .20, very weak; .20–.39, weak; .40–.59, moderate; .60–.79, strong; .80–1, very strong), the correlations between the affect evoked by the accident scenario and trust in SDV, between the negative affect evoked by SDV and trust in SDV, between the negative affect evoked by SDV and perceived severity, were moderate. Other correlations were weak or very weak.

Table 1. Mean, standard deviation (SD), and correlation between constructs.

Construct	Mean	SD	1	2	3
1. Affect evoked by scenario	4.50	1.99			
2. Negative affect evoked by SDV[a]	6.50	2.06	$-.17^{**}$		
3. Trust in SDV[a]	4.73	2.21	$.47^{***}$	$-.46^{***}$	
4. Perceived severity	6.50	1.97	$-.16^{**}$	$.45^{***}$	$-.26^{***}$

Note: ** $p < .01$; *** $p < .001$. [a] "SDV" was replaced by the human name "Anda" in the anthropomorphic condition.

Next, the multivariate analysis of variance (MANOVA) was run to examine the effects of SDV type and accident cause and their interaction effect. None of the main effects of SDV type ($p = .81$) and accident cause ($p = .57$) and their interaction effect ($p = .61$) were significant. Following separate univariate ANOVA revealed similar results. Thus, we did not run mediation analysis. Table 2 shows the mean of the four constructs in the four different treatments.

Table 2. Mean values of constructs in different treatments.

Construct	Normal SDV ($n = 136$)		Anthropomorphic SDV ($n = 124$)	
	Others-caused ($n = 61$)	Self-caused ($n = 75$)	Others-caused ($n = 62$)	Self-caused ($n = 62$)
Affect evoked by scenario	4.85	4.32	4.47	4.42
Negative affect evoked by SDV	6.48	6.57	6.31	6.62
Trust in SDV	4.57	4.73	4.90	4.73
Perceived severity	6.52	6.60	6.48	6.37

Finally, we run an ordinal least square (OLS) regression analysis of perceived severity and regarded as SDV type, accident cause, affect evoked by accident scenario, negative affect evoked by SDV, and trust in SDV as predictors. In this OLS regression model, negative affect evoked by SDV was the only significant predictor of perceived severity (Table 3).

Table 3. Regression analysis of perceived severity.

Predictor	*B*	*SE*	*t*	*p*
SDV (anthropomorphic = 1)	−0.13	0.22	−0.58	.56
Accident cause (self = 1)	−0.12	0.22	−0.55	.58
Affect evoked by scenario	−0.09	0.06	−1.37	.17
Negative affect evoked by SDV	0.41	0.06	6.77	< .001
Trust in SDV	−0.01	0.06	−0.22	.82

Note: *B*, unstandardized coefficients; *SE*, standard error.

4 Discussion and Conclusions

Our major result is that simple anthropomorphic features (human name, male photo) and accident cause (caused by self and caused by others) did not induce different impacts on participants' perceived severity of traffic accidents involving SDV, trust in SDV, and affect evoked by the accident information and by the SDV technology. Thus, adding simple anthropomorphic features did not play a role in affecting participants' perception of the accident and affective responses (affect and trust). Our participants seemed impervious to the effects of anthropomorphism.

Previous studies [9, 10] investigated the effect of anthropomorphic features on trust in the context of autonomous driving and reported mixed results to some degrees. Waytz et al. [9] designed three driving conditions, human, agentic (similar to our normal SDV), and anthropomorphic, and measured participants' self-reported trust before they experienced a simulated accident caused other drivers and their behavioral trust (e.g., similar to relax, measured by heart rate change and startle evaluated by independent raters) afterward. Waytz et al. found that participants in the anthropomorphic condition did not report higher trust before the accident, but had higher behavioral trust after the accident, than those in the agentic condition. Lee et al. [10] found that human appearance did not affect participants' cognitive trust (similar to the trust in our study) in the unmanned driving agent; but, it increased participants' affective trust (similar to affect in our study). Combining these results and our result, we argue that the function of the simple anthropomorphic design in creating people's self-reported trust needs more empirical investigations, at least in the case of autonomous driving.

We investigated the psychological predictors of perceived severity of the SDV accident. Three psychological predictors (affect evoked by the accident scenario, negative affect evoked by SDV, and trust in SDV) significantly correlated with perceived severity. However, negative affect evoked by SDV was the only significant predictor of perceived severity in the OLS regression model. According to the affect heuristic [14], people's affect associated with a technology will influence their evaluation of the technology [12]. Thus, our finding might be in line with the affect heuristic.

Several limitations should be noted. First, we did not check the manipulation of the simple anthropomorphic design. Thus, we cannot rule out the possibility that our simple anthropomorphic design might not lead participants to perceive higher anthropomorphism from the anthropomorphic SDV. If so, it could in part explain the non-significant differences between participants' perceptions and affective responses in the anthropomorphic SDV and the normal SDV. An anthropomorphic design that involves attributing essential human characteristics (i.e., a humanlike mind, thinking, abilities of reasoning, and feeling) to the agent [9] would have higher impacts. Further inquiry may consider how the strong anthropomorphic design affects people's perception of SDVs. Second, we did not consider changes in affective responses (affect evoked by SDV and trust in SDV) before and after participants were given the accident scenario information. These changes might finely reveal the potential differences in participants' perception and judgment between these two SDV types.

References

1. WHO (2015) Global status report on road safety 2015. World Health Organization (WHO), Geneva, Switzerland
2. Dhillon B (2007) Human reliability and error in transportation systems. Springer, London
3. Bansal P, Kockelman KM, Singh A (2016) Assessing public opinions of and interest in new vehicle technologies: an austin perspective. Transp Res Part C Emerg Technol 67:1–14

4. Anderson JM, Kalra N, Stanley KD, Sorensen P, Samaras C, Oluwatola OA (2016) Autonomous vehicle technology: a guide for policymakers. RAND Corporation, Santa Monica
5. Liu P, Yang R, Xu Z (2018) How safe is safe enough for self-driving vehicles? Risk Analysis (accepted)
6. Kalra N, Paddock SM (2016) Driving to safety: How many miles of driving would it take to demonstrate autonomous vehicle reliability? Transp Res Part A Policy Pract 94:182–193
7. Favarò F, Eurich S, Nader N (2018) Autonomous vehicles' disengagements: Trends, triggers, and regulatory limitations. Accid Anal Prev 110:136–148
8. Teoh ER, Kidd DG (2017) Rage against the machine? Google's self-driving cars versus human drivers. J Saf Res 63:57–60
9. Waytz A, Heafner J, Epley N (2014) The mind in the machine: Anthropomorphism increases trust in an autonomous vehicle. J Exp Soc Psychol 52:113–117
10. Lee J-G, Kim KJ, Lee S, Shin DH (2015) Can autonomous vehicles be safe and trustworthy? Effects of appearance and autonomy of unmanned driving systems. Int J Hum-Comput Interact 31(1):682–691
11. Duffy BR (2003) Anthropomorphism and the social robot. Robot Auton Syst 42(3):177–190
12. Siegrist M, Sütterlin B (2014) Human and nature-caused hazards: The affect heuristic causes biased decisions. Risk Anal 34(8):1482–1494
13. de Visser E, Monfort SS, McKendrick R, Smith MAB, McKnight PE, Krueger F, Parasuraman R (2016) Almost human: Anthropomorphism increases trust resilience in cognitive agents. J Exp Psychol Appl 22(3):331–349
14. Slovic P, Finucane ML, Peters E, MacGregor DG (2004) Risk as analysis and risk as feelings: Some thoughts about affect, reason, risk, and rationality. Risk Anal 24:311–322

Visual Behaviors and Expertise in Race Driving Situation

Takaaki Kato(✉)

Keio University, 5322 Endo, Fujisawa-shi, Kanagawa 252-0882, Japan
tiger@sfc.keio.ac.jp

Abstract. It is generally considered that experts are typically more accurate and quicker in their responses and generally employ fewer fixations and longer duration. In motorsports, very few attempts have been made at perceptual-motor skills of racing drivers. We analyzed visual search activities of racing drivers in the simulated racing driving situation. A total of 15 subjects took part in the experiment and they all had normal or corrected-to-normal vision. Two expert racecar drivers were competing in national and international races at that moment. Two intermediate drivers had experienced in race driving on the motor racing circuits. Others had no race experience. The results indicated differences between experts, intermediates and novices. Experts' vehicle speed was higher than others. It can be seen that experts held the brake pedal position earlier and smaller time when they approached the corner, then stepped on the accelerator earlier and reached the maximum throttle position faster at the middle or apex of the corner. RMS of Steering wheel acceleration reveled that experts moved steering more variable while cornering. Proportion of eye movements types showed fixations of experts were more than intermediates and novices and, experts seldom blink and tended to set their line of sight towards and beyond the tangent point during the corner. SD of horizontal eye angle of experts were smaller than intermediates and novices. Findings which reveals expert racing driver used a systematic visual search strategy and control a car efficiently are discussed from not only expert performance approach but also empirical perspectives.

Keywords: Race driving · Expertise · Eye movements

1 Introduction

Expert performers are almost routinely able to cope with severe constraints and can consistently demonstrate superior performance. In recent decades, numerous studies of visual search behaviors in sports have been reported, methodologies such as eye movement recording that can be used to identify the mechanisms which mediate experts' superior performance in perceptual-motor skills, and the expert performance approach presents a descriptive and inductive approach for the systematic study of expert performance. It is generally considered that experts are typically more accurate and quicker in their responses and generally employ fewer fixations and longer duration.

S. Bagnara et al. (Eds.): IEA 2018, AISC 823, pp. 245–248, 2019.
https://doi.org/10.1007/978-3-319-96074-6_26

In motorsports, very few attempts have been made at perceptual-motor skills of racing drivers. We analyzed visual search activities of racing drivers in the simulated racing driving situation.

2 Methods

2.1 Subjects

A total of 15 subjects took part in the experiment and they all had normal or corrected-to-normal vision.

Two expert racecar drivers (age 22 & 40) were competing in national and international races at that moment. Two intermediate drivers (age 24 & 22) had experienced in race driving on the motor racing circuits. Others (mean age 21.09 (SD 1.96)) had no race experience.

2.2 Driving Environment

The experiment was conducted in race driving simulator (Project CARS) which was used for racecar driver training. Subjects are received oral instructions regarding the driving and asked to repeatedly drive the same circuit course which is one of the most famous race track had been used for Formula one race in Japan. The circuit, with a length of 5,380 m, consisted of 15 corners. See Fig. 1 for an overview of the circuit layout and the corner radii for the various corners.

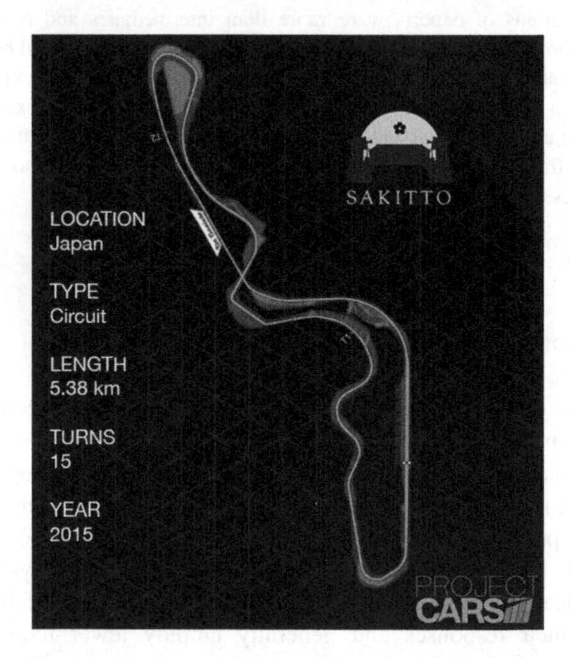

Fig. 1. Overview of the circuit layout and the corners.

2.3 Apparatus

During driving, the subject's eye movements were measured with the lightweight eye tracker at 50 Hz using Tobii Pro Glasses 2.

2.4 Data Analysis

The horizontal component of eye movements, steering wheel movements, brake and throttle maneuvers when driving in corners were calculated from the fastest lap data for each participant. Particularly, the first corner (C1) was focused on analyze in this study. One of expert's data was excluded because of different car conditions.

3 Results

3.1 Best Lap and Vehicle Speed

The results indicated differences between experts, intermediates and novices. The average of best lap time was 1:48.520 for experts, 1:54.612 for intermediates, and 2:14.718 for novices.

3.2 Vehicle Control

Even in the corners, experts' vehicle speed was higher than others. It can be seen that experts held the brake pedal position earlier and smaller time when they approached the corner, then stepped on the accelerator earlier and reached the maximum throttle position faster at the middle or apex of the corner. SD of Steering wheel velocity and RMS of Steering wheel acceleration reveled that experts moved steering more variable while cornering (see Table 1).

Table 1. Means of vehicle control of each groups (standard deviations in parentheses).

Vehicle control	Experts	Intermediates	Novices
Lap time on C1 (s)	9.149	9.304	9.681
Car velocity before C1 (km/h)	230.00	190.00(2.83)	168.91(33.75)
Car velocity after C1 (km/h)	150.00	135.50(2.12)	116.27(20.14)
SD of steer velocity (deg/s)	20.96	19.85(8.63)	20.00(9.38)
RMS of steer acceleration (deg/s^2)	31.74	29.44(12.58)	28.54(15.08)
SD of horizontal eye angle (deg)	1.80	2.08(1.06)	3.55(1.93)

3.3 Eye Movements

Figure 2 shows proportion of eye movements types. It can be seen that fixations of experts (94.98%) were more than intermediates (92.97%) and novices (85.73%) and, experts seldom blink (0.63%). The results from eye tracking data indicated that experts in particular tended to set their line of sight towards and beyond the tangent point during the corner. Consequently, SD of horizontal eye angle of experts (1.80) were smaller than intermediates (2.08) and novices (3.55) (see also Table 1).

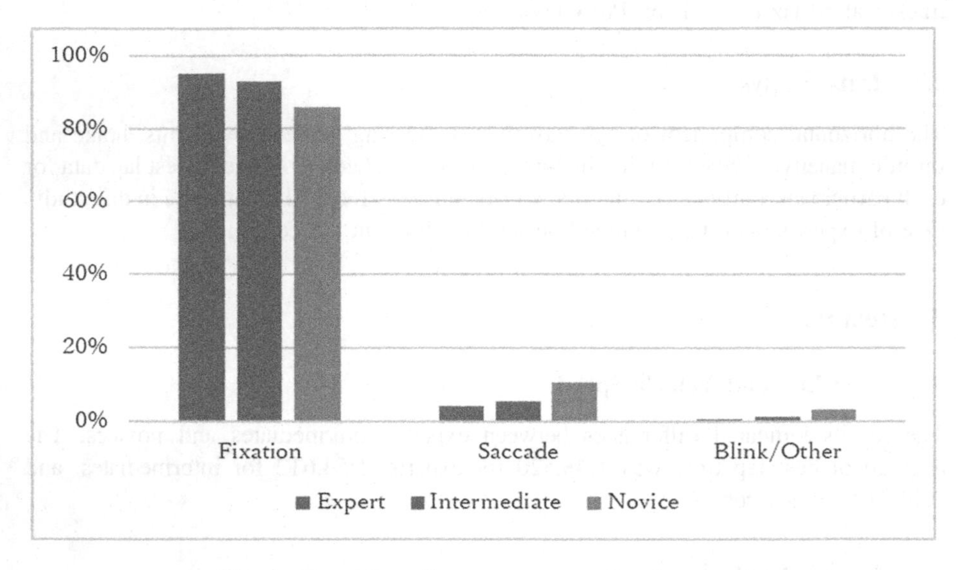

Fig. 2. Proportion of eye movements of each groups.

4 Conclusions

Recent study has demonstrated that racing drivers showed faster lap times, higher steering activity, and a more variable gaze behavior while cornering [1, 2]. This study has identified differences between expert race drivers, intermediate race drivers and novices, and particularly experts moved their eyes quietly as well as in other sports. Still, small sample size did not allow to make significant differences. Further research needs to clear not only in experimental situation, but from empirical perspectives.

References

1. Land MF, Tatler BW (2001) Steering with the head: The visual strategy of a racing driver. Curr Biol 11:1215–1220
2. van Leeuwen PM, de Groot S, Happee R, de Winter JCF (2017) Differences between racing and non-racing drivers: A simulator study using eye-tracking. PLOS one 12(11):e0186871

The Efficacy of Eye Blink Rate as an Indicator of Sleepiness: A Study of Simulated Train Driving

Hardianto Iridiastadi(✉)

Institut Teknologi Bandung, Bandung 40132, Indonesia
hiridias@vt.edu

Abstract. Despite the steady decline in the number of train incidents in Indonesia, rail safety remains an important national issue. Sleepiness has been cited as a major contributing factor in previous rail accidents, and the ability to detect sleepiness while a machinist is performing his job is of importance. The present investigation aimed at determining if eye blink rate (EBR) could be used in determining the degree of sleepiness during train driving tasks. A group of 12 male subjects were asked to drive a train simulator for 4 h in the morning, with sleep durations of 2, 4, and 8 h were allotted the night before the experiment. The driving task was fairly monotonous, with one stop (train station) for every two hours. A second group was also asked to perform the same tasks, but the driving condition was more dynamic (train stopped every 20 min). A high definition camera was mounted in front of the subjects, and recorded the entire face of the subjects continuously throughout the experiment. Rates of eye blink were determined every 20 min, resulting in 12 data points throughout the experiment. Similarly, scores of Karolinska Sleepiness Scale (KSS) was used to assess perceived sleepiness. Results of this experiment demonstrated that frequency of eye blink tended to increase, but in somewhat inconsistent fashion. KSS scores, on the other hand, increased consistently throughout the experiment. It was concluded here that it was fairly difficult to assess sleepiness based merely on raw blink rate data.

Keywords: Sleepiness · Eye blink rates · Train driving · Simulator Karolinska Sleepiness Scale

1 Introduction

Rail operations in Indonesia have become much more complex, particularly due to the rapid developments of railway networks in all major islands and double-track constructions in the island of Java. Hours of operations have also been expanded to 24 h. In majority of large cities, both light rail transit (LRT) and mass rapid transit (MRT) have been introduced in the past five years. While statistics shows that the rates of railway incidents have declined steadily in the past decade, the rapid development of railway system could increase the likelihood of future train incidents.

National papers and the authorities have pointed out that fatigue, particularly sleepiness, as a major contributing factor in many railway incidents not involving level

S. Bagnara et al. (Eds.): IEA 2018, AISC 823, pp. 249–255, 2019.
https://doi.org/10.1007/978-3-319-96074-6_27

crossings. Employing Human Factors Analysis and Classification System (HFACS) for rail sector, Iridiastadi and Ikatrinasari [1] also reported fatigue and sleepiness as an important cause of previous railway incidents. Currently, however, the Indonesian Railways Company (PT KAI) only administers a few tests prior to duty assignment (of a train driver), including blood pressure, heart rate, and body temperature measurements. Fatigue occurred while driving the train has not been addressed, and the management usually assumes that each driver always comes prepared before their duty.

Research has shown that certain eyelid (and oculomotor) parameters can be used in evaluating fatigue development during driving tasks [2]. These include saccadic peak velocity, pupil diameter, and saccade amplitude, and eyelid characteristics, such as blink frequency, blink durations, and eye closure [2–4]. Blink frequency, in particular, can be easily calculated and determined. However, the use of this parameter in evaluating sleepiness has received mixed results. Not only do blink rates correlate with sleepiness, they are also associated with changes in work demands [5].

The objective of this study was to evaluate if blink rate frequency can be of any value in the assessment of sleepiness during train driving. The job of a train driver tends to be relatively monotonous, a condition that can also result in lower cognitive demands [6] and induce sleepiness. In this study, patterns of eye blink frequency were characterized during simulated train driving.

2 Methods

2.1 Experimental Procedures

A total of 24 male subjects aged 21 years old (SD = 0.95) were recruited and asked to drive a train simulator in the morning. The subjects were college-aged engineering students with similar physical and educational background as actual train driver trainees. They did not have any reported medical conditions, and were asked not to smoke nor drink coffee on the day of the experiment.

Twelve of the subjects drove the simulator for four hours continuously (train speed was 15–75 km/h) in a relatively monotonous condition (train stopped every 2 h). The other half drove in more dynamic condition (train stopped every 20 min). All the subjects had to follow speed limits that changed randomly. The subjects were required to sleep for 2, 4, or 8 h prior to the experimental day in a bedroom close to the laboratory. An experimenter slept in the same bedroom to ensure that the subjects had the required amount of sleep duration. To minimize the chronic effects of sleep deprivation, a minimum of 6 days were allotted between these sleeping conditions.

A high-definition camera was mounted underneath the simulator screen, with the purpose of recording the entire face (particularly the eyelids) of the subjects throughout the duration of the experiment. Eye blink rate (EBR) was determined every 20 min (performed off-line). A red light was positioned adjacent to the camera; it was turned on every 20 min and the subjects were required to write down their perceived rating of sleepiness using the Karolinska Sleepiness Scale/KSS [7].

2.2 Data Analysis

Two sets of data were collected in this experiment, including eye blink rates (EBR) and KSS scores. The experiment was a mixed design, in which sleep duration was a within-subject factor (with 3 levels), while the task condition was a between-subject factor (with 2 levels). The Kruskal-Wallis test was used in determining the main and interaction effects. A p-value of 0.05 was used in determining the significance of all statistical tests.

3 Results

During the monotonous conditions, the initial EBR were around 41, 46, and 52 blinks per minute (bpm) for the 2, 4, and 8 h of sleep, respectively (Fig. 1). While the EBR pattern for the 8 h sleep duration tended to be constant throughout the experimental session, the 2 and 4 h of sleep conditions resulted in increasing trends, with peak at around 60 bpm toward the end of the experimental session. Both of these conditions resulted in a decreased EBR at the end of the session (the average of roughly 55 bpm).

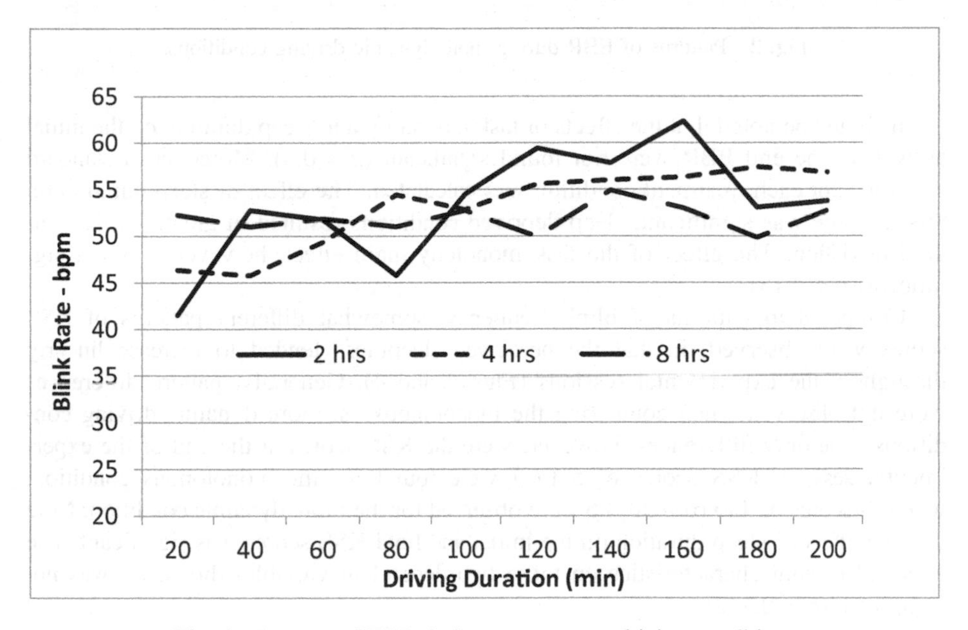

Fig. 1. Patterns of EBR during monotonous driving condition.

Slightly different patterns of EBR were found during the more dynamic driving scenarios (Fig. 2). Subjects with 8 h of sleep were associated with an initial EBR of 48 bpm, a slight increase toward the 3rd hour (a peak of 53 bpm), and a final EBR of 42 bpm. For the 2 and 4 h of sleep conditions, the initial EBR were 35 and 40 bpm, respectively. Both conditions resulted in increased EBR (a peak of 45 and 50 bpm for the 2 and 4 h sleep durations, respectively). For both condition, a decline in EBR was observed toward end of experimental sessions.

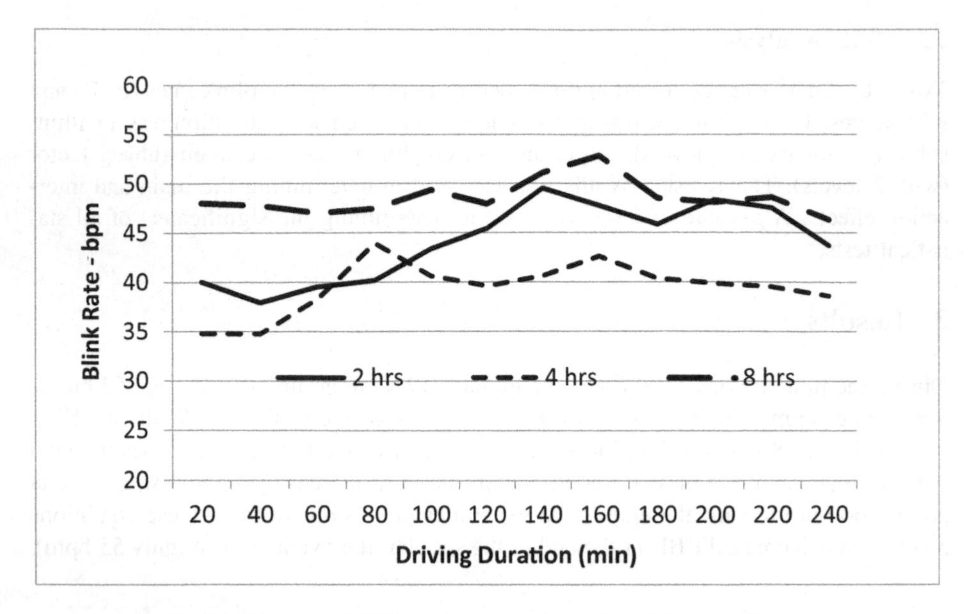

Fig. 2. Patterns of EBR during more dynamic driving conditions.

It should be noted that the effects of task monotony and sleep duration on the initial as well as the end EBR were not found significant ($p = 0.3$). Moreover, a standard deviation for each treatment condition was calculated. The effect of sleep duration on this measure was significant; sleep deprived conditions resulted in greater EBR standard deviation. The effect of the task monotony main effect, however, was not significant ($p = 0.13$).

Compared to patterns of blink frequency, somewhat different patterns of KSS scores were observed, in that the perceived sleepiness tended to increase linearly throughout the experimental sessions (Figs. 3 and 4). Generally, pattern differences were not observed when comparing the monotonous vs. more dynamic driving conditions. The only differences, however, were the KSS scores at the end of the experimental session. KSS scores of 5 to 7 were found for the monotonous condition, whereas scores of 4 to (roughly) 6 were observed for the more dynamic condition. Note that the effect of sleep duration on the initial and final KSS scores was significant. The effect of driving characteristics on these two dependent variables, however, was not significant ($p = 0.122$).

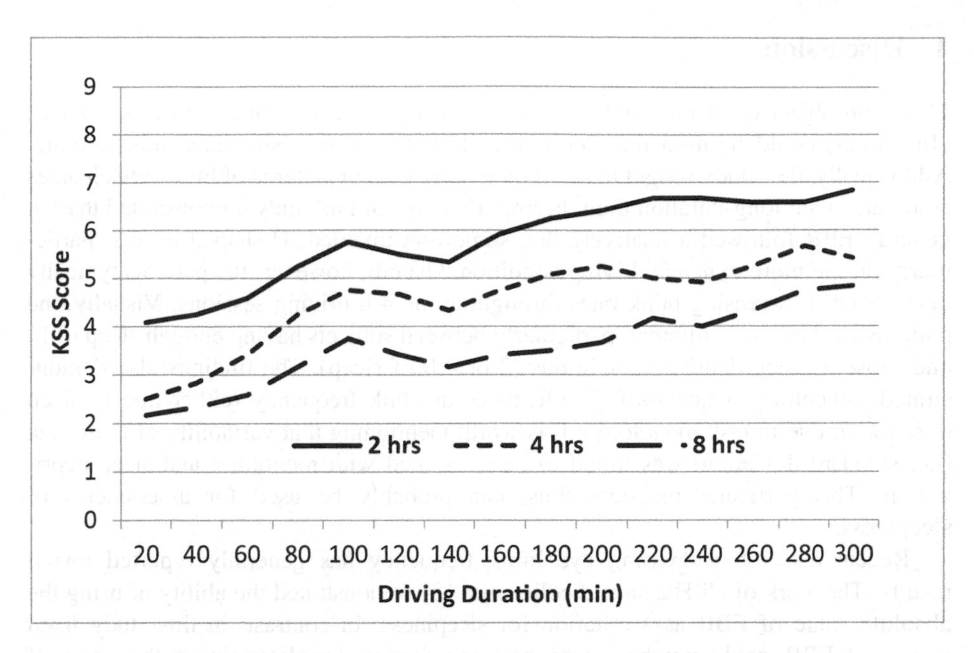

Fig. 3. Changes in KSS scores during monotonous driving task.

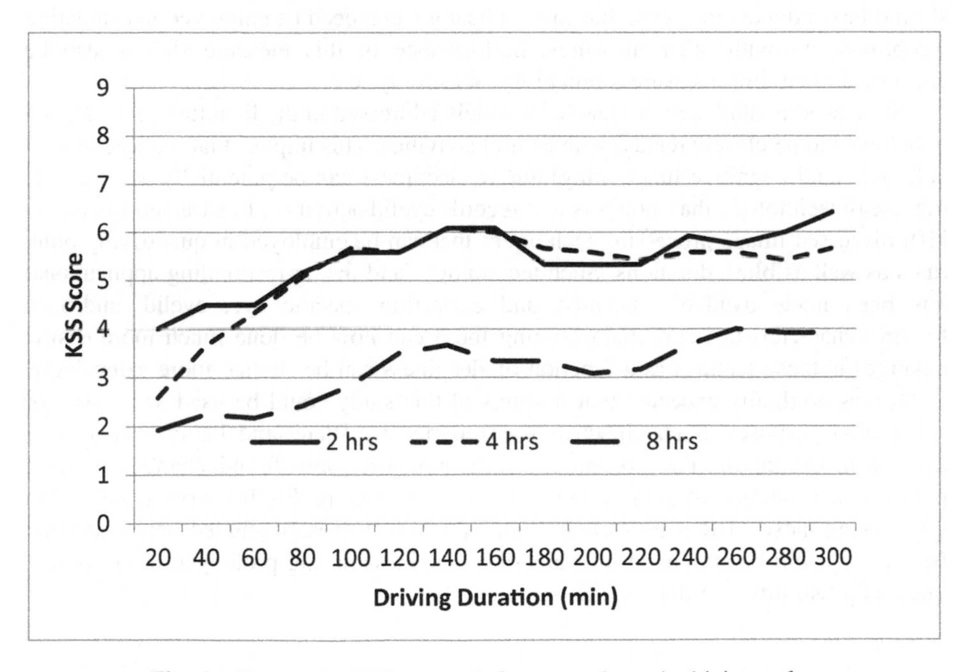

Fig. 4. Changes in KSS scores during more dynamic driving tasks.

4 Discussion

The main objective of this study was to evaluate if a feature of eyelid changes (i.e. blink rates) could be used in determining sleepiness during simulated train driving. Additionally, this study sought to see if there were certain patterns of blink rate changes associated with long-duration train driving. Findings of this study demonstrated that, in general, EBR followed a relatively flat, somewhat inverted, U-shaped curves, particularly during more dynamic driving condition. Overall, however, the patterns typically demonstrated increasing blink rates throughout the 4-h driving sessions. Visually, the patterns seemed to be different, particularly between subjects having enough sleep (8 h) and those in sleep deprived conditions (2 or 4 h of sleep). The findings also demonstrated difficulties in quantifying differences in blink frequency (either due to sleep duration or due to task monotony). It is worth mentioning that variability of EBR data (i.e. standard deviation) was found to be associated with monotony and sleep deprivation. This particular measure, thus, can probably be used for assessments of sleepiness.

Recent research addressing eye blink frequency has generally reported mixed results. The work of Ul-Husna and colleagues [8] demonstrated the ability of using the absolute value of EBR as a criterion for sleepiness. In contrast, in this study fixed values of EBR could not be established as criterion for determining the onset of sleepiness. EBR variability, however, might be used for this purpose. Further research should be conducted to ensure that such indicator can indeed be employed in evaluating sleepiness. As with other measures, performance of this measure should also be addressed, including measure's reliability, sensitivity, etc.

Spontaneous blink activity has been widely addressed in the literature [e.g., 9], and is believed to be closely related with central activities. This implies that reduced central activities (and cognitive functioning) due to sleepiness can be potentially assessed via the use of technology that monitors and records eyelid activities. Fuletra and Bosamiya [10] discussed image processing techniques that can be employed in quantifying blink rates as well as blink durations. Such technology (and the corresponding applications) has been made available recently, and extracting specific eye, eyelid, and face features/characteristics and characterizing these can now be done much more easily. Changes in these features as a function of sleepiness can be studied more extensively.

It was originally expected that findings of this study could be used as a basis for scheduling purposes. Such objective was reported by Wang and Peing [11] in their driving investigation. The present research, however, only found changes in EBR patterns as a function of sleep duration and task monotony. Further work is needed in quantifying these EBR patterns (e.g., data variability). It is concluded here that blink frequency could be employed in evaluating sleepiness during prolonged train driving, but this possibility warrants additional research.

References

1. Iridiastadi H, Ikatrinasari ZF (2012) Indonesian railway accidents: utilizing human factors analysis and classification system in determining potential contributing factors. Work 41:4246–4249
2. Schleicher R, Galley N, Briest S, dan Galley L (2008) Blinks and saccades as indicators of fatigue in sleepiness warnings: looking tired? Ergonomics 51:982–1010
3. Abe T, Nonomura T, Komada Y, Asaoka S, Sasai T, Ueno A, dan Inoue Y (2011) Detecting deteriorated vigilance using percentage of eyelid closure time during behavioral maintenance of wakefulness tests. Int J Psychophysiol 82:269–274
4. Di Stasi LL, Renner R, Catena A, Canas JJ, Velichkovsky BM, Pannasch S (2012) Towards a driver fatigue test based on the saccadic main sequence: a partial validation by subjective report data. Transp Res Part C 21:122–133
5. Gao Q, Wang Y, Song F, Li Z, Dong X (2013) Mental workload measurement for emergency operating procedures in digital nuclear power plants. Ergonomics 46(7): 1070–1085
6. Dunn N, Williamson A (2012) Driving monotonous routes in a train simulator: the effect of task demand on driving performance and subjective experience. Ergonomics 55(9): 997–1008
7. Miley AA, Keclund G, Akerstedt T (2016) Comparing two versions of Karolinska Sleepiness Scale (KSS). Sleep Biol Rhythms 14:257–260
8. Ul-Husna A, Roy A, Paul G, Raha MK (2014) Fatigue estimation through face monitoring and eye blinking. In: International conference on mechanical, industrial and energy engineering, 140214-1–140214-5, Bangladesh
9. Cruz AAV, Garcia DM, Pinto CT, Cechetti SP (2011) Spontaneous eyeblink activity. Occular Surf 9(1):29–41
10. Fuletra JD, Bosamiya D (2013) A survey on driver's drowsiness detection techniques. Int J Recent Innov Trends Comput Commun 1(11):816–819
11. Wang L, Peing Y (2014) The impact of continuous driving time and rest time on commercial drivers' driving performance and recovery. J Saf Res 50:11–15

Information Content of a Route Guidance System Based on the Characteristics of Elderly Drivers

Akifumi Tsuyuki[1(✉)] and Tatsuru Daimon[2]

[1] Graduate School of Science and Technology, Keio University, Yokohama, Japan
tsuyuki.a@keio.jp

[2] Faculty of Science and Technology, Keio University, Yokohama, Japan

Abstract. Landmarks and clues outside the car that elderly drivers can easily recognize from inside the car and on a car navigation system during route navigation were extracted. In a laboratory survey, we measured the degree of awareness of logos by type of business. We then conducted a driving simulator experiment to verify the response rate, reaction distance, and line of sight direction according to the business type and store, vehicle location, shop location, and presence/absence of signboards that had good results in the laboratory survey set as variables. Results show that a vehicle navigation system should ideally use clue information outside the car that drivers can view without largely averting their eyes from the front; use clue information that is presented at a large visual angle; and use clue information that is conspicuous in color and shape and known by the elderly driver.

Keywords: Navigation system · Route guidance · Clue information

1 Introduction

The numbers of traffic accidents and fatal accidents have been decreasing in Japan. However, the number of accidents involving elderly drivers has not been decreasing. In addition, the elderly population will exceed 30% of the total population in 2025 in Japan. It is thus necessary to implement measures that prevent traffic accidents involving elderly drivers.

Route guidance provided by car navigation systems can support the driving of the elderly. Studies on route guidance often sketch a map [1]. This is a method of letting the cognitive map, which is drawn in the head, depict when moving in the space. Information and strategies used for route guidance are related to urban structures that depend on, for example, the regionality of countries and buildings.

Recently developed route guidance systems use dynamic information obtained from outside the vehicle to provide drivers with more information than ever before. Meanwhile, it is considered that a decline in cognitive functionality makes it difficult for elderly drivers to collect information and perform cognitive processing smoothly. Therefore, elderly drivers have difficulty in processing car navigation landmark information.

S. Bagnara et al. (Eds.): IEA 2018, AISC 823, pp. 256–266, 2019.
https://doi.org/10.1007/978-3-319-96074-6_28

The present study extracts the requirements of clue information used by the route guidance of a car navigation system that elderly drivers can easily spot to facilitate the matching of clue information outside the vehicle with information displayed by the car navigation system.

2 Laboratory Survey

The present survey investigated the extent of elderly driver's prior knowledge of landmarks used in actual car navigation. It also investigated what knowledge is in the logos of stores and examined the logos of stores that are candidates of landmarks in the experiment using the driving simulator.

2.1 Survey Method

A questionnaire survey was conducted for elderly drivers and young drivers about stores and store logos that are often observed while driving. As shown in Table 1, representative stores were selected from seven types of business, namely convenience stores, gas stations, banks, car dealerships, restaurants, fast food outlets, and apparel stores, and the store logos were presented to the participants.

Table 1. Business/store logos presented in the laboratory survey

Business type	Store
Convenience Store	7-Eleven, Lawson, Family Mart, Sunkus, Ministop
Gas Station	Eneos, Shell, Idemitsu
Bank	Sumitomo Mitsui Banking Corporation, Mizuho Bank, MUFG Bank
Car Dealership	Toyota, Nissan, Honda
Restaurant	Gusto, Denny's, Ohsho,
Fast Food Restaurant	Yoshinoya, Sukiya, Kentucky Fried Chicken, McDonald's, Mos Burger
Apparel Store	Uniqlo, Aoyama, Aoki

We asked the questions listed in Table 2 for each store logo. Question 2 was only answered if the participant recognized the logo in Question 1. The level of awareness of the logo was obtained in Question 2 and averaged across participants. Meanwhile, the percentage of correct answers was obtained according to the accuracy of each subject's selection of the logo. For Questions 3 and 4, we set out 45 logos on the question sheet for the participant to match to stores. To avoid replying to the next question, we divided the questionnaire for each question except Q1 and Q2 and provided the next questionnaire when the participant's answer was complete.

Table 2. Subjective evaluation of logos in the questionnaire

Question	Store
1st Question Sheet	Question 1: Do you know the logo of this store? Answer 1: "I know" or "I do not know" Question 2: Only answer if you answered "I know" in Question 1. How do you know the logo? Answer 2: # I can image the color and shape exactly # I can choose the exact logo if I watch some images # I understand that it is the shop if I see the logo
2nd Question Sheet	Question 3: Please select the logo that represents each store *For example, we enclose part of the store name with the color of the letter or present an incomplete logo (a) convenience store (b) car dealership (c) bank
3rd Question Sheet	Question 4: Please select the logo that represents each store *I presented a logo used in an actual environment (a) convenience store (b) car dealership (c) bank

2.2 Participants

Participants were grouped as elderly drivers and young drivers. The elderly driver group comprised 11 elderly men aged 65 to 77 years and having an average age of 72.4 years who held regular driving licenses. The young driver group comprised three young men aged 21 to 22 years and having an average age of 21.3 years.

The elderly drivers had more than 36 years' driving experience each and drove more than once a week. The young drivers had driving experience of 1 to 2 years and drove about once a week (two people) and about once a month (one person).

2.3 Measured Items

Table 3 explains how the level of knowledge and percentage of correct answers were determined.

Table 3. Measurement item and its explanation

Measured item	Explanation
Level of Knowledge	The level of knowledge is either "I know" or "I do not know" in Question 1. An answer of "I do not know" gets a score of zero. Question 2 is answered if the answer to Question 1 is "I know".
Percentage of Correct Answers	The number of participants who selected the correct logo (hereinafter referred to drivers who gave the right answer) is counted for each store and the percentage of correct answers is calculated as follows. *Percentage of correct answers (%) = Total number of drivers who gave the right answer / Total number of drivers in each group × 100*

2.4 Results

Figures 1 and 2 show that elderly drivers tended not to know the logos, especially those of fast food restaurants. However, they had higher levels of knowledge of the logos of gas stations and car dealerships. The knowledge level for the logos of convenience stores strongly depended on the store.

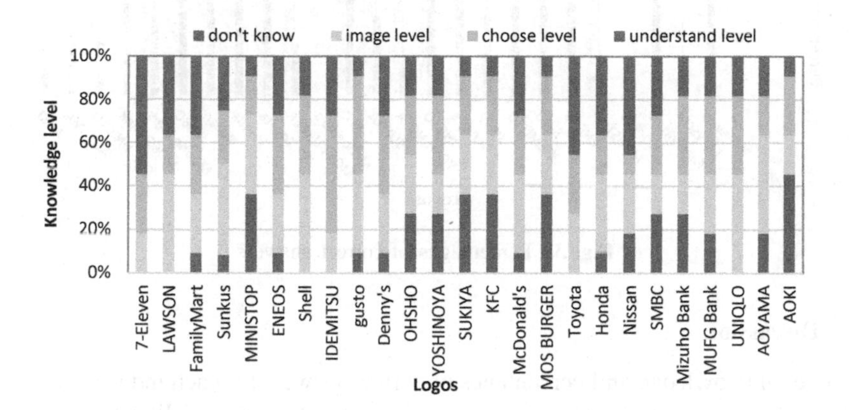

Fig. 1. Level of knowledge of logos for elderly drivers

Figure 3 shows that the percentages of correct answers for gas stations, car dealerships, and several convenience stores were high for both elderly and young drivers, while those for convenience stores and fast food restaurants strongly depended on the store. Overall, the percentages of correct answers given by elderly drivers tended to be lower than the percentages of correct answers given by young drivers.

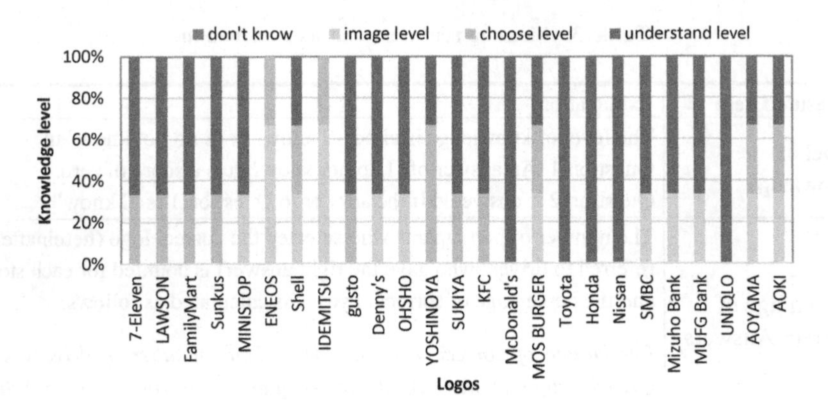

Fig. 2. Level of knowledge of logos for young drivers

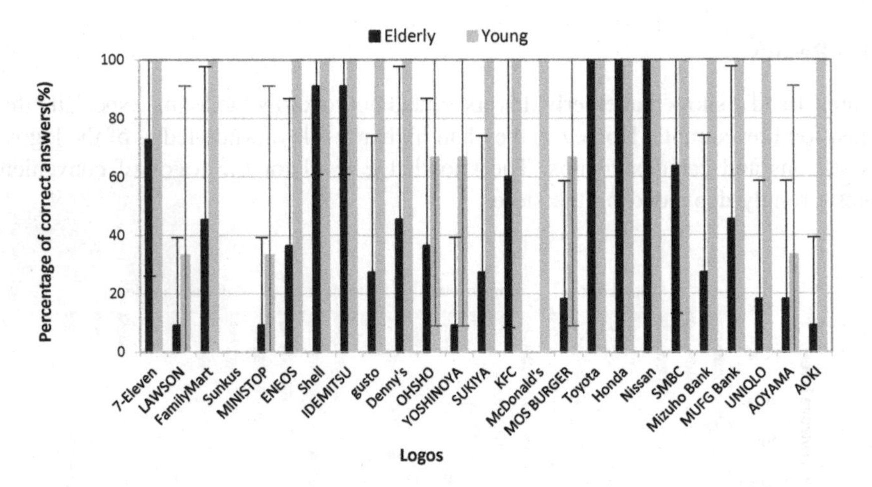

Fig. 3. Percentages of correct answers

2.5 Discussion

The level of knowledge and percentages of correct answers for each industry was found to depend on not only the number of stores but also the drivers' lifestyle habits. For elderly drivers, gas stations and car dealerships are of the industry directly related to driving and cars with which they are familiar. It is considered that, coupled with their long driving experience, their long-term memory of gas stations and car dealerships is as high as that of young drivers. Meanwhile, the knowledge level and percentage of correct answers for convenience stores and fast food restaurants were low because elderly drivers visit these stores less often.

It is thus considered that although young drivers have sufficient knowledge of well-known stores used in current car navigation systems, it is possible that elderly drivers do not recognize the stores displayed by such systems.

3 Driving Simulator Experiment

3.1 Experimental Equipment

The experimental environment is presented in Fig. 4. The environment consisted of a vehicle cabin, three (150-in.) screens surrounding the cabin, and a liquid-crystal-display projector. In addition to the presentation of the driving image, the tracking task was projected onto the front screen. Levels of operation of the steering wheel, accelerator pedal and brake pedal were imported into the computer, and the display of the tracking task was controlled according to the operation level.

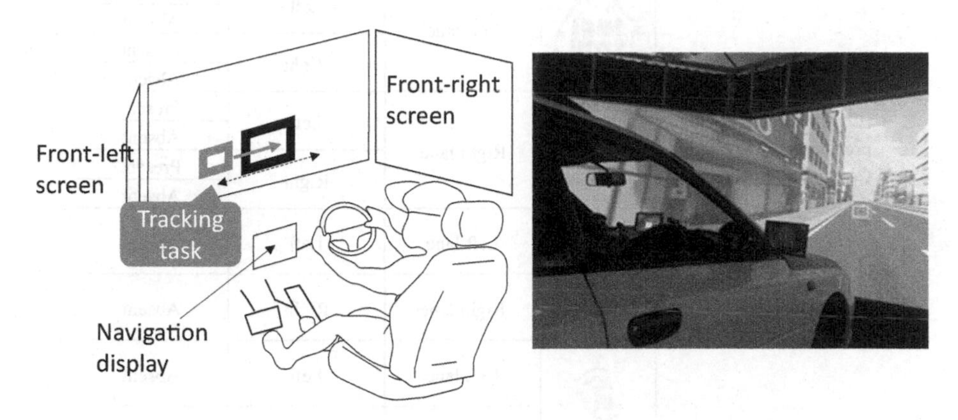

Fig. 4. Outline of the experimental equipment and overview of the driving simulator

For the driving image, we used a UC-win software (FORUM 8) to create a four-lane straight road and buildings such as stores.

As a means of presenting logos for route guidance, a 7-in. display (hereinafter referred to as the navigation display) was installed at the top of the installation panel.

3.2 Participants

The same participants as in the laboratory survey participated.

3.3 Experimental Method

A tracking task was projected to direct the line of sight of the participant forward. Participants were instructed to follow irregularly moving tracking targets by steering within a certain range (Fig. 4).

We imposed a secondary task to search for specific landmarks in the driving situation while doing the tracking task. Table 4 gives the simulation conditions.

Table 4. Business/store logos presented in the driving simulator experiment

Business type	Logo	Lane position	Store location	Signboard
Convenience Store	SEVEN&i HOLDINGS	Left lane	Left	Present
				Absent
			Right	Present
				Absent
		Right lane	Left	Present
				Absent
			Right	Present
				Absent
(Continuation from the previous page) Convenience Store	MINI STOP	Left lane	Left	Present
				Absent
			Right	Present
				Absent
	MINI STOP	Right lane	Left	Present
				Absent
			Right	Present
				Absent
Gas Station		Left lane	Left	Absent
		Right lane	Right	Absent
	IDEMITSU	Left lane	Left	Absent
		Right lane	Right	Absent
Fast Food Restaurant		Left lane	Left	Present
				Absent
			Right	Present
				Absent
		Right lane	Left	Present
				Absent
			Right	Present
				Absent
	KFC	Left lane	Left	Present
				Absent
			Right	Present
				Absent
		Right lane	Left	Present
				Absent
			Right	Present
				Absent

3.4 Measured Items

The level of knowledge and percentage of correct answers were determined as described in Table 5.

Table 5. Measurement item and its explanation

Measured item	Explanation
(Continuation from the previous page) Response Rate Response Rate	Whether the stores presented by the navigation display were correctly discovered in the driving situation was judged. Participants were requested to verbally state whether they found each store on the left or right side of the road. It was judged that the participant was "able to react" only when the right or left position answer was correct. In addition, the timing of the pressing of the brake pedal was measured from the time when the logo was displayed and a beep tone from the navigation display. The reaction rate is calculated as follows. *Response rate = Number of stimuli reacted under the same condition / Number of stimuli presented under the same condition* × *100*
Reaction Distance	On the basis of the timing of the pressing of the brake pedal, the distance between the vehicle and the landmark was calculated. The distance is thus defined as the distance required for the response. In other words, the farther the point at which the brake pedal was pressed is from the landmark position, the earlier the landmark was discovered.
Azimuth Angle and Character Size (Visual Angle)	We defined the azimuth angle and character size (visual angle) created when the participant found the landmark properly in the driving situation. The azimuth angle is the angle formed by lines connecting the center of the store and the driver's eye position, and the center of the tracking task and the driver's eye position. It is an index for evaluating how far the store is from the driver's eye position at the time of reaction. 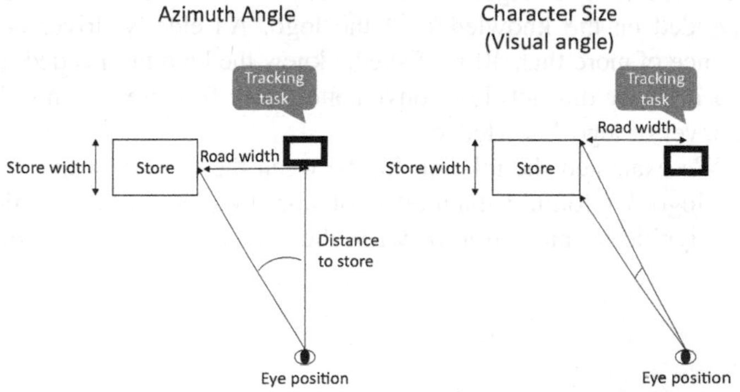 Azimuth Angle Tracking task Store width Store Road width Distance to store Eye position Character Size (Visual angle) Tracking task Road width Store width Store Eye position

3.5 Experimental Procedure

The objectives of the research, contents of the experiment, protection of personal information and the process of withdrawing consent were explained in writing and verbally, and informed consent was obtained from each participant. We then explained the experiment and operation method. We allowed participants to practice the experiment and operation and confirmed that they were sufficiently familiar with the experiment in the opinion of the experimenters and the participants themselves. After measuring the control conditions, the experimental conditions were measured. The order of presenting the experimental conditions was changed for each participant. Participants were able to take a break as appropriate.

3.6 Results

The results of the reaction rate and the reaction distance for all parameter settings, such as the vehicle position and store location, are presented in Figs. 5 and 6. The response rate of the elderly drivers tended to be lower than that of the young drivers, while the reaction distance was shorter for the elderly drivers; significant differences were obtained in t tests.

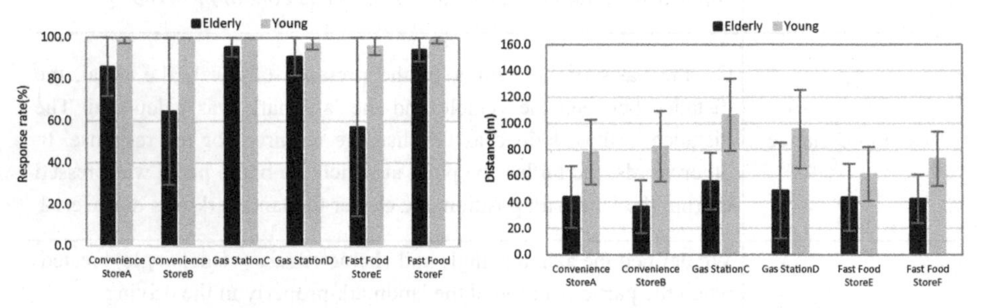

Fig. 5. Average response rate for each store **Fig. 6.** Average reactive distance for each store

We studied the effect of logo knowledge obtained in the laboratory survey on the reaction distance. We focused only on the results for the elderly because young drivers knew all the logos. Figure 7 shows that, for fast food restaurants, the reactive distance depended on the knowledge of the logo. An elderly driver could spot a logo at a distance of more than 40 m if she/he knew the logo but needed to be nearer to find the logo if she/he did not. For convenience store B, there was no difference according to the level of logo knowledge.

We examined the relationship between the vehicle position and store location for each logo. We found a main effect of store location for both participant groups, as well as a significant interaction between the vehicle position and store location.

Fig. 7. Average reaction distance by awareness (for elderly drivers only, excluding the stores responded as "know all of them")

3.7 Discussion

Level of Logo Knowledge

Clue information with higher knowledge tends to contribute to a higher reaction rate and earlier finding. This is considered to be due to the process of checking the logo presented on the navigation display by referring to the driver's long-term memory and then searching for the corresponding feature or structure outside the vehicle. Therefore, the logo that exists in the driver's long-term memory is quickly and accurately recognized. However, if the logo does not exist in the elderly driver's long-term memory, the elderly driver stores the logo shown by the navigation display in her/his short-term memory and collates the matching features and structures outside the car. The process is considered to disturb the smooth finding of logos, leading to a poor response rate and shorter reaction distance.

In addition, the response rate and reaction distance of elderly drivers were found to be lower than those of young drivers in most situations. This is considered to be due to the elderly's lifestyle and limited area of movement in addition to the elderly's poor ability to switch attentional resources. In other words, the elderly drivers' limited area of activity and limited knowledge of logos, such that they are only aware of the logos of stores near their homes, explains their poor performance. Conventional navigation systems show many landmarks with which drivers are not familiar depending on the generation of the user. The present results suggest that even stores generally well known throughout the country may not be useful to elderly people depending on static conditions, such as the elderly's lifestyle and area of activity, and dynamic conditions, such as the driving situation.

Position

The response rate was higher for stores located on the left. This is because vehicles are driven on the left side of the road in Japan and drivers usually pay attention to stopped vehicles and pedestrians on the left as part of safe driving.

Character Size (Visual Angle)

When a vehicle is driven in the left lane, stores on the left are found almost parallel to the direction of travel of the vehicle and have less opportunity to be viewed. A logo on the right side of the road is therefore more observable. When a vehicle travels in the left lane and the store is on the left, the character size can be increased if the sign is oriented vertically against the driving direction, as is commonly done. In this respect, gas stations with large store areas generally scored highly in terms of the correct answer rate and reaction distance.

4 Conclusion

The present study found that vehicle navigation systems can benefit from using clue information outside the car for which drivers do not need to largely avert their line of sight from the front; using clue information that presents at a large character size; and using clue information that elderly drivers know and is conspicuous in terms of color and shape.

Reference

1. Lynch K (1960) The image of the city. MIT Press, Cambridge

A Team Drives the Train: Human Factors in Train Controller Perspectives of the Controller-Driver Dynamic

Anjum Naweed(✉)

Appleton Institute for Behavioural Science, Central Queensland University, 44 Greenhill Rd, Rockhampton, SA 5034, Australia
anjum.naweed@cqu.edu.au

Abstract. Signal passed at danger events (SPADs) impact safety-risk on rail networks, despite the introduction of novel technologies aimed at addressing their cause and effect. Much of the rail safety literature has had a tendency to focus on activities within the cab, placing a spotlight on "errors" within the train driving role. However, a train is not propelled by a single person—is it is propelled by a tightly-coupled team where driving and train controlling activities are distributed but must work in concert. This study set out to understand how controllers perceive the controller-driver dynamic, and how these perspectives impact upon SPAD-risk. Interviews were conducted with 35 train controllers from 6 rail organisations across Australia and New Zealand. Data were collected using the SITT forward scenario simulation method and analysed using conventional content analysis. Eleven different perspectives were identified, ranging in type and varying by frequency, each with implications for the strength of the coupling in distributed cognition between the controller and driver roles and with implications for SPAD-risk. How these perspectives may influence controller-driver dynamics are illustrated using sample scenarios from the data. The findings emphasise key dimensions of the teaming factors in the movement of trains and illustrate how the underlying values and philosophies in different train controlling cultures influence safety. Findings are discussed in the context of obtaining a holistic and more informed model of train driving.

Keywords: Train controllers · SPAD-risk · Teamwork

1 Introduction

In conventional rail networks, a signal passed at danger or 'SPAD' happens when a train goes past a red signal.[1] It is like a car going through a red light, but trains take much longer to stop and cannot manoeuvre to avoid collisions. For this reason, the risks are greater and impact to service more severe. Although most SPADs are contained before real harm occurs, in Australia, each SPAD can cost an organization up to

[1] The term "SPAD" is used to encompass any scenario where a train has encroached into an area it has no authority and can therefore include non-signalled environments. SPADs are most common in setting which use traditional driving/multi-aspect signalling practices but can also happen in advanced rail networks with sophisticated signalling systems.

S. Bagnara et al. (Eds.): IEA 2018, AISC 823, pp. 267–277, 2019.
https://doi.org/10.1007/978-3-319-96074-6_29

$110,000 in loss of service, missed shipping windows, timetabling disruption, follow-on impact to other traffic, post-SPAD investigation, retraining, and fines [1]. The costs are invariably greater for SPADs that result in derailments and collisions, and in the worst-case of scenarios, loss of life. The UK Ladbroke Grove SPAD in 1997 cost 31 lives, the US Chatsworth SPAD in 2008 cost 25, and in 2015, a SPAD east of Lucknow in Northern India killed at least 30 [2]. In Australia, publicly available SPAD statistics hint at ~1,000 SPADs happening annually [3, 4], and although the country has managed to avoid calamity, they are costing the economy many millions of dollars annually [1]. Whilst SPADs are a problem now, increasing rail traffic and maximised capacities mean they have the potential to be an even bigger problem in the future.

SPADs are frequently blamed on the individual by way of distraction or errors in decision making [7], compelling a view that the driver is the sole agent responsible for when they happen (see Fig. 1). In practice, it is a 'team' that drives the train where the driver and the controller functions work as a distributed system. For this reason, research is starting to view SPADs less from the individual and more from a systems perspective and qualifying it as a "wicked problem" [8], meaning they are being linked with cultural challenges and contradictory requirements that are incomplete, hidden, and difficult to recognize [9].

Fig. 1. Two examples of posters hanging in rail organizations warning drivers against SPADs.

1.1 Towards Understanding the Train Controller-Driver Dynamic

Non-technical human factors in SPADs are associated with the effects of time pressure, patterns of station dwelling, increases in concurrent workload (e.g. team communications during safety critical periods), and priming effects (i.e. expectation bias) from routine [5–7]. Many of these are defined by the interconnect shared dynamic between the driver and the 'controller' – a term which varies from place to place (e.g. signaler, dispatcher) but describes those who interact with the train driver remotely and control train movements. The train driver and train controller work in a close dynamic, and although each has his or her own discrete functions, it is actually a team dyad that operates the train, with the driver and controller forming part of a tightly coupled, distributed system. However, the synergy in this team may be disrupted and/or distorted by conflicting views on service delivery and a power differential that renders the driver subservient to the controller's productivity goals [7].

Previous rail research [e.g. 5, 7, 8] has sought to understand how train drivers view their relationship with signals, revealing that they are contextualised in a number of different ways. For example, the relationship is described in ways that assigned primacy, such as *"top priority"* or in terms of feelings about the quality of the relationship, such as *"respect"*; they are also personified interpersonally, with descriptions such as *"my colleague"* and *"my best friend"*, or in terms of spiritual ethics or divine law, such as *"my religion"*, and *"God"* [7, 8]. These findings point to a strong, highly intimate and often complex dynamic underlying the relationship between the driver and the signal, revealing much about the driver-train-signal system. However, very little is known about the 'driver-train-signal-controller' system. Given that controllers effectively manipulate signals to communicate corresponding train operation, this presents a gap.

The Joint Cognitive Systems theory [9] implies an intimate and dyadic coupling between drivers and controllers at a functional level, such that they are both parts of a joint cognitive system. Thus, instead of looking at the 'train-driver-signal' system in isolation, it is important to view it in terms of the 'train-driver-signal-controller' system and enjoin these parts as a whole cognitive system. If this is the case, then the way that train controllers and train drivers relate to one another at an interpersonal level could create barriers or destabilising factors that impact how the system functions, and ultimately, train safety. It is important to develop a better understanding about the driver-controller dynamic.

While much research has developed a good understanding on how train drivers view SPAD-risk and their relationship with train controllers, we still do not have even a basic understanding of how controllers themselves perceive how SPADs are caused, and how they see themselves in relation to the train driver. Developing an understanding of the controller-driver dynamic from the perspective of the controlling role is fundamental for understanding the non-technical dimensions and intricacies of the system.

1.2 Aims and Objectives

The aim of this research was to develop a better understanding of the controller-driver dynamic, in order to be able to inform a more holistic model of SPAD-risk, capturing the nuances of these roles. The objective of this particular study was to disambiguate the relationship between controllers and drivers by identifying how drivers were viewed from the controller perspective. The research was undertaken using the following research question, *"how do train controllers perceive their relationship with train drivers, and how does this perspective impact upon SPAD-risk?"* This paper then, presents preliminary findings and corresponding ideas.

2 Methods

2.1 Research Design

This study drew on a qualitative phenomenographic approach to obtain rich insight into the decision process, experiences, perceptions and opinions of participants [10, 11].

Semi-structured one-to-one interviews were used to elicit knowledge and support the application of a method called the Scenario Invention Task Technique (SITT) [5, 12]; a generative simulation task, the SITT combines principles of multi-pass retrospective inquiry from the Critical Decision Method (by Klein, Calderwood and Macgregor [13]) and story-telling approach in the Rich Picture Data method [14] to externalize decision-making in complex work.

The SITT, and very similar processes, have been successfully applied in the context of naturalistic decision making in a number of complex domains such as rail [5, 15, 16] and healthcare [17], and involve participants creating challenging scenarios and stories specific to their work with the aid of illustrations, schematic drawings, representations and/or work plans to assist with articulation.

2.2 Participants and Recruitment

A total of 35 train controllers (all male) ranging from 29–66 years of age (M = 46.9, SD = 9.78) took part in the study. Participants were recruited from 6 rail organizations across Australia, including Perth (WA), Brisbane (QLD), Sydney (NSW), Melbourne, (VIC), and in New Zealand (Wellington). Most participants had more than 10 years of experience (M = 18.78, SD = 11.98). Organisation support was initially acquired from a national Australasian SPAD Group, and participants were subsequently recruited using contacts from each of these organizations.

2.3 Procedure

After providing informed consent and completing a demographical information card, participants were interviewed in two parts using the semi-structured protocol shown in Fig. 2. Questions included closed-format responses, open-ended questions to encourage consideration and presentation of new ideas, and cued questions calling for relatively short answers in a narrow range. Closed-format responses were used to obtained data about the controller-signal relationship (*"in one or a few short words, describe your relationship with the railway signal"*), and the controller-driver dynamic (e.g. *"in one or a few short words, describe how you see your relationship with train drivers"*).

The SITT was applied over three steps in the second part of the interview, commencing with an instruction for the participant to: *"Invent a scenario where even the most experienced train controller inadvertently decreases rail safety"* An increase in SPAD-risk was provided as an example, but was not necessary; scenarios such as derailments or collisions could also be created that did not necessarily feature SPADs. Participants were encouraged to explain the scenario during construction and created it using felt markers and A3-sized paper.

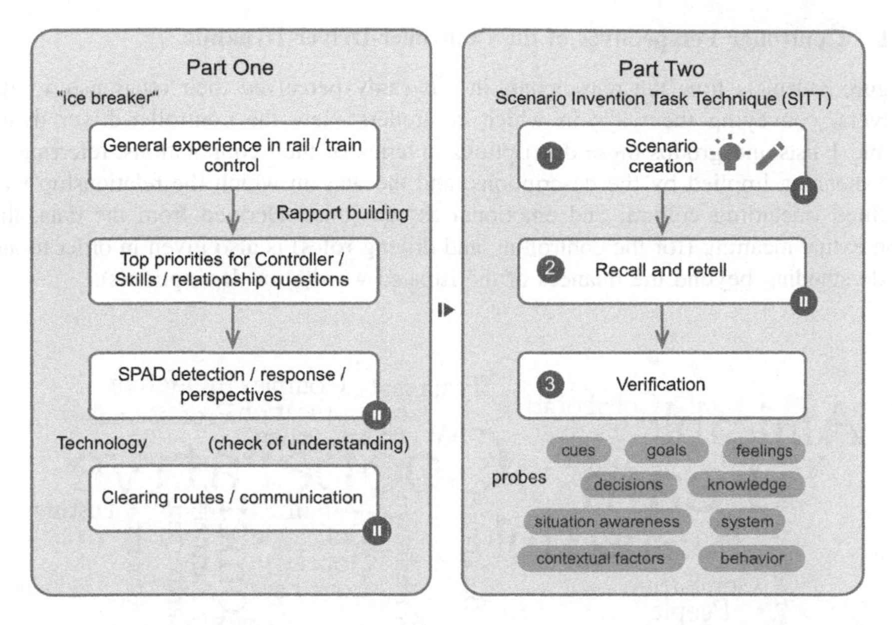

Fig. 2. Overview of data collection protocol.

2.4 Ethical Considerations

The protocol was piloted prior to commencement. Identifying data were redacted at the time of transcribing to ensure anonymity and confidentiality. The study was approved by the University's Human Research in Ethics Committee (Approval no. H17/03-047).

2.5 Data Analysis

Voice recordings were transcribed verbatim. Scenario data from the SITT was inductively analysed (i.e. without using pre-formulated categories) using conventional content analysis [18], in order to identify and create links between key scenario features described during the SITT, and all data collected in the first part of the protocol. For some aspects of the protocol, such as the closed-format responses, analysis was used to determine the connotations and contextual meanings. Coding occurred semantically from description to interpretation, with codes undergoing refinement into categories.

World clouds were utilized during analysis as a word summarization research tool [19] to visualize outcomes for closed-format questions, by distilling text down to those words that appeared with highest frequency.

3 Results and Discussion

In the first section of the results, findings associated with controller-driver dynamic are presented, followed by illustrative examples associated with specific scenarios that lend insights into how controller-driver dynamics impact upon SPAD-risk.

3.1 Controller Perspectives of the Controller-Driver Dynamic

Figure 3 depicts how the participants in this study perceived their relationship with drivers, conveying the ways in which controllers view the controller-driver dyad. Table 1 lists and groups these descriptions in terms of their connotations, referring to the meaning implied by the descriptions, and the way in which the relationship was defined (including cultural and emotional associations). Derived from the data, the contextual meaning (for the controlling and driving roles) is also given in order to aid understanding beyond the nuances of the isolated word(s) or description(s).

Fig. 3. Word cloud visualizing how train controllers characterized their relationship with train drivers.

Table 1. Connotations and contextual meanings of descriptions used to describe the Controller-Driver relationship.

	Description of relationship	Connotations (The controller-driver dynamic was defined through…)	Contextual meaning
1.	*Fair* *Good*	Quality of the relationship	Driver seen as equal, treated well
2.	*Respectful*	Feeling of the quality of the relationship, stemming from underlying admiration	Controller has optimistic acceptance of driver behaviour
3.	*Trust*	Strength of the relationship	Reliance on driver and confident expectation of driving role
4.	*Complex*	Interconnectedness and opacity in relationship	What controllers and drivers do is hard to understand
5.	*Withdrawn*	Separation and distance in relationship	Drivers and controllers are co-located; drivers do not want to communicate with controllers and controllers do not want to communicate with drivers

(*continued*)

Table 1. (*continued*)

	Description of relationship	Connotations (The controller-driver dynamic was defined through…)	Contextual meaning
6.	*Partnership* *Partners* *Colleague*	Functional dynamic(s)	Controllers and drivers are associated and have joint interests
7.	*Team* *Cooperative* *Work with them* *Collaborative* *Communication*	Assessment of combined action	Emphasis of coordinated effort required to achieve goals/relationship between controller-drivers is marked by cooperation
8.	*Customer*	Commoditised, transactional nature of the relationship	Controller is delivering a service to drivers, emphasis is on rail business model
9.	*Backbone* *Leave them alone*	Chief supporting mechanism of the system	Drivers and controllers are important, and must be protected
10.	*Combative* *Testing* *Love-hate* *Amicable* *Hopeless*	Interpersonal quality, simultaneous or alternating emotions	Controllers and drivers' perspectives vary, there is a predisposition to contend/disagree with one another, disputes may or may not be resolvable
11.	*People* *Indifferent* *Operators*	Impartiality	Drivers and controllers are both a "cog in the machine"

As can be seen from Fig. 3 and Table 1, participants chose to emphasize their relationship with train drivers in different ways, defining it according to its quality, function, feelings, connectedness, and so on. From the word cloud, it can be seen that descriptions such as *"cooperative"*, *"team,"* *"amicable"*, *"respectful"*, and *"customer"* were recurrent ways of describing the controller-driver relationship across participants, with *"cooperative"* and *"team"* the most prevalent. Defining the controller-driver dynamic through assessment of combined action, and in the context of coordinated effort for goal-attainment was therefore a dominant view. Nevertheless, relationship dynamics were defined in many other ways. When viewing the relationship in terms of its quality, it was generally positive (e.g. *"fair"*, *"good"*), and where this was associated with feelings, evoked optimistic acceptance of behavior (i.e. *"respectful"*). The *"trust"* descriptor was one way of qualifying this by emphasizing a reliance upon and confident expectation of a driver operating the train in a way that befitted their own expertise.

The nature of the controller-driver dynamic was also described as *"complex"*, simultaneously implying an interconnectedness of many moving parts, but also a system that was dense and hard to understand easily. In some ways, the complexity in the relationship was further elaborated by descriptors that leant insight into its interpersonal qualities; descriptions such as *"combative, "testing", "love-hate", "hopeless"* and even *"amicable"* belied feelings that while controller and driver perspectives varied, there was an inherent predisposition for contention and conflict.

Some of the controller-driver dynamic perspectives were highly contrasted. For example, one perspective was to be completely impartial, where descriptions of the two roles described drivers and controllers as little more than cogs in a machine (i.e. *"people", "indifferent", "operators"*), whereas another described them as the chief supporting mechanism of the whole rail system (*"backbone", "leave them alone"*).

3.2 Controller-Driver Dynamic Factors Influencing SPAD-Risk

Figure 4 depicts a SPAD scenario created by a participant who viewed the controller-driver dynamic as *"withdrawn"* (see Table 1) emphasizing co-location and the view that train drivers do not want to communicate with train controllers and vice versa. In this scenario, the driver of Train 1–A has yet to clear the section and is approaching Signal 1. In the meantime, the driver of a following train, Train 2–A, has just had a SPAD two signals away, at Signal 3. Upon detection of the SPAD, the area controller elects to ring the train controller to advise of the SPAD (in this setting, those who interacted with signals and control the train were called 'area controllers' and were overseen by an overarching controller). Partway through this, the area controller notices that train 2–A has in fact departed to Signal 2. The area controller promptly contacts the driver of Train 2–A, instructed to stop the train immediately and then informed that they have had a SPAD—but not before the driver has a second SPAD at Signal 2.

Fig. 4. Multi-SPAD scenario depicting the driver of a train (2–A) continuing on to a SPAD (at Signal 2) after having already had a SPAD at the previous signal (Signal 3). *Notes:* top portion of scenario is a "mimic board", presenting an abstracted view of how the scenario appears to the Area Controller on their panel; handwriting replaced by typescript to preserve anonymity.

In the scenario shown in Fig. 4, the area controller made an implicit assumption that Train 2–A had stopped after the initial SPAD, but moreover, that the driver knew they had a SPAD. In this scenario, the controller defined the dynamic with the driver in a manner where they believed the drivers did not want to communicate with controllers (indeed the driver did not call the controller), and therefore elected to notify their train controller (i.e. overarching controller) first. Accurate detection of the train coming to a stop was compounded by mimic board design (see top half of Fig. 4), which presented an abstracted view where 'blocks' reflected train progress but did not reflect train position in real-time. Key questions here were, why did the area controller call the overarching controller first? Did the driver know they had has a SPAD, and if so, why did the driver not call the controller?

Figure 5 depicts a SPAD scenario created by a participant who viewed the driver as a *"customer"* in their dynamic. This scenario features single line running, where all trains share the same corridor and crossing is controlled via loops. The scenario also features a "priority train"—a train designation that controllers must favor in terms of expediency and prioritize for movement (due to business agreements). In this scenario, the controller routes the priority train (no. 217) into the loop because they assess that more time will be lost trying to maneuver a much larger train into the loop than a smaller one. The larger non-priority train (no. 220) has only a 25 m leeway into the loop as opposed to the priority train, which has 350 m. However, in this scenario, the non-priority train driver, who is very used to being given priority and seeing green signals, has a SPAD at the turning signal.

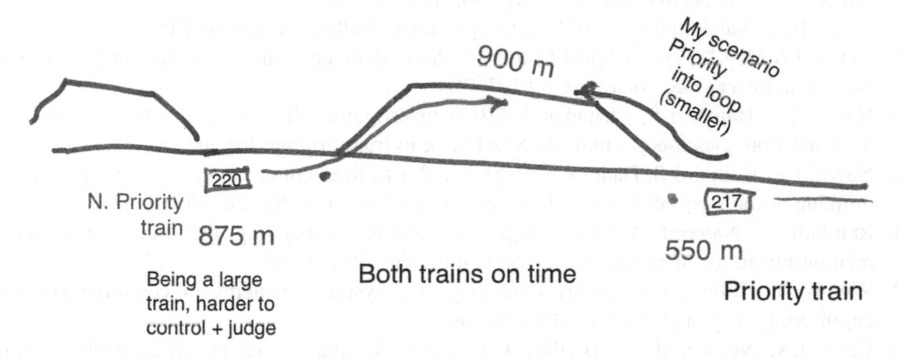

Fig. 5. SPAD scenario depicting a priority train (no. 217) having a SPAD after being routed to into go into a loop. *Note*: handwriting replaced by typescript to preserve anonymity. (Color figure online)

In the scenario shown in Fig. 5, the controller defined their dynamic in a way that commoditized the driver and emphasized service delivery in the context of the business model. The underlying assumption made here was that drivers drive to signals, when it would be more accurate to say that they drive to habit and expectation [7]. A scenario such as a this, where drivers are routinely given green signals, invariably provides ideal error-producing, ergo SPAD-producing conditions. To mitigate opportunity for error, a controller less inclined to relate with the driver as a "customer" and more as a team-member or partner, may make a call to the priority train driver beforehand to advise them of the maneuver beforehand.

4 Conclusions

The findings presented here lend some empirical credence to a view that different controller perspectives of the controller-driver dynamic inform, and engender, a specific kind of engagement, which invariably impacts the system, and can potentially influence SPAD-risk. The findings also provide an indication of the sort of factors that may be analysed to develop a better understanding of SPADs. The next steps in this research is to build on the preliminary findings with a more complete analysis of data, and map specific archetypes of train controller engagement. These data may then be used to develop a more informed 'controller-signal-driver-train' system model and pave the way for strategies that can embrace diversity in different perspectives whilst optimizing system safety and performance.

References

1. Naweed A, Trigg J, Cloete S, Allan P, Bentley T (in press) Throwing good money after SPAD? Exploring the cost of signal passed at danger (SPAD) incidents to Australasian rail organisations. Saf Sci. https://doi.org/10.1016/j.ssci.2018.05.018
2. Singh H, Hanna J (2015) India train derailment: 30 killed, 50 injured, 20 March
3. ATSB (2012) Australian Rail Safety Occurrence Data, 1 July 2002 to 30 June 2012. Canberra, ACT: ATSB, 2012 Contract No.: RR-2012-010
4. Indep Trans Saf Regulator (2011) Transport safety bulletin: Focus on SPAD
5. Naweed A (2013) Psychological factors for driver distraction and inattention in the AU and NZ rail industry. Acc Anal Prev 60:193–204
6. Naweed A, Rainbird S, Chapman J (2015) Investigating the formal countermeasures and informal strategies used to mitigate SPAD risk in train driving. Ergonomics 58(6):883–896
7. Naweed A, Rainbird S, Dance C (2015) Are you fit to continue? Approaching rail systems thinking at the cusp of safety and the apex of performance. Saf Sci 76:101–110
8. Rainbird S, Naweed A (2016) Signs of respect: embodying the train driver–signal relationship to avoid rail disasters. Appl Mobilities 2(1):50–66
9. Woods D, Hollnagel E (2006) Joint cognitive systems: patterns in cognitive systems engineering. Taylor & Francis, Boca Raton
10. Curry LA, Nembhard IM, Bradley EH (2009) Qualitative and mixed methods provide unique contributions to outcomes research. Circulation 119(10):1442–1452
11. Ferroff CV, Marvin TJ, Bates PR, Murry PS (2012) A case for social constructionism in aviation safety and performance research. Aeronutica 2(1):1–12
12. Naweed A, Balakrishnan G (2014) Understanding the visual skills and strategies of train drivers in the urban rail environment. Work 47(3):339–352
13. Klein G, Calderwood R, Macgregor DG (1989) Critical decision method for eliciting knowledge. IIEE Trans Syst Manuf Cybern 19(3):462–472
14. Monk A, Howard S (1998) Rich picture: a tool for reasoning about work context. Interactions 2(5):21–30
15. Naweed A, Balakrishnan G, Bearman C, Dorrian J, Dawson D (2012) Scaling generative scaffolds towards train driving expertise. In: Anderson M (ed) Contemporary ergonomics and human factors 2012: proceedings of the international conference on ergonomics & human factors 2012. CRC Press, Blackpool, p 235

16. Filtness AJ, Naweed A (2017) Causes, consequences and countermeasures to driver fatigue in the rail industry: The train driver perspective. Appl Ergon 60:12–21
17. O'Keeffe VJ, Tuckey MR, Naweed A (2015) Whose safety? flexible risk assessment boundaries balance nurse safety with patient care. Saf Sci 76:111–120
18. Hsieh H-F, Shannon SE (2005) Three approaches to qualitative content analysis. Qual Health Res 15(9):1277–1288
19. Viegas FB, Wattenberg M, Feinberg J (2009) Participatory visualization with wordle. IEEE Trans Vis Comput Graph 15(6)

Driver's Cardiac Activity Performance Evaluation Based on Non-contact ECG System Placed at Different Seat Locations

Rahul Bhardwaj[1] and Venkatesh Balasubramanian[2(✉)]

[1] Haria Seating Systems Limited, Hosur, India
rb@haritaseating.com
[2] RBG Lab, Engineering Design Department, IIT Madras, Chennai, India
chanakya@iitm.ac.in

Abstract. Electrocardiography (ECG) is known to be a reasonable measure of driver fatigue. In this study, we have estimated the cECG performance while placing it on seat base and seat back of the driver seat. Ten male licensed volunteers participated in this study for the duration on the one hour on the simulator. cECG electrodes were place at the seat back and other set of cECG electrodes were placed at the seat base. cECG signals were acquired from both seat back and seat base and it was correlated with the conventional ECG system. Based on Magnitude square coherence (MSC) analysis, it was observed that all the ECG signals acquired from different source had good coherence with ECG ($p > 0.05$). It was observed that the cECG signals acquired from the seat base shown good coherence as compared to the signals acquired from seat back. Perspiration effect reveled that the signals from the seat base were more consistent and reliable as compared to seat back signal ($p > 0.05$). However, combined cECG from seat back and seat base was better than using a single source for sensor.

Keywords: Signal correlation · Capacitive ECG · Ubiquitous monitoring Simulator driving · Driver performance

1 Introduction

According to a recently published report by the Ministry of Road Transport and Highways [1], in India, road accidents were the cause of death in about 150,785 of accidental deaths in 2016, and this is 31% (480,652) of total traffic accidents. This is a marginal increase of about 3.1% compared to the previous year and trend of death due to road accidents has been observed to be increasing from year 2003 to 2016. It has also been reported that 84% of accidents are due to driver error. Driver fatigue is one of the major causes of the road accidents globally [2, 3] and it contributes about 10–20% of total road accidents [4]. Accidental death due to driver fatigue may be different as many times driver doesn't experience the effect of fatigue [4].

Lower in performance of the driver may be interpreted as driver fatigue which leads in lower the response time, slower cognitive response and diminished physiological

S. Bagnara et al. (Eds.): IEA 2018, AISC 823, pp. 278–285, 2019.
https://doi.org/10.1007/978-3-319-96074-6_30

arousal [5, 6]. There are two types of fatigue: physical fatigue and mental/cognitive fatigue and these two complement each other.

For the sake of driver safety, driver fatigue detection system is important in transportation industry. So far, many studied have been performed on simulator and on-road to estimate the driver fatigue using numerous techniques. Driver drowsiness detection system is mainly based on subjective and objective parameters. Subjective based parameters based on subject's oral response whereas, objective based parameter is based on physiological response from subjects using sensors and systems. In objective parameters, three measures have been considered to monitor driver drowsiness: vehicle based parameters [7], behavioral based parameters [8] and physiological based parameters [6].

Several researches have been done so far based on physiological parameters such as, ECG, EMG, and EEG etc. Despite of clinically proven of all these techniques, these have certain limitations such as, wire obtrusion, requirement of conductive gel, hair removal and more importantly, it would be unrealistic to develop a product containing all these limitations.

First non-contact ECG system was proposed by [9], which was based on the principle of capacitively coupled electrocardiography (cECG). cECG systems have been used in several applications such as, like office chair [10, 11], incubators [12], exercise assistance [13], wheelchair [14] and car [15]. Researchers have found that the features extracted from the ECG signals reflect state of the driver. Heart rate variability (HRV) is considered to be reliable indicator of driver drowsiness [16]. While recording cECG signals, in general, cECG electrodes are fixed at the seat back. Because of driver movement while driving, ECG signals are being distorted due to forward/backward movement of driver, shifting gear and controlling the steering wheel.

In this study, we have estimated driver fatigue using cECG system while fixing electrodes at the seat back and seat base. To validate the cECG signals, we have used mean squared coherence (MSC) to correlate the ECG and seat back cECG signals. Similarly, ECG and seat base cECG signals. Effect of perspiration on cECG signals has also been estimated using signal to error ratio (SER).

2 Materials and Methods

2.1 Subject Details

Ten male volunteers participated in this study with their average height 1.72(±0.07) meters, weight 64.8(±10.5) kgs and age 23(±3.1) years. All the subjects participated in this study were right handed. Handedness was checked using Edinburgh Handedness inventory (Oldfield 1971). It was self report from all the subjects that they all possesses normal hearing, vision and was free from any type of cardiac disorder. All the participants were requested to take at least 8 h of sleeping rest a day prior to the experiment and they were all advised not to consume tea, coffee or any type of stimulant 5–6 h before the experiment. A written consent was taken from every subject who participated in the experiment.

2.2 cECG System Design

In this study, two cECG electrodes and one right-leg driven ground electrode have been used. Thin flexible copper plates were used due to its good conductivity (6.30×10^7 S/m). Active electrodes were places 25 cm above the seat base and each electrode was 8 cm away from symmetry line of the seat [17]. Placements of active electrodes on seat back and seat base are shown in Fig. 1. In this system, copper plate and driver's body act as parallel plate of capacitor whereas, cloth act as a dielectric (Eq. 1).

$$C = \frac{\varepsilon_0 \varepsilon_r A}{d} \tag{1}$$

Where, A - Electrode's surface area, d - Distance between electrode and skin, ε_r - Dielectric constant of the material used (cloths) and ε_0 - Vacuum permittivity.

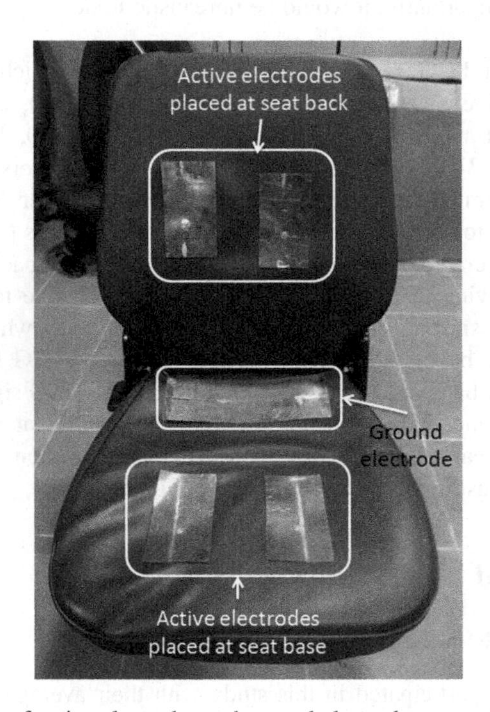

Fig. 1. Placements of active electrodes and ground electrode on seat back and seat base.

A high input impedance operational amplifier with unity gain was used which had input resistance and capacitance of $10^{13}\,\Omega$ and 1 pF respectively. To amplify the signal and achieve high input impedance, an instrumentation amplifier was used. To filter the acquired ECG and cECG signal, a Butterworth 4th order band pass filter and band stop filter with the frequency range of 0.05–40 Hz and of 49–51 Hz respectively with the total gain of 500. Complete setup of the processing of the cECG is shown in Fig. 2.

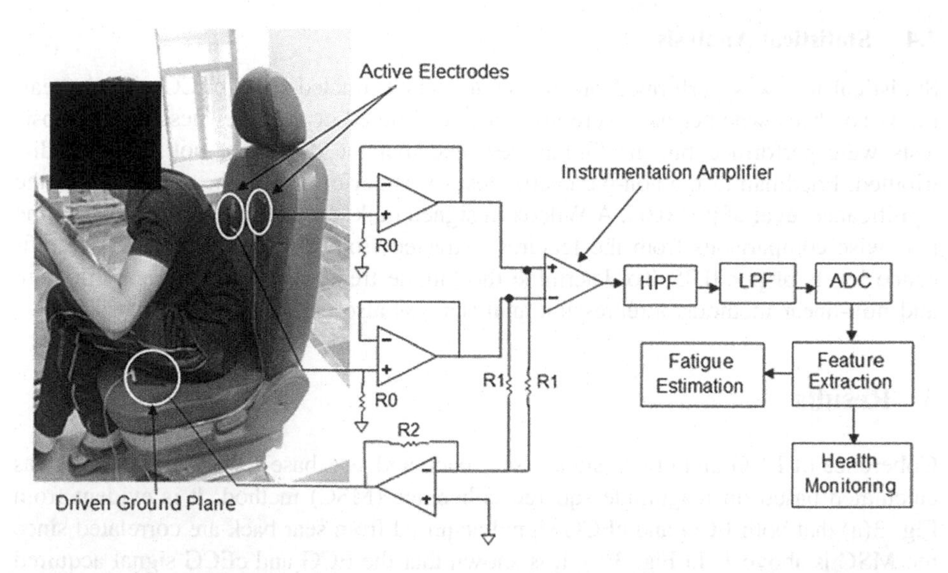

Fig. 2. Diagram represents location of the active cECG and driven ground plane electrodes in the seat. It shows the signal acquisition, processing and feature extraction of cECG signal.

2.3 Magnitude Squared Coherence and Signal to Error Ratio

Magnitude squared coherence (MSC) was performed to find the similarity between two signals based on their frequency content [18, 19] (Eq. 2).

$$|\gamma_{uv}(\Omega)|^2 = \frac{|S_{uv}(\Omega)|^2}{S_u(\Omega)S_v(\Omega)} \tag{2}$$

Where, $\gamma_{uv}(\Omega)$ represents the coherence function of ECG and cECG signal (seat back and seat base), $S_{uv}(\Omega)$ denotes cross power spectral estimate of ECG and cECG signals (seat back and seat base), $S_u(\Omega)$ represents the power spectral density of ECG signals and $S_v(\Omega)$ represents the power spectral density of cECG signals (seat back and seat base). Correlation between two signals at a given frequency Ω is proportional to the magnitude squared coherence value. Two signals are considered to be correlated if MSC between two signals is one. Similarly, two signals are considered to be uncorrelated if MSC between two signal is zero [19].

Signal strength and quality over a period of time due to perspiration was examined based on signal to error ratio (SER) (Blanco-Velasco et al. 2008) from the Eq. 3.

$$\mathrm{SER} = \frac{\sum_{t=0}^{L-1} x^2(t)}{\sum_{t=1}^{L-1} [x(t) - \hat{x}(t)]^2} \tag{3}$$

Where, $\hat{x}(t)$ is the cECG signal (seat back and seat base) at 1st, 5th, 10th, 15th, 20th, 25th, 30th, 35th, 40th, 45th, 50th and 55th min and $x(t)$ is the cECG signal at 60th min.

2.4 Statistical Analysis

Statistical test was performed on the parameters extracted from cECG signal (heart rate). To check whether data were normally distributed or not, skewness and Kurtosis tests were performed but the finding revealed that the data were not normally distributed. Friedman test, a non-parametric test - was performed on the cECG data at the significance level of $p < 0.05$. A Wilcoxon signed rank test was performed to verify the pair wise comparisons from the features extracted from cECG signals at the significance levels of $p < 0.05$. To determine the fatigue trend based on frequency domain and non-linear methods, a regression analysis was also performed.

3 Results

Coherence of ECG and cECG signal (seat back and seat base) of all the subjects was calculated based on magnitude squared coherence (MSC) method. It is evident from Fig. 3(a) that both ECG and cECG signal acquired from sear back are correlated since the MSC is above 1. In Fig. 3(b), it is shown that the ECG and cECG signal acquired from sear base are also correlated since the MSC is above 1. It was found that the seat the MSC of ECG and cECG of seat base are higher than ECG and cECG of seat back. A Wilcoxon signed rank test revealed that there were significant differences ($p < 0.05$) at 1st, 5th, 20th and 30th Hz in MSC acquired from seat back and significant differences ($p < 0.05$) at 1st, 5th, 10th and 15th Hz in MSC acquired from seat base.

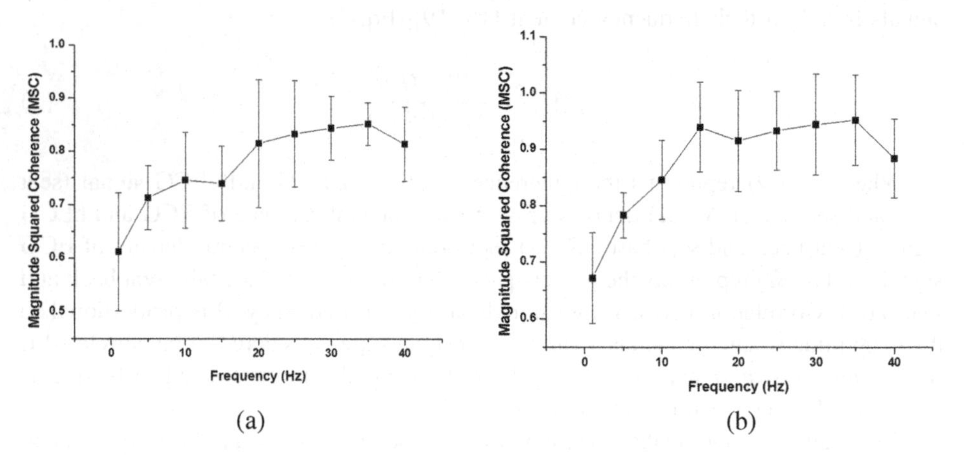

Fig. 3. Magnitude squared coherence (MSC) magnitude were plotted against frequency content of ECG and cECG electrodes (a) seat back and (b) seat base signals. MSC was estimated at every 5 Hz frequency. It was observed that MSC of ECG and cECG showed good coherence.

Figure 4 shows the result of signal to error ratio (SER) of cECG signal acquired from seat back and seat base. SER of cECG signal was estimated at every 5 min interval during 60 min of simulated driving task. SER values of the cECG signals were observed to be increasing on the progress of driving task. A Wilcoxon signed rank test revealed that there were significant differences ($p < 0.05$) at 1st, 10th, 25th, 35th and 45th min in SER from seat back and significant differences ($p < 0.05$) at 1st, 15th, 20th and 45th min in SER acquired from seat base.

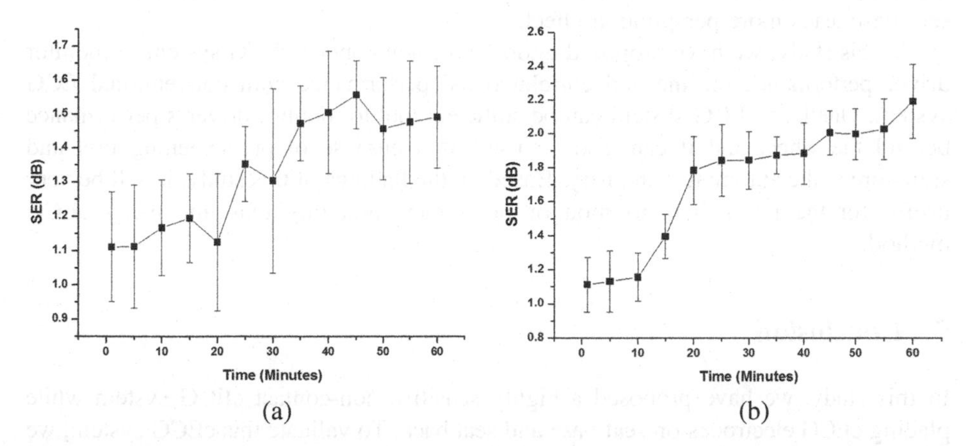

Fig. 4. Signal to noise ratio (SER) was plotted against 60 min of simulated driving while placing cECG electrodes at (a) seat back and (b) seat base. SER was calculated at every 5 min of time interval. It was observed that SER was increasing on the progress of simulated driving task due to perspiration.

4 Discussion

In this study, we have attempted to establish the correlation between ECG and cECG signals acquired from seat back and seat base. In addition to that, effect of perspiration on cECG signal quality during simulated driving session has also been estimated. Signal quality of ECG and cECG signals in this study has also been established based on magnitude squared coherence method (MSC). It is evident from the Fig. 3(a) that most of the mutual frequencies of ECG and cECG (seat back) were close to one ($77.42 \pm 4.91\%$) which represents that there is good correlation of the ECG and cECG signals [18, 19]. Similarly in Fig. 3(b) that most of the mutual frequencies of ECG and cECG (seat base) were close to one ($87.41 \pm 3.71\%$) which represents that better correlation of the ECG and cECG signals (seat base) than ECG and cECG signals (seat back).

Coherence function has been used mainly in neurological and cardiac investigations. Coherence function was applied to electroencephalography (EEG) signal processing as linear measure by [20]. The authors measured the linear synchronization using coherence method and revealed that common frequency for two selected EEG channels is in range 1 Hz–10 Hz.

Signals quality acquired from cECG (seat back and seat base) were observed to be getting better as time progressed (Fig. 4). This would have happened due to the effect of perspiration which leads to increase in dielectric strength of the cloth several times. Decrease in electrode-skin impedance happens due to moistening of underlying skin by sweat-gland activity with the passage of time [21]. It is evident from the Fig. 4(a) that the SER obtained from cECG (seat back) is fluctuating and less consistent as compared to the cCEG acquired from seat base. There are mainly two reasons of better ECG acquired from seat base; (i) less movement happens on the seat base by the driver while driving, (ii) Consistent contact of driver's thigh with the cECG electrode placed at the seat base leads more perspiration effect.

In this study, we have proposed unobtrusive non-contact cECG system to monitor driver performance on-line and correlated its' performance with conventional ECG system. One lead cECG system can be sufficient tool to monitor driver's performance behind the wheel and it can also be used as a purpose of pre-screening tool and signature in the automotive industry. Based on the findings of this study, it will be very useful for the researchers to monitor heart rate variability real-time using cECG method.

5 Conclusion

In this study, we have proposed a highly sensitive non-contact cECG system while placing cECG electrodes on seat base and seat back. To validate this cECG system, we recorded ECG signal from conventional ECG system along with cECG and both acquired signals were compared based on magnitude square coherence and it was found that both the signals have good coherence and similarity based on frequency content. However, effect of perspiration was also noticed in cECG signals which occur due to contact of cECG electrode and subject's body for a long period time. It was found that the signal quality was getting better with progressing driving time and it was quantified based on signal to error ratio. It was also found that the ECG signal quality was better that were acquired from seat base.

Acknowledgment. The authors would like to show gratitude to all the members of Rehabilitation Bioengineering Group (RBG) at Indian Institute of technology Madras, India, and other volunteers for their participation in this study.

References

1. Road accidents in India - 2015, Ministry of Road Transport and Highways, Government of India (2016)
2. Connor J, Whitlock G, Norton R, Jackson R (2001) The role of driver sleepiness in car crashes: a systematic review of epidemiological studies. Accid Anal Prev 33(1):31–41
3. Åkerstedt T, Ingre M, Kecklund G, Anund A, Sandberg D, Wahde M, Kronberg P (2010) Reaction of sleepiness indicators to partial sleep deprivation, time of day and time on task in a driving simulator–the DROWSI project. J Sleep Res 19(2):298–309
4. Karrer K, Roetting M (2007) Effects of driver fatigue monitoring–an expert survey. International conference on engineering psychology and cognitive ergonomics. Springer, Heidelberg, pp 324–330
5. Tzamalouka G, Papadakaki M, Chliaoutakis JE (2005) Freight transport and non-driving work duties as predictors of falling asleep at the wheel in urban areas of crete. J Saf Res 36 (1):75–84

6. Jagannath M, Balasubramanian V (2014) Assessment of early onset of driver fatigue using multimodal fatigue measures in a static simulator. Appl Ergonom 45(4):1140–1147
7. Forsman PM, Vila BJ, Short RA, Mott CG, Van Dongen HP (2013) Efficient driver drowsiness detection at moderate levels of drowsiness. Accid Anal Prev 50:341–350
8. Zhang Z, Zhang J (2010) A new real-time eye tracking based on nonlinear unscented Kalman filter for monitoring driver fatigue. J Control Theor Appl 8(2):181–188
9. Richardson PC (November 1967) The insulated electrode: a pasteless electrocardiographic technique. In: 20th Annual Conference on Engineering in Medicine and Biology, vol 9, pp 15–17)
10. Lim YG, Kim KK, Park S (2006) ECG measurement on a chair without conductive contact. IEEE Trans Biomed Eng 53(5):956–959
11. Aleksandrowicz A, Walter M, Leonhardt S (2007) Ein kabelfreies, kapazitiv gekoppeltes EKG-Messsystem/Wireless ECG measurement system with capacitive coupling. Biomed Tech 52(2):185–192
12. Kato T, Ueno A, Kataoka S, Hoshino H, Ishiyama Y (August 2006) An application of capacitive electrode for detecting electrocardiogram of neonates and infants. In: Engineering in Medicine and Biology Society, EMBS 2006. 28th Annual International Conference of the IEEE. IEEE, pp 916–919
13. Lee YD, Chung WY (2009) Wireless sensor network based wearable smart shirt for ubiquitous health and activity monitoring. Sensor Actuator B 140:390–395
14. Postolache O, Girao P, Joaquim J, Postolache G (2009) "Unobstrusive heart rate and respiration rate monitor embedded on a wheelchair,". In: Proceedings of the IEEE 4th International Workshop on Medical Measurements Application, Cetraro, pp 83–88
15. Wartzek T, Eilebrecht B, Lem J, Lindner H-J, Leonhardt S, Walter M (2011) ECG on the road: Robust and unobtrusive estimation of heart rate. IEEE Trans Biomed Eng 58 (11):3112–3120
16. Mulder L (1992) Measurement and analysis methods of heart rate and respiration for use in applied environments. Biol Psychol 34:205–336
17. Jung SJ, Shin HS, Chung WY (2012) Highly sensitive driver health condition monitoring system using nonintrusive active electrodes. Sens Actuators B Chem 171:691–698
18. Bendat JS (1971) Analysis and Measurement Procedures. RANDOM DATA, 407
19. Bendat JS, Piersol AG (1980) Engineering applications of correlation and spectral analysis. Wiley-Interscience, New York, p 315
20. Quiroga RQ, Kraskov A, Kreuz T, Grassberger P (2002) Performance of different synchronization measures in real data: a case study on electroencephalographic signals. Phys Rev E 65(4):041903
21. Geddes LA, Valentinuzzi ME (1973) Temporal changes in electrode impedance while recording the electrocardiogram with "dry" electrodes. Ann Biomed Eng 1(3):356–367

Mental Workload and Performance Measurements in Driving Task: A Review Literature

Totsapon Butmee(✉), Terry C. Lansdown, and Guy H. Walker

Heriot-Watt University, Edinburgh, UK
{tb19,T.Lansdown,g.h.walker}@hw.ac.uk

Abstract. The concept of human mental workload in the field of human factors and psychology has a long history with important applications in the aviation and automotive industries. The main objectives of this literature review are defined the 'mental workload' term and determined mental workload measurement methods in driving task. Mental Workload is a complex concept and it is difficult to define this term. It has no a universal accepted definition. Mental workload level cannot be detected directly, however, it has found that relates to limitation of individual internal resources to accomplish the task, and also involves a multi-dimensional variable. Previously, several studies have been indicated that mental workload relate with operators' performance, task demand and mental resource supply. Extremes (underload or overload) mental workload can degrade operators' performance. Several assessment methods have been proposed for investigating mental workload. They can be performed in experimental or operational settings. There are seven selection criteria to select the most appropriate methods. These include sensitivity, diagnosticity, intrusiveness, implementation requirement, operator acceptance, selectivity and bandwidth and reliability. Dozens of Mental workload measurement techniques have been developed and categorized into three main groups. (i) *Subjective rating,* which were categorized into unidimensional and multidimensional. NASA-TLX, SWAT, RSME and MCH are the famous examples of subjective-based techniques. (ii) *Performance measures* are divided into primary task and secondary task measures. Primary task measures are capable of discriminating the resource competition between individual differences. For example, speed instability, distance headway instability, lateral position from road centerline, lane excursion, time spent out of lane can be widely used to represent the driver primary performances. In secondary task measures are more diagnosticity than primary task measures and subjective measures. Correct response, time response of additional secondary task are a well-known examples of secondary task performance measures in driving research. Additionally, (iii) *Physiological techniques* also have high sensitivity in measurement, but results from these methods can easily are confounded by other external and extraneous interference. Measures of Eye Functions have been frequently used if compared with other Physiological techniques. It can be argued that the combined methods are recommend cooperatively to predict human mental workload.

Keywords: Mental workload measurements · Performance measurements
Driving task

S. Bagnara et al. (Eds.): IEA 2018, AISC 823, pp. 286–294, 2019.
https://doi.org/10.1007/978-3-319-96074-6_31

1 Introduction

Driving task has been defined as numerous highly dynamic tasks occurring during a changing environment [1]. The driving complexity has been reported that depends on several elements including *road design* (city roads vs. rural roads vs. motorways), *traffic flow* (low density vs. high density), and *road layout* (straight vs. curves, even vs. inclined, junction vs. no junction) [2]. Driving is not only a physical task (e.g., applying force on steering wheel and pedals), but also a visual and mental task [3]. In the term of mental task, the driver's brain deals with millions of sensory such as visual, verbal and other daily stimuli, particularly in a changing environment of driving task. Too much information needs to be proceed along with a limited resource of driver. These situations can affect to high demand of driver information processing system and can link to the high rate of vehicle colliding [1]. Mental workload issue has been concerned since the last four decade in different research fields [1, 4, 5]. Especially in driving research, extremes (underload or overload) mental workload can degrade drivers' performance and increase collisional rate. For example, it has been reported that too many in-vehicle information, activities or systems such as navigation system, conversation with a friend on hands-free cell phone, using a speech-to-text interfaced e-mail were associated with distraction effect and information overload during driving [6, 7]. This is also supported by Lansdown, Brook-Carter and Kersloot [8], multiple in-vehicle systems can significantly impose higher mental workload. On the other hand, engaging prolong with automated vehicle or monotonous underload task has been also reported that it turn driving tasks into a vigilance decrement task [1], attentional loss and performance decrements [9].

Evaluating mental workload is an important component of system design and analysis [10]. Especially, in driving task, drivers' mental workload assessments could be helpful in improving driving tasks to reduce the number of road accident [3]. Mental workload measurement has a long history from aviation research. In driving research, previous studies have used the same concept of aviation to investigate driver's mental workload. The main objectives of this review are to define metal workload term in driving task and summarize the results of recently conducted studies on mental workload measurement criteria and techniques in driving research. The review proceeds with definitions following with conclusion and discussion of the related findings

2 Definitions

The concept of Human Mental Workload in the field of human factors and psychology has a long history with important applications in the aviation and automotive industries. However, it is difficult to define this term [11]. At the present, it has no clear definition and universal accepted term of Mental Workload [11–14]

From review of a number of publications, there are various views that have been proposed to understand the term mental workload. For example, O'Donnell and Eggemeier [5] defined the term of workload as a 'portion of the operator's limited capacity actually required to perform a particular task'. In more general term, it has been suggested that mental workload can be described by the amount of human 'information processing capacity' which is spent for task performance [15]. An interesting definition, De Waard [1] presented three key terms to understand mental workload. First, 'task demand' is required to reach the goals and sub-goals by operator's performance. Second, 'workload' is the proportion of the capacity that is allocated for task demand. It is reacted by task demand. Third, 'effort' includes 'state-related effort' which is exerted to maintain an optimal state for task performance, and 'task-related effort' is exerted in the case of controlled information processing. A new perspective by Young et al. [11] explained mental workload as a multidimensional construct, and it is described by 'task' (e.g. demand and performance), 'operator' (e.g. skill and attention) characteristics, and the environmental context. Thus, mental workload is the result of an interaction between task demands and individual characteristics [3]. A literature review by Xie and Salvendy [14] summarized main tenets of mental workload including: amount of *mental work (effort)* to complete a task; it cannot be detected directly; it involves the depletion of *individual internal resources* to accomplish the task; it is a *multi-dimensional* variable, i.e. time load, mental-effort load, psychological-stress load; it can be influenced by many factors. In driving research domain, mental workload was specifically defined as the effort to maintain the driving task within a subjective safety zone [16]

As can be seen that mental workload is related to attentional demand on human's information process. Driving task, for example, drivers usually have higher mental workload under complex dual –task, in comparison with driving only due to the driver has to invest more attentional resources to meet the task demand. Extremes or sub-optimal (both underload and overload) mental workload can degrade operators' performance. This is the significant issue that should be considered particularly in driving task due to it can lead to driving performance degradation and road accident.

3 Literature Search Methods

The literature search was conducted using Google Scholar with using one of six following keywords: "mental workload measurements" "performance measurements" "mental workload assessments" "performance assessments" "mental workload evaluations" "performance evaluation" combined with one of four following keywords: "driving task" "automobile driving" "car driving" "vehicle driving".

4 Mental Workload and Performance Measurements

4.1 Measurement Criteria

As mentioned above, the results from workload measurements can predict operator's performance. The main reasons of mental workload measurement for quantifying the mental cost which is used to perform a certain task in order to predict operator and systems' performance. However, there are concerns regarding the application of workload measurements. Several criteria have been proposed for selecting and developing measurement techniques.

There are five selection criteria to select the most appropriate methods. *Sensitivity* is the capability to discriminate the nature of the workload which is imposed by task(s) demand [5] or reflect changes in workload [1, 17]. This is the most important criteria for detect changing of workload. Some studies require the main source of workload. Thus, *diagnosticity* is an important criteria. Diagnosticity can be defined as the capability of technique to identify the specific source of workload or locus of demand [1], e.g., perceptual versus central processing versus motor resources [5] not to others. Moreover, *intrusiveness* is considered, particularly additional secondary task measures and physiological technique might interfere the primary task. Intrusiveness refers to the degree the method interferes with the performance of a primary task [13]. Ease versus complexity of the measurement procedure is considered as an *implementation requirements* criteria. For example, the training of operators or the need of specific equipment for data collection and analysis were referred to implementation requirement. Particularly, in real world situation such as on-road driving condition, implementation requirement becomes the important criteria [1]. *Operator acceptance* is defined as operator's willingness to follow the instructions of techniques and actually utilize a particular technique [5]). It was suggested that a mental workload measurement method should be simple, showing understanding and directness [13]. *Selectivity and Bandwidth and reliability* were extended as an additional criteria by Longo [13]. Generally, a useful mental workload assessment technique should has 'high sensitivity, a high bandwidth, low intrusiveness on primary task and high reliability, as well as showing concurrent and convergent validity' [13]. Each measurement technique has its own advantages and disadvantages and, there are appropriate for different contexts. It has been recommended here that using combination of different measures of mental workload to get the most accurate assessment [17] and more comprehensive assessment than using only one technique [7].

4.2 Mental Workload Measurement Techniques

Dozens of mental workload measurement techniques have been developed and categorized into three main groups [5, 11, 17, 18] including: (i) subjective measurement; (ii) performance-based measurement; and (iii) physiological measurement.

Subjective measurements have been designed to indicate the perceptions of the operator on mental workload experienced after an experiment or across a task overall. The main advantages of subjective self-report measures are reducing application cost, high sensitivity to underload and overload situations [13], all of these techniques do not interfere with primary task [19], and this is an easiest method of assessing workload [17] However, difficult to discriminate between physical workload and mental workload, and operator disability for detecting internal changes are claimed as the main disadvantages of subjective methods [17]. Subjective rating scales have been categorized into two groups [13, 17, 19] including: (i) *unidimensional scale*, it has only one dimension e.g., Modified Cooper-Harper (MCH) [20], Subjective Workload Dominance technique (SWORD) [21] and Rating Scale Mental Effort (RSME) [22]; and (ii) *multidimensional scale*, there are more complex dimensions, more time consuming and more diagnostic e.g., NASA-Task Load Index (NASA-TLX) [4] and Subjective Workload Assessment Technique (SWAT) [23]. NASA-Task Load Index (NAS-TLX) is well-know and frequently used to measure driver's mental workload [11]. However, the related limitation of subjective methods is recall bias with the short-term memory of the operator, so workload ratings should be completed as soon as possible after task performance [5].

Performance measures of mental workload are used to quantify how well an operator is performing the particularly task(s). According to various studies, in driving research, performance measures have been categorized into two categories including *direct measurements* and *indirect measurements* [5, 15, 19]. Direct measurements focus on performance of the main or *primary task* such as lateral position, longitudinal position, speed controlling [19] steering and car following [15]. In addition, *indirect dual-task measurements* point out secondary task performance such as manual response to a stimulus presented in the visual or auditory field [19] and peripheral detection task (PDT) [15]. Both primary and secondary task measures have been accepted to use moderately in driving research [17]. However, primary task measurements are not sensitive to change in low workload situation [17]. Moreover, one of limitations of primary-task techniques is that they may be limited with respect to diagnostic capability [5, 24]. For secondary task measures, intrusiveness is the main limitation of secondary-task performance techniques, most techniques usually require mock-up simulators or operational equipment and some operator training.

Physiological measures are based on the concept that increased mental demands can lead to increased physical response from human body [17]. Several point of views were proposed to categorize physiological mental workload measures. Physiological measures have been categorized into three main major categories including: eye, brain and heart related measures and other measures such as skin and muscle activity [24, 25]. Another point of view O'Donnell and Eggemeier [5] sub-divided physiological into four classes: brain, eye, cardiac and muscle function. In addition, Miller [17] recommended five physiological measure including: cardiac; respiratory; eye; speech and brain activity. After reviewed of several studies, we summarized the examples of physiological measures which have been accepted widely to assess driver's mental workload in Table 1. It seems to be that eye blink measurements such as blink rate and blink duration have been used in several driving research with most accurate for visual workload. However, a common drawback is that most physiological measures are required special equipment as well as trained operators with technical expertise to utilize these equipment and interpret the data [24]. In addition, the signals from physiological measures might interfere by other factors, so it might misleading indicator of mental workload.

Since several techniques have been developed for measuring human mental workload. Miller [17] presented the consideration criteria to select the appropriate methods. The decision tree by Miller considers about Interval of collection, Obtrusive, Form of gathering data, time consideration, sensitivity, cost of implementation, and reliability in Table 2.

Table 1. The examples of physiological measures

Physiological measures	Examples
1. Eye behaviour measurement	Blink rate, blink duration, pupil diameter and Electrooculogram (EOG)
2. Measures of cardiac functions	Heart Rate (HR), Heart Rate Variability (HRV), Inter-Beat-Interval (IBI) and Electrocardiogram (EKG)
3. Measures of brain functions	Electroencephalogram (EEG)
4. Measures of muscle functions	Electromyogram (EMG)
5. Other	Electrodermal Activity (EDA) and Hormone Levels

Table 2. Decision tree to choose the mental workload techniques

Considerations	Type of data	Methods
Interval of collection	Continuous	Physiological Primary performance measures Secondary performance measures
	During	Unidimensional rating
	After	Multidimensional rating
Obtrusive	Obtrusive	Brain measures Respiratory measures Secondary performance measures
	Unobtrusive	Subjective Other physiological Primary performance measures
Form of gathering data	Verbal	Unidimensional rating
	Written	Multidimensional rating
	Machine gathered	Physiological Primary performance measures Secondary performance measures
Time consideration (only subjective measure)	Yes	Unidimensional rating
	No	Multidimensional rating
Sensitivity	High	Unidimensional rating Multidimensional rating Brain Measures
	Medium	Secondary performance measures Primary performance measures Cardiac measure Eye measures
	Low	Other physiological

(*continued*)

Table 2. (*continued*)

Considerations	Type of data	Methods
Cost of implementation	High	Brain measures Respiratory measures Eye measures
	Moderate	Multidimensional rating Most primary performance measures Most secondary performance measures Other physiological
	Low	Unidimensional rating
Reliability	High	Unidimensional rating Multidimensional rating Brain Measures Eye activity
	Medium	Secondary performance measures Other physiological
	Low	Primary performance measures

5 Discussion

The purposes of this review are to define mental workload term in driving task and summarize the results of recently conducted studies on mental workload measurement criteria and techniques in driving research. As a result of the review, numerous definitions have been proposed mental workload. Based on a different point of view, in driving task, mental workload is related to attentional demand on driver's information process due to the driver has to invest their attentional resources to meet the task demand. From analysis of a number of publication, it seems reasonable to suggest that, a useful mental workload assessment technique should has 'high sensitivity, a high bandwidth, low intrusiveness on primary task and high reliability. Various empirical techniques have been reported in the literature as confirmed assessed of driver's mental workload. In the literature review dealt with three types of mental workload measurement: subjective, performance and physiological measurements. In subjective measures, NASA-TLX multidimensional rating has been used commonly and well-known in driving research. It should be noted that performance-based measurements of driver's mental workload might be the most frequently used methods to indicate driver's mental workload. However, it has been claimed that both primary and secondary task measures have been accepted to use moderately in driving research [17]. Many tools have been accepted to represent driver's mental workload in physiological approaches. Especially, eye behavior measures have been frequently used assessment tools in driving domain. However, it has no excessive conclusions to confirm which method is the most appropriate between eye blink behaviors and pupillometry. Although several mental workload measurement techniques have been developed more than 40 years, no one has been accepted as the best technique. Thus, it should be noted that the combined methods are recommend cooperatively to predict human mental workload [7, 17]. Much research on driver's mental workload has been required. Especially, in the near future, autonomous systems have been predicted to replace the human roles. Human factor research should be investigated excessively beside with the developing of these sophisticated automobile technologies.

References

1. de Waard D (1996) The measurement of drivers' mental workload. Groningen University, Traffic Research Center, Netherlands
2. Paxion J, Galy E, Berthelon C (2014) Mental workload and driving. Front Psychol 5:1–11
3. Marquart G, Cabrall C, de Winter J (2015) Review of eye-related measures of drivers' mental workload. Procedia Manuf 3:2854–2861
4. Hart SG, Staveland LE (1988) Development of NASA-TLX (Task Load Index): results of empirical and theoretical research. Adv Psychol 52:139–183
5. O'Donnell RD, Eggemeier FT (1986) Workload assessment methodology. In: Boff KR, Kaufman L, Thomas JP (eds) Handbook of perception and human performance. Cognitive Processes and Performance, vol. 2. John Wiley and Sons, A wiley-interscience publication, United States, pp 42-1–42-49
6. Piechullaa W, Mayserb C, Gehrkec H, Königc W (2003) Reducing drivers' mental workload by means of an adaptive man–machine interface. Transp Res Part F: Traffic Psychol. Behav 6(4):233–248
7. Strayer DL, Turrill J, Cooper JM, Coleman JR, Medeiros-Ward N, Biondi F (2015) Assessing cognitive distraction in the automobile. Hum Factors 57(8):1300–1324
8. Lansdown TC, Brook-Carter N, Kersloot T (2004) Distraction from multiple in-vehicle secondary tasks: vehicle performance and mental workload implications. Ergonomics 47 (1):91–104
9. Körber M, Cingel A, Zimmermann M, Bengler K (2015) Vigilance decrement and passive fatigue caused by monotony in automated driving. Procedia Manuf 3:2403–2409
10. DiDomenico A, Nussbaum MA (2008) Interactive effects of physical and mental workload on subjective workload assessment. Int J Ind Ergon 38(11):977–983
11. Young MS, Brookhuis KA, Wickens CD, Hancock PA (2015) State of science: mental workload in ergonomics. Ergonomics 58(1):1–17
12. Cain B (2007) A review of the mental workload literature. Defence Research and Development Toronto, Canada
13. Longo L (2015) A defeasible reasoning framework for human mental workload representation and assessment. Behav Inf Technol 34(8):758–786
14. Xie B, Salvendy G (2000) Review and reappraisal of modelling and predicting mental workload in single-and multi-task environments. Work Stress 14(1):74–99
15. Brookhuis KA, van Driel CJ, Hof T, van Arem B, Hoedemaeker M (2009) Driving with a congestion assistant; mental workload and acceptance. Appl Ergon 40(6):1019–1025
16. Boer ER (2001) Behavioral entropy as a measure of driving performance. In: Proceedings of the first international driving symposium on human factors in driver assessment, training and vehicle design, pp 225–229
17. Miller S (2001) Workload measures. National Advanced Driving Simulator, Iowa City
18. Cantin V, Lavallière M, Simoneaub M, Teasdale N (2009) Mental workload when driving in a simulator: effects of age and driving complexity. Accid Anal Prev 41(4):763–771

19. da Silva FP (2014) Mental workload, task demand and driving performance: what relation? Procedia Soc Behav Sci 162:310–319
20. Cooper GE, Harper Jr, RP (1969) The use of pilot rating in the evaluation of aircraft handling qualities (No. AGARD-567). Advisory Group for aerospace research and development Neuilly-Sur-Seine, France
21. Vidullch MA, Ward GF, Schueren J (1991) Using the subjective workload dominance (SWORD) technique for projective workload assessment. Hum Factors 33(6):677–691
22. Zijlstra FRH (1993) Efficiency in work behaviour: a design approach for modern tools
23. Reid GB, Nygren TE (1988) The subjective workload assessment technique: a scaling procedure for measuring mental workload. Adv Psychol 52:185–218
24. Rehmann AJ (1995) Handbook of human performance measures and crew requirements for flightdeck research (No. CSERIAC-ACT-350). Crew System Ergonomics Information Analysis Center Wright-Patterson AFB OH
25. Ryu K, Myung R (2005) Evaluation of mental workload with a combined measure based on physiological indices during a dual task of tracking and mental arithmetic. Int J Ind Ergon 35 (11):991–1009

The Influence of Non-driving Related Tasks on Driver Availability in the Context of Conditionally Automated Driving

Jonas Radlmayr(✉), Fabian Marco Fischer, and Klaus Bengler

Chair of Ergonomics, Technical University of Munich,
Boltzmannstr. 15, 85748 Garching, Germany
jonas.radlmayr@tum.de

Abstract. This study looked at the effect of three different non-driving related tasks (NDRT) (Surrogate Reference Task (SuRT), n-back task and a motoric task), instructed or free engagement into these tasks and the resulting take-over performance in two different take-over situations after conditionally automated driving.

We conducted a study with 53 participants in a static driving simulator. Participants were split into three groups and each group was assigned one of the three NDRT's. Each participant per group experienced two different take-over situations twice, totaling in four take-overs. In addition, prior to a take-over, participants were either instructed to engage in the task assigned to their group or could choose their NDRT's freely.

Dependent variables to assess driver availability were percent eyes on road (PEOR), standard deviation of the horizontal gaze dispersion (HGD) and blink frequency and changes in the center of pressure and contact area (COP) in the seat. To analyze take-over performance, we looked at gaze reaction time, take-over time, time to collision (TTC), standard deviation of lateral position (SDLP), longitudinal and lateral accelerations and subjective ratings.

Results showed significant changes for the different NDRT's for the driver availability variables. The type of instruction did not show differences in the take-over performance, while we saw significant differences between the two different take-over situations.

We concluded that the influence from different take-over situations is high, while differences in driver availability can be measured well with the right sensors, but do not lead to different take-over performances.

Keywords: Automated driving · Non-driving related tasks · NDRT
Take-over performance

1 Introduction

With current vehicles already allowing Level 2 automated driving [1], the introduction of Level 3 or conditionally automated driving [1] can be foreseen for the near future. In Level 3, the automation takes over the dynamic driving task and drivers can engage in non-driving related tasks since they do not have to monitor the system anymore. In case

S. Bagnara et al. (Eds.): IEA 2018, AISC 823, pp. 295–304, 2019.
https://doi.org/10.1007/978-3-319-96074-6_32

a system limit is reached, drivers are responsible to take-over after the request to intervene (RtI) [2] and continue driving manually. Thus, any changes to drivers' availability [2] to take-over due to engagement into NDRT's is of interest for ensuring safety and comfort. Previous studies already looked at effects of NDRT's on take-over performance [3], but did not measure changes to the driver availability directly. NDRT's can potentially show negative effects due to over- or underload [4], or positive effects in case they establish a medium level of arousal [5]. To quantify changes to drivers' availability to take-over, we utilized eye-tracking and seat pressure mats. The horizontal gaze dispersion allows insight on the tracking behavior and can be understood as indicator for the knowledge of the current traffic situation [6]. In addition with the blink frequency, which is sensitive to different kinds of workload (visual versus cognitive) we established an understanding of the effects of the NDRTs on the driver availability [7]. For evaluation of the take-over performance, we relied on well-established metrics found in [8].

This study aims to answer the following research questions:

- Do different modalities of NDRT's effect drivers' availability to take-over and can this be measured using eye-tracking and seat pressure mats?
- Do different NDRT's translate into different take-over performances evaluated in two take-over situations?

2 Method

2.1 Sample

Fifty-three participants experienced the study in the static driving simulator of the chair for ergonomics. Three participants were excluded from analysis due to simulator sickness or technical issues. The individual plots in the results chapter will show the exact number of data sets considered in case data from specific situations could not be used due to technical issues. Table 1 shows the available data sets of the three groups.

The sample had a mean age of 31.7 years (SD = 15.6 years) and included 24 females. All participants had a valid drivers' license, normally distributed experience concerning the average mileage per year. Seventy percent of participants had experienced conditionally automated driving in a simulator, but only three participants more than one time.

Table 1. Sample composition of the three groups

Group	Surrogate reference task (SuRT)	N-back task	Motoric task
	$n = 17$	$n = 17$	$n = 16$

2.2 Experimental Setup

The vehicle mockup consisted of a BMW 6-series convertible and allowed a high immersion with a 180° field of view horizontally. Simulation software was SILAB.

The design of experiment consisted of a mixed design with three independent variables. Table 2 shows the three independent variables. The between factor type of NDRT split the sample into three groups.

Table 2. Mixed design of experiment with one between factor and two within factors.

Factor	Type of non-driving related task (between)	Instruction (within)	Take-over situation (within)
	Surrogate reference task (SuRT) (visual-motoric NDRT)	Free	"Accident" (High complexity) [9]
	N-back task (cognitive NDRT)	Instructed	"Construction site" (low complexity) [9]
	Shape-sorter ball (motoric NDRT)		

One group engaged into the surrogate reference task [10] which is a visual-motoric task that shows many small circles on a tablet screen arranged in columns. One of the small circles is slightly larger than the rest and must be identified by the participants and selected. The tablet was mounted in the center stack of the vehicle mockup.

The second group engaged into the n-back task [11], which is a cognitive task that features a voice reading a number sequence over the speakers. Participants were asked to repeat the numbers depending on the chosen difficulty of the task. This means that a 2-back task consists of the voice reading two additional numbers before participants can reply the first one and so on. This induces cognitive workload that increases if the number associated with the n-back increases. The feasibility of these two standardized tasks has already been shown to be applicable to analyze a standardized task load from NDRT's in the context of conditionally automated driving [3].

The third group engaged into a motoric task that "was represented by a shape-sorter ball used for young children to build up visual thinking and dexterity. The ball with 16 shapes was stored in a bag with holes to put hands through and placed on the passenger seat when not in use. For task handling, participants had to put the hands inside the bag and fit the shapes through the corresponding holes. As there was no possibility of visual control, eyes could remain on the scenery" [12]. Figure 1 shows the surrogate reference task and the motoric task. Since the n-back task consisted of the auditory modality and induced cognitive workload, Table 3 shows the execution of the n-back task as 2-back task.

The other two factors instruction and take-over situation were within factors. Participants experienced the two different take-over situations "accident" and "construction site" after approximately five minutes of automated driving each. To allow a combination with the factor instruction, participants had to take-over a total of four times, two times in the same situation. Every take-over situation succeeded a short interval of automated driving, in which participants were either instructed to engage into the specific NDRT, depending on the group they were in or could freely behave.

The free behavior was put under certain constraints: participants should not deactivate the automation but could freely engage into either doing nothing, or the three different NDRT's. All NDRT's were canceled when a RtI was issued and a take-over situation followed.

Fig. 1. The left figure shows the SuRT in the center console of the vehicle. The right figure shows the shape-sorter ball that was put into a bag with openings when used during the experiment.

Table 3. Sequence of a 2-back task as it was implemented.

Trial manager/computer voice	2	4	5	1	8
Participant	–	–	2	4	5

The take-over situations differed in their overall criticality. Participants in the "construction site" had to stabilize the vehicle while participants in the "accident" situation had to execute a lane change maneuver. The design of the take-over situations and the resulting overall criticality followed the classification of take-over situations for conditionally automated driving in [9]. The "accident" situation represents a high criticality while the "construction site" can be understood to be of low to medium criticality. Figure 2 shows the two take-over situations.

At the time of the RtI, both take-over situations could not be perceived by the participants because the system limits as shown in Fig. 2 were either blocked from view due to preceding traffic or by having it appear after a small crest. The RtI was

Fig. 2. The left figure shows the situation "accident" with the ego lane being blocked by an accident. The right figure shows the construction site where participants had to stabilize the vehicle. Both screenshots were taken right before the system limit was reached. The RtI was issued 233 m/7 s time budget prior to the system limit.

designed to be a double beep with 70 dB at approximately 2600 Hz together with a change in the automation status displayed in the cluster display. The traffic density was designed to be random traffic with 20 vehicles per kilometer throughout the entire experiment and the take-over situations.

2.3 Measures

The acquisition of relevant metrics can be split into two different intervals to allow a better understanding. The driver availability during the intervals with active automation was measured using a three-camera eye-tracker from SmartEye, seat pressure mats from xSensor and a motion tracking software using a Microsoft Kinect v2 camera. Data from the motion-tracking could not be used for analysis due to limited space inside the vehicle mockup rendering all motion tracking data unfeasible.

The second interval consisted of the time between the RtI was issued and participants passing the system limit and represented the take-over performance. Vehicle dynamics could be analyzed using data acquired by the simulation software. Table 4 gives an over view of the specific metrics used to quantify driver availability and take-over performance.

2.4 Procedure

The experiment lasted approximately a total of two hours. Participants answered a demographic questionnaire in the beginning, before a detailed instruction of the simulator, the automation and the experiment followed. Participants could familiarize themselves with the vehicle dynamics in the simulator and experience the automation in addition with a generic take-over situation to avoid first-exposure results.

The track itself was a three lane German Autobahn, the automation adapted the speed of the ego vehicle up to 120 km/h, executed lane change maneuvers and could be deactivated by steering, braking or using the automation button again. The intervals of

Table 4. Overview of the dependent variables split into driver availability metrics and take-over performance metrics.

Driver Availability metrics during the automated drive	Take-over performance metrics during the take-over situation
Eye-Tracking – Percent eyes on road (PEOR) – Standard deviation of horizontal gaze dispersion (HGD) – Blink frequency Seat pressure mats (both seat and backrest) – Average pressure – Variance of maximum pressure – Center of pressure (COP) – Contact area Motion Tracking (unfeasible data due to limited space in the vehicle mockup)	Eye-Tracking – Gazes to mirror Time and quality aspects – Gaze reaction time – Take-over time (TOT, begin of maneuver by either steering >2° or using a pedal >10%) – Min. time to collision (TTC) – Max. lateral acceleration – Min. longitudinal acceleration (brake acceleration) – Standard deviation of lateral position (SDLP, only for construction site)

automated driving lasted approximately 4–5 min each before a RtI was issued. After each take-over situation, participants were asked to activate the automation again and were asked a few questions concerning their subjective rating of the current take-over situation and performance. The experiment ended after each participant had experienced the four take-over situations and were asked to exit the Autobahn at the next stop. After a final questionnaire the trial ended.

3 Results

The analysis was conducted in SPSS and the level of statistical significance was set to $\alpha = .05$ if not corrected otherwise.

3.1 Changes Concerning Driver Availability

For the analysis of the PEOR, one participant from the SuRT group was excluded since the eye-tracker produced corrupted data due to excessive make-up. All frames, for which the eye-tracker reported a quality of tracking below 70% were excluded from analysis. The changes represent the duration of eyes on road (through the windshield) during engagement into the NDRT's compared to intervals of automated driving were participants did not engage into a NDRT. A Levene-Test showed significant results for the SuRT data and a succeeding F_{max}-test yielded 18.77. Consequently, the α-level was adjusted to .025. An ANOVA showed significant differences between the SuRT group and the other two groups and a strong effect size, $F(2,47) = 50.3$, $p < .001$, $\eta^2 = .68$. Figure 3 shows the plot of the data and illustrates the significant results.

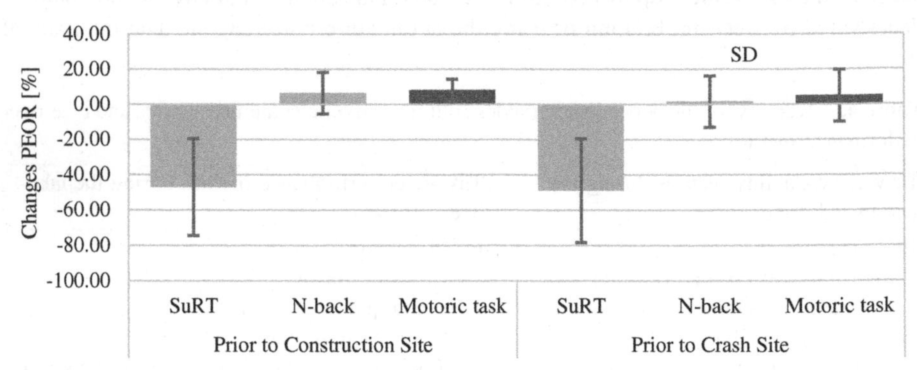

Fig. 3. Changes in the percent eyes on road/glances through the windshield. $n_{motoric} = 16$, $n_{SuRT/n\text{–}back} = 17$.

The analysis of the standard deviation of horizontal gaze dispersion was only conducted for the motoric task and the n-back task, since horizontal tracking was very limited if instructed to engage in the visual demanding SuRT. Results showed no differences between the motoric task and the n-back task but a significant difference compared to no engagement into a task, $F(1,30) = 39.75$, $p < .001$, $\eta^2 = .57$.

Blink frequency was analyzed identically compared to the HGD to show the changes when engaging into a NDRT versus non-engagement. Results showed significant changes between the SuRT and the motoric task, $F(2,38) = 7.86$, $p = .001$, $\eta^2 = .29$, but also significant differences between the two take-over situations $F(1,38) = 5.07$, $p = .03$, $\eta^2 = .12$.

In addition to the eye-tracking data, data from the seat pressure mats was also analyzed. Results were split into data from the seat and from the backrest. We looked at the changes in the COP between the interval without NDRT versus engagement into the task. The mean COP during no task was calculated as baseline and for every frame during engagement into the NDRT, we looked at the distance from the COP to the baseline COP.

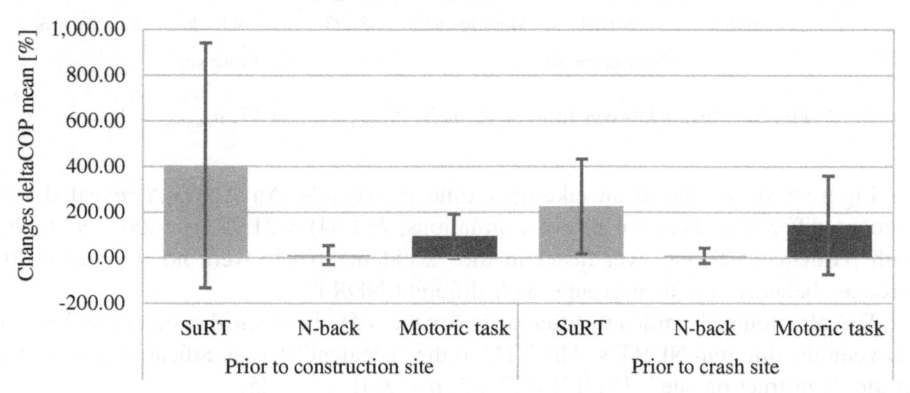

Fig. 4. Mean changes during engagement in the NDRT's during automated driving for the backrest. $n_{motoric/n\text{-}back} = 16$, $n_{SuRT} = 17$.

There were no significant changes for the seat, but the backrest showed significantly higher changes for the SuRT group compared to the n-back task and the motoric task, $F(2,46) = 9.30$, $p < .001$, $\eta^2 = .29$. Figure 4 shows the mean changes of the COP compared to the baseline COP in percent.

Analyzing data for the contact area, results from the COP were confirmed. There were no significant changes for the change in contact area for the seat, but significant changes for the backrest with the SuRT group showing smaller contact areas compared to the n-back group and the motoric task group, $F(2,46) = 12.52$, $p < .001$, $\eta^2 = .35$.

For this analysis, the α-level was adjusted to .025, since F_{max}-test was 15.78. Significant results for the backrest but not for the seat were also confirmed when looking at the maximum pressure. The change in mean pressure showed the opposite picture with significant changes for the seat but not for the backrest.

3.2 Take-Over Performance

The analysis of take-over performance will only look at situations that were preceded by instructed engagement into tasks. The free behavior intervals introduce a high variance and would not allow analysis of group differences.

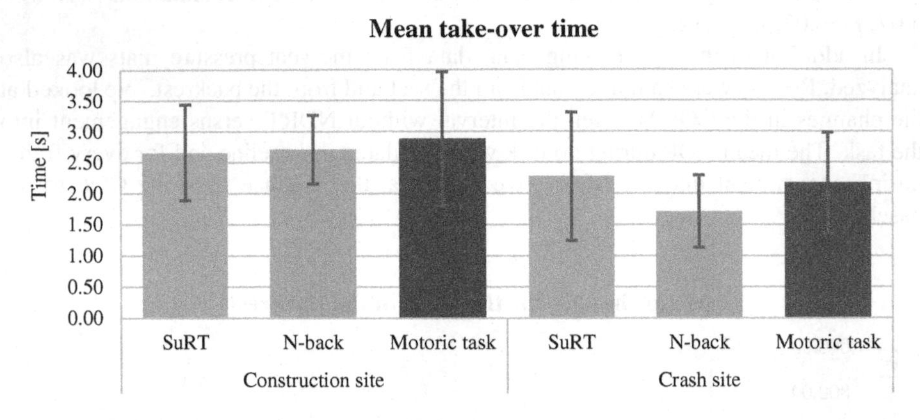

Fig. 5. Mean take-over time in seconds. $N_{SuRT/n\text{-}back} = 17$, $n_{motoric} = 16$.

Figure 5 shows the mean take-over time in seconds. An ANOVA revealed significant differences between the two situations, $F(1,64) = 21.95$, $p < .001$, $\eta^2 = .26$, with participants taking over faster in the "accident". There were no significant differences between the three groups with different NDRT's.

We also found significant differences for the TTC between the situations but not between the different NDRT's. The TTC in the "accident" was significantly lower than in the "construction site", $F(1,47) = 38.82$, $p < .001$, $\eta^2 = .45$.

The SDLP was only analyzed for the construction site due to the necessary lane change maneuver in the "accident". Results revealed no significant changes between the different NDRT groups.

We also analyzed the minimum longitudinal acceleration or brake acceleration and the maximum lateral acceleration. We found significantly lower long. accelerations for the "construction site", $F(1,64) = 42.44$, $p < .001$, $\eta^2 = .40$. The lateral accelerations also showed significant differences, with the "accident" having higher lateral accelerations than the "construction site", $F(1,64) = 282.66$, $p < .001$, $\eta^2 = .82$. Both types of accelerations did not show any significant differences between the three groups of different NDRT's.

Subjective ratings of the complexity, criticality and the available time budget revealed participants to significantly perceive the "accident" to be more critical, more complex and more urgent compared to the "construction site".

A qualitative analysis of the free instruction results revealed, that participants tend to engage into NDRT's during automated driving rather than doing nothing and would switch between different NDRT's rather than engaging into only one.

4 Discussion

Results show that the use of eye-tracking and seat-pressure mats allows the detection of changes in driver availability to some extent. The HGD cannot be used to differentiate between different modalities, but allows the detection of engagement into a NDRT in general. Blink frequency also shows significant changes between the NDRT's but also the situations. This either shows the influence of the track or more likely the large individual differences between participants. The significant results for the blink frequency should be viewed critically.

Participants react significantly faster in the "accident" situation, which can be attributed to the higher overall criticality of the "accident" situation compared to the "construction site" adding to a perceived urgency [8]. The higher criticality is punctuated by looking at the TTC, the accelerations and the subjective rating of the two situations. The lateral accelerations are within expectation, since the "construction site" does not feature a lane change maneuver.

Results from the free behavior should be viewed critically, since we offered standardized NDRT's and did not allow realistic tasks or activities. Nonetheless we conclude that participants take up on the offer of engaging into NDRT's in conditionally automated driving.

5 Summary

Results from this experiment are in line with previous findings but offer an additional, more detailed assessment of changes of driver availability during automated driving. Concluding, different modalities of NDRTs do not seem to effect take-over performance in a critical way. Contrary, the influence of different criticality of the take-over situations is revealed and is consisted with findings from the overall scope of research. Eye-tracking and seat pressure mats offer a promising way of assessing changes in driver availability even though, in this experiment, they did not result in changes of the take-over performance accordingly.

Acknowledgment. This work results from the joint project Ko-HAF - Cooperative Highly Automated Driving and has been funded by the Federal Ministry for Economic Affairs and Energy based on a resolution of the German Bundestag.

References

1. SAE Standard J3016 (2016) Taxonomy and definitions for terms related to on-road motor vehicle automated driving systems, Surface vehicle recommended practice
2. Marberger C, Mielenz H, Naujoks F, Radlmayr J, Bengler K, Wandtner B (2017) Understanding and applying the concept of "Driver Availability" in automated driving. In: Stanton NA (ed) Advances in Human Aspects of Transportation. Proceedings of the AHFE 2017 International Conference on Human Factors in Transportation, 17–21 July 2017, The Westin Bonaventure Hotel, Los Angeles, California. Springer, Cham, pp 595–605

3. Radlmayr J, Gold C, Lorenz L, Farid M, Bengler K (2014) How traffic situations and non-driving related tasks affect the take-over quality in highly automated driving. In: Proceedings of the human factors and ergonomics society annual meeting, vol 58, pp 2063–2067
4. Stanton NA, Young MS (1997) Automotive automation: investigating the impact on drivers' mental workload. Int J Cogn Ergon 1:325–336
5. Neubauer C, Matthews G, Langheim L, Saxby D (2012) Fatigue and voluntary utilization of automation in simulated driving. Hum Factors 54:734–746
6. Victor TW, Harbluk JL, Engström JA (2005) Sensitivity of eye-movement measures to in-vehicle task difficulty. Transp Res Part F Traffic Psychol Behav 8:167–190
7. Recarte MÁ, Pérez E, Conchillo Á, Nunes LM (2008) Mental workload and visual impairment: differences between pupil, blink, and subjective rating. Span J Psychol 11: 374–385
8. Gold ÇG (2016) Modeling of take-over performance in highly automated vehicle guidance. Dissertation, Technische Universität München, München
9. Gold C, Naujoks F, Radlmayr J, Bellem H, Jarosch O (2017) Testing scenarios for human factors research in level 3 automated vehicles. In: Stanton NA (ed) Advances in human aspects of transportation. Proceedings of the AHFE 2017 International Conference on Human Factors in Transportation, 17 – 21 July 2017, The Westin Bonaventure Hotel, Los Angeles. Springer, Cham, pp 551–559
10. ISO14198:2012 (2012) Road vehicles - ergonomic aspects of transport information and control systems. Calibration tasks for methods which assess driver demand due to the use of in-vehicle systems, vol ISO/TS 14198:2012. International Organization for Standardization, Switzerland, p 16
11. Kirchner WK (1958) Age differences in short-term retention of rapidly changing information. J Exp Psychol 55:352
12. Gold C, Berisha I, Bengler K (2015) Utilization of drivetime–performing non-driving related tasks while driving highly automated. In: Proceedings of the Human Factors and Ergonomics Society Annual Meeting. SAGE Publications, Los Angeles, pp 1666–1670

A Systems Analysis of the South African Railway Industry

Jessica Hutchings[1(✉)] and Andrew Thatcher[2]

[1] Transnet Centre of Systems Engineering (TCSE), University of the Witwatersrand, Johannesburg, South Africa
jessica.hutchings@wits.ac.za

[2] Psychology Department, University of the Witwatersrand, Johannesburg, South Africa

Abstract. Railway occurrences in South Africa remain high, despite occurrence investigations conducted by the various organisations. A failure to thoroughly investigate the underlying causes for incidents may be a possible reason for the number of recurrences. A systems analysis of the South African railway industry illustrates the systemic factors within the railway system that influences the effectiveness of the occurrence investigation process. Systemic factors refer to challenges, pressures, frustrations or obstacles that contribute to the complexity of railway accident investigations. Rasmussen's (1997) Risk Management Framework was operationalised for the South African railway system. A qualitative multi-method approach was adopted in this research. Methods included a print media analysis of 133 reported railway accidents, governance document analyses, 23 semi-structured interviews with railway investigators, 4 observations during actual inquiries and analyses of railway occurrence reports. The data were compared and verified against each other using triangulation. An Accimap was used to graphically illustrate the complexity of the investigation process. This systems analysis tool highlights that the system of accident investigations is indeed a complex system in its own right, and not just the rail accidents themselves. The rail investigation system in South Africa can be described as a broken 'system of systems'. Deficiencies and complexities in the system of accident investigations limits the effectiveness of the entire investigation process from achieving its objectives - that is to learn from such events - offering an explanation for why railway safety trends remain unchanged in South Africa.

Keywords: Rail · Accident investigations · Systems

1 Railway Safety in South Africa

Rail plays an important role in the sustainability of transport in South Africa. However, for rail to be sustainable, safe railway operations are required. Regardless of whether accidents and incidents result in fatalities and/or injuries, these can result in significant costs to businesses. Therefore the financial, moral, and legal impacts of accidents and incidents needs to be managed effectively. One such way of managing accidents and incidents is through effective and independent investigations. Effective investigations

S. Bagnara et al. (Eds.): IEA 2018, AISC 823, pp. 305–313, 2019.
https://doi.org/10.1007/978-3-319-96074-6_33

are important as the purpose is to identify what happened and to implement remedial actions to prevent a recurrence.

Railway *occurrences* are defined in South Africa by the Railway Safety Regulator (RSR) as railway accidents or incidents that are either due to operational or security events. These may result in fatalities and/or injuries. Occurrences, accidents, events and incidents are used interchangeably in this article. *Railway safety performance* is defined by the number of operational occurrences, number of security incidents such as theft and vandalism, and accidents, fatalities and injuries.

In South Africa, there are on average 4500 railway occurrences annually resulting in fatalities, injuries, damage to rolling stock and the environment. Railway safety performance trends in South Africa have fluctuated in the last nine years indicating that there is a definite need to establish why this is the case. This is particularly important given that interventions by the Regulator and Operators to improve railway safety performance are provided during investigations. This raises the question as to whether these are effective and implemented, or whether it is the actual investigation process that is ineffective in addressing the real reasons for why an accident occurred, with similar events repeating. Figure 1 illustrates the safety performance trends in South Africa as reported by the RSR (2017).

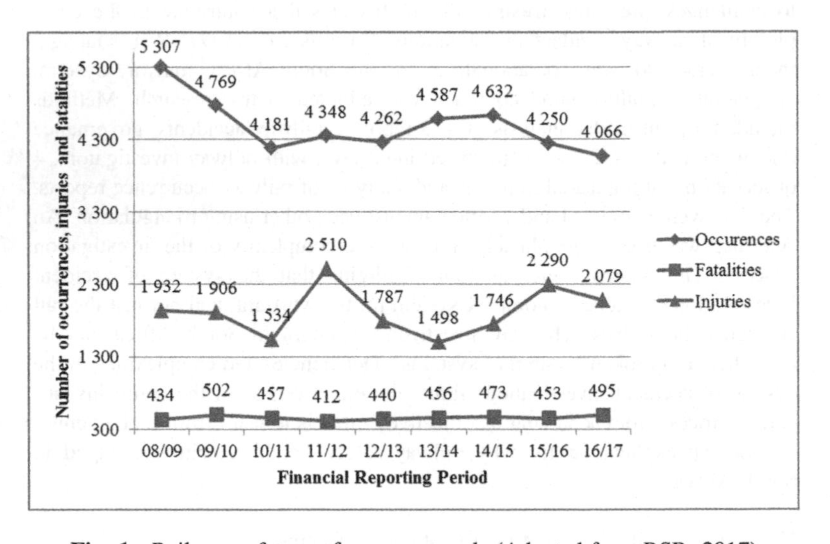

Fig. 1. Railway safety performance trends (Adapted from RSR, 2017)

Figure 1 demonstrates that safety performance trends have oscillated, with a slight decrease in numbers since 2016. Given the added impact of occurrences on the economic growth of South Africa and the status of safety performance in Fig. 1, there is a need to understand how effective interventions are within the railway industry in addressing critical occurrences such as derailments, collisions and level crossing incidents. The objective of this research was to focus on the actual investigation process of railway occurrences rather than on why the accidents occur. It was hypothesised that the actual investigation process of accidents is a contributor to the repeated number of similar occurrences. This was based on the premise that thorough accident investigations should result in a reduction of occurrences as the actual causes of the events are identified and recommendations are implemented. The latter of course need to be reliable, accurate and achievable. Therefore it was proposed that systemic factors inherent within the bigger rail socio-technical system have an impact on the effectiveness of the investigation process and the findings and recommendations that emerge from these. Systemic factors are defined as the challenges, pressures, frustrations and obstacles inherent within the investigation system creating its complexity. The effectiveness of the investigation process refers to the accuracy, reliability, quality, validity and objectivity of the actual investigation process, its findings and recommendations.

A systems analysis tool, an Accimap was developed during this research (Hutchings 2017). Adopting a systems approach allows for a holistic and comprehensive identification of the external and internal parts or components within and between organisations, their interactive nature, feedback and interdependence, in addition to the role that the environment plays in the system. All of these factors were considered in the development of the Accimap. The identification of the systemic factors influencing the effectiveness of the investigation process (the outcome of the Accimap) will not only benefit the South African railway industry but will also contribute to accident investigation theory by providing an alternative and novel approach. In developing the Accimap, Rasmussen's (1997) Risk Management Framework was operationalised to the South African railway context. This formed the basis for the final Accimap illustrating the systems analysis of the South African railway industry.

2 Rasmussen's (1997) Risk Management Framework

Rasmussen's (1997) Risk Management Framework provided the theoretical framework for this research and was used to illustrate the South African railway system. Figure 2 illustrates Rasmussen's framework, a popular systemic accident analysis method that has also been used to analyse rail accidents (Underwood and Waterson 2014).

The levels in Rasmussen's (1997) Risk Management Framework served as sources from which the data was collected for this research. In order to contextualise Rasmussen's (1997) Risk Management Framework to suit the South African railway context, adjustments to the levels were necessary. Rasmussen (1997) states that the exact number of levels and references of each level can vary depending on the system being studied. In this research the *Public* was included as a level in the system of accident investigations. Rasmussen's (1997) Risk Management Framework does not include the public as an actual level but makes reference to public opinions and pressures influencing the system (dynamic force) at a Government level. Public opinion regarding railway safety was deemed as an important actor in the bigger rail socio-technical system and not just an outside pressure. This is because the public in South Africa have a constitutional right to safe and secure transport. The public were represented by way of the media. A media analysis was conducted to determine if railway accidents are reported on, and if so, this would then inform the public about the state of

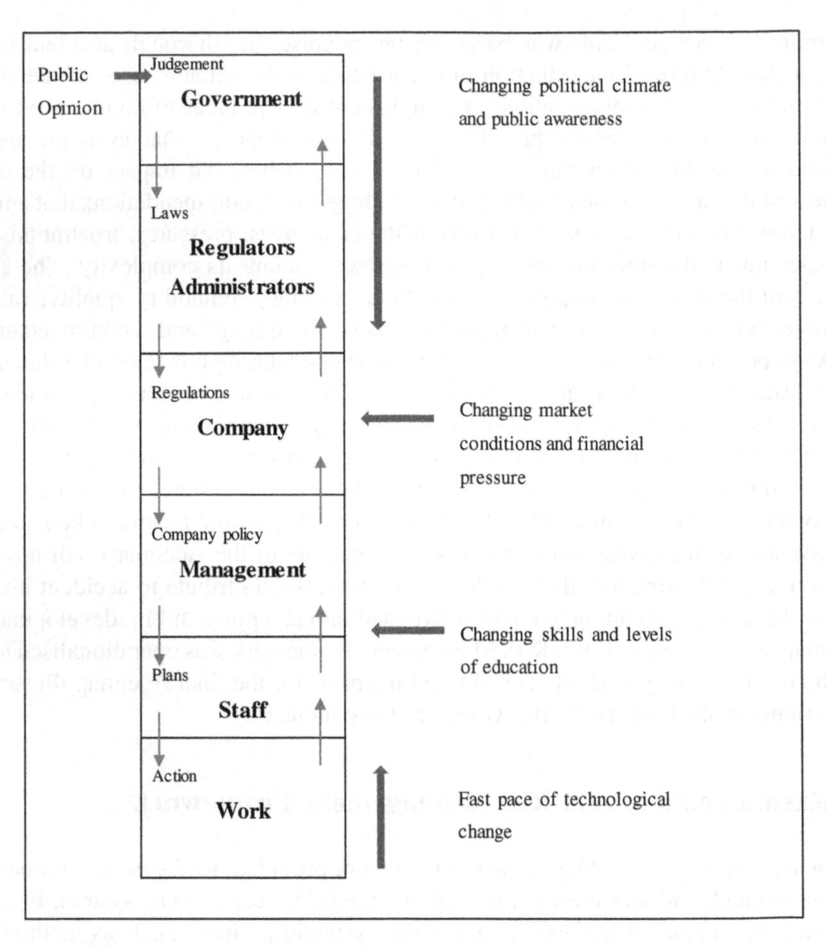

Fig. 2. Rasmussen's (1997) Risk Management Framework (Adapted from Rasmussen and Svedung 2000, p. 11)

railway safety. Furthermore, if the public are not adequately informed about the state of railway safety then the public may not be an influential dynamic force. The role of society, represented by the public is acknowledged as an external pressure that can force the Government to introduce legislation and control the practices of safety (Rasmussen 1997). The purpose of including the public as its own level in this study rather than an outside pressure, is because the author believes that the role of the public is far greater than just an external pressure especially in South Africa. The socio-economic conditions and the historical context have resulted in many people relying on the South African Government to provide safe and reliable transport after years of underinvestment, segregation and the deterioration in rail infrastructure. As a result of an increase in urbanisation this has placed greater demands on the rail transport system as a mode of travel to get to work (Department of Transport, 2015). The inclusion of society as part of the system of interest can therefore play a more significant role in influencing the systems performance. An additional reason for including the public (represented by the media) as a level on its own is because the media influences the severity and level of an occurrence investigation conducted by the Regulator and Operators. Governance documents reviewed at the Regulator and Operator revealed that if an occurrence receives media attention and public outcry that this increases the level of severity of the inquiry and the investigation team. In other words, the occurrence is deemed more serious than may have originally been classified. The public are therefore an important decision maker in the accident investigation process.

The use of Rasmussen's framework in accident causation generally refers to the *Staff* level as operational front line staff, for example a train driver or signaller involved in the accident. In the investigation process of accidents there were a few difficulties applying this framework, particularly at this level and the Management level. In this study looking at the investigation process, the Staff level refers to the individuals who conduct investigations. However, the investigators who formed part of the data collection were also *Management*, as was identified in the Operator who did not have appointed investigators, but by virtue of being a Manager anyone in this role could investigate accidents. Given the skills and competencies required to be an investigator, this raised the question as to whether the Managers are equipped with such and how this would impact of the effectiveness of the investigation process. This study identified that the skill, competencies and training of investigators was indeed a systemic factor influencing the effectiveness of the investigation process. Most of the investigators interviewed had received little to no accident investigation training, investigated because of their seniority or management level in the organisation and acknowledged the lack of competencies and training to be a skilled investigator. In terms of the *Work* level in the system of accident investigations, this refers to the actual work of conducting an investigation or inquiry.

3 Methods

A qualitative multi-method approach was adopted in this research (Hutchings 2017). Methods included a print media analysis of 133 reported railway accidents, governance document analyses, 23 semi-structured interviews with railway investigators, 4 observations during actual inquiries and analyses of railway occurrence reports (see Fig. 3). The data were compared and verified against each other using triangulation. The advantages of adopting qualitative research using a multi-method design provided the researcher with an in-depth understanding of the railway system and the intricacies within a 'system of systems'.

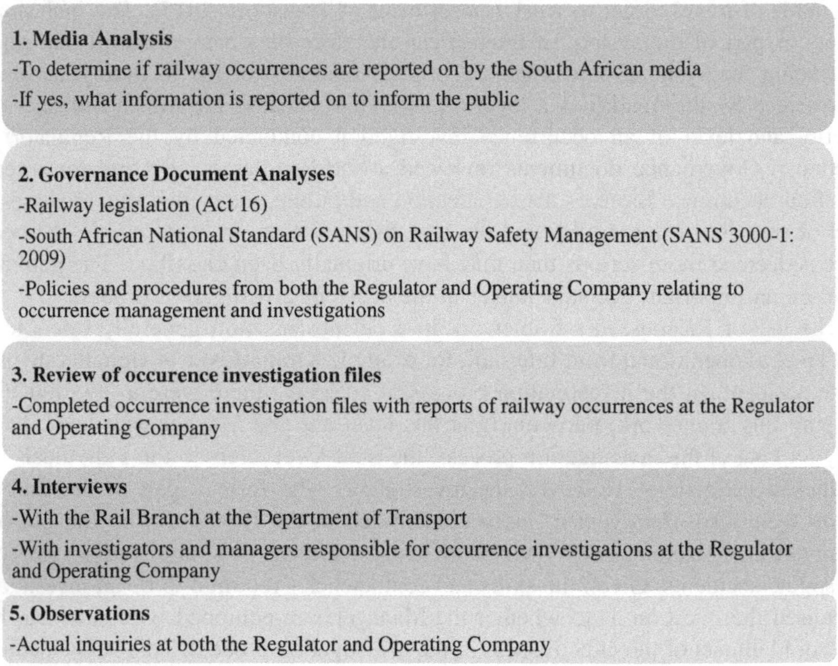

1. Media Analysis
-To determine if railway occurrences are reported on by the South African media
-If yes, what information is reported on to inform the public

2. Governance Document Analyses
-Railway legislation (Act 16)
-South African National Standard (SANS) on Railway Safety Management (SANS 3000-1: 2009)
-Policies and procedures from both the Regulator and Operating Company relating to occurrence management and investigations

3. Review of occurence investigation files
-Completed occurrence investigation files with reports of railway occurrences at the Regulator and Operating Company

4. Interviews
-With the Rail Branch at the Department of Transport
-With investigators and managers responsible for occurrence investigations at the Regulator and Operating Company

5. Observations
-Actual inquiries at both the Regulator and Operating Company

Fig. 3. Methods and sources of data used in this research

4 A Systems Analysis of the Investigation Process of Accidents

An Accimap was used to graphically illustrate the complexity of railway accident investigations across the entire system. Rasmussen and Svedung (2000) identified the need to graphically represent the causal flow of accidents based on the principles of Rasmussen's (1997) Risk Management Framework. The Accimap is a systems analysis tool that is traditionally used in accident causation analysis to structure the socio-technical system behind an accident, the preconditions, the functions of the different system levels involved, and how they contributed to an accident (Rasmussen and Svedung 2000). In this study the Accimap was used in a different context to illustrate decision making during normal work for all the levels of the railway system and the influence of the stressors found in modern dynamic society impacting on the *investigation process of accidents* and not the actual accident itself. Multiple contributing or systemic factors were mapped onto the levels of the South African rail socio-technical system illustrating the context in which accident investigations take place. The outcome of the Accimap highlights all the factors impacting on the effectiveness of the investigation process contributing to a recurrence of accidents (see Fig. 4).

This systems analysis tool highlighted that the accident investigation process is an example of a complex system in its own right, and not just the rail accidents themselves. The Accimap highlighted how the media, representing the public, are an

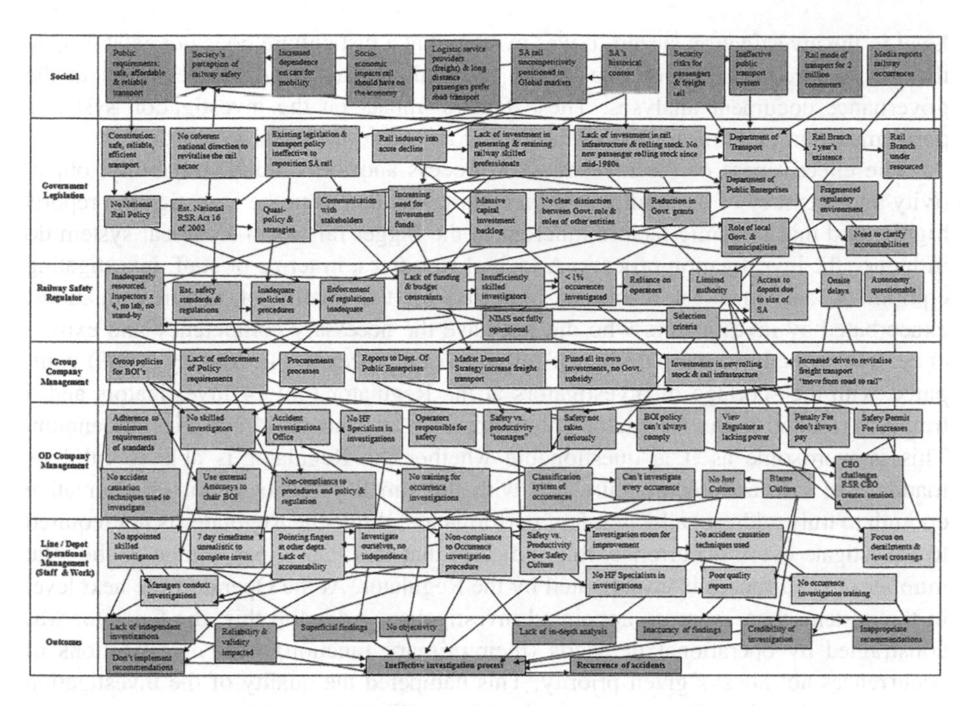

Fig. 4. An Accimap of the investigation system

important decision maker in the system of accident investigations as they are able to influence the seriousness given to an investigation at both the Regulator and Operator. However, the media analysis illustrated that although railway accidents are reported, the number of reports were substantially fewer than the number of occurrences reported annually by the Regulator. Therefore the public (and the media) are not acceptably informed about the state of railway safety in order to affect pressure on the South African Government to improve railway safety.

The Department of Transport (representing the Government) in the investigation system plays no role in the actual investigation of occurrences, and it was reported that in general it does not actively participate in railway safety unless the type of occurrence results in media attention and a number of fatalities or injuries for example in level crossing events. The data revealed a number of level crossing events, and a number of repeated occurrences at the same level crossing with fatalities. The power of the Department of Transport to ensure safe and reliable road and rail transport is questionable given this, the high number of railway occurrences, and the failure of Government to develop a National Rail Policy in South Africa. The consequence of the latter has resulted in generic legislation that is not well understood at the highest levels of the system. These are examples of systemic factors at a Government level that influences the levels below. Constraints and challenges at this level in the investigation system impacted on the next level of the system, the Regulator. The results demonstrated very little vertical integration between the different levels below the Government level, and also up to the Public level. At the both the Regulator and Company level in the investigation system it was evident that a just culture does not exist despite this being a requirement of one of the national railway standards reviewed in the governance document analyses. These factors impact on the investigation system's performance in addition to that of the bigger rail socio-technical system.

The effectiveness of the investigation process and the validity, reliability, objectivity and accuracy of the findings and recommendations from the investigation reports highlighted that systemic factors inherent in the bigger rail socio-technical system do influence the investigation process. A lack of resources, in terms of staff, investigating equipment, and financial resources impacted on the quality of the investigations; exacerbated by investigators who did not have the necessary competency and experience. The annual average number of railway occurrences in the country (4500) compared with the shortage of investigators at the Regulator (only 4 investigators and 4 trainees) resulted in the Regulator only being able to conduct 34 inquiries per annum. This is worrisome as it is questionable whether the Regulator is able to fulfil its mandate of safe and secure railways, with this small percentage not representative enough to truly address railway safety performance. While the Regulator is not required to investigate every occurrence, this indicates that less than a percentage of the total number of occurrences is investigated by the Regulator. At the Operator, the next level in the system, there were no appointed investigators. Added to this, the Operator was constrained by operational demands (their primary function) with investigations of occurrences not always given priority. This hampered the quality of the investigation process and therefore the accuracy of the reports. Challenges such as time constraints, pressures and demands from higher up in the system, no accident causation methods being used, shortage of skilled investigators, absence of investigator training, non-compliance to governance documents, lack of investigating equipment, time delays in commencing with inquiries after an occurrence impacting on witnesses memory to recall events, financial costs of using external specialists which limits the number of inquiries, and a lack of a just culture within the entire system at each level all impacted on the effectiveness of the investigation process from preventing recurrences. These factors influence the investigation process from being thorough and effective in trying to really identify what happened. By implication the recommendations that are suggested therefore act as a 'band aid' as the symptoms are only identified. The appropriateness of the recommendations in addressing the actual causes of the events is also questionable and is a contributing factor to recurrences.

At all of the levels of the system from the Government to the Operator level, similar findings were demonstrated. The systems analysis showed how the relationships within and between the different levels in the bigger rail socio-technical system played a role in influencing the effectiveness of the investigation process.

5 The South African Rail Investigation System Is a Broken 'System of Systems'

The systems analysis revealed that the ability of the occurrence investigation process to effectively address railway safety performance is influenced by decisions makers and events from all levels of the bigger rail socio-technical system. It can be confirmed that constraints and challenges higher up in the system influence the actual process of conducting investigations with a greater impact on the lower levels of the system where it is mandatory to investigate all railway accidents. The bigger rail socio-technical system influences its nested systems of which this research revealed is the system of accident investigations. The accident investigation process is an example of a complex system. The systemic factors highlight how the bigger rail socio-technical system has a number of challenges, constraints and pressures that negatively influence the performance of the accident investigation system and therefore its effectiveness in identifying what happened and what can be done to prevent recurrences. These factors contribute to the accident investigation systems complexity. The rail investigation system in South Africa can be described as a broken 'system of systems' with the disjointed parts in the bigger rail socio-technical system contributing to its complexity. Deficiencies and complexities in the system of accident investigations limits the effectiveness of the entire investigation process from achieving its objectives - that is to learn from such events - offering an explanation for why railway safety trends remain unchanged in South Africa.

Note to the Reader

The results of this research were based on information developed from data collected at a particular time from certain organisations - with their full support - and in the interest of process improvement. The author recognises that some of the findings may since have been addressed or are in the process of being addressed by these organisations where the data were collected. They were provided with a copy of the Doctoral thesis that this paper was developed from. The organisations are aware of the findings and have reiterated their commitment to improving railway safety by considering the implementation of the corrective actions and initiatives at a strategic level to address the results.

References

Department of Transport (2015). http://www.transport.gov.za/documents/11623/21629/NationalRailPolicyGreenPaper.pdf/ea72ecab-2990-41cd-a0dd-f5e7a2c3e82a. Accessed 31 Mar 2018

Hutchings J (2017) Systemic factors in the investigation of South African railway occurrences (Doctoral dissertation). http://wiredspace.wits.ac.za/handle/10539/23846

Railway Safety Regulator (2018). https://rsr.org.za/Documents/State%20of%20Safety%20Reports/RSR%20State%20of%20Safety%20Report%202016-17.compressed.pdf. Accessed 31 Mar 2018

Rasmussen J (1997) Risk management in a dynamic society: a modelling problem. Saf Sci 27 (2/3):183–213

Rasmussen J, Svedung JRI (2000) Proactive risk management in a dynamic society. Swedish Rescue Services Agency, Karlstad

Underwood P, Waterson P (2014) Systems thinking, the Swiss cheese model and accident analysis: a comparative systemic analysis of the grayrigg train derailment using the ATSB, AcciMap and STAMP models. Accid Anal Prev 68:75–94

Effect on Mode Awareness When Changing from Conditionally to Partially Automated Driving

Anna Feldhütter[1(✉)], Nicolas Härtwig[1], Christina Kurpiers[2], Julia Mejia Hernandez[2], and Klaus Bengler[1]

[1] Chair of Ergonomics, Technical University of Munich, Boltzmannstr. 15, 85748 Garching, Germany
anna.feldhuetter@tum.de
[2] BMW AG, Knorr Strasse 147, 80788 Munich, Germany

Abstract. Future vehicles will combine different levels of capable driving automation characterized by varying responsibilities for users. This development will lead to an increase in system complexity which poses the risk of confusing the driver. Based on the theory of proactive interference, we hypothesize that the users' mode awareness suffers especially when changing from Level 3 "Conditional Automation" to Level 2 "Partial Automation". Consequently, a mode transition intermitted by a short phase of manual driving acts as a countermeasure for a loss of mode awareness. Assumptions were tested in a driving simulator study with 45 valid participants. Mode awareness was operationalized by the visual attention towards driving-relevant areas and a qualitative analysis of an interview. Results indicate that in partial automation, visual attention does not deteriorate due to a lack of mode awareness, but rather to the development of overreliance arising from the experience with a very reliable partially automated system.

Keywords: Mode awareness · Partial automation · Conditional automation Visual attention · Overreliance

1 Introduction

Due to the technical advances in recent years, automated vehicles will be present in road traffic within the next few years. As a result of progressive development towards full driving automation, vehicles will feature increasingly capable automation levels. However, more complex automated systems might pose the risk of confusing drivers and increase the probability of mode-related errors [1, 2]. Therefore, automated systems need to be designed in such a way that minimizes confusion with regard to the automation mode. In this study, we investigate a countermeasure to avoid mode confusion with the aim of contributing to the reliable operation of different levels of automation in one vehicle.

Driving Modes and Levels of Automation. Different levels of driving automation have specific requirements with respect to their operational environment (e.g. infrastructure, climate, etc.), determining their scope and allotting varying responsibilities to

S. Bagnara et al. (Eds.): IEA 2018, AISC 823, pp. 314–324, 2019.
https://doi.org/10.1007/978-3-319-96074-6_34

their users. According to the taxonomy of the Society of Automotive Engineers (SAE) [3], recently released driving automation systems in production vehicles can be classified as Level 2 "Partial Automation" (PA). On this level, the automation performs both lateral and longitudinal vehicle guidance [3]. When it exceeds its scope, PA fails, also without notifying the driver (silent system boundary) in order to prevent too much reliance on such warnings. Therefore, the user is considered to be the driver at all times and is expected to complete the dynamic driving tasks (DDT) by continuously monitoring the vehicle behavior in the respective traffic situation. The next evolutionary step is to temporarily release the user from all driving-related responsibilities. On Level 3 (Conditional Automation, CA), the system comprises the entire range of DDT, and thus the user can engage in non-driving-related activities (NDRA). CA recognizes its system limits and timely requests the driver to intervene if it becomes necessary [3]. With CA, the user is considered to be the fallback level, which means he/she is expected to be receptive to the system's requests to intervene and to timely respond accordingly. Both automation levels, PA and CA, are limited to a specific scope and therefore will not be available everywhere and at any time. Outside of this scope, only lower automation modes or manual driving (MAN) will be available.

Mode Awareness and Mode Errors. The aviation sector, where cockpit automation already started back in the 1980s, quickly faced mode awareness problems. Sarter and Wood [2] stated that more complex automated systems might pose the risk of confusing drivers and therefore provoke erroneous behavior. Such incorrect driver behavior is referred to as "mode error", and occurs when the driver is not aware of the currently activated automation mode and its functional capability or assumes a different mode is currently activated [2, 4], in other words, if there is a lack of mode awareness. A review of several incidents and even accidents in the aviation sector caused by a lack of mode awareness revealed that the main reasons for the occurrence of mode confusion and the resulting erroneous behavior were poor system feedback, high complexity of the automated system, and insufficient mental models on the part of the pilots [2, 4]. Furthermore, pilots had problems recognizing the activated mode or system-initiated mode changes as well as the corresponding impact on the aircraft [4]. The automotive domain is now facing similar challenges as the aeronautic domain. Implementing different levels of capable automation modes into one vehicle increases the complexity of the system. In each automation mode, drivers have to keep in mind which tasks are taken over by the system and which they have to remain responsible for. Here, an adequate human-machine interface (HMI) is needed to ensure that the user is aware of the activated mode. Therefore, mode confusion and lack of mode awareness are identified as risk factors for the safe operation of automated vehicles and need to be further examined [5]. What is considered a mode error is determined by the current mode of a driving automation system and the corresponding responsibilities of a human operator. SAE definitions provide implications for the driver behavior required for safe operation at a specific level of automation [3]. While driving in PA, the driver must monitor the system at all times, meaning that he/she has to verify that the vehicle is behaving as it is supposed to depending on the given situation (surrounding vehicles/traffic, speed, road conditions, etc.). Therefore, insufficient monitoring behavior is considered to be a mode error in PAD.

Interference Theory. Interference theory describes memory processes related to forgetting. Memory traces of newly and previously encountered stimuli interfere with one another [6]. They potentially impede the retrieval of both old (retroactive interference) and new (proactive interference) memories [7]. Interference between stimuli occurs especially when stimuli are conceptually similar. Consistently separating conceptually similar stimuli by a conceptually different stimulus might reduce interference [8].

Due to the very similar driving experience when driving alternatingly with PA and CA could pose the risk that user can temporarily not distinguish between the two modes, especially considering the use case when driving on the right lane on a freeway with low traffic density and no tactical maneuvers are required, for instance overtaking slower vehicles. However, the ability of the systems to detect system boundaries or malfunctions and the corresponding responsibility of the user differ significantly. Thus, changing between PA and CA is assumed to be particularly prone to causing reduced mode awareness.

Research Questions. Based on interference theory, we assume that mode awareness in partially automated driving (PAD) is reduced and the probability of mode errors occurring increases if the previous driving mode was conditionally automated driving (CAD). Forcing users to drive manually for a defined time interval between CA and PA separates conceptually similar experiences by a conceptually distinct experience. Accordingly, it is expected to inhibit interference effects and reduce the probability of mode errors occurring by emphasizing the users' current role. We examined these questions in the context of a driving simulator study.

2 Method

2.1 Participants

In total, 50 participants took part in this study. All participants were recruited from among BMW employees. All participants had to meet the criterion that they were not experts in vehicle automation and were asked about this prior to the experiment. Five participants were excluded from data analysis due to unrealistic behavior in the driving simulator. The remaining sample consisted of 45 participants (35 males, 10 females) with a mean age of 39.04 years ($SD = 5.98$). Most participants ($N = 30$) stated that they drove 5,001–20,000 km per year.

2.2 Experimental Design

This driving simulator study aims at examining the effect of driving phase and transition condition on mode awareness. The simulated test track was located on a three-lane freeway with low traffic density. Participants were randomly assigned to the two transition conditions (intermitted, immediate). Figure 1 shows the test track of this experiment with its driving phases: All participants started with 10 min of manual driving (MAN1), followed by 10 min of PAD (PAD1) and subsequently by 20 min of CAD. Next, participants assigned to the immediate transition condition changed directly from CAD to a second 10-min phase of PAD (PAD 2), while participants in the intermitted transition condition drove 3 min manually (MAN2) between CAD and PAD2. In both PAD phases, the test track and the surrounding traffic were designed in such a way that overtaking maneuvers were not necessary and no system limits or failures occurred. Thus, we designed the PA very/maximum reliable in our experiment. After finishing PAD2, a malfunction occurred conforming to the capability of an SAE PA: on a straight road section, the vehicle slowly drifted to the right without notifying the driver, due to a tar joint which was mistaken for a lane marking. If the driver does not timely intervene, the vehicle deviated from the right lane or even from the road.

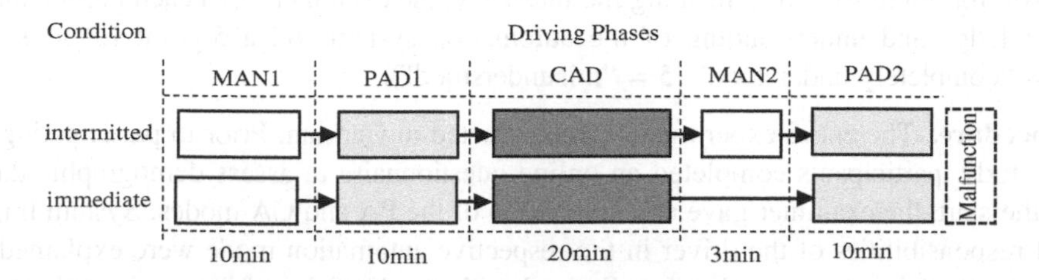

Fig. 1. Test track with the immediate and the intermitted conditions.

Assessment of Mode Awareness. To assess mode awareness, we used a combination of objective metrics, an examiner rating, and a semi-structured interview. All components are described in the following.

Mode errors occur due to a loss of mode awareness. In PAD, insufficient monitoring behavior is considered a mode error. SAE standards do not explicitly specify a sufficient monitoring behavior for PAD. However, system boundaries and malfunctions might occur anytime and might not be detected by the system. Hence, it is crucial for safe operation that the drivers' monitoring behavior be comparable to manual driving, and so MAN was considered the baseline. According to Bubb, Vollrath, Reinprecht, Mayer, and Körber [9], visual perception is the main sensory channel for managing the driving task. Therefore, we operationalized monitoring behavior by the visual attention ratio towards driving-relevant areas (road and driving environment ahead, the instrument panel, and all three mirrors). Attention ratio is defined as the percentage of time that a participant glanced into the area of interest (AOI) in relation to the total duration of each driving phase. Furthermore, attention ratio towards the presented NDRA was also measured for examining visual attention in more detail. In addition, the percentage of participants who reacted to the malfunction in time and prevented the vehicle from deviating from the lane/road was assessed. For each driving phase (except MAN2), the examiner rated the participants' mode awareness on an 11-point scale from 0 ("no mode awareness at all") to 10 ("perfect mode awareness"), based on observing the participants' monitoring behavior, reaction towards mode change requests, and usage of the respective mode features. The semi-structured interview used for examining the mental model of participants with respect to their understanding of the system consisted of questionnaires and open questions. On a 7-point Likert scale (1 = "not at all as expected"; 7 = "completely as expected"), the participants rated whether the system behaved as expected with regard to the malfunction at the end of the test track.

Furthermore, participants were presented with five statements about the responsibilities in the respective modes which could be answered with “correct” or “false”. Additionally, participants were asked to describe the systems’ functionality and their own responsibilities in their own words. Moreover, a questionnaire with a 7-point Likert scale from 0 (“strong disagreement”) to 6 (“strong agreement”) was completed, which contained six items for self-rating the overall mode awareness and six items to evaluate the HMI concept (auditory and visual feedback). Greater values indicate a higher level of mode awareness or a more positive perception of the HMI. Depending on whether participants engaged in the NDRA in PAD, they were asked if they had any specific reason for doing so. After finishing the interview, the examiner rated each participant’s knowledge and understanding of the automation systems on a 5-point Likert scale (1 = “completely understood”; 5 = “misunderstood”).

Procedure. The entire experiment was conducted in German. Prior to participating in the study, participants completed an online questionnaire to assess demographic data. At the start, the examiner gave an explanation of the PA and CA modes. System limits and responsibilities of the driver in the respective automation mode were explained at length. Participants were also familiarized with the NDRA. Afterwards, participants completed a training drive to become familiar with the simulation and the automation system. Before starting the actual test track, eye-tracking equipment was calibrated. The completion of test track took approximately 50–55 min. Lastly, the examiner conducted the 20-min debriefing, including the semi-structured interview and all questionnaires.

2.3 Apparatus

Driving Simulator, Automation and Eye-Tracking. The study was conducted in a static driving simulator at the BMW Research and Innovation Center. The BMW 5-Series half-vehicle mockup was surrounded by five projectors producing a front view of about 200°. All three mirrors could be used. Two speakers behind the front seats and two speakers in front played road and engine noises. The driving simulator was equipped with active steering. The automation modes PA (“Autobahnassistent”) and CA (“Autopilot”) were implemented, conforming to the SAE taxonomy. Participants were instructed that PA informs drivers about upcoming detected system boundaries. However, not all system boundaries and malfunctions would be detected. Therefore, they must supervise the system at all times. CA detects all system boundaries and drivers would be timely requested to intervene if necessary. Participants can engage in other activities but have to be receptive to taking over control if needed.
Gaze behavior was assessed using the Dikablis Essentials head-mounted eye-tracking system (by Ergoneers), with a sampling frequency of 50 Hz.

Non-driving-Related Activity. In front of the central information display, a tablet was mounted that ran an ordinary user-paced quiz game (single choice questions with four answer options). We chose this NDRA because it requires cognitive resources and is still interruptible. For each correct answer, participants won points. To evoke a realistic motivation to play the game, the participants were informed that the top ten players in terms of total score would be notified via e-mail. Participants were instructed that they may (but did not have to) play the game whenever they considered it to be appropriate in the respective driving situation and whenever they felt comfortable doing so.

Human-Machine Interface. The different automation modes were indicated by green (PA) and blue (CA) symbols in the instrument panel (IP) (Fig. 2) as well as by an approximately 120 cm-long LED stripe mounted across the bottom of the windshield (Fig. 2) glowing continuously in the corresponding color of the symbol. Availability of a higher-level automation mode (PA or CA) was indicated to the driver by a single acoustic gong and a text in the IP. Automation modes were activated with separate buttons on the steering wheel labeled "Assist" (PA) and "Auto" (CA). A third button ("Off") deactivated all automation. There was a two-level cascade for downgrading to a lower-level automation mode determined by the system (CAD to PAD/MAN). First, a single acoustic gong and a text in the IP was triggered. Additionally, the symbols in the instrument cluster and the LED stripe started to flash at 1 Hz. If the mode change was not confirmed by the driver, the second step triggered a sharp double beep and caused the symbols in the instrument cluster and the LED stripe to flash at 2 Hz.

Fig. 2. Left: Instrument panel indicating active PA ("Autobahnassistent"); Right: LED stripe indicating the active automation mode by illumination in the respective color (here: PA). (Color figure online)

3 Results

Monitoring Behavior and Reaction to the Malfunction. Monitoring behavior was assessed by the visual attention towards driving-relevant areas in the different driving phases and under different transition conditions. MAN2 was too short for a suitable comparison. When driving manually, most of the participants showed a high attention ratio with mean values >88%. In PAD1, attention ratio decreased to mean values of about 51%, and in CAD to mean values of about 17%. In PAD2, attention ratio rebounded to about 33%. Consistently, the attention ratio towards the tablet where the NDRA was presented evolved reversely in each phase (Fig. 3). In MAN1, the mean attention ratio (tablet) was very low (approx. 2%). In CAD, participants showed high visual attention (approx. 78%) towards the quiz game. Figure 3 indicates that more attention was allocated towards the tablet in PAD2 (about 60%) than in PAD1 (about 40%).
Statistically, we analyzed the effect of driving phase and transition condition on attention ratio with linear a mixed-effects model. We specified participants as random effects. This allowed us to account for participants' interindividual differences in disposition to monitoring behavior (intraclass correlation; ICC). Table 1 shows an

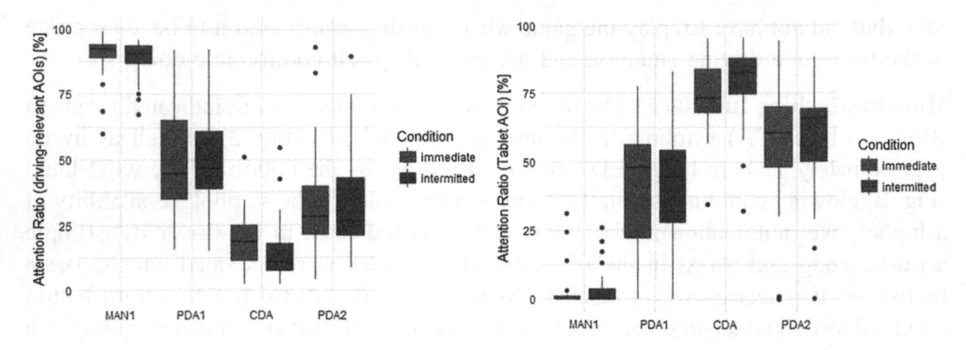

Fig. 3. Attention ratio towards driving-relevant AOIs (left) and towards the tablet (right) separated by condition. Whiskers represent upper/lower hinges ±1.5*IQR.

Table 1. Summary of mixed-effects model

	Attention ratio		
	β	CI	p
Fixed Parts			
(Intercept)	89.21	82.45–95.97	<.001
PAD1	−38.37	−46.27 – −30.47	<.001
CAD	−70.59	−78.49 – −62.69	<.001
PAD2	−55.07	−62.97 – −47.18	<.001
Intermitted condition	−1.60	−11.49 – 8.30	.752
PAD1: intermitted	0.88	−10.69 – 12.44	.882
CAD: intermitted	−1.08	−12.64 – 10.48	.855
PAD2: intermitted	0.17	−11.40 – 11.73	.978
Random Parts			
σ^2		194.879	
$\tau_{00,\ vp}$		90.520	
N_{vp}		45	
ICC_{vp}		0.317	
Observations		180	
R^2		.841	

Note. Estimates (β); confidence intervals (*CI*); p-values (p) with Kenward-Roger approximation; between-group variance (σ^2); within-group variance ($\tau_{00,\ vp}$); number of groups (N_{vp}). intraclass correlation (ICC_{vp})

overview of the model parameters. All four driving phases are significant predictors of attention ratio. Relative to an estimated intercept for driving manually of 89% (SE = 3.45; $p < .001$), attention ratio on average decreased by 38% points (SE = 4.03; $p < .001$) when driving in PAD1, by 71% points (SE = 4.03; $p < .001$) when driving in CAD, and by 55% points (SE = 4.03; $p < .001$) when driving in PAD2. Pairwise comparison reveals significant lower attention ratio in PAD2 than in PAD1 (t (129) = 5.78; $p < .0001$), irrespective of condition. No significant difference in the visual attention towards driving-relevant areas can be found between PAD2 in the immediate condition and PAD2 in the intermitted condition ($t(132) = 0.28$; $p = 1$). Therefore, a change between CAD and PAD2 intermitted by manual driving did not affect the monitoring behavior.

When analyzing the participants' reactions to the malfunction, we found that only seven participants (16%) intervened before the vehicle left the lane. 25 participants (56%) reacted when two wheels of the vehicle had left the lane, and 12 participants (29%) did not notice the vehicle's lane deviation until it had completely left the lane or the road.

Examiner Rating, Questionnaire and Interview. The examiner rating of the participants' mode awareness reveals largely appropriate behavior in MAN1 and CAD. In both PAD phases, the participants' mode awareness was rated lower during PAD1 (approx. 7.5) and PAD2 (approx. 6). The questionnaire for self-assessing overall mode awareness yields a mean value of 4.5 ($SD = 1.0$), indicating that participants felt they had a rather good overview of the system's behavior and functionality. The evaluation of the HMI items shows a largely positive perception of the HMI ($M = 4.5$, $SD = 0.9$). However, many participants mentioned that they rated the HMI more negatively due to a lack of auditory feedback for the malfunction. The participants' rating with regard to whether the system behavior in terms of the malfunction was as expected shows that for 21 participants (47%), the system had not (at all) behaved as expected (score 1 or 2, $M = 3.36$, $SD = 2.0$). Only three participants (7%) stated that the system had behaved completely as expected. The reasons stated for regarding the malfunction as unexpected were a lack of auditory feedback (51%) and that participants had not expected such an extensive failure in such a simple situation (31%). All five statements about the automated driving systems were evaluated correctly by 32 participants (71%). 12 participants (27%) correctly evaluated 4 out of 5 statements. Only one participant answered incorrectly. While the two statements regarding the CAD mode were always answered correctly, participants were less certain about the three statements concerning the PAD mode. Here, 12 participants (27%) believed they could unrestrictedly engage in NDRA.

The reasons stated as to why the participants played the quiz game rather than monitoring the PA were reliance on the system (58%), uncertainty concerning the automation mode (33%), and boredom due to the passive monitoring task (29%). The examiner rating of the participants' knowledge about the automated driving system reveals that 24 subjects (53%) had completely understood the systems' functionality, the system limitations, and their own responsibility, 16 participants (36%) had largely understood the systems, and five participants (11%) at least partially understood.

4 Discussion and Conclusion

In this study, we examined the effect of driving phases and transition condition on the mode awareness in PAD. We assumed that driving in conditionally automated mode right before driving in partially automated mode would lead to a loss of mode awareness resulting in a decrease in monitoring behavior. We found a significantly reduced attention ratio towards driving-relevant areas in PAD2, where the preceding automation mode was CA, as compared to PAD1, where the previous mode was manual driving. Consistently, the attention ratio towards the tablet where the NDRA was provided increased in PAD2. That means that in PAD2, participants increasingly neglected their monitoring task and played the quiz game more intensely. As a consequence, only one quarter of the participants could prevent the vehicle from leaving the lane when a malfunction occurred. The interview revealed that one third of the participants considered the two modes to be difficult to distinguish between due to their similarity. These results indicate that a lack of mode awareness might have played a role in PAD2. However, the interview also showed than half of the participants named greater trust as a reason for their increased playing behavior in PAD2. This overreliance was developed, since no system failure occurred except for the one at the end of the track. This is in accordance with the research of Lee and See [10], who found that trust in an automated system depends on the experience gained with the automated system. Furthermore, the interview and the examiner rating revealed that the participants generally had a good awareness and understanding of the two automation modes and could largely recall their respective responsibilities. Nevertheless, participants seemed to lack a deeper understanding of what an actual unnoticed system failure or a silent system boundary could lead to, since the malfunction – especially without any warning – was completely unexpected by most of the participants. This might have further promoted the development of excessive reliance. Overreliance – and not the lack of mode awareness – being potentially one main reason for reduced monitoring behavior might also explain why intermediate manual driving was not an effective measure to regain a sufficient level of monitoring behavior after changing from CAD to PAD2.

Limitations and Implications for Further Research. Due to the research question of this study, we implemented a PA with high reliability (close to ideal), which could have caused or at least promoted the development of overreliance. Furthermore, the design of the test track was not conducive to remind participants of their respective responsibility concerning their monitoring task (no overtaking maneuvers required during phases of PAD) and, additionally, the HMI concept did not provide any assistance to remind the drivers about their current responsibility (e.g. hands-off times or eyes-off times in PAD were not limited). Furthermore, the transition from CAD to PAD was designed very unobtrusively which would also not be implemented in real systems. Both the system design and the test track are considered the worst case and are unlikely to occur in the future use case. Future investigations should examine whether overreliance also occurs in a more realistic system design when more system boundaries occur and propose countermeasures to avoid overreliance from developing. For instance, the consequences of a (silent) system limit could be pointed out to the participants more clearly by having them experience an example of a specific situation during the training drive.

Furthermore, the problem of overreliance is probably not restricted to the use case of changing between PA and CA. Instead, overreliance can potentially also arise in the event of prolonged uninterrupted phases of only PAD if no malfunction or system boundaries occur. This is also the result of the National Transportation Safety Board's report on the causes of a fatal crash of a Tesla Model S 70D [11].

Finally, data were gathered in a driving simulator study and the transfer of findings into the real world is limited. It is questionable whether the strong decrease in monitoring performance in PAD compared to MAN would occur in real traffic.

5 Conclusion

In a driving simulator study, we examined whether mode awareness in PAD is reduced if the previous driving mode was CAD. We found reduced monitoring behavior and more intense engagement in the NDRA during PAD, indicating a decrease in mode awareness. However, results of the interview revealed that most of the participants had a rather good understanding of the system and the poor monitoring behavior was mainly due to the development of excessive reliance on the partially automated system. However, overreliance could have been artificially promoted, since, for the purpose of this study, a very capable PA with no malfunctions was designed. Future research will address the development of suitable countermeasures to prevent overreliance.

References

1. Feldhütter A, Segler C, Bengler K (2017) Does shifting between conditionally and partially automated driving lead to a loss of mode awareness? In: Stanton NA (ed) Proceedings of the AHFE 2017. International Conference on Human Factors in Transportation, Los Angeles, pp 730–741
2. Sarter NB, Woods DD (1995) How in the world did we ever get into that mode? Mode error and awareness in supervisory control. Hum Factors. https://doi.org/10.1518/001872095779049516
3. Society of automotive engineers (2016) taxonomy and definitions for terms related to driving automation systems for on-road motor vehicles. SAE International, Warrensdale, PA (SAE Standard J3016_201609)
4. Sarter N (2008) Investigating mode errors on automated flight decks. Illustrating the problem-driven, cumulative, and interdisciplinary nature of human factors research. Hum Factors. https://doi.org/10.1518/001872008X312233
5. Martens MH, van den Beukel AP (2013) The road to automated driving: dual mode and human factors considerations. In: 2013 16th international IEEE conference on intelligent transportation systems (ITSC 2013), The Hague, Netherlands, 6–9 October 2013. IEEE, Piscataway, pp 2262–2267. https://doi.org/10.1109/itsc.2013.6728564
6. Eysenck MW (2004) Psychology. An International Perspective. Psychology Press, Hove

7. Kane MJ, Engle RW (2000) Working-memory capacity, proactive interference, and divided attention. Limits on long-term memory retrieval. J Exp Psychol Learn Mem Cogn. https://doi.org/10.1037/0278-7393.26.2.336
8. Underwood BJ (1969) Attributes of memory. Psychol Rev. https://doi.org/10.1037/h0028143
9. Bubb H, Vollrath M, Reinprecht K, Mayer E, Körber M (2015) Der Mensch als Fahrer. In: Bubb H, Bengler K, Grünen RE, Vollrath M (eds) Automobilergonomie. Springer Fachmedien Wiesbaden, Wiesbaden, pp 67–162
10. Lee JD, See KA (2004) Trust in automation: designing for appropriate reliance. Hum Factors. https://doi.org/10.1518/hfes.46.1.50_30392
11. National transportation safety board (2017) Collision between a car operating with automated vehicle control systems and a tractor-semitrailer truck near Williston, Florida, 7 May 2016. Highway Accident Report NTSB/HAR-17/02. National Transportation Safety Board, Washington, DC. https://www.ntsb.gov/investigations/AccidentReports/Reports/HAR1702.pdf

Ergonomic Approach of the Influence of Materials and the User Experience in the Interior of Automobiles

Talita Muniz Ribeiro(✉) and Jairo José Drummond Camâra(✉)

Universidade do Estados de Minas Gerais ED/UEMG, Antonio Carlos Avenue, Belo Horizonte, Minas Gerais 7545, Brazil
talitamr@gmail.com, jairo.camara@uemg.br

Abstract. A new paradigm emerges around automobiles: People's expectations regarding transport and mobility are changing. New ways of interacting and relating to the vehicle emerged. This is mainly due to the significant increase of imbedded technology and connectivity as well as the updating of ADAS systems increasingly present in vehicles. In this context, automotive design needs to be rethinked. In this way, car and transport users in general are being impacted and the automotive industry is being forced to revise its business model.

This new scenario highlights the importance of areas such as ergonomics and user experience, essentials for automakers to invest in new projects development. Vehicle ergonomics needs to adapt to these new needs of industries and user, aiming at future trends: connectivity and interaction. How to design vehicle's interiors that are more and adapted to the new demands of both, automotive industry and market and provide a better customer experience of these vehicles? The purpose of this article is to demonstrate a macro vision of how the relation of materials, finishes and technologies influence the experience inside a vehicle. Although there is a consolidated understanding of vehicle ergonomics and user experience, the introduction of this knowledge into industrial activities should be further explored and partnerships with universities and the industry should be encouraged so that automakers can be prepared for this new paradigm where transportation becomes a service.

Keywords: Vehicle ergonomics · User experience · Materials

1 Introduction

It is notable the significant changes that the transportation in general has been suffering in recent years. The digital revolution that is becoming increasingly present in the vehicles, this emphasizes the importance of increasing the performance of the design areas within the automakers regarding ergonomics and user experience, which have become essential contents for the automakers to invest in the development of new automotive projects.

Bhise (2011) sustain the importance of developing more research and design tools for ergonomic vehicle evaluations to design future vehicles more efficiently due to technological advances and changes in drivers' expectations regarding the capabilities

S. Bagnara et al. (Eds.): IEA 2018, AISC 823, pp. 325–331, 2019.
https://doi.org/10.1007/978-3-319-96074-6_35

of current functionalities in vehicles, in addition to change the perception of quality will require automotive designers to expand their knowledge and create new concepts for the automotive area.

To design a user-centered vehicle, Norman (2008) affirm that it is essential to know who those users are, their needs, characteristics and behavior. In this context the selection of materials is an important step to please the customer and ensure a good design. claims about the importance of a user centric design professional, this author concern "Serving clients is a means of relieving frustration and confusion, the feeling of helplessness. Make them feel they are in control and empower them."

In this manner both areas vehicular ergonomics in partnership with the user experience should inform all the development team, problems related not only to the posture and reachability of users inside the vehicle, but also if the use of certain material, shape, color and texture, hinders readability or may cause annoyances that interferes with the drivability or interaction of the user in the vehicle.

This paper is the result of first impressions, from an initial research to the development of a work plan related to a study proposal for a future master's thesis. The notes were based in remarks, and professional experience within an automaker, specifically in product ergonomics area. In summary, the theoretical basis of this work was made from bibliographical and documental researches about the trends and developments practiced in the automotive industry, especially related on vehicle ergonomics, materials, market trends, and the experience primarily aimed at car users.

2 Development

2.1 Vehicle Ergonomics

According to Rozestraten (2006), one of the first contributions that the ergonomics field provided to the vehicle development in the twentieth century was the implementation of the closed body, thus creating the car or cabin. The car has been receiving several ergonomic improvements since the 1950s, such as the closed body attachment, improvements in driver's visual field, windows and mirrors, habitability, seat comfort, the addition of improved onboard technology the communication and understanding of the users regarding vehicle experience. Previously, drivers needed more information about the vehicle, especially those related to its functionalities and speed. In this way, the cabins correspond to every internal part of the vehicle intended mainly for the steering assembly and the accommodation of the passengers.

2.2 The Current Automotive Context

> *"The global automotive industry is experiencing the first steps of a transformation, in a dimension it has never experienced before, and it is being require to redesign its business in all spheres. New and more agile processes, new entrants in the supply chain, new consumer demands, new business models, new partnerships and new competitors are indicators that nothing will be as before"* KPMG (2016).

According to the PWC - Global Site for Practical Strategies for World Industry (2016), the automotive industry has seen a significant increase in the use of digital artifacts and new materials due mainly to the numerous technological advances of recent years, and it is foreseeing the continuous employment increase due this main development trends for the automotive sector; connected cars, shared cars, and autonomous vehicles.

The ABMD - Brazilian Association of Data Marketing (2015) states that by the year of 2017 about 23 million cars were connected to the Internet, and that this number is expected to increase to 152 million by the end of 2018. Several automakers are already aware and are investing in the use of technological devices with greater connectivity inside the vehicles, developing projects that use services related to the Internet of Things, which still conform to ABMD from mobile applications, new multimedia centers and even insurance based on the actual use of the vehicle.

All this technological device improves the interaction of the user inside the vehicle, allow a more efficient human machine interaction, assisting in its monitoring and decision-making process. In this way, interfaces such as displays, cluster, head-up displays, controls, ADAS and infotainment in general play an important role. All these systems and devices give the user new possibilities of relating to the car, helping it in the control and stability of the vehicle promoting a positive experience.

2.3 The Influences of Materials in the Cockpit

Another important point when talking about ensuring good experiences in the internal environment of the vehicle refers to the application of materials and finishes that interfere directly in the well-being of the driver and passengers. The materials present in the cockpit should guarantee good tactile sensations, ensure good visibility and anti-reflective and anti-glare qualities, especially in the components that must transmit messages to users such as displays, instrument panel, etc.

Analyzing and proposing the most stable and suitable materials with the characteristics mentioned above that should be used in the vehicle cabin are also part of the duties of a vehicle ergonomist. Parallel to the increased technology and its implications, the use of materials, finishes and textures can result in the reduction of distractions and stress of the driver of the vehicle. Emphasizing this importance, the author Bhise (2011) defines the analysis of quality craftsmanship and sensorial perceptions of the brand as one of the specialized technologies that contribute to the improvement of the characteristics of the vehicle.

In this way the evaluation of the selection and application of materials in the cockpit guarantees more harmony, better tactile and visual sensations of the controls, sharpness and smoothness felt during its maneuverability and operation. Another point raised Bhise (2011) refers to internal lighting that should ensure the readability of screens, ease of localization of functionalities and visual effects. All these questions allied with technology make the "being" in the vehicle much more pleasant and satisfying.

"The number of possible futures that can be introduced in future vehicles will increase fast with advances in technologies. The ergonomics engineers will need to consider many issues during designing the systems and deciding whether the benefits that can be claimed would be indeed realized and the costs and disadvantages with their introduction can be minimized." BHISE (2011)

The FAURENCIA (2016) related in your Press Kit for Paris Motor Show 2016 with the tittle "Transforming the Driving Experience" your concept about cockpit of the future relative about material application and technology. They believe that new technologies: making the cockpit more connected, adaptable and predictive.

"In the Intuition cockpit, Faurecia has developed innovation to convert wood, aluminum, plastic or so on conventional controls are replaced with touch sensitive, capacitive integrated switches into the decorative surface, such as in the Deco Control Alu control panel. When these switches are used to turn on air conditioning, for example, they provide a haptic and visual response through vibration and illumination indicating the activation of the heating or cooling system. These same smart surfaces can bring ambient light to the cockpit."

Reiterating what was previously said, some measures related to color and textures application were observed, such as overflush spare parts with finishes or large details, clear colors or reflective finishes, or of high gloss like chrome inside the car cabin that can reflect the sunlight coming in through the windows and windshield. The highlight of pieces with this type of finish, can interfere with the driver's drivability and readability, especially on the devices that uses it in critical areas such as instrument panel, navigator, rear-view mirrors or even in the windshield.

In this way, ergonomic analysis and studies are extensive, evaluate and inform the project team, all problems that in a certain way interferes with comfort and handling, which also covers the use of certain technology, material or finishing, which can cause reflections such as areas of intense brightness that may obscure the driver's visibility, whether the driver must properly visualize the road ahead of him or her, the rear-view mirrors, dashboard and any other navigation components.

2.4 The New Demands and Possibilities

The concept of future vehicles is closely connected with future mobility. we experienced one of the greatest moments of rupture in the traditional automotive industry, we are seeing arise this kind of possibilities: autonomous cars, shared cars and the reformulated electric cars. Habitability, of cockpit is a good premise for conquering consumers, but other factors have also become so important as to win over future consumers: price and friendly interaction and consumption.

To illustrate this statement, Carter for IQ (2018) says that it is the consumer who drives the changes in the automotive sector, not the companies. According to him, consumers perceive new technologies as added value for their lives and if they still aggregate low they become the ones chosen, even if these benefits come from companies that are on the verge of leaving the market.

"The benefit to the consumer is what drives demand, and those in their way will usually fall. Take Uber, a company that despite riding roughshod over incumbent taxi providers and city laws, provided a cheaper, cleaner, more efficient service to their customers with less friction points. Customers demanded Uber and eventually most cities rewrote bylaws to include them. Industry and technology change is thus very difficult to block if fueled by customer demand." CARTER for IQ (2018)

2.5 The Future Vehicle Experiences

The technical journalist Pahsey for IQ Automotive (2018) reports that one of the major challenges for the automotive industry and its stakeholders in this new scenario is the integration of technology and the development of human-machine interface capability, once autonomous vehicles technology will gradually assume the task of driving the car, which makes the user less responsible for controlling the car and more willing to perform other activities made viable inside the vehicle by virtue of the connectivity and advanced computer program.

The technical journalist also stresses the importance of automotive displays, which have made great strides in a short period of time, the development of these digital artifacts will be focused on the development and improvement of the Human Machine Interface (HMI), prioritizing to provide pertinent information to the driver in an agile and easy format, so that reaction time is faster and reduced distraction, but also make the environment of the vehicle conducive to the transition between the modes of manual and autonomous driving. All these advances should expand further in the next five years as the entire automotive industry moves toward autonomous driving.

"Looking further into the future, standalone vehicles will free the driver to focus on other tasks through displays in the vehicle. This will free designers to produce increasingly innovative and futuristic concepts that go beyond the cockpit and absorb the entire interior of the car." PAHSEY for IQ Automotive (2018)

In addition, Barat et al. (2017) stresses that it is still necessary a lot of developments in Human Machine Interaction systems like displays, Head-up displays and controls.

"These needs are really pulled by consumer electronics market and the smart, connected and autonomous vehicle. To fill the gap with consumer electronics market and to be able to have dedicated solutions for automotive market, the key points are affordable and limited-size systems, compliance with automotive standards (long-term reliability), compatibility with driving tasks, false detection and misuses management, multimodal interactions and the proposition of relevant automotive use-cases." Barat et al. (2017)

Therefore, according to these authors if all these items called by him from key points are taken into consideration in parallel with a design methodology oriented to a user experience, the future vehicles will propose more attractive and up-to-date interiors, useful, with more really needed resources, with a friendly interaction.

3 Final Considerations

The car is highly connected to its context, it in this way, understanding this context is important to analyze what experiences are important inside the future vehicles. We are at the beginning to experience a new way as we relate to vehicles as we have never seen before, all this new technologies and new possibilities related in this paper give other choices and other possibilities to relate with cars, as we are already experiencing, such as electric cars, autonomous cars and shared cars. In this way the industry should focus its efforts on producing mobility services, aiming to guarantee clear visual language, security, privacy to this user with new habits and needs.

Due this new digital landscape around cars, it is necessary to design cars and vehicle systems that allows much more than moving from one point to the other, providing positive and desirable experiences to drivers and passengers during the use of the vehicles.

Extending the application of materials in the automotive sector such as nanotechnology and biosensitive materials, as well as the increasingly digitalization and automation of the vehicle makes this product more and more hybrid and multitasking, going further to become autonomous and independent of human direction. All these transformations allow a vast field of studies for the development of new research on usability, materials and automotive consumer experience, this change requires automakers to seek new ways to design and implement digital interfaces across all vehicle segments.

Due to what was exposed in this paper, it is possible to verify that although there is a consolidated knowledge about vehicle ergonomics and user experience, it is necessary to expand the performance of these areas in the industrial activity of the traditional automotive industry.

The automaker that does not update itself and keep linked to old concepts, standards and procedures will not be able to remain in the future market. The change must begin with the process development, vehicle design teams involved with the automotive process should be multidisciplinary, and the focus should be on the user experience.

It is necessary to foster partnerships with both technology companies and universities in the new projects, as these will generate new research and developments that will allow automotive projects more appropriate to the user of new automotive era, the autonomous era, with friendlier interactions where the focus is mobility and not to generate a new vehicle.

References

ABMD – Brazilian Associacion of Data Marketing (2015) Carros Conectados criam novo Ecossistema baseado na Internet das Coisas. http://abemd.org.br/noticias/carros-conectados-criam-novo-ecossistema-baseado-na-internet-das-coisas. Accessed 28 Mar 2018

Barat D, Fromion, A, Féron S, Le Guen L, Lainé V (2017) Automotive HMI: present uses and future needs. PSA Group, Engineering and Quality Division, Velizy-Villacoublay, France, Invited Paper per Society for Information Display, vol 48, no 1, Book 1: Session 26: Future of Automotive Displays and HMI, June 2017 https://doi.org/10.1002/sdtp.11633. Accessed 21 Mar 2018

Bhise VD (2011) Ergonomics in the automotive design process, 1st edn. Taylor&Francis Group, Boca Raton

Carter J (01 March 2018) for Automotive IQ (2018) Online community for the automotive professionals. https://www.automotive-iq.com/autonomous-drive/articles/who-stole-my-company-coming-disintermediation-automotive-new-mobility. Accessed 21 Mar 2018

FAURECIA Homepage (2016) The Presskit for Paris Motor Show 2016. http://www.faurecia.com/en/news/faurecia-paris-motor-show-exhibit-features-innovative-technologies-transform-driving-experience-20092016. Accessed 06 Apr 2018

KPMG, International Cooperative – (Audit, Tax e Advisory Services) (2016) Global Automotive Executive Survey. https://assets.kpmg.com/content/dam/kpmg/pdf/2016/01/gaes-2016.pdf. Accessed 07 Apr 2018

Norman DA (2008) Design Emocional: Por que adoramos (ou detestamos) os objetos do dia a dia, 1st edn. Rocco, Rio de Janeiro Tradução Ana Deiró

Pawsey C (20 March 2018) for Automotive IQ (2018) Online Community for the automotive professionals. https://www.automotive-iq.com/autonomous-drive/articles/displays-and-autonomous-driving-challenges-and-solutions-us. Accessed 21 Mar 2018

PWC – Global Site for Practical Strategies for World Industry (2016) Auto industry trends. https://www.strategyand.pwc.com/trends/2016-auto-industry-trends. Accessed 21 Mar 2018

Rozestraten RJA (2006) A Ergonomia Veicular do Século XX. Psicol pesqui trânsito **2**(1) http://scielo.bvs-psi.org.br/pdf/ppet/v2n1/v2n1a07.pdf. Accessed 25 Feb 2018

The Effects of Food Packaging on Driving Performance When Eating While Driving

Swantje Zschernack(✉) and Chloe Bennett

Rhodes University, Grahamstown 6140, South Africa
s.zschernack@ru.ac.za

Abstract. Background: Eating and drinking while driving are common forms of secondary tasks and are perceived as a lower risk by many drivers and received less attention in terms of safety of driving. **Aim**: Little is known about food packaging on the safety of driving; therefore, the aim of the study is to understand the effects of food packaging on driving performance when eating while driving. **Method**: The study compares two different types of packaging during driving, unpacking and driving and eating and driving in a simulator setting. A total of 12 participant ranging from 20–30 years Rhode University students were recruited. For each packaging condition participants were required to perform four activities (pure driving, unpacking of meal, eating of meal and a second session of pure driving) while driving and throughout tracking deviation and perceived control were measured. Each driving activity lasted for 3 min with 3 min breaks in between. **Results:** The results on the effects of food packaging on driving performance found that there were no statistical differences found between the packaging conditions ($p > 0.05$) for driving performance (tracking deviation), however there was statistical difference found between the packaging conditions for perceived control ($p < 0.01$). There were also statistical significances between the different activities ($p < 0.05$), highlighting unpacking of meal been the worst, followed by eating of meal and then the pure driving conditions. **Conclusion:** This study suggested that food packaging does not influence driving performance when eating while driving; meaning that driving is as dangerous with cardboard packaging as to paper wrapper packaging. the study suggests that people perceived driving performance differently.

Keywords: Distraction · Eating and driving · Packaging · Driving safety

1 Background

People on a regular basis perform in a wide variety of multitasking activities when they drive which causes them to divert their attention away from the primary task of driving [13]. This can potentially degrade driving performance and have serious consequences for road safety [13]. Attention refers to a mental focus, serious concentration (cognition process) for smooth process of information as well as information processing capacity of an individual [14]. Driving is primarily a complex, multitask activity that requires full attention at all times; any minor distraction or slip could lead to one causing

S. Bagnara et al. (Eds.): IEA 2018, AISC 823, pp. 332–342, 2019.
https://doi.org/10.1007/978-3-319-96074-6_36

accidents [14]. Distraction in driving has been a concern for road safety professionals and other stakeholders for many years [14].

Distracted driving occurs when a driver redirects his or her attention away from driving to another task or activity resulting in inattention [14]. Distracted driving disturbs the attention needed when driving safely thereby resulting in driving performance been compromised [14]. Distractions also make drivers less able to see potential problems and properly react to them [14]. Distracted driving affects reaction time, lane positions, crash risks, and gaze on the road as well as driving safely [12]. Driver distraction is considered one form of driver inattention and is also considered a contributing factor in over half of inattention crashes [12, 15]. Manual distraction has been shown to have negative effects on driving performance, often leading to decrease in driving speed, increase in collision, increases in lane position deviation as well as changes in glance behaviour [1].

1.1 Effects of Eating and Driving

Stutts et al. [13] highlighted that eating and drinking while driving causes similar proportions of road crashes, accidents and injuries as that to cell phone use for example 1.7% versus 1.5% respectively. Distracting behaviours such as eating and texting while driving have shown to negatively affect driving performance, as it has direct effects on motorists that engage in these behaviours behind the wheel and for the safety of other road users [4]. Since there are no statistics on eating while driving available in South Africa, data from the US are presented. According to the National Highway Traffic Safety Administration (NHTSA), car accidents due to distracted driving kill more than 9 people and also injure approximately 1,060 other individuals every day and about one out of every car accident that resulted in injuries was caused by at least one distracted driver (Riggs, Bailey, Gupton, & Cravens, 2013). People who eat and drive increase the odds of an accident by 80% and 65% of near miss accidents [6]. Manual transmission vehicles double the chances of driving crashes caused by food consumption. Fast foods such as burgers, chips, tacos, soup and many more ranked the worst foods to eat behind the wheel [6].

ExxonMobil found that over 70% of drivers admit to eating while driving behind the wheel and 83% drink beverages while driving [6]. Yet eating and drinking while driving is often perceived as a behaviour that is safe to when driving a car [4]. In a UK based study it was found that while eating and driving a driver's reaction time increased by 44% and drinking behind the wheel reaction time increased by 22% [11]. Drivers who eat or drink while driving are also 3.6 times more likely to be in an automobile crash in comparison to drivers who do not eat or drink while driving [8].

One of the main reasons why eating and driving might represent safety risks to the driver is because it prevents drivers from keeping both hands on the steering wheel, it takes drivers eyes away from the road, prevents drivers from noticing changes in the road conditions or important road signs and it can also slow down drivers reaction times [2]. Eating and driving represents three types of distractions, these include manual, cognitive and visual distractions [1]. Drivers must remove their hand from the steering wheel to manipulate another object (manual), engage in a task requiring a mental workload in addition to thinking about driving as well as glance away from the road and redirects his or her attention to eating [1]; for example, drivers must unwrap food packaging, hold the food with at least one hand, apply flavouring and complete other activities while using a vehicle which makes eating while driving a dangerous activity. Eating and driving has also shown to increase crashes during critical incidents such as pedestrian and vehicle collision [15].

Eating while driving often occurs more frequently when people are running late and with the number of fast food drive-thru in most dangerous driving areas such as highways, the task of driving has been normalized by our society [5]. Multinational fast food companies' supply food that could be easily eaten while driving and in their advertising made people believe that it is all right to eat in our cars while driving [5]. The fast food drive-thru has also made it easier and accessible and encourages people especially drivers to eat most fast food while steering the wheel of a car which could all impact driving performance [5].

1.2 Food Packaging and Driving Performance

Food packaging serves many important purposes, to protect food products from outside influences and damage as well as to carry and contain the food. Some packaging is also needed for safe and efficient transportation [9]. Food packaging plays a vital role in ease of access and handling of food [9]. Types of packaging material include, paper, paperboard/cardboard, glass, aluminium, plastic and many more. The most commonly used food packaging for fast food industries with drive-thru are paper and paperboard/cardboard packaging [9]. There is no literature on the influence food packaging has on driving performance.

2 Methods

2.1 Experimental Design

This research project aims to understand the effects of food packaging on driving performance when eating while driving. This study compares two different types of packaging during driving, unpacking and driving and eating and driving in a simulator setting. It follows a repeated measure design, in which each participant performs all conditions.

There are two independent variables, namely type of packaging (cardboard and paper wrapper) and the activities with the packaging and meal consisting of individual food items such as a burger, chips and canned beverages (see Fig. 1). The two independent conditions namely the cardboard and paper wrapper packaging were compared. For each condition (cardboard and paper wrapper packaging) participants were required to perform four activities (pure driving 1, unpacking of the meal, eating of the meal and pure driving 2) while throughout driving dependent variables were measured. The pure driving 1 and 2 activities require participants to drive normally following the white line as accurately as possible presented on a screen in front of them. The second activity requires participants to unpack the meal, taking the burger and fried chips out of either the cardboard or paper wrapper packaging as well as open the cold beverage

Fig. 1. Fast food meal and packaging (cardboard packaging on the left and paper wrapper packaging on the right).

and putting the straw into it while driving. The third activity requires participants to consume the food while driving.

The order of the cardboard and paper wrapper conditions was randomised (half of the participants performed the test firstly with the cardboard packaging while the other of the participants began with the paper wrapper packaging), However the order of the activities was not, because eating before unpacking is not possible. Instead a second pure driving condition has been included to identify a possible learning effect that might mask the effect of packaging. Noteworthy, the study aims to look at the effects of the type of packaging on driving performance but does not compare the impact different types of foods have on driving performance.

In total twelve participants (4 males, 8 females) were recruited, their ages ranging from 20–30 years. Volunteering participants were recruited from students around Rhodes University and had to have a valid driver's license and a minimum of three years driving experience as anything less than 3 years driving experience can overestimate the results. Participants had to have good eye sight and no food allergies. Participants with visual impairment were required to wear contact lenses.

2.2 Dependent Variables

Tracking Deviation

Tracking deviation considers the driving ability of the participants'. This continuous tracking task reflected behavioural aspects of motorists (attention, and performance allocation) [7]. It highlighted the consistency or inconsistency to remain on the middle white line, with the tip of the yellow triangle (the front of the vehicle) as accurately as possible. Any distraction will result in a swerve away from the line demonstrating less vehicle control. The average deviation is calculated from the target line in meters and the accuracy of driving on the target, a percentage will be given. A low fidelity driving simulator was used to measure driving performance variable such as tracking deviation. A low-fidelity driving simulator was used to provide a controllable, repeatable environment which allows individuals to explore these potentially risky conditions in an ethical and safe context [14]. Previous studies have shown that a low fidelity simulator is suitable in measuring tracking deviation/lane keeping performance [15]. The driving simulator was presented by a curved road with a yellow triangle at the bottom of the screen indicating the vehicle. The speed on the simulator remained constant at 4 km/hr throughout the testing sessions.

Perceived Control
The perceived control is a subjective measure used to measure driver's estimation of perceived control as well as driving skills. A five point Likert scale was used with scores ranging from one being poor or worst driving ability to five being excellent or best driving ability.

2.3 Experimental Setup

The low fidelity driving simulator vehicle consisted of a light motor vehicle with a non-force feedback steering wheel on a metal dash board and a car seat which could be adjusted according to the anthropometry of the participant. Each participant sat behind the wheel of the simulator vehicle facing the white projector screen (1350 mm × 1030 mm) placed in front of them where the road scene of the simulator was projected. A tracking task using a simplified road scene was shown on the white projector. It required participants to use the steering wheel to direct the static yellow arrow representing the bonnet of a vehicle on the moving white line as accurately as possible (Fig. 2).

Fig. 2. The simulator software representing the visual scene that is projected on to the white projector screen (Color figure online)

2.4 Procedure

The experimentation was conducted in the laboratory in the Department of Human Kinetics and Ergonomics at Rhodes University. After obtaining ethical clearance from the Human Kinetics and Ergonomics Ethics Committee participants were recruited.

Participants were required to attend an introduction and habituation session in the HKE department. For the testing session Participants were required to attend two testing sessions and were tested individually. For each condition (cardboard and paper wrapper packaging) participants were required to perform four activities (pure driving1, unpacking of meal, eating of meal and pure driving 2) while driving. The pure driving 1 and 2 activities required participants to drive normally following the white line as accurately as possible. The second activity required participants to unpack the meal, taking the burger and fried chips out of either the cardboard or paper wrapper packaging as well as open the cold beverage and putting the straw into it while driving. The third activity required participants to consume the food while driving. Participants did not unpack and eat the meal against time. Each driving activity lasted for 3-min with 3-min breaks in between, in order to set up for the next condition and for the participants to complete the perceived control questionnaire (Fig. 3).

Fig. 3. Participants during the unpacking (left) and eating activity (right)

Statistical analysis of data collected was performed using Microsoft Office Excel (Windows 7, 2010, USA) and STATISTICA 12.0 software programme. Two-factorial measures analyses of variances (ANOVAs) using the factors activity (eating, unpacking, pure driving) and packaging (cardboard, paper) were conducted to assess the differences between the conditions. A confidence level of *($p < 0.05$)* was used to show significance. Sex was considered as co-variate, but as no differences between male and females were found the results are being presented for the entire sample

3 Results

3.1 Tracking Deviation

The descriptive statistics of the tracking deviation is illustrated in Fig. 4 indicating the mean and standard deviation values for the different activities for each condition.

For the purpose of the study the higher the mean value for tracking deviation the worse the participants' vehicle controls. Participants performed worst in both packaging conditions during the unpacking of the meal activity (the mean value for driving deprivation tripled in comparison to pure driving 1 activity), followed by eating of the meal activity which doubled in mean value in comparison to pure driving 1 activity and then following this was the pure driving activities having better driving performance (Fig. 4).

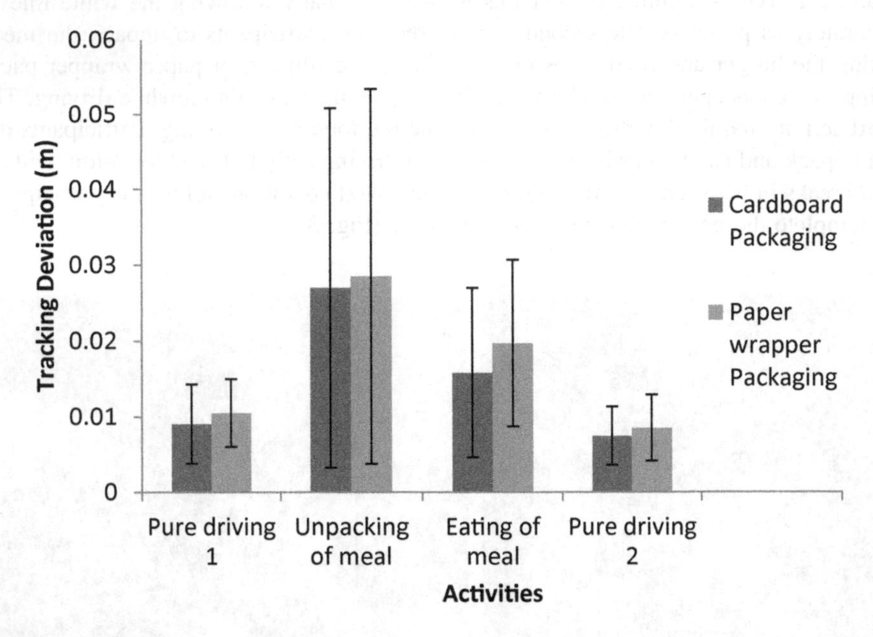

Fig. 4. Average tracking deviation for all activities for each condition (error bars indicate standard deviation, n = 12).

Two-factorial measures analyses of variances (ANOVAs) using the factors activity (eating, unpacking, pure driving) and packaging (cardboard, paper) showed statistical significances between the activities ($p < 0.05$). There was no effect of packaging nor an interactional effect.

3.2 Perceived Control

For the purpose of the study the higher the value (ranging from 1–5) the better the participant's estimation of perceived control and driving skill, whereas the lower the value the worst the participant estimation of perceived control and driving skill. Participants perceived to have the least vehicle control during the unpacking of meal activity, followed by eating of the meal activity and then the pure driving 1 and 2 activities for both packaging conditions (Fig. 5).

Two-factorial measures analyses of variances (ANOVAs) using the factors activity (eating, unpacking, pure driving) and packaging (cardboard, paper) revealed statistical significances between the conditions ($p < 0.01$). There was no effect of packaging nor an interactional effect between the conditions and activities ($p > 0.05$).

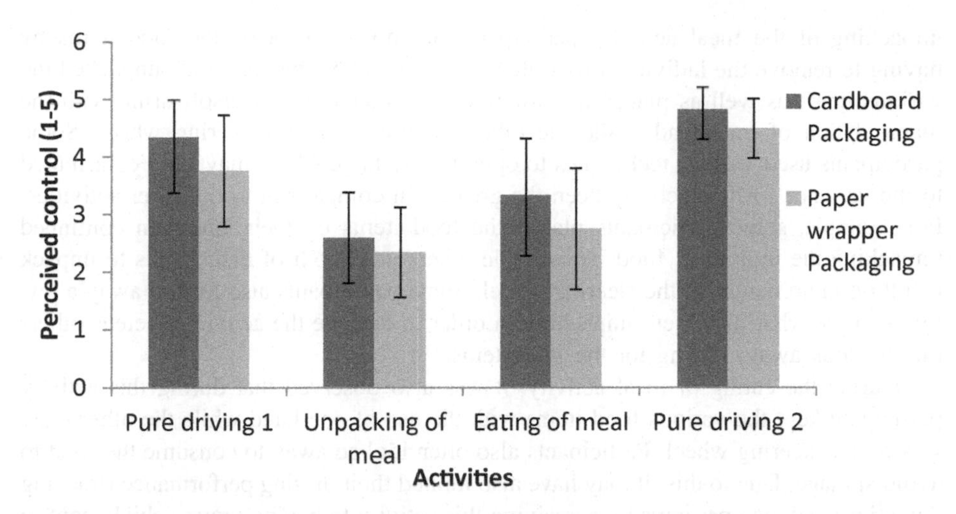

Fig. 5. Average score values and standard deviation (error bars) measured for all activities for each condition (rated between 1 = poor control and 5 = excellent control, n = 12).

4 Discussion

The task of eating and driving in a vehicle can result in visual and physical distraction thus contributing to increase risk of accidents [4]. With many individuals still eating and drinking behind the wheel and many fast food drive-thru's making it easier and accessible for people especially drivers to eat fast food while steering the wheel of a car, the task of driving has been normalized by our society and can contribute to reduced driving performance [5]. The main purpose of this study was to determine the effects of food packaging on driving performance when eating while driving and whether the packaging of food would make a difference in driving performance, thus improving driving safety.

The principle findings from this study suggest that food packaging does not influence driving performance. This means that driving is as dangerous with cardboard packaging as to paper wrapper packaging. Interestingly however, is that the study found that people perceived driving performance differently. The possible reasons for this are, participants perceived that the cardboard packaging was easier to handle while driving because they did not have to worry about the paper wrapper while driving. It was also perceived as easier to unpack while driving. However, participants perceived that they drove worst in paper wrapper condition but in fact they are not. This might have affected their confidence in driving, however in terms of driving performance (tracking deviation) there were no significant differences found between cardboard packaging and paper wrapper packaging.

Unpacking was associated with reduced simulated driving performance (tracking deviation) because the activity introduced multiple distractions that may have limited the driving performance and thereby participants also perceiving this activity to having worst vehicle control and driving ability for example it was denoted as fair. During the unpacking of the meal activity, participants attempted to locate the food items by having to remove the individual food items from the paper bag and then unpacked the various items as well as placed a straw into the canned beverage/soft drink with the manipulation of one hand while the other hand was on the steering wheel. Some participants used various techniques to open the package which may have contributed to the variance for this activity been the greatest in comparison to the other activities. For example, some participants placed the food items on their laps then continued unpacking the individual food items while others used both of their hands to unpack resulting in no hands on the steering wheel. Some participants also looked away a few times to see what they were unpacking in order to execute the activity whereas others did not look away, feeling for the food items.

During the eating of meal activity, it was also observed that during this activity participant kept the various food items with the use of one hand while the other hand was on the steering wheel. Participants also often looked away to consume the food to avoid spillage. Due to this, it may have also limited their driving performance (tracking deviation) and also participants perceiving this activity to having worst vehicle control and driving ability. Previous research has highlighted that eating of the meal contributes to incidences of inattention in lane change [3, 13]. Eating behind a wheel leads to increases in collision as well as lane position deviation [1]. A similar study also highlighted that distracted driving behaviour such as eating while driving appears to negatively impact driving measures of lane position control and reaction times [4]. The study also showed that those who were drinking a beverage behind the wheel were 18 percent more likely to have poor lane control. For example, the act of sipping on a soda beverage behind a wheel, participants found it difficult to remain in the center of the lane. This can be hazardous if a vehicle is attempting to pass you and you accidentally swerve into their lane [4]. These finding were consistent with the findings of the current study. Interestingly however in terms of subjective measure, previous research indicated that drivers perceived that the task of eating behind the wheel to be a more acceptable behaviour and are therefore less likely to identify the associated risks or modify eating behaviours behind the wheel [[12, 13, 16]). Therefore, current findings do not support previous works. Given the findings of the current study greater awareness on the impact of unpacking and eating while driving and driving safety are required.

The significant differences found between pure driving 1 and pure driving 2 activities for perceived control could be explained by the fact that participants felt more confident afterwards highlighting a possible learning effect considering that the unpacking and eating when they are already learning it increased the assuredly of these differences. However, this was not the case, as for performance while driving, pure driving 1 was not significantly different to pure driving 2 meaning that there was no possible learning effect that masked the effect of packaging. In terms of the effect of interactions, no differences were found for the interactional effect between the conditions and activities, meaning that the packaging type did not affect any possible changes between the activities. Therefore, driving performance was not made safer by putting the individual food items in different packaging types.

The study acknowledges that there were limitations in terms of equipment and protocol used in the study. In terms of equipment use, with the use of a low-fidelity simulator it may evoke unrealistic driving behaviour of drivers that eat and drink as it imitates an automated vehicle and does not imitate a manual vehicle making it easier for participants to complete. This may underestimate the behaviour of eating and drinking while driving on driving performance variables in a real scenario. The simulator software also does not represent road features such as traffic signs, road signs, billboards, stop streets, pedestrians and does not require participants to use both the brake and accelerator to stop and start a vehicle that take place in the real life setting

5 Conclusion

This study suggested that food packaging does not seem to influence driving performance when eating while driving, meaning that driving is as dangerous with cardboard packaging as to paper wrapper packaging. Interestingly however, is that the study found that people perceived driving performance differently. This might have affected their confidence in driving, but in terms of their driving performance which is tracking deviation it did not differ.

The unpacking of the meal activity highlighted the worst performance this may be explained by the fact that participants used various techniques to open the package which may have contributed to the variance for this activity been the greatest in comparison to the other activities. Following this was eating of the meal. These behaviours lead drivers neglecting to put both hands on the steering wheel which could have contributed to increased exposure and risk of accidents as well as participants perceiving these activities to having worst vehicle control and driving ability.

References

1. Alosco ML, Spitznagel MB, Fischer KH, Miller LA, Pillai V, Hughes J, Gunstad J (2012) Both texting and eating are associated with impaired simulated driving performance. Traffic Inj Prev 13(5):468–475
2. Beissmann T (2012) Eating, drinking while driving slows reaction times. http://www.caradvice.com.au/171581/eating-drinking-while-driving-slows-reaction-times-study. Accessed 4 Apr 2016
3. Campbell BN, Smith JD, Najm WG (2002) Examination of Crash Contributing Factors Using National Crash Databases. (Report No. DOT-VNTSC-NHTSA-02-07). Volpe National Transportation Systems Center, Cambridge, MA
4. Irwin C, Monement S, Desbrow B (2015) The influence of drinking, texting, and eating on simulated driving performance. Traffic Inj Prev 16(2):116–123
5. Jakle JA, Sculle KA (1999) Fast food: roadside restaurants in the automobile age. The John Hopkins University Press, Baltimore
6. Locher J, Moritz O (2009) Eating while driving causes 80% of all car accidents, study show. http://www.nydailynews.com/new-york/eating-driving-80-car-accidents-study-shows-article-1.427796. Accessed 4 Mar 2016
7. Louw T (2013) An investigation into control mechanisms of driving performance resource depletion and effort regulation. Unpublished Master of Science thesis, Rhodes University, Grahamstown, South Africa

8. Lytx (2014). Lytx Data Finds Three Dangerous Activities You May Be Doing While Driving Every Day. http://www.lytx.com/press-releases/lytx-data-finds-three-dangerous-activities-you-may-be-doing-while-driving-every-day. Accessed 4 Apr 2016
9. Marsh K, Bugusu B (2007) Food packaging—roles, materials, and environmental issues. J Food Sci 72(3):39–55
10. Siu J (2013) Eating while driving deadlier than texting behind the wheel, study shows. http://www.autoguide.com/auto-news/2012/05/eating-while-driving-deadlier-than-texting-behind-thewheel-study-shows.html. Accessed 4 Mar 2016
11. Stutts JC, Reinfurt DW, Staplin L, Rodgman EA (2001) The role of driver distraction in traffic crashes. AAA Foundation for Traffic Safety, Washington, DC
12. Stutts J, Feaganes J, Reinfurt D, Rodgman E, Hamlett C, Gish K, Staplin L (2005) Driver's exposure to distractions in their natural driving environment. Accid Anal Prev 37(6): 1093–1101
13. Tlhoaele K (2013) Understanding the multitask process of eating while driving and its potential effect on driving performance. Unpublished Honours thesis Rhodes University, Grahamstown, South Africa
14. Young K, Lee JD, Regan MA (2008) Defining driver distraction. In: Young K, Lee JD, Regan MA (eds.) Driver distraction: theory, effects, and mitigation, pp. 31–40. CRC Press, Florida, USA
15. Young MS, Mahfoud JM, Walker GH, Jenkins DP, Stanton NA (2008) Crash dieting: the effects of eating and drinking on driving performance. Accid Anal Prev 40(1):142–148
16. White MP, Eiser JR, Harris PR (2004) Risk perceptions of mobile phone use while driving. Risk Anal 24(2):323–334

Anthropomorphism: An Investigation of Its Effect on Trust in Human-Machine Interfaces for Highly Automated Vehicles

Erik Aremyr, Martin Jönsson, and Helena Strömberg(✉)

Chalmers University of Technology, 41296 Gothenburg, Sweden
helena.stromberg@chalmers.se

Abstract. Trust has been identified as a major factor in relation to user acceptance of Highly Automated Vehicles (HAV). A positive correlation has been suggested between increased trust and the use of anthropomorphic features in interfaces. However, more research is necessary to establish whether this is true in an HAV context. Thus, the aim of this study was to investigate how trust in HAVs is influenced by HMI design with different degrees of anthropomorphism: baseline, caricature, and human. Ten subjects participated in an in-vehicle trial to test the designs. The results showed no significant difference in levels of trust between conditions. Instead, it was found that anthropomorphism may affect user acceptance indirectly through its effect on perceived ease of use and usefulness. The findings imply that designers must be cautious when using anthropomorphism and consider adaptability and customisability to incorporate new and diverse user needs associated with the use of HAV.

Keywords: Highly Automated Vehicles · Human – Machine Interaction
Anthropomorphism

1 Introduction

The development of Highly Automated Vehicles (HAV) is rapidly progressing and within a few years, several of the major car manufacturers are expected to release their first generation. Automation in vehicles is predicted to improve road safety, decrease the need for parking space and add comfort to the driver. However, to be able to reap these benefits, the HAV technology will need to be accepted and utilised by users. According to the Automation Acceptance Model, AAM [1], user acceptance of automation depends on trust and task-technology compatibility, in addition to perceived usefulness, perceived ease of use and external variables. Trust has also been identified empirically as an important precursor to HAV acceptance [2, 3].

Trust in automation in general has been seen to be affected by the anthropomorphism of the automated system [4–6]. Anthropomorphism has also been suggested as a design feature for increasing trust in HAVs in particular [7, 8]. Waytz and colleagues [8] describe the *act of anthropomorphising* a nonhuman as not simply being the act of attributing superficial characteristics such as a face to an agent, but rather attributing essential human characteristics such as a humanlike mind to it. It is by identifying these

S. Bagnara et al. (Eds.): IEA 2018, AISC 823, pp. 343–352, 2019.
https://doi.org/10.1007/978-3-319-96074-6_37

characteristics that one can, without producing a replica of a human being, design an interface that is perceived as humanlike. In their study, Waytz et al. [8] conclude that anthropomorphic features such as name, gender, and voice, used in vehicle-driver communication, correlated positively with increased trust in the vehicle's ability to perform competently. They suggested that by blurring the line further between human and nonhuman by utilising anthropomorphism in interfaces, the driver's willingness to trust technology in place of humans would increase. Further studies using humanoid robot as co-drivers to represent anthropomorphic conditions [9, 10] also report increased trust in comparison to control conditions.

However, studies concerning the link between anthropomorphism and trust in HAV have so far been mainly simulator studies [8, 9, 11] and the transferability of results to real-life conditions on the road is not certain. Additionally, indications point to more variables that interact to create trust in HAV. For instance, perceived competence of the system was indicated to be more important than anthropomorphism [11] and the interaction of anthropomorphic features together with system personality and transparency create varying trustworthiness [7]. So, although research from different areas seem suggest a positive correlation between anthropomorphism and trust, further research is needed to understand the link in an HAV context. Thus, a study was conducted with the aim of investigating how trust in HAV is influenced by interface designs based on different degrees of anthropomorphism in a vehicle moving on the road. Based on the previous research, it was predicted that an increased degree of anthropomorphism would lead to an increased level of trust in the vehicle. The objective was to make suggestions on how to design interfaces that support an appropriate level of trust.

2 Method

In order to investigate the connection between trust and anthropomorphism in interface design, three different designs with varying degree of anthropomorphism were designed and tested in a within-subjects test on a track using a Wizard-of-Oz setup.

2.1 Conditions

Three tested interface designs were constructed based on three desired points on a scale of anthropomorphism from machine-like to human-like. The intention was to create two conditions on the opposite ends of the anthropomorphic spectrum with one condition somewhere in between. In doing so, the least anthropomorphic condition would most closely resemble current automotive interfaces and act as a baseline. The three points were adapted from Zhang and colleagues [12] and described as: Non-anthropomorphic agents; Agents designed with certain anthropomorphic features as well as intentional deviations from the human "prototype" and; Agents highly resembling the human "prototype". These three points were translated into three interface designs (Fig. 1), named Baseline, Caricature, and Human respectively.

The interfaces were carefully crafted to differ in degree of anthropomorphism but communicate the same information. The information communicated was the vehicle's awareness of its surroundings, the vehicle's intended actions and the reason behind them. To provide this information, the designs all consist of two halves:

- Left half (constant across conditions), using "the Ring Concept" [13] to communicate awareness and symbols for intended actions. When an object is in the vehicle's vicinity, parts of the grey ring change colour to yellow or red depending on distance to the object. This part was included to give participants the same understanding of the vehicle's intentions to the same extent between conditions.
- Right half (three different degrees of anthropomorphism). This half gave the reasons for the vehicle's actions (e.g. "avoiding obstacle on the left-hand side"). Baseline uses text and icons to display messages and a two-tone earcon for sound. Every time a message is displayed, it is preceded by the earcon. Caricature consists of icons and an animated caricatured representation of a human face named "Kim", as well as a two-tone earcon and computer-generated voice that reads messages. Human consists of icons and a video recording of a human named "Isak" speaking the messages, as well as the two-tone earcon.

Fig. 1. The three conditions used in the study. The ring concept is shown to the left and the conditions are shown to the right. From top to bottom: baseline, caricature, and human.

Common for all conditions is that they make use of both visual and auditory displays. This is because the Human and Caricature conditions both need a visual representation to achieve a higher degree of anthropomorphism. Anthropomorphism is not all about appearance but is rather a result of a combination of humanlike attributes. A face helps to increase the perception of the virtual agent having an identity. Furthermore, by using both modalities in all conditions the differences between the conditions not related to anthropomorphism were minimised.

2.2 Participants

Test participants were recruited using the University's lists of voluntary user study participants. Eleven participants were selected after screening for age and gender, corresponding to a future population of drivers of HAV. Participants one to ten answered all questions needed for the quantitative analysis. Of the eleven participants, six were male and five were female, all between 25–68 years old (mean = 47), with driving experience ranging from less than two years to over 20 years.

2.3 Design and Procedure

The test was conducted with a within-subjects design, as this allowed collection of as many observations as possible from a relatively small number of participants, while minimising the effect of individual differences among the participants. During the test, participants were exposed all three conditions in a randomised order.

The user test was carried out on a test course (approx. 400 m) inside a fenced-off area to which no other traffic had access. A scenario with safety critical events was created be able to measure trust. The situations were avoiding collision with (1) a moving obstacle in the form of a pedestrian crossing the road unexpectedly and (2) a static obstacle in the form of a traffic cone in the way of the vehicle's projected path. These events were chosen as they were deemed critical enough to affect a driver's trust formation, as implementable in the test setting.

A right-hand driven Volvo V40 was used to represent an SAE level 5 autonomous vehicle in a Wizard-of-Oz setup. A real vehicle was chosen over a simulator as a real vehicle is more likely to invoke a feeling of the participants being in a safety-critical situation - possibly affecting trust - and thereby increasing the study's ecological validity. The interface was displayed on a 9.7-in. tablet mounted on the dashboard on the passenger's side (see Fig. 2), and audio from a speaker placed between the driver and the passenger. To create the illusion of autonomy in the vehicle, a blind was mounted between driver and passenger, so that they could not visually confirm that a person was driving the vehicle. A test leader was placed in the backseat and controlled the interface animations and initiated the auditory displays, using a laptop connected to the participant's display. Participants also had a green button mounted next to the tablet, which they could press to start the test.

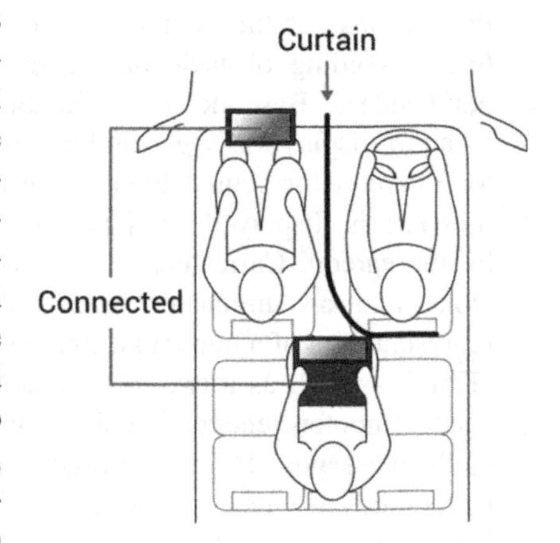

Fig. 2. Wizard-of-Oz setup.

Procedure. Participants first received an introduction to the study, including a consent form, information about the possibility to end the test at any time by pressing the green button, and explanation of the interface structure. They were also told that the blind was used so that the person sitting on the other side of the blind would not disturb them during the test. Having completed the introduction, the participants were taken on an initial warm-up lap, in which no condition was applied, to accustom them to the interface structure and the vehicle's driving style. Then, during the actual test, the vehicle was driven three laps with one condition active during each lap. Each lap consisted of two encounter locations. In order to minimise the effects of the participants predicting the vehicle's actions between the laps, the vehicle avoided the traffic cone on different sides depending on the active condition. In addition, the pedestrian shifted encounter location between each lap.

Data Collection. Video was used to record the complete test using three cameras, focusing on the road ahead, on the test participant's face, and on the test participant, screen, and button from behind. This setup was used in order to record the test participants' physical reactions to events on the road ahead as well as their answers to the interview questions. After experiencing each condition, the participants completed questionnaire (all on 1-7 Likert scales), that assessed:

- *Trust.* Trust was measured using five of the seven items from [14], chosen based on their relevance to the study. The test participants also answered brief interview questions regarding positive and negative opinions of each condition.
- *Perceived Ease of Use* was measured with three items chosen to assess how well participants understood the information given to them through the interface and if they found the interface unnecessarily complex. Items came from [14] and SUS [15].
- *Anthropomorphism* was measured as a manipulation check, using four items from the Godspeed questionnaire series (Godspeed I) [16], to assess whether the conditions were perceived to convey different degrees of anthropomorphism.

After completion of the whole test, a short semi-structured interview was conducted. During the interview, the participants were asked to rank the conditions according to how much they liked them and trusted them with the help of pictures of each condition as mediating objects. The interview also collected participants' opinions of the features of the different conditions, and about how autonomous the test participant thought that the car was. At the end of the test, the participants were rewarded with a cinema ticket for their contribution.

2.4 Analysis

The correlation between the degree of anthropomorphism and the level of trust in the system was tested using a non-parametric statistical test, a Wilcoxon Signed-Rank Test [17] with an applied continuity correction, which was performed pairwise on all concepts ($\alpha = .05$). The rest of the questionnaire data was summarised and plotted. The audio recordings of the interviews were transcribed and a thematic analysis was conducted, creating an affinity diagram. For each of the themes found in the affinity diagram, an analysis of the implications for HMI design was conducted.

3 Findings

3.1 Effects on Trust of the Anthropomorphic Features

The integrated results of five items measuring trust show that the conditions did not differ notably. On a 7-point Likert scale, Baseline (mean = 5.12, s.d. = 1.51) scored highest, followed by Human (mean = 5.06 s.d. = 1.19), and Caricature (mean = 4.66 s.d. = 1.39). Statistically, no significant differences between any of the concepts could be found ($\alpha = .05$). However, when asked to rank the conditions in terms of which they trusted the most Human and Baseline were ranked first by five persons each, while Caricature only by one (see Fig. 4), indicating that the participants to some degree judged the conditions differently in terms of trust.

3.2 Other Factors Building Trust in the Vehicle

From the interviews it was clear that participants had assessed their trust in the vehicle based on other factors than the anthropomorphic features. The main factor was how the vehicle behaved, for instance when turning and braking, which was consistent across conditions. A majority of the participants indicated that they felt completely safe for the entire duration of test, largely based on the driving behaviour of the "vehicle" and the way it managed to handle the events in the scenario. The content delivered through the interface rather than the appearance of it was also a frequently mentioned factor leading to trust. For instance, when asked about whether or not he experienced differences in trust depending on the three conditions, participant 10 said that: *"No, I don't think so. I think more about this [the Ring Concept], where I can see what it [the vehicle] sees"*. The score for how well participants understood the system was equally high for all conditions (see Fig. 3). That the participants understood the system through the content provided by the interface appears to have affected their rating of trust more than the representation through anthropomorphic features.

3.3 The Experience of the Anthropomorphic Features

Human (0.73) scored higher than the Caricature condition (0.32) on the anthropomorphism scale, which shows that the two conditions successfully represented different degrees of anthropomorphism with the intended relation (Fig. 3). However, Baseline (0, 41) scored higher than the Caricature condition, despite the fact that Baseline contained no anthropomorphic features. Interview data suggests that Caricature had anthropomorphic attributes that were (too) far from the human "prototype" to trigger the perception of anthropomorphism. In addition, Caricature was the least liked condition, and least trusted, because of its annoying voice, "toy-like appearance" and that participants found this type of representation to hold little value for the user experience. For instance, participant two said "*... [the eyes] are an unnecessary detail that draws your attention. It is hard not to look at them*". Eyes and the sensation of being stared at was frequently mentioned in the interviews (10 out of 11), especially in response to the Human condition. For instance, participant ten said "*...it was a little unpleasant when someone was staring at you the whole time. I know that no one is actually staring at* *me, but you are still programmed by evolution to react to a human gaze"*. There was a strong agreement that the staring sensation was both distracting and unpleasant, and thus led to a negative experience of the HAV.

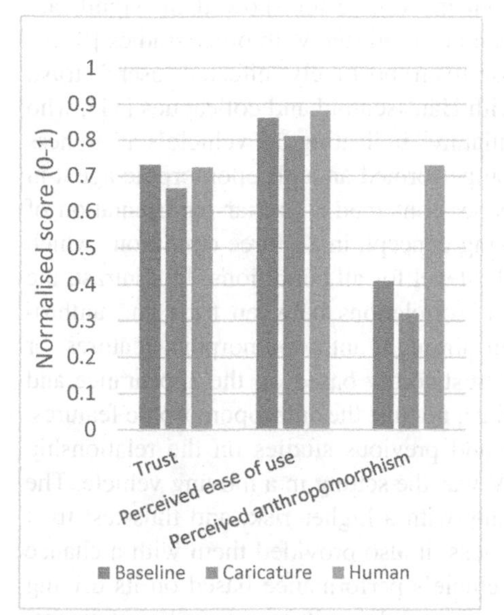

Fig. 3. Normalized score for the items measuring trust, perceived ease of use, and perceived level of anthropomorphism.

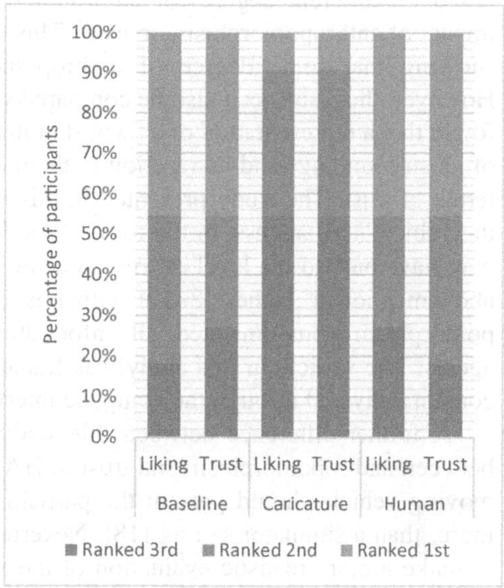

Fig. 4. Ranking of the conditions against each other, in terms of how much the participants liked them, and how much they trusted them.

The visual representation as an agent was not perceived as entirely negative however, and five participants commented on the benefits of having a focal point. The visual representation of an agent drew their attention, and the speech messages delivered by the agent helped them focus on the message. Participant nine highlighted the difference between Human and Baseline: *"If there is a human, you can let go of everything else and one can focus on it. But with text [Baseline], one's thoughts may wander off and that is not the point"*. A highly anthropomorphic agent represented visually and through audio may thus create an intense focal point useful for capturing the driver's attention and delivering information. However, participants (8 out of 11) also discussed the suitability of the design solutions in relation to their circumstances and situational need for information - *"what I like here [Baseline] is that the information is available, but it is not forced upon me. If I consider the car to be in control, I do not need it to constantly tell me what it is doing"* (participant one). Suggested adaptations included adaptations to the state of the external environment, including the need for driver intervention due to driving related activities, but also adaptation to the state of the cabin, including their awareness level and the overall noise level. Thus, the use of anthropomorphic agents as a focal point may need to be adapted to the message being delivered.

4 Discussion

The aim of the study was to investigate how trust in HAV is influenced by interfaces based on different degrees of anthropomorphism. The result showed no significant impact of anthropomorphism on trust. This can be contrasted with other studies [8–10] showing that some degree of anthropomorphism positively affected users' trust. However, the results can also be compared with Häuslschmid and colleagues [11], who found that a representation of a "world in miniature" indicating the vehicle's awareness of its surroundings and its reaction to them outperformed an anthropomorphic agent in terms of trust. The study presented in this paper contained a similar representation of the vehicle's awareness, in the form of the Ring concept, in all three conditions which may have pushed the level of trust to a similar level for all conditions. In contrast, the abovementioned studies [8–10] with positive correlations between trust and anthropomorphism communicated all information through anthropomorphic features or agents. The vehicle in this study was found trustworthy based on the appearance and content delivered through the complete interface, not just the anthropomorphic features.

A further difference between this study and previous studies on the relationship between anthropomorphism and trust in HAV was the setting in a moving vehicle. The moving vehicle should present the participants with a higher risk, and thus test trust more, than a simulator setting [18]. Nevertheless, it also provided them with a chance to make a more realistic evaluation of the vehicle's performance based on its driving behaviour. Since this behaviour was calm and managed the situations well, participants put high levels of trust in the vehicle based on this aspect alone. However, it can also be discussed whether the test failed to invoke a feeling of being in a safety-critical situation since it was conducted in a closed off area with a relatively slow-moving vehicle. Together, these two aspects, the content and the vehicle behaviour, indicate that the vehicle may not need to be embodied as an agent in order for trust to grow.

Instead, the appearance of the interface, including the anthropomorphic features, may instead affect how useful and easy-to-use the interface is deemed. The anthropomorphic features used in the conditions were not perceived as central elements of the interaction, rather as ways of embodying the information. They were thus evaluated as separate features (gimmicks), seen to bring more or less value to the use experience, depending on their adaptation to the situation and personal preference of the participants. Caricature was for example perceived as unsuited to the safety-critical situation of driving, drawing unnecessary attention, as well as hard to use with its annoying voice, which affected the participants willingness to trust it. These findings emphasise the importance of the feedback loop in AAM [1]; going from the evaluation of ease-of-use and usefulness in actual system use to trust, over the direct impact of trust on perceived ease-of-use and usefulness in the formation of the intention to use. Anthropomorphic features may have a role to play in the design of interfaces that induce trust in HAV, but the usefulness of the features and usability they provide need to be carefully balanced in the design. Failing to consider the suitability of the interface communication style to the need for driver intervention and driver preferences may result in an HMI which is unsatisfactory to use, making it less likely that users will adopt the technology.

Further research appears necessary to establish the links between trust, user acceptance and HAV, especially taking the vehicle behaviour and complete information communication into consideration. It should also be noted that this study is small in comparison to e.g. Waytz and colleagues [8], who conducted a large-scale simulator study using a between-subjects design, while this study consisted of a small-scale test in a real vehicle using a within-subjects design. The small scale and sample size of this study means that a significant correlation between anthropomorphism and trust cannot be ruled out. However, the interview data and the additional factors found contributing to the formation of trust support the indication that anthropomorphic features lead to higher trust in HAV.

4.1 Conclusion and Implications

In conclusion, the study presented there failed to find a link between use of anthropomorphism and trust in HAV. While the sample size is too small to claim that there is no such link, the findings indicate that there several factors combining to influence trust and override the effects of anthropomorphism, including the presence of easy-to-understand information about the vehicle's awareness and actions, the performance and style of the vehicle's driving behaviour and how well the information given is adapted to the situation. The study thus indicates that the interrelation between the vehicle's driving behaviour and the interface's content and appearance should be studied further to be able to make assured claims about trust.

In terms of interaction design, an implication of the study is to *use anthropomorphism with caution,* and consider the quality, utility, and timing. If used properly, with the task at hand in mind, anthropomorphic features may act as focal point to direct driver attention and facilitate learning in the initial interaction with HAV. On the other hand, if used improperly, for instance by including staring eyes, anthropomorphic features may lead to driver distraction. Successful integration of anthropomorphic features into HMI in HAV requires a balance between attention and distraction created by a strong focal point and recognizing the interaction between the vehicle's total communication to the driver.

References

1. Ghazizadeh M, Lee JD, Boyle LN (2012) Extending the technology acceptance model to assess automation. Cogn Technol Work 14(1):39–49
2. Choi JK, Ji YG (2015) Investigating the importance of trust on adopting an autonomous vehicle. Int J Hum-Comput Int 31(10):692–702
3. Schoettle B, Sivak M (2014) A survey of public opinion about autonomous and self-driving vehicles in the US, the UK, and Australia. Report. https://deepblue.lib.umich.edu/handle/2027.42/108384
4. Green BD (2010) Applying human characteristics of trust to animated anthropomorphic software agents. State University of New York at Buffalo
5. Pak R, Fink N, Price M, Bass B, Sturre L (2012) Decision support aids with anthropomorphic characteristics influence trust and performance in younger and older adults. Ergonomics 55(9):1059–1072

6. de Visse EJ, Krueger F, McKnight P, Scheid A, Smith M, Chalk S, Parasuraman R (2012) The world is not enough: trust in cognitive agents. In: Proceedings of the Human Factors and Ergonomics Society Annual Meeting, vol 56, No 1, pp 263–267
7. Hoff KA, Bashir M (2015) Trust in automation integrating empirical evidence on factors that influence trust. Hum Factors 57(3):407–434
8. Waytz A, Heafner J, Epley N (2014) The mind in the machine: anthropomorphism increases trust in an autonomous vehicle. J Exp Soc Psychol 52(5):113–117
9. Kraus JM, Nothdurft F, Hock P, Scholz D, Minker W, Baumann M (2016) Human after all: effects of mere presence and social interaction of a humanoid robot as a co-driver in automated driving. In Adjunct Proceedings of AutomotiveUI 2016 Adjunct. https://doi.org/10.1145/3004323.3004338
10. Lee J-G, Kim KJ, Lee S, Shin D-H (2015) Can autonomous vehicles be safe and trustworthy? effects of appearance and autonomy of unmanned driving systems. Int J Hum-Comput Int 31(10):682–691
11. Häuslschmid R, von Bülow M, Pfleging B, Butz A (2017) Supporting trust in autonomous driving. In: Proceedings of the 22nd International Conference on Intelligent User Interfaces (IUI 2017). https://doi.org/10.1145/3025171.3025198
12. Zhang T, Zhu B, Lee L, Kaber D (2008) Service robot anthropomorphism and interface design for emotion in human-robot interaction. IEEE International Conference on Automation Science and Engineering, 23–26 Aug 2008, Paper presented at the 2008
13. Ekman F, Johansson M (2015) Creating appropriate trust for autonomous vehicles. M.Sc Thesis. Chalmers University of Technology
14. Helldin T, Falkman G, Riveiro M, Davidsson S (2013). Presenting system uncertainty in automotive UIs for supporting trust calibration in autonomous driving. In: Proceedings of AutoUI 2013. http://dx.doi.org/10.1145/2516540.251655
15. Brooke J (1996) SUS - A quick and dirty usability scale. Usability Eval Ind 189(194):4–7
16. Bartneck C, Kulić D, Croft E, Zoghbi S (2009) Measurement instruments for the anthropomorphism, animacy, likeability, perceived intelligence, and perceived safety of robots. Int J Soc Robot 1(1):71–81
17. Lowry R (1998) Concepts and Applications of Inferential Statistics. 1st edn. Subchapter 12a. The Wilcoxon Signed-Rank Test
18. Deniaud C, Honnet V, Jeanne B, Mestre D (2015) The concept of “presence” as a measure of ecological validity in driving simulators. J Interact Sci 3:1. https://doi.org/10.1186/s40166-015-0005-z7

In the Right Place at the Right Time? A View at Latency and Its Implications for Automotive Augmented Reality Head-Up Displays

Matthias Walter(✉), Tim Wendisch, and Klaus Bengler

Chair of Ergonomics, Technical University of Munich,
Boltzmannstr. 15, Garching, Germany
matthias.mw.walter@tum.de

Abstract. In the presented study, latency between physical event and displayed content is considered in order to provide a quantifiable criterion. In a driving simulator the influence of a contact-analog lane marker which was subjected to different stages of latency (17 ms, 50 ms, 100 ms) is examined. In total 43 participants took part in this experiment. A detection response task was conducted to evaluate the subjects' reaction times and cognitive workload [1]. Usability was assessed by applying the system usability scale [2]. Changes in latency have a significant influence on stress and usability. Specifically latencies over 50 ms have a negative effect on the dependent variables. Results suggest that latencies of up to 50 ms are still considered acceptable in terms of usability as evaluated in the implemented use case.

Keywords: Head-up display · Augmented reality · Contact analog Latency · Driving simulator

1 Introduction

The driving task has changed over time by an increase of assistance and information systems in the vehicle. While these systems are designed to enhance safety and comfort, the user is still tasked with supervising them [3]. Onc tcchnical solution to control the complexity of visual information is the Head-up Display. Information can be displayed in the driver's primary field of view by projecting a virtual image over the bonnet of the car. This way the driver can monitor relevant information without head movement and with minimal focal accommodation or eye movement. Consequently distraction and reaction times to unexpected traffic events are reduced [4–7].

A possible next step in the development of Head-up Displays is to integrate Augmented Reality (AR). AR is the superimposition of the real world with virtual objects [8]. Similar to the conventional HUD, information in an ARHUD is displayed in the driver's primary field of vision, yet it is directly superimposed to or linked with real objects. As already shown AR efficiently guides the driver's attention and improves the detection of objects [9, 10]. Utilizing the spatial link between virtual information and physical world can help to reduce cognitive workload and reaction times in critical traffic events and increase the understanding of advanced driver assistance systems.

S. Bagnara et al. (Eds.): IEA 2018, AISC 823, pp. 353–358, 2019.
https://doi.org/10.1007/978-3-319-96074-6_38

The ARHUD technology has the potential to improve how information is presented in the automotive environment and to improve road safety, yet no marketable solution has been presented so far. Numerous technical challenges must be overcome in order to provide an acceptable user experience. Conventional Head-up Displays sold in today's production cars fit in the constricted room between dashboard and steering column. ARHUDs on the other hand provide an increased field of view and virtual image distance and therefore need to incorporate more and larger mirrors or lenses. Hence, packaging poses a more serious challenge for ARHUDs which occupy 4 to 10 times the volume. Secondly, sensors, e.g. GPS, and data, e.g. maps, in cars cannot provide the accuracy augmented reality applications depend on. Pfannmüller, Walter and Bengler [12] showed that deviations of 6 m from the optimal position reduced usability significantly while driving. But the geometrical distance between a virtual and a real object cannot be determined in the field yet.

In the current driving simulator study we therefore investigated the impact of three different levels of latency within an augmented reality lane departure warning on cognitive workload and usability.

2 Method

2.1 Participants

Forty-fife subjects took part in this experiment. 60.5% of participants reported to drive more than 5,000 km per year. One participant had to be excluded due to simulator sickness and another due to technical difficulties with the simulator. The remaining sample consisted of 8 women and 35 men. On average participants were 25.88 (SD = 5.93) years old and ranged from age 20 to 52. They held their driving license for 8.40 years (SD = 3.48).

Furthermore two subjects had to be excluded from analysis of the detection response task (DRT). One had trouble in using the device resulting in a > 99% error rate. The second lost the tactile stimulus during the experiment.

2.2 Driving Simulator

The study took place in the driving simulator at the Chair of Ergonomics, Technical University of Munich. It consists of a BMW 640i mock-up, surrounded by a 270° view on three front screens and three rear screens (side mirrors and rear-view mirror). The SILAB simulation environment was used. The augmented reality lane departure warning integrated into the simulation environment (see Fig. 1).

Three levels of latency were defined for the augmented reality lane departure warning: (a) 17 ms, (b) 50 ms, and (c) 100 ms. Using an industry standard automotive HUD display with a refresh rate of 60 Hz this translates into (a) 1, (b) 3 and (c) 6 frames delay. Participants were instructed to expect an experimental HMI consisting of lane markings visible on both sides of the car during the whole experiment (see Fig. 1). The subjects experienced the three stages of latency in a randomized order.

Fig. 1. Positioning error resulting from 17, 50 and 100 ms latency (left to right)

To determine which road conditions are particularly critical for the experiment, we conducted a field study. For this purpose we drove through various curves and intersections in the Munich area, then analyzed the vehicle data analytically. We calculated the difference between virtual content and real environment resulting from yaw rate and curve radius. Fast driven curves on country roads and rapid lane changes on the highway resulted in the highest deviations. Therefore, the route was a combination of a country road and a highway section. On the rural section participants drove through six curves. Two curves had a radius of 400 m (lightly bent curve), two a radius of 290 m (medium bent curve) and two a radius of 120 m (sharply bent curve). On the highway section the subjects drove through four construction sites forcing them to change the lane. They did two lane changes on a straight and two more while cornering. In summary the subjects experienced 5 different situations, 3 differently bent curves, one lane change on a straight and one while cornering.

Reaction times were measured with a tactile detection response task. As shown by Engström, Åberg and Johansson [13] the DRT can be used to evaluate cognitive workload. In the present study a button was fixed on the index finger of the subjects' preferred hand. A vibration element was taped to the back of the participants' other hand. A tactile stimulus was given and the subjects clicked the button against the steering wheel.

We measured using the Systems Usability Scale (SUS) [2]. Participants rated each of the 10 items on a 5-point Likert scale with the endpoints fully disagree and fully agree, resulting in an overall score between 0 and 100.

2.3 Procedure

Participants gave written, informed consent, filled out a demographic questionnaire and familiarized themselves with the driving simulator and the DRT. They drove the above described route of six curves (two lightly, two medium and two sharply bent) and four lane changes (two on a straight, two while cornering) three times, i.e. at every latency level.

The order of conditions was randomized to minimize learning effects. The detection response task was conducted nonstop during each drive. Participants filled out the above-described questionnaires after each route, i.e. three times.

3 Results

To analyze the impact of latency in the visual augmentation on cognitive workload repeated measures 3 × 5 (Latency [17 ms, 50 ms, 100 ms] × Situation [lightly bent curve, medium bent curve, sharply bent curve, lane change straight, lane change curve]) ANOVAs were calculated for reaction times and errors of the DRT separately. To analyze the impact of latency on usability a repeated measures ANOVA was calculated for the overall SUS score.

The significance level of the statistical analysis was .05. Data of two participants were not evaluated due to technical difficulties and simulator sickness. Another two participants were excluded from the analysis of the DRT (see. 2.1). In total data of forty-three participants were used to evaluate the impact of latency on usability and those of forty-one subjects were used to evaluate its impact on reaction time and DRT error rate.

Latency had no significant effect on reaction times, $F(2,80) = 0.134$, $p = 0.875$. Nor had it a significant effect on the DRT error rate over all situations, $F(2,80) = 1.804$, $p = 0.171$. Interestingly an increase in latency had a significant effect on the error rate while performing a lane change task, $F(2,80) = 3.744$, $p = 0.028$. Higher latency led to an increased error rate (*M17 ms* = 4.89, *M50 ms* = 5.91, *M100 ms* = 10.00). Post-hoc *t*-tests (Bonferroni-corrected) revealed significant differences between 17 ms and 100 ms (see Fig. 2).

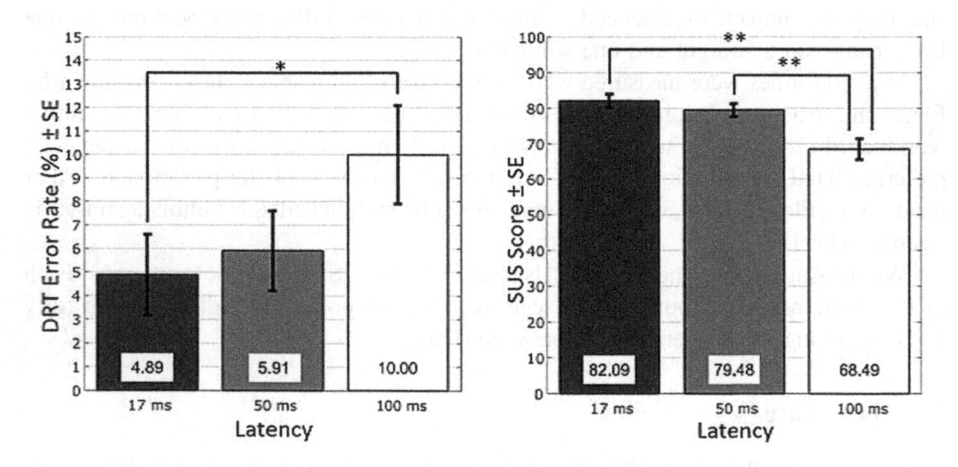

Fig. 2. DRT error rate (lane change; left) and SUS scores (right) depending on latency; * $p < .05$, ** $p < .001$ (post-hoc t-tests, Bonferroni-corrected).

Latency had a significant effect on usability (SUS score), $F(2, 84) = 21.52$, $p < .001$. An increased latency led to a reduction in usability scores (*M17 ms* = 82.09, *M50 ms* = 79.48, *M100 ms* = 68.49). Post-hoc *t*-tests (Bonferroni-corrected) revealed significant differences in SUS scores between 17 ms and 100 ms, and between 50 ms and 100 ms (both $p < .001$, see Fig. 2).

4 Discussion

The manipulation of latency had no significant impact on the cognitive workload as represented by DRT reaction times and error rate over all situations. Only while performing a lane change task the subjects made significantly more errors in reacting to the DRT stimulus. The tactile DRT was chosen to separate visual and tactile stimuli. Therefore it is possible that the tactile DRT was not sensitive enough to detect differences due to variations in latency. These findings require further investigation and should be validated using visual stimuli for the DRT.

A significant impact of latency on SUS usability scores was revealed. Especially the drop from 82 points at the lowest (17 ms) to 68 points at the level of highest latency (100 ms) represents a decline in usability from excellent to only marginally acceptable [2]. This coincides with the findings of Pfannmüller [14] who found that imprecision in positioning virtual navigation clues leads to reduced usability. Furthermore it shows the importance of fast and accurate sensors for augmented reality applications and customer demands. Furthermore the signal transfer times, especially in an automotive context with multiple bus systems and gateways must not be neglected.

5 Conclusion

The results suggest a negative impact of increased latency of AR information in an automotive ARHUD on cognitive workload and usability ratings, underlining importance of high frequency and low latency sensors. As some technological restrictions will not be overcome in the foreseeable future, research should focus on developing robust and error-tolerant concepts for information display in the ARHUD within the technological boundaries. Furthermore, future soft- and hardware architectures need to accommodate latency as a primary factor for ARHUD usability.

Acknowledgment. The authors would like to thank Mr. Tobias Schumm and Mr. Johannes Salzberger from Audi AG for their continuing support and expertise during this experiment.

References

1. Bubb H, Bengler K, Grünen R, Vollrath M (2015) "Automobilergonomie" (ATZ/MTZ-Fachbuch). Wiesbaden: Springer Vieweg. http://search.ebscohost.com/login.aspx?di-rect=true&scope=site&db=nlebk&AN=959251
2. Brooke J (1996) SUS-a quick and dirty usability scale. Usability Eval Ind 189:194
3. Bengler K, Dietmayer K, Farber B, Maurer M, Stiller C, Winner H (2014) Three decades of driver assistance systems: review and future perspectives. IEEE Intell Transport Syst Mag 6:6–22
4. Gish KW, Staplin L (1995) Human factors aspects of using head up displays in automobiles: a review of the literature. Washington DC
5. Kiefer RJ (1998) Defining the "HUD benefit time window". Vision in vehicles - VI. Derby, England, North Holland, Amsterdam, New York, pp 133–142

6. Kiefer RJ (1999) Older drivers' pedestrian detection times surrounding head-up versus head-down speedometer glances. Vision in vehicles - VII, Marseille, France. Elsevier, Amsterdam, New York, pp 111–118
7. Horrey WJ, Wickens CD, Alexander AL (2003) The effects of head-up display clutter and in-vehicle display separation on concurrent driving performance. Proc Hum Fact Ergon Soc Annu Meet 47:1880–1884
8. Azuma RT (1997) A survey of augmented reality. Presence-Teleop Virt. 6:355–385
9. Yeh M, Wickens CD (2001) Display signaling in augmented reality: effects of cue reliability and image realism on attention allocation and trust calibration. Hum Factors 43:355–365
10. Rusch ML, Schall MC, Gavin P, Lee JD, Dawson JD, Vecera S, Rizzo M (2013) Directing driver attention with augmented reality cues. Transp Res Part F Traffic Psychol Behav 16:127–137
11. Pfannmüller L, Walter M, Bengler K. (2015) Lead me the right way?! The impact of position accuracy of augmented reality navigation arrows in a contact analogue head-up display on driving performance, workload, and usability. In: Proceedings 19th triennial congress of the IEA, Melbourne
12. Engström J, Johansson E, Östlund J (2005) Effects of visual and cognitive load in real and simulated motorway driving. Transp Res Part F Traffic Psychol Behav 8(2):97–120. https://doi.org/10.1016/j.trf.2005.04.012
13. Pfannmüller L (2017) "Anzeigekonzepte für ein kontaktanaloges Head-up Display", Dissertation, Technische Universität München

HMI of Autonomous Vehicles - More Than Meets the Eye

Helena Strömberg(✉), Lars-Ola Bligård, and MariAnne Karlsson

Chalmers University of Technology, SE 412 96 Gothenburg, Sweden
helena.stromberg@chalmers.se

Abstract. Cars are becoming increasingly automated and intelligent and will soon be able to drive on their own. The new intelligent technology will mean that communication between driver and vehicle will and must change. However, much of the research in autonomous vehicle interaction still revolves around the traditional GU interfaces and modes. To open up the full potential of interactive possibilities and allow for the creation of interfaces that can enable effective and satisfactory communication between driver and vehicle, this paper will present and argue for a holistic framework to aid analysis and design of human – vehicle interaction. The framework is based on four types of interactive surfaces. The first is the explicitly designed interfaces of today, the second is the interior design of the vehicle as a whole, the third is the implicit information included in the vehicle's movement pattern, and the fourth is the interactive technology brought into the vehicle. The framework focuses on the interaction related to the operation of the vehicle (not in-car entertainment) and on the vehicle information output.

Keywords: Autonomous vehicles · Interactive surfaces · Vehicle design

1 Introduction

With the current technological development, cars are becoming increasingly automated and will soon be able to drive on their own. Autonomous vehicles, i.e. vehicles that self-driving, are considered to bring increased safety and efficiency to the transport system and are marketed with the promise that the user can perform other activities while the vehicle drives. However, to reap these benefits, the vehicles will need to be accepted, both as a form of personal transportation and as a part of the traffic environment. To become accepted, autonomous vehicles have to be perceived as useful, as well as safe and competent [1]. The vehicles will need to be observed to perform as intended and as expected by potential users, and not pose any danger to people inside or outside the vehicle.

Once a user has taken the step to use an autonomous vehicle, acceptance of that particular vehicle is created in interaction with it, in terms of how useful, easy to use and safe it is perceived to be. The user will continuously evaluate the performance and experience of the vehicle to build acceptance and trust [2]. If the performance and experience meet expectations, acceptance is likely established and the user share their

S. Bagnara et al. (Eds.): IEA 2018, AISC 823, pp. 359–368, 2019.
https://doi.org/10.1007/978-3-319-96074-6_39

positive evaluation with others. If not, the user may not only dismiss the particular vehicle but dismiss autonomous vehicles altogether. The interaction between human and vehicle is thus central to build acceptance of autonomous vehicles in general.

However, research has so far focused interaction between human and automated vehicles, i.e. partial self-driving and on handover situations where the human will need to take over in case of failure (e.g. [3–5]). Few studies have looked at the more overarching questions of interaction, like how the driver perceives the performance of the vehicle and experiences normal operation. As the autonomy and increased agency of the vehicles will involve a shift in the type of information that needs to be exchanged between human and vehicle, it is essential to clarify how the preconditions for interaction have changed in comparison to manually driven vehicles. Thus, the aim of this paper is to present and argue for a holistic framework to aid analysis and design of interaction between human and autonomous vehicle. The framework focuses on the interaction related to the operation of the vehicle (i.e. not e.g., in-vehicle entertainment) and on how the vehicle communicates information.

2 Interaction in the Autonomous Vehicle

The new intelligent technology introduced in vehicles will cause a shift in the division of driving labour and the associated communication needs between user and vehicle. The former driver (from here on denoted as user when referring to the user of an autonomous vehicle) will move from performing the actual driving (forming intentions such as “turn left” or “avoid pedestrian”) to a managing position, deciding on strategic directions (such as “go to this destination”, “park close to the entrance” or “prioritize speed over comfort”) (targeting level, cf. Extended Control Model, ECOM, [6]) and letting the vehicle execute the actions as it knows best. This means moving from making active decisions on a tracking level and relinquish the task of keeping aware of the surroundings on the regulating level, but still monitor the location of the vehicle related to the destination on the monitoring level.

2.1 Information Content

The shift from manoeuvring to monitoring the vehicle has effect on the interaction pattern with the vehicle. Instead of the driver continuously giving input to the vehicle through pedals and steering wheel and actively interpreting driver-vehicle system performance based on, e.g. the sound of the engine and the state of the traffic environment outside of the vehicle, the user will take a more passive role and evaluate performance by assessing the decisions made by the vehicle based on the output from the vehicle. To be able to assess the vehicle’s performance, the user will need to interpret the vehicle’s actions, intentions and awareness: Is it turning or swerving? Will it choose the highway route? Has it detected the person at the zebra crossing? Thus, the design of the information flow from the vehicle increases in importance.

Currently, information in cars is designed to enable users to make decisions on tracking and regulating levels by providing detailed insight into the status of the vehicle. Such information includes current speed, rpm, fuel level and indicators concerning the "health of the vehicle" (need for service, errors, warnings etc.). There is also information from systems such as eco-driving support, lane keep assist, and GPS-navigation to support the driver's actions. In autonomous vehicles, the user will instead need information to judge the vehicle's performance on monitoring and targeting levels. The information should give the user possibility to understand what the vehicle is doing and can include information about:

- Status of the trip, including estimated time to destination, planned route, and possibly reasons for the chosen route
- The vehicle's intentions in the immediate future, including upcoming actions, their reasons, and "strategies" (I intend to drive slowly because of wet surface)
- The vehicle's awareness about its vicinity, including what it sees, if it has identified whether it is dangerous, in harm's way, safe or unidentified.
- If some kind of action is necessary from the user.

However, depending on the design decision about how involved the user should be in the operation of the vehicle, the amount of information needed varies. If the user should not be involved at all it could be sufficient with the possibility to set the destination and an indicator that "all is alright". If the user is to be more engaged in the performance and individual traffic decisions, information will have to include more of the items on the list.

2.2 Interface Format

As is the case with drivers today, the user will take in information from multiple sources and judge performance of the vehicle based on this. Such sources include explicitly designed interfaces, side-effects of other systems (e.g. sound of the engine and the movement of the car) and the traffic environment outside of the car. However, increasing automation and the decoupling of the user from the driving task have implications for the user's readiness to take in information from these various sources.

One important aspect is that automation offers the user the possibility to engage in other activities than driving, such as looking at the view, checking emails or taking a nap [7]. This will in turn mean a decreased receptivity to information from the vehicle. Users will not attend to the vehicle as they have been, or to the surroundings. They may also be disturbed by information from the vehicle that is perceived as unnecessary or delivered in a way that is not adapted to the new circumstances.

Another very important aspect is that the user will experience the movement of the vehicle in a new way. Passengers, in contrast to drivers, are more susceptible to motion sickness as they do not have foreknowledge of the vehicle motion [8, 9] and often have a disturbed sightline which conflicts with the experienced motion. Passengers without direct control of the vehicle have also been found to have a higher motion sensitivity [10] and more attuned to the movements of the vehicle, which could have both negative and positive implications.

2.3 Implications for Interaction Design

Together, the shifts described present challenges for the design of the interaction between vehicle and user:

- There is an increased emphasis on the user's ability to judge the information output from the vehicle, rather than the user's ability to input commands.
- The interface will need to present new types of information (intentions, awareness) difficult to communicate through traditional interface designs.
- The traditional driver interface placement will no longer be a natural focus point and may not be looked at.
- The user will attend to other "stimuli": devices, movement, activities.

Because of these challenges, a traditional automotive graphical user interface will no longer be as effective as it is today. Nevertheless, much of the research on autonomous vehicle interaction still revolves around the traditional modes. To open up the full potential of interactive possibilities and allow for the creation of interfaces that can enable effective and satisfactory communication between user and autonomous vehicle, we propose a framework of interactive possibilities that takes into account which types of information that need to be conveyed as well as the multitude of ways in which to do it. The purpose of the framework is to provide a structure that supports the understanding of today's design and facilitates future designs by presenting a structure for the interaction.

3 The Interactive Surfaces of a Vehicle

This section presents a framework for understanding the possibilities available to vehicle developers and interaction designers to convey relevant information about the performance of the vehicle to the user: the vehicle's information output. The information can be communicated across four interactive surfaces. Only one of these surfaces is today designed to explicitly communicate information to the driver, but the driver retrieves information from the other sources as well. Each of the surfaces is presented below, with examples of how their information-carrying potential and role in interaction are affected by automation.

3.1 Surface 1: Today's Explicitly Designed Driver - Vehicle Interfaces

In today's vehicle there is a large amount of information provided to the driver to enable and facilitate the use of the vehicle in a safe and efficient manner. This information is conveyed through explicit interfaces whose sole purpose is to convey information.

Today, the type of information conveyed through these interfaces concerns primarily the status of the vehicle (see Sect. 2). The information is placed on the dashboard, centre stack, and windshield [cf. 11] and is usually represented in visual formats, with digital or analogue dials, indicator lights, digital screens, and as auditory signals..

With the changes brought on by automation, the type of information necessary to provide to the user also changes; from status to performance. Using the same type of dials and indicators as before becomes challenging, as the presentation of information on trip status, vehicle intentions, and awareness appears to require more complex representations. Ideas include representing the entire vehicle system as an agent [12] or dividing it into several different graphical representations [13]. In addition, users will not be paying attention with the same frequency to these interfaces, which means that they have to be understood at a glance, or designed so that the user feels compelled to spend a longer time understanding the representation when needed.

Together, this indicates that today's explicitly designed interfaces will not be able to play the same role as they do now. It is however unlikely that interfaces with the sole purpose of conveying information are removed from the vehicle, but they may have to be combined with other interactive surfaces. They will remain useful for information of occasional interest (but not critical information) and for information that cannot be conveyed in any other way.

3.2 Surface 2: The Interior Design of the Vehicle as a Whole

The second interactive surface is the interior of the vehicle as a whole.

Today, the interior surrounds the driver and comprises aspects such as seating design and comfort, lighting, climate, and additional features such as additional screens, pockets, and folding tables. The interior includes also the sound environment, filtering sound from the vehicle and surroundings as well offering entertainment in the form of e.g. music. However, the interior design of the vehicle is not used to convey information about the operation and the performance of the vehicle. Instead, the interior of the vehicle is designed to convey experience aspects, branding, and to give the driver and passengers comfort and good vehicle experience, while not distracting the driver.

With the changes brought on by automation and the disengagement of the driver from driving, the user will likely attend more attention to the state of the interior, which opens it up as a space for communication. There is also more freedom to design the interior in innovative ways when the demand for forward-facing seating and undisturbed views of the outside are no longer. The interior design could be used to inform the driver of changes in vehicle status and direct attention to trip status, such as making it brighter when arriving at a destination to wake the user up or shifting seating arrangement to highlight approaching an unautomated stretch of road. The opportunity to reconfigure the vehicle physically to denote its status has been suggested in less automated vehicles, such as pulling the steering wheel back when the vehicle has taken over control (e.g. Volvo Concept 26). Lighting has also been tested for communicating vehicle intentions, including informing drivers about intended manoeuvres by illuminating certain areas of the car (e.g. braking and turning [14]). Similar ideas have been tested for directing drivers' attention towards other road users [e.g. 15, 16], but could also be used to convey the vehicle's awareness of e.g. bicyclists and pedestrians. These types of interfaces have been positively evaluated as unobtrusive and intuitive [15] which indicates that the types of designs might not be useful for more critical information.

Together, this indicates that while there is room for innovative solutions to use interior design for explicit interaction, the best use of this surface needs to be further explored. Ideas from literature suggest that it can be used to grab and direct attention if designed in one way or form an unobtrusive background information depending upon how it is designed.

3.3 Surface 3: The Implicit Information in the Vehicle's Movement Pattern

A large part of the information that the driver collects from the vehicle is given by the vehicle's reaction to the driver's commands and its traction to the road and relation to other vehicles. Using this information, the driver is able to judge the vehicle's progress. In an autonomous vehicle, the user will still react to this source of implicit information, and it can be designed to explicitly communicate vehicle performance, intentions, etc. This type of vehicle movement information is today used by passengers to judge the performance of a driver, and therefore represents great potential for communicating both intentions and awareness of the vehicle in a natural and intuitive way.

Today, the information obtained from the vehicle's movement pattern contributes to the driver's understanding of their own operation of the vehicle – did the vehicle perform the intended action, is it safe to continue in the same way, am I getting closer to the goal? For a passenger, the movement of the vehicle serves as a window into the control, awareness, state and intentions of the driver. By comparing the outside context (vehicles, roads, traffic situations) to the driver's behaviour the passenger can get some insight into the capacity of the driver. A passenger can also inspect the state of the drivers in this scenario to gain even more insight into their current performance.

With the changes brought on by automation, this type of insight can be transferred to the new context. However, in automated driving, the state of the driving system is not as easy to assess as the behaviour of a human driver, but the vehicle movement patterns provide a powerful window of insight. This type of information is beginning to be explored in terms of communicating intentions of the vehicle to other road users [17]. Studies of the potential of this interactive surface in informing the user of the vehicle's intentions and awareness have begun to emerge as well, for example for communication of intent [18] and trust in automation [19]. The movement of the vehicle has been found to affect acceptance of automation in other ways, e.g. uncomfortable acceleration and deceleration in automated driving mode has led users to change back to manual driving [20]. Furthermore, lane positioning and lateral steering behaviour appear important in conveying the capabilities of the system [21, 22]).

Together, this indicates that it should be possible to actively design the vehicle's movement patterns to include explicit communication to the user. There are however many questions that need to be explored to get there. For example, how should movements be designed to effectively convey intentions and awareness without being uncomfortable, nauseating or annoying? And how can movement patterns be created that are transferable across culture, situations, and contexts?

3.4 Surface 4: The Interactive Technology Brought into the Vehicle

The fourth interactive surface is not part of the actual vehicle but consists of the nomadic devices that drivers and passengers bring into the vehicle, e.g. smartphones and tablets. It is very likely that the users of autonomous vehicles will direct their attention towards these types of devices, as this is the case already today. Since vehicles can connect to these devices, they represent a potential fourth surface of communication worth exploring.

Today, these devices are brought into the car mainly for entertainment and for communication with the outside world, but they can also be used as aids in the operation of the vehicle, mainly as navigation aids. Some car manufacturers have designed applications that can communicate with the vehicle when the vehicle is not in use. Such applications can show the status of the car and let the user communicate with the car from a distance. For example, Volvo-on-Call allows the user to set destination and climate control settings, but also log trips with extra data such as fuel consumption and other car brands e.g. BMW and Audi have similar applications.

With the changes brought on by automation, users can (acceptably and legally) start using nomadic devices during the driving of the vehicle. These devices may become the thing that users of autonomous vehicles spend the most time interacting with during a trip and thus what their attention is focused on. They therefore have great potential to be used to communicate messages to the user that need immediate attention.

Together, this indicates that one should actively consider and design the interaction with nomadic devices so that it fits with the other three surfaces and not counteracting the design as a whole.

4 Discussion

The framework proposed in this paper highlights the many different sources of information from a vehicle that are available, and how increasing automation affects which type of information that will be, or could be, transmitted across each of four surfaces. The framework and the examples of shifts brought on by automation demonstrate that while the traditional type of graphical, driver-centric user interface will lose some of its relevance, there are many interactive surfaces that before only carried implicit information and that now will unfold as potential explicit communication surfaces.

However, by unwrapping these new surfaces, many new questions are raised. These will need to be explored further in both research and creative design to understand how the surfaces can best be used to effectively and efficiently communicate relevant information to the user. First and foremost, studies need to be conducted on how to address one of the new roles of the human: the passenger role. Building too much on the old experiences of the driver may create lock-in effects, and result in that the wrong type of information is presented, and in a way not adapted to the new situation. In addition, imagining only one user as before will be problematic when moving away from driver-centric user interfaces. When using the whole vehicle as an interaction space all of the vehicle's occupants will experience the same information flow, but are they equal recipients of the information, or should it be directed at a main user?

In relation to each of the surfaces, which type information content that best fits with which surface is an important question. The timing of different information will likely also vary depending on the type of information and the surfaces across which it is communicated. Additionally, each surface can be used in multiple ways, opening up for new innovative forms of interfaces. However, there are already examples of research using the same type of format (light strips around the cabin) to convey different types of information (awareness of surroundings and intended action) with seemingly equally positive reception from users. This indicates that standardization across brands will be necessary to avoid user confusion, especially if users should be able to switch between vehicles. (something important in the envisioned car sharing schemes).

For the surfaces currently not in use for explicit communication to the driver, it is also important to further investigate how they can be utilized without impairing the main functionality of the surface, i.e. how to maintain the comfort and usefulness of the interior design, the safety and efficiency of the vehicle movement patterns, and the original functionality of the devices. Poor designs emphasising the communication potential over the initially purposed may cause everything from annoyance (interior design and devices), to nausea (movement), to danger for the user and/or other road users.

4.1 Combinations and Coherence Across Surfaces

For the occupant of the vehicle, the information gained across the four surfaces is not perceived as separated but together forms the experience of the autonomous vehicle and its performance. This presents the opportunity to combine information elements on several surfaces to create a clear message. Combinations have powerful potential, combining cabin state and displays can draw attention to more complex communication, or the sense that all is alright can be strengthened by a calm defensive driving behavior combined with a calm interior and muted displays.

However, it also creates the risk that information obtained from different surfaces is perceived as contradictive. Offering a calm driving behavior and interior while the explicit user interface tries to alert the user to take action may not produce the desired effect to attract the user's attention. A key challenge is thus how to design all four surfaces as coherent interaction. The framework presented here lays the foundation to recognize all four surfaces, which should increase the probability that the vehicle's interaction design space is considered as a whole and lessen the risk that parts are designed in isolation. Nevertheless, at the moment the framework does not provide guidance for how to create good combinations or ensure that contradicting design decisions are avoided.

Currently the most important advice to practitioners is to have a strategy for the creation of a coherent interaction. It needs to consider what the overall message is that should get across to the user in the specific moment, and how that information should be divided across surfaces, in which combinations and in which order. The strategy also needs to cover how any new piece of information should be integrated into the existing information across all surfaces so that the introduction of this new information is either adapted to the state of the current vehicle experience, or that the whole vehicle experience is changed in alignment with the new information.

Working this way will bring disciplines not previously involved in vehicle–user communication into the design process; disciplines who may not have realised their role in users' information intake. New multidisciplinary collaborations within vehicle manufacturers will be required, and potentially more challenging problems will result in a need for new methods and tools,

4.2 Further Work with the Framework

While the framework can be a guide in understanding the risk for unintended contradictions in the information from the vehicle to its user, it does not cover potential conflicts in the information obtained by the user inside the vehicle and road users outside the vehicle. The interactive surfaces (now and in the future) provide other road users with information about the driver's or vehicle's performance, awareness and intentions. The framework can be extended to cover information across these same four surfaces to road users outside of the vehicle, so that potential conflicts can be avoided and the vehicle's performance can be interpreted equally well from 'the inside' and 'the outside'. Additionally, the framework currently only covers the vehicle's output, but the user's input possibilities (such as controlling climate or setting target destination) could be equally opened up with increasing automation and divided across surfaces. These surfaces are however likely to be different than the four identified here.

4.3 Concluding Remarks

The human's role in driving will change to more of a passenger role, and with it the interest in information content from the vehicle and the receptivity to different formats of communicating information. While this unlocks more available communication channels; the four interactive surfaces described in the framework, it will also lead to more work than today for designers in identifying which the communicative aspects of the vehicles are and how they will be interpreted, what is interpreted as information about performance, and what should be utilised or covered up to give the vehicle a unified, coherent, safe and easily interpreted expression.

References

1. Choi JK, Ji YG (2015) Investigating the importance of trust on adopting an autonomous vehicle. Int J Hum Comput Interact 31(10):692–702
2. Ghazizadeh M, Lee JD, Boyle LN (2012) Extending the technology acceptance model to assess automation. Cogn Technol Work 14(1):39–49
3. Morgan PL et al (2018) Manual takeover and handover of a simulated fully autonomous vehicle within urban and extra-urban settings. In: Advances in intelligent systems and computing, pp 760–771
4. Mirnig AG, Stadler S, Tscheligi M (2017) Handovers and resumption of control in semi-autonomous vehicles: what the automotive domain can learn from human-robot-interaction. In: ACM/IEEE international conference on human-robot interaction

5. McCall R et al (2016) Towards a taxonomy of autonomous vehicle handover situations. In: Proceedings of the AutomotiveUI 2016 - 8th international conference on automotive user interfaces and interactive vehicular applications
6. Hollnagel E, Woods DD (2005) Joint cognitive systems: foundations of cognitive systems engineering. CRC Press, Boca Raton
7. Payre W, Cestac J, Delhomme P (2014) Intention to use a fully automated car: attitudes and a priori acceptability. Transp Res Part F Traffic Psychol Behav 27(PB):252–263
8. Wada T (2017) Motion sickness in automated vehicles. In: Proceedings of the 13th international symposium on advanced vehicle control, Munich
9. Rolnick A, Lubow RE (1991) Why is the driver rarely motion sick? The role of controllability in motion sickness. Ergonomics 34(7):867–879
10. Pretto P et al (2014) Variable roll-rate perception in driving simulation
11. Kern D, Schmidt A (2009) Design space for driver-based automotive user interfaces. In: Proceedings of the 1st international conference on automotive user interfaces and interactive vehicular applications, AutomotiveUI 2009
12. Häuslschmid R et al (2017) Supporting trust in autonomous driving. In: Proceedings of the 22nd international conference on intelligent user interfaces. ACM, Limassol, pp 319–329
13. Ekman F, Johansson M, Sochor J (2016) To see or not to see-the effect of object recognition on users' trust in "automated vehicles". In: ACM international conference proceeding series
14. Löcken A, Heuten W, Boll S (2016) Autoambicar: using ambient light to inform drivers about intentions of their automated cars. In: Adjunct proceedings of the AutomotiveUI 2016 - 8th international conference on automotive user interfaces and interactive vehicular applications
15. Löcken A, Heuten W, Boll S (2015) Supporting lane change decisions with ambient light. In: Proceedings of the 7th international conference on automotive user interfaces and interactive vehicular applications. ACM, Nottingham, pp 204–211
16. Pfromm M, Cieler S, Bruder R (2013) Driver assistance via optical information with spatial reference. In: 16th international IEEE conference on intelligent transportation systems (ITSC 2013)
17. Habibovic A et al (2017) Let's communicate: how to operate in harmony with automated vehicles. In: Khal M (ed) Special report: advances in automotive HMI. Automotive World Ltd
18. Brown B, Laurier E (2017) The trouble with autopilots: assisted and autonomous driving on the social road. In: Proceedings of the conference on human factors in computing systems
19. Aremyr E, Jönsson M, Strömberg H (2018) Anthropomorphism: an investigation of its effect on trust in human-machine interfaces for highly automated vehicles. In: 20th congress international ergonomics association, Florence
20. Dikmen M, Burns C (2017) Trust in autonomous vehicles: the case of tesla autopilot and summon. In: 2017 IEEE international conference on systems, man, and cybernetics, SMC 2017
21. Merat N, Lee JD (2012) Preface to the special section on human factors and automation in vehicles: designing highly automated vehicles with the driver in mind. Hum Factors 54 (5):681–686
22. Price MA et al (2016) Psychophysics of trust in vehicle control algorithms. SAE Technical Papers, April 2016

Understanding Situation Awareness Development Processes Through Self-confrontation Interviews Based on Eye-Tracking Videos

Léonore Bourgeon(✉), Vincent Tardan, Baptiste Dozias, and Françoise Darses

French Armed Forces Biomedical Institute, BP73, 91220 Brétigny-sur-Orge, France
leonore.bourgeon.pro@gmail.com

Abstract. The importance of Situation Awareness (SA) for effective decision making and safety in risky and complex socio-technical systems is closely linked to the methods used to evaluate it. Many of the existing methods are based primarily on measures of SA quality. However, in the context of the design of a training program to improve the SA skills of teams operating nuclear submarines, it became apparent that it was necessary to develop a method to account for the cognitive processes underlying the development of SA. To achieve this, we held self-confrontation interviews with operators, who were asked to explicit their own actions through eye-tracking video recordings. These interviews allowed operators to verbalize their mental representations and reasoning, while the eye-tracking video highlighted the allocation of visual attention. This method has been put in practice with eight submarine team leaders and revealed several points of interest. First, it makes it possible to take into account Endsley's three levels of SA: the eye-tracking video addresses the capture of visual information, while the operator's explanation captures information from other sensory channels as well as their understanding and expectations of the situation. Furthermore, the risk of memory lapses is overcome by eye-tracking video anchoring; data reliability is improved by the supplementary explanations provided by the operator; and data are enriched by access to operators' emotional experience. This method provides some interesting insights, especially in the context of professional training design.

Keywords: Situation Awareness · Measures · Dynamic decision making Sub-marines

1 Introduction

Situation awareness (SA) is a key challenge for operators working in risky and complex socio-technical systems. They must adapt to situations that are constantly evolving and in which decision-making is associated with important safety issues. Endsley's model [1, 2] describes SA in three interdependent levels following information processing mechanisms: (1) perception of information; (2) comprehension of the situation; and

S. Bagnara et al. (Eds.): IEA 2018, AISC 823, pp. 369–378, 2019.
https://doi.org/10.1007/978-3-319-96074-6_40

(3) projection of how the situation will evolve. According to this model, the development of SA is directly influenced by the goals and expectations of operators, together with individual factors (experience, knowledge, skills) and system-related factors (automation, interface design, complexity).

Many methods designed to evaluate SA have been developed in order to measure SA quality. However, they are regularly criticized for their lack of validity and reliability [3, 4]. In addition, they do not capture the processes that are involved in developing SA or the factors that may influence them.

The need to highlight the cognitive processes involved in developing SA emerged during a project to develop a training that aims at improving SA skills for French nuclear submarine operators. Given this context, we propose a new method for evaluating the processes involved in developing SA, based on self-confrontation interviews that draw upon eye-tracking videos of operators in a simulator.

In this paper, we first present a brief summary of existing SA measures, we then describe our method and illustrate its implementation together with its contributions through a study conducted with operators of French nuclear submarines.

2 SA Evaluation Methods

The many SA evaluation methods can be divided into three main classes:

2.1 Based on Direct Evaluation of the SA Product

These methods make it possible to assess the quality of the operator's SA by measuring the difference between the actual situation and the situation as it is perceived by the operator.

These measurements can be objective, thanks to *probe techniques* that consist in provide queries on SA, either by freezing the situation during the simulation (the *freeze probe technique*) or in real-time during the activity (the *real-time probe technique*) or after the activity has ended. The most commonly used *freeze probe technique* is SAGAT (Situation Awareness Global Assessment Technique) [5]. This consists in designing queries on the situation encountered by the operator by examining the three SA levels described in the Endsley model [1]. These methods have the disadvantage of potentially interfering with the activity. This is because either the activity is interrupted to provide SA queries (freeze technique), or questions are asked as the activity is ongoing, thus adding a further task for the operator. Techniques that are based on queries that are provided *a posteriori* have the disadvantage of relying upon what the operator remembers. They can therefore potentially present memory biases.

Subjective measures are also used to evaluate SA product: *self-rating* or *observer-rating techniques*. In these cases, it is the level of SA that is perceived, either by the operator him or herself, or by an expert observer of the activity. However, this type of method is often criticized for the fact that it seems impossible for an individual to be aware of all of their mental representations, all the more so after the activity. This criticism is even more pertinent in the case of third-party evaluation.

2.2 Based on Indirect Measures of SA Product

In this type of method, SA evaluation is implicitly based on another, more easily measurable, indicator of activity, notably, performance measures. The assumption is that if performance is successful, then the operator's SA was good.

However, this link does not always exist, and the execution of an action adapted to the situation that is encountered may also result from an incorrect awareness of the situation. For example, in a study of the influence of emotional competence in medical residents faced with a simulated unexpected medical emergency, some participants implemented the relevant therapeutic action even though their SA was incomplete: having failed to diagnose the cause of the patient's life-threatening distress, they decided to try a 'default' solution because they did not know what the correct treatment option was [6].

2.3 Based on the Analysis of Process Indices

This type of method aims to evaluate operators' SA based on the analysis of process indices during their activity. These are mainly eye-tracking and verbal protocol analysis techniques. Eye-tracking makes it possible to determine the direction and time of fixation of the gaze on specific zones in the workspace. The percent of time fixating on an important clue or event is calculated as an indicator of SA quality, with expectation of a positive correlation between fixation time and SA quality [7]. Verbal protocol analysis techniques are based on the *think aloud* procedure or verbal exchanges during the activity. The analysis of verbal exchanges within teams makes it possible to link the intention of the message with a level of SA: a more homogeneous distribution of verbal messages across the three SA levels is associated with a better quality of SA [8, 9].

The value of this type of method is that it makes it possible to evaluate SA as it develops during the activity and without interference. However, one of the main criticisms of this type of method is not always reliable, since it rarely includes operators' explanations. This is typical of the *looked-but-failed-to-see* phenomenon [3] or, for verbal exchanges, it can be the difference between what is said and what is thought.

2.4 Synthesis of SA Evaluation Methods

It seems to us that existing methods to evaluate SA are insufficient to highlight processes that are involved in SA development and suffer from many biases.

Nevertheless, the analysis of gaze direction remains very relevant in understanding SA level 1 (i.e. processes that support the perception of information useful for understanding the situation). However, to avoid the *looked-but-failed-to-see* phenomenon, it is necessary to supplement this by showing operators their eye-tracking videos as this gives them an opportunity to explain their strategy for gathering visual information.

On the other hand, to access processes that allow the operator to understand the situation and evaluate its evolution (SA levels 2 and 3), none of the existing methods seem sufficiently precise. Therefore, we adopted the individual self-confrontation interview method, which is used in ergonomics as a tool: (1) to access the subjective experience and the cognitive processes underlying the individual's action; and (2) to compensate for memory lapses [10].

Yet, while this method is commonly used with external perspective videos, the originality of our method is to use internal perspective videos thanks to eye-tracking videos. Thus, our method should provide a more accurate insight into the three SA levels, but also highlight the factors influencing SA development.

3 Description of the Proposed Method

3.1 Advantages to Confront Operators with Traces of their Own Actions with Eye-Tracking Videos

Confronting operators to traces of their own actions, in the form of video recordings, has long been used in ergonomics to describe activity in a way that is as faithful as possible to the individual's experience [11]. The main value of the exercise is to access parts of the activity that cannot be observed, or that the individual is not fully aware of: stream of actions, thoughts (e.g. interpretations, expectations or even assessments), sensory perceptions and emotions [12].

Access to this, subjective experience is achieved through special techniques, called *re-situating interviews* that help to guide the interviewee. They aim at reviving the individual's memory of the situation and prevent them from rationalizing or revising their experience. Questions can be asked to guide the interviewee in recalling their experience, while taking care to avoid influencing the answers. Thus, a major advantage compared to other techniques is to minimize the influence of the expectations of the experimenter, which is a common problem in query methods, for example.

The self-confrontation interview based on eye-tracking videos is of particular interest in evaluating SA because: (a) it minimizes the risk of forgetting without interfering with the activity; (b) it makes it possible to understand the determinants of the actions of operators; (c) it provides evidence of the allocation of visual attention while minimizing the *looked-but-failed-to-see* phenomenon.

Providing eye-tracking videos during self-confrontation interviews is already being successfully used in the medical and sports field as a debriefing technique with the aim of improving the performance of operators [13, 14].

3.2 Illustration of the Implementation of the Method with Submarine Operators

We implemented our method in the context of a project that aimed to develop the SA skills of French nuclear submarine operators. We chose to focus on the *diving-safety* team leader, whose main role is to establish and maintain SA in order to meet both operational and safety objectives.

Participants. Eight French nuclear submarine *diving-safety* teams participated in this exploratory study. These teams consisted of three operators:

- The *Chief Petty Officer* (CPO) is the team leader. He or she is responsible for the movements and safety of the boat. To this end, their role is to supervise and coordinate the activity of team members as a function of the operational orders issued by the command. He or she is responsible for the decisions taken by the team;
- The *Diving Safety Board Operator* (DSBO) must continuously monitor the status of technical equipment through dedicated interfaces;
- The *Helmsman* (HM) is responsible for steering the submarine and monitoring diving parameters using dedicated interfaces.

The CPO may also interact face-to-face with a representative of the command: the *Officer of the Watch* (OW) who is responsible for the operations of the submarine, and who relays operational orders to the CPO (see Fig. 1).

Fig. 1. Shot of a French nuclear submarine *diving-safety* team

The CPO's actions are, therefore, principally focused on obtaining information from operators concerning the state and position of the submarine. On the basis of this information, they develop a representation of the situation that allows him or her to make decisions in response to operational needs while ensuring the safety of the boat.

CPOs were 33 years old on average (SD = 3 years) with an average experience of 10 submarine missions (SD = 3) and a range of 4 to 14. Their experience as CPO ranged from 0 to 3 missions.

Recording Activity During the Simulation. Observations were recorded during team training sessions in a full-scale simulator. All teams had to manage the same situation corresponding to the most difficult scenario of the training session, in which various risks arose simultaneously: risks related to the state and position of the submarine (pump damage affecting the weight of the boat and air leak) and risks related to external threats (shallow water and presence of ships on the surface).

The CPO was provided with portable *ASL Mobile Eye* eye-tracking glasses composed of a monocular sensor sampled at 30 Hz. These glasses are very light (78 g) and did not obstruct operators. These glasses are also equipped with a microphone that captures verbal exchanges in high quality.

In addition, four cameras located inside the simulator recorded the activity of all team members. All of the latter were equipped with lapel microphones that provided a back-up of what was said in the case of a doubt.

***A Posteriori* Self-confrontation Interviews with Eye-Tracking Videos.** The eye-tracking videos started with the same event for all participants. This allowed us to compare possible strategies as a function of the events that occurred.

Participants viewed the eye-tracking video of their own activity and the experimenter stopped it either as soon as a new area of interest was fixated, during a new visual path, or when a particular event occurred (see Fig. 2).

Fig. 2. Screenshot of an eye-tracking video from the point of view of a CPO. The direction of the gaze is represented by the red circle. (Color figure online)

The goal was for the CPOs to describe and explain their visual behaviour and actions, verbalize their representation of the situation and their reasoning. To do this, the interviewer guided him or her via questions that reflected the three levels of SA, such as: Level 1 - "*At that moment, what caught your attention?*", "*What were you looking at?*", "*Were you listening to the conversation that was going on at the time between the other operators?*", Level 2 - "*At that point, what did you understand about the situation?*", "*What was your aim?*", "*What were your intentions?*", Level 3 - "*What consequences did you think your decision would have on how the situation would unfold?*".

Interviews lasted an average of 46 min (SD = 20 min).

3.3 Contributions of the Self-confrontation Interview Technique Based on Eye-Tracking Videos

The analysis of the eight interviews highlighted six key points:

- **Understanding level 1 SA through the identification of the allocation of visual attention**

By watching the recording of the direction of his or her gaze thanks to the eye-tracking video, the operator could confirm the information that he or she perceived while explaining the determinants of this information capture. For example, CPO5 spontaneously described the information visually detected from the HM interface, followed by an explanation of this information capture that highlights a process of cooperation through the need to perform a cross check.

CPO5—"I look at the rate of climb, I look at the rudder angle, rate of climb, forward rudder angle. Depending on the order we give to the front rudder, which we have to force more or less, we can know in relation to the rate of climb whether we are heavier or lighter. It's a very good indicator at very low speed. So, he [HM5] *tells me this, I trust it. I'm also doing my own little* [visual] *scan. Okay, that's good. We both agree."*

- **Understanding level 1 SA through the allocation of attention on other sensory channels**

When operators work in teams where information gathering is an important part of their activity, attention is not only allocated to the visual channel. The auditory channel is just as important, as each operator has to listen to the verbal exchanges between other members of the team to maintain their SA. Onboard submarines, even other channels are also used (kinaesthetic or olfactory channel). This information is not necessarily expressed verbally during the activity, but the self-confrontation interview can very easily facilitate its subsequent verbalization. In the following excerpt, the CPO explained that he focused his auditory attention on two operators and simultaneously was searching for visual information on a third interface, all with the same purpose: to ensure that his order to pump has been correctly followed.

Interviewer—"There, you gave an order to the DBSO, while you watch what the HM is doing. Are you listening to the DBSO while you watch the…?"

CPO6—"Yes, when I ask him to do something [the DBSO], *for example what I just asked, to pump* 300 l, *I look at what the HM does. I listen to the Commander who was behind at the same time and I also listen to the DBSO. Without looking at him, I know he just asked for it, so I'm waiting for the start of pumping. The person below is going to tell him: it's okay, I'm starting to pump. At the same time, above the cockpit, you actually have a small rate of climb indicator. So, looking at the sensor that's up there, I can see whether we're pumping or not."*

This example clearly shows a process of development of SA, seen in the combination of the order that is given, and the capture of visual and auditory information that aims to monitor the unfolding situation and check that the CPO's expectation of the consequences of pumping on the rate of ascent is correct or not.

- **Understanding SA levels 2 and 3 through explanations of the reasoning underlying observed behaviours**

The explanations given during interviews enable operators to verbalize their understanding of the situation, their reasoning, and their expectations. These phases in the elaboration of SA are very important to record as they are often poorly verbalized during the activity. Indeed, many implicit pieces of information are shared between members of an expert team, which leads operators to give orders directly after information sharing, without verbalizing SA phases 2 and 3 (understanding and anticipating the situation). In the following example, the interview captured the processes of the development of SA through observations of the difference between what the CPO expected would happen and what occurred in reality.

> *CPO3—"I'm actually looking at the HP air collector to see if it's consistent with the fact that we're taking in air. And then I look at the atmospheric pressure, which should have dropped, because we're pumping air on board. And there, actually, we do have a fall at the beginning but it's gone up again. Which is completely incoherent. So there, from then on, I know there's an air leak on board."*

- **Compensating for memory lapses thanks to eye-tracking video anchoring**

In expert activities, a lot of information is processed automatically at a low level of awareness, as soon as it is perceived. It can be difficult for the operator to recall and verbalize this mental activity. Watching videos of one's own activity from an internal perspective, as with eye-tracking videos, can serve as an anchor and help to compensate for memory lapses. In the following excerpt, the CPO describes an automated visual scan of the instrument panel. At first, he does not seem to remember why he was looking at this area of the interface, which illustrates the low level of awareness of this action. Viewing the eye-tracking video enabled him to reconstruct this visual strategy.

> *Interviewer—"So there, you just heard his message. And there, are you looking at something?"*

> *CPO5—"Yes, I don't know what it is anymore. I don't remember. I know that... well yes, I check that the air gauge has really fallen and... what am I doing? I'm taking advantage of it to check the whole instrument panel."*

- **Improving data reliability thanks to supplementary explanations provided by the operator**

Data obtained as the activity unfolds may lack reliability, since operators may look at pieces of information without treating them or share judgment of the situation with other team members but not agree with what he or she is saying. Thanks to a confidence climate between interviewer and interviewee, supplementary explanations given by operators may allow to detect this difference and hence improve reliability.

The example below illustrates how difficult CPOs can find it to express their representation of the situation faced with a superior with whom they disagree. Although the CPO has agreed to a speed command given by the OW, the CPO does not in fact agree with this command but expresses the fact that he feels unable to say anything.

> *CPO5—"I suggested 2 knots to him* [OW5], *he asks me for 1 knot. I agree 1 knot. But really, I already know that I'll definitely lose immersion, because my weight isn't regulated, I'm too heavy for such a slow speed. (…). Then, the boat loses immersion and we sink. In fact, I said it out loud at one point. I have to justify myself by saying: here, we're sinking, and I need a front engine 3. But then, when I say that, it's already too late."*

Interviewer— "When you say you know that 1 knot is a problem, you can't say it to him?

CPO5—"I know, yes, but I can't say it to him."

- **Access to emotional experience**

Emotional experience is a subjective feeling experienced by an individual. This feeling is not necessarily expressed. The self-confrontation interview can stimulate this expression if it is useful for understanding the activity.

In the following example, the CPO describes the stress felt when faced with a recurring problem that he cannot deal with effectively. This feeling of stress may explain a strategy of seeking a solution for this minor damage, to the detriment of seeking solutions for the more significant damage encountered during the scenario.

Interviewer—"Are you trying to deal with the damage?

CPO7—"of the oil station, which in the end is not something that's a huge problem, but as it came back fairly regularly throughout the platform, after a while, it's fairly stressful to see the same problem all the time and to see that it doesn't change, that we can't find a solution. (…). It often happens in reality that we have degraded stations, and then we get used to it. But here, we have a defect, the problem with the oil station doesn't go away and that adds a bit of stress because I ask myself: have I done everything that I should have? Did I do everything I could to deal with this problem so that we could improve the situation, or not?"

4 Conclusion

Understanding the cognitive processes underlying the development of SA through self-confrontation interviews based on eye-tracking videos provides some interesting insights: direct access to operators' information-gathering processes on the different sensory channels, explanation of their understanding of the situation, and their expectations regarding how the situation will unfold. In this sense, it appears to us that this method is more accurate and more reliable than the usual techniques based on queries, although a detailed study should be run to test this. The fact that the operator is not disturbed during the activity and that the risk of memory lapse is limited thanks to the anchoring provided by eye-tracking videos means that this method seems promising, in particular in the context of the design of SA training.

Acknowledgements. The authors would like to thank all participants and the staff of the submarine navigation school who made it possible to implement this exploratory study.

References

1. Endsley MR (1995) Towards a theory of situation awareness in dynamic systems. Hum Factors 37:32–64
2. Endsley MR (2015) Situation awareness misconceptions and misunderstandings. J Cogn Eng Dec Making 9(1):4–32
3. Salmon PM, Stanton NE, Walker GH, Jenkins D, Ladva D, Rafferty L, Young M (2009) Measuring situation awareness in complex systems: comparison of measures study. Int J Industr Ergon 39:490–500
4. Endsley MR (2015) Final reflections: situation awareness models and measures. J Cogn Eng Decis Making 9(1):101–111
5. Endsley MR (1995) Measurement of situation awareness in dynamic systems. HumFactors 37:65–84
6. Bourgeon L, Bensalah M, Vacher A, Ardouin JC, Debien B (2016) Role of emotional competence in residents' simulated emergency care performance: a mixed-methods study. BMJ Qual Saf 25:364–371
7. Moore K, Gugerty L (2010) Development of a novel measure of situation awareness: the case for eye movement analysis. In: Proceedings of the human factors and ergonomics society 54th annual meeting, 1650–1654
8. Lee SW, Park J, Ar Kim, Seong PH (2012) Measuring situation awareness of operation teams in NPPs using a verbal protocol analysis. Annal Nucl Energy 43:167–175
9. Tardan V, Bourgeon L, Darses F (2016) How do team leaders elaborate situation awareness?: an exploratory study in nuclear submarine simulator. In: Proceedings of the european conference on cognitive ergonomics, Article No 21. ACM, New York
10. Mollo V, Falzon P (2004) Auto- and allo-confrontation as tools for reflective activities. Appl Ergon 35:531–540
11. Cahour B, Licoppe C(2010) Confrontations with traces of one's own activity. Revue d'anthropologie des connaissances 4(2): a-k (2010)
12. Cahour B, Salembier P, Zouinar M (2016) Analyzing lived experience of activity. Le Travail Humain 79(3):259–283
13. O'Meara P, Munro G, Williams B, Cooper S, Bogossian F, Ross L, Sparkes L, McClounan M (2015) Developing Situation Awareness amongst nursing and paramedicine students utilizing eye tracking technology and video debriefing techniques: aproof of concept paper. Int Emerg Nurs 23:94–99
14. Omodei MM, McLennan J, Withford P (1998) Using a head-mounted video camera and two-stage replay to enhance orienteering performance. Int J Sport Psychol 29:115–131

Experimental Study on the Effects of Air Flow from Cross-Flow Fans on Thermal Comfort in Railway Vehicles

Hiroharu Endoh[1(✉)], Shota Enami[1], Yasuhiko Izumi[1], and Jun Noguchi[2]

[1] Railway Technical Research Institute, 2-8-38 Hikari-cho, Kokubunji-shi, Tokyo, Japan
endo.hiroharu.10@rtri.or.jp

[2] East Japan Railway Company, 2-479 Nisshin-cho, Kita-ku, Saitama, Japan

Abstract. This study examines the effects of the periodic wind variations generated by cross-flow fans to offset the sensation of warmth under realistic conditions of occupant density and temperature change in commuter trains in Japan. Subjective experiments were performed in a stationary commuter vehicle at a rolling stock depot. The results of the study showed that (1) the temperature at which the subjects felt a neutral thermal sensation on average was 24.6 °C in the fan-off condition and 27.5 °C in the fan-on condition, (2) the temperature at which 20% of the subjects felt discomfort due to warmth was 24.3 °C in the fan-off condition and 27.6 °C in the fan-on condition, (3) the relations between air temperature and subjective evaluations under an occupant density of 0.3 m^2 per person were almost the same as those under an occupant density of 0.2 m^2 per person and (4) the draft risk model substantially overestimated the discomfort due to coldness at temperatures over 22 °C.

Keywords: Thermal comfort · Commuter train · Cross-flow fan Draft

1 Introduction

Thermal environment is an important factor affecting passenger comfort in railway vehicles. In Japan, in addition to the cooling and heating units, cross-flow fans are installed in the ceiling as supplementary fans in almost all commuter trains to circulate the air in a cabin and provide beneficial cooling to passengers, especially during the hot and humid summer seasons. Efficient use of cross-flow fans for supplementary cooling is important for both thermal comfort and energy conservation.

In a thermal environment wherein the thermal sensation of occupants is close to neutral, air movement may cause local thermal discomfort. This unwanted air movement is called a draft. The percentage of dissatisfied occupants due to a draft is expressed by the draft risk model, and the thermal comfort standard sets room air speed limits low according to this model [1]. However, recently, large field studies on office buildings found that in warm, neutral and slightly cool conditions, occupants accepted and rather preferred air speeds higher than those predicted by the draft risk model [2].

S. Bagnara et al. (Eds.): IEA 2018, AISC 823, pp. 379–388, 2019.
https://doi.org/10.1007/978-3-319-96074-6_41

The findings concerning office buildings may be applied to the thermal environment in commuter trains. However, there are some differences in the characteristics of the thermal environment in an office building and in a commuter train. Compared to an office building, in a commuter train, the occupant density is higher, especially during rush hours, and the rate of temperature change is higher, as shown in Fig. 1 [3], due to the rapid changes in thermal loads, such as the number of passengers in a cabin and hot air entering from outside. Furthermore, each passenger experiences a periodic variation in the wind because the cross-flow fans in the ceiling swing at a constant frequency to provide an air flow for as many passengers as possible. Whilst some studies have investigated the effects of periodic wind variation on thermal comfort under constant temperature conditions [4], the effects under a high occupant density and a high rate of temperature change, as observed in commuter trains, have not been reported.

This study examines the effects of the periodic wind variation generated by cross-flow fans to offset the sensation of warmth under realistic conditions of occupant density and temperature change in commuter trains in Japan. In this study, subjective experiments were performed in a stationary commuter vehicle at a rolling stock depot. In the experiments, the temperature was changed to simulate the changes shown in Fig. 1 and the occupant density was changed to simulate some typical densities in commuter trains in Japan. The effects of periodic wind variations generated by cross-flow fans were examined by analysing the subjects' ratings of thermal sensation and comfort and comparing them with those predicted by the draft risk model.

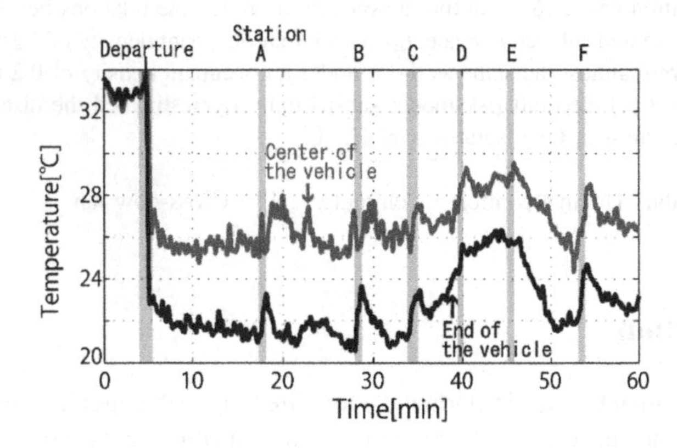

Fig. 1. An example of temperature changes measured in a commuter train [3].

2 Method

2.1 Experimental Period and Subjects

The experiments were conducted in a stationary commuter vehicle on 22 August 2017 (first experiment) and 28 August 2017 (second experiment). Each day, 31 people (15 males and 16 females) participated as subjects. Different people participated on 22 August 2017 and 28 August 2017. The characteristics of the subjects are listed in Table 1.

Table 1. Characteristics of the subjects.

Average (SD)	First experiment			Second experiment		
	Male	Female	Total	Male	Female	Total
Number	15	16	31	15	16	31
Age	29.4(9.3)	32.9(8.7)	31.2(9.0)	32.5(9.5)	33.5(11.2)	33.0(10.2)
Height [cm]	174.0(5.4)	160.4(4.0)	166.9(8.3)	173.9(8.0)	160.1(5.2)	167.0(9.7)
Weight [kg]	72.5(12.5)	55.7(11.2)	63.8(14.4)	70.0(13.3)	51.7(7.3)	60.8(14.1)

2.2 Outline of the HVAC Installation

Figure 2 shows an outline of the heating, ventilation and air-conditioning (HVAC) equipment in the commuter vehicle used in the experiments. A cooling unit is mounted on the roof of the vehicle. The conditioned air (cooling air) passes through the air duct in the ceiling and moves from the cooling unit towards the end of the vehicle. The conditioned air is then blown down through the air outlet ports arranged in 2 rows at equal intervals. The cross-flow fans are installed in the ceiling at 6 positions per vehicle, as shown in Fig. 2 (see overhead view). The cross-flow fans suck in air from the cabin and blow it into the cabin again whilst swinging at a constant frequency. Note that the airflow from the cross-flow fans is not directly connected with the cooling air from the cooling unit. The temperature of the air flowing from the cross-flow fans is almost the same as that of the cabin.

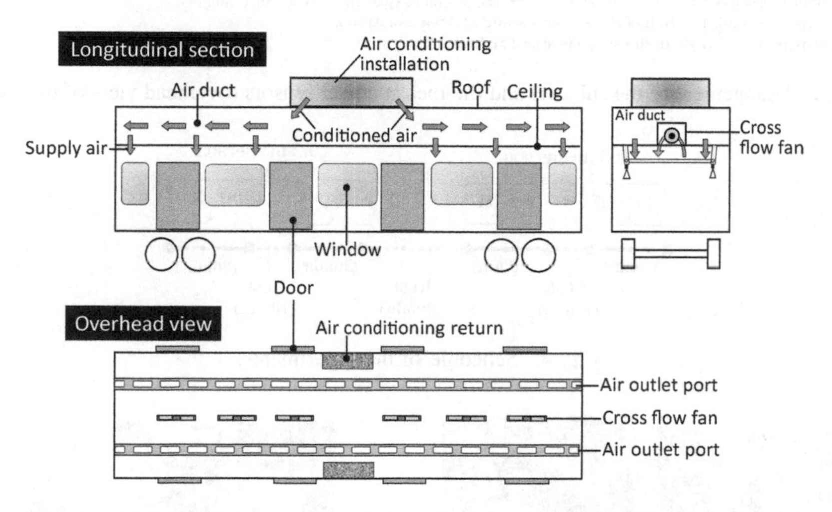

Fig. 2. Air-conditioning system of the vehicle used in subjective experiments.

2.3 Experimental Conditions

In a regular service, the cooling intensity of air conditioning and the air speed of the cross-flow fans are controlled automatically with reference to the interior temperature, exterior temperature and total weight of passengers. However, it was impossible to produce the desired experimental conditions under this automatic control within a limited amount of time and a limited number of subjects. Therefore, in the experiments, the temperature changes shown in Fig. 1 and the air speed from the cross-flow fans were controlled by manually adjusting the cooling intensity of air conditioning and turning the cross-flow fans on and off.

All subjects wore the same clothing (half-sleeve shirts, trousers, underwear and socks) and stood near a cross-flow fan. They were exposed to a temperature that increased from approximately 24 °C to 28 °C and then decreased to approximately the initial temperature again. The experiment was performed under 2 airflow conditions (fan-on and fan-off conditions) and 2 occupant density conditions (0.3 and 0.2 m^2 per person). The occupant density was controlled by adjusting the number of subjects standing in area 1 and area 2 shown in Fig. 3. A schedule of the experiments is shown in Fig. 4. The experimental environment is shown in Fig. 5. To decrease the effect of solar radiation through the windows on the subjects, all windows were covered by screens during the experiments.

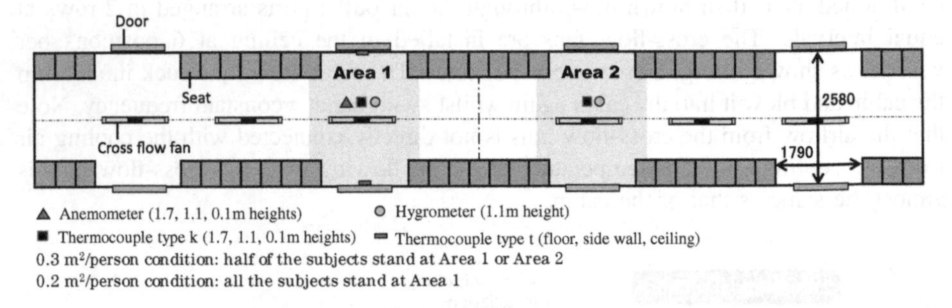

Fig. 3. Arrangement of the subjects and the measurement sensors (overhead view of the cabin).

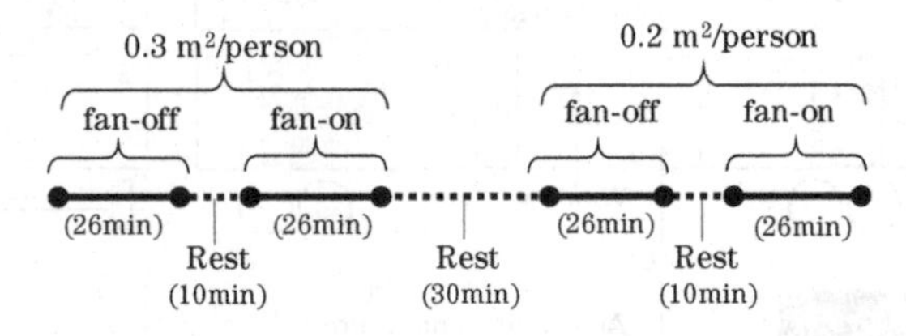

Fig. 4. Schedule of the experiments.

(a) Occupant density of about 0.3 m^2/person

(b) Occupant density of about 0.2 m^2/person

Fig. 5. Typical arrangement of the subjects standing near a cross-flow fan.

2.4 Measurements and Method

The following measurements were performed:

- Air temperature and air speed at heights of 0.1, 1.1 and 1.7 m above the floor
- Surrounding surface temperature at the centre of each surface (floor, side walls and ceiling)
- Relative humidity and global temperature at a height of 1.1 m above the floor

Figure 3 shows the arrangement of the sensors and subjects. The details of the measurement sensors used in the experiments are summarised in Table 2.

Table 2. Measurement sensors.

Measurement items	Measurement sensor	Sampling time
Air temperature (vertical distribution)	Thermocouple type k φ0.2 mm (LR9692, Hioki)	5 s
Relative humidity	Hygrometer (TR3110, T&D)	5 s
Surface temperature	Thermocouple type t φ0.32 mm	1 s
Air velocity	Anemometer (6542, Kanomax)	1 s
Globe temperature	Globe ball: φ75 mm Thermocouple type k φ0.2 mm (LR9692, Hioki)	1 s

The subjective evaluation included thermal sensation, wind sensation and degree of satisfaction. The evaluations were repeated at 2-min intervals. The evaluation scales are listed in Table 3, and the questionnaire form is shown in Fig. 6. The subjects responded with both the local and overall sensation for each subjective item other than the degree of satisfaction.

Table 3. Scale of the subjective items.

Thermal sensation	cold -4	slightly cold -3	cool -2	slightly cool -1	neutral 0	slightly warm +1	warm +2	slightly hot +3	hot +4
Wind sensation	not feel 0	slightly feel +1	feel +2	very feel +3					
Degree of satisfaction	"satisfied"	"dissatisfied" → ☐Cold ǀ ☐Hot							

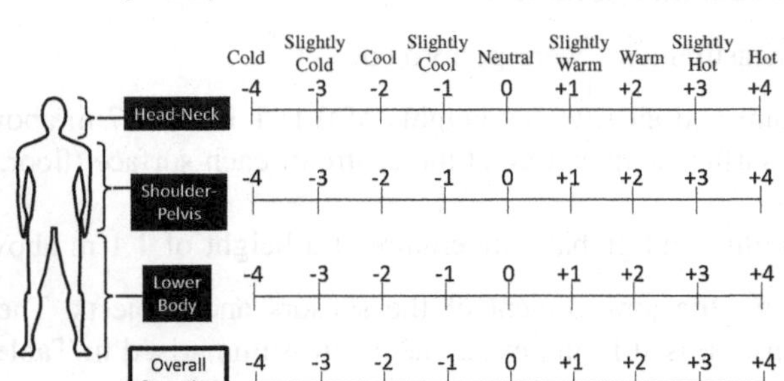

Fig. 6. The questionnaire form.

3 Results and Discussions

Figure 7 shows examples of the time-series variations of temperature and air velocity measured in the second experiment. The air temperature increased and decreased in a range of around 24 °C–28 °C in each condition. The air speed was less than 0.5 m/s at all heights in the fan-off condition. In the fan-on condition, cyclic air speeds with a period of approximately 15 s were observed. Their peak velocities were around 1.2 m/s at 1.7 m, 0.8 m/s at 1.1 m and 0.3 m/s at 0.1 m, whereas their mean velocities were around 0.5 m/s at 1.7 m, 0.4 m/s at 1.1 m and 0.2 m/s at 0.1 m. The peak and mean air velocities at each height were almost the same for occupant densities of 0.3 m^2 per person and 0.2 m^2 per person. The relative humidity was in the range 50%–70% during the experiments.

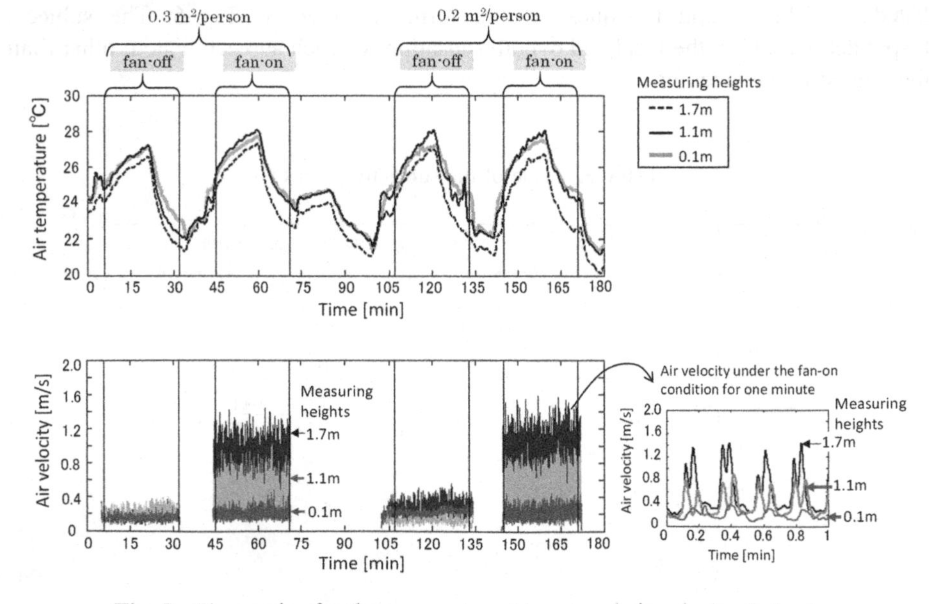

Fig. 7. Time series for the temperature (above) and air velocity (below).

Figure 8 shows examples of the time-series variations of the subjective evaluations: (a) mean thermal sensation, (b) mean wind sensation and (c) percentage of dissatisfied subjects. In the fan-on condition, the subjects felt the wind mainly on their upper bodies [Fig. 8(b)]. Compared to the fan-off condition, the warmth sensation [Fig. 8(a)] and the percentage of dissatisfied subjects [Fig. 8(c)] were substantially lower. For example, for an occupant density of 0.2 m^2 per person, the reported mean thermal sensation was approximately warm and the percentage of dissatisfied subjects due to warmth was approximately 70% at the maximum temperature in the fan-off condition. The mean thermal sensation decreased to neutral to slightly warm, whilst the percentage of

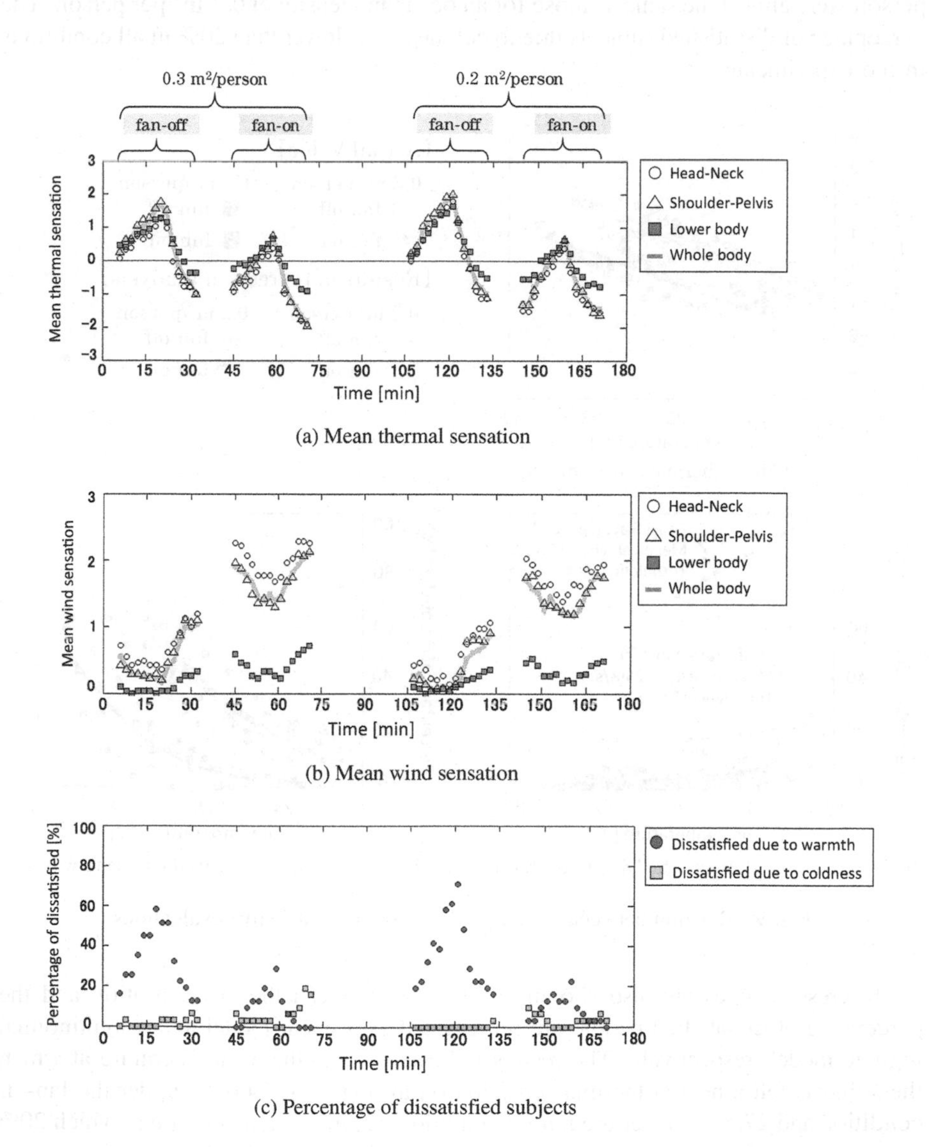

Fig. 8. Time series of subjective evaluation.

dissatisfied subjects decreased to 20% in the fan-on condition. The thermal effects of the airflow from the cross-flow fans for an occupant density of 0.2 m^2 per person were almost the same as those for an occupant density of 0.3 m^2 per person.

Figure 9 shows the relation between air temperature and the subjective evaluations. The horizontal axis is the temperature averaged over the heights of 0.1, 1.1 and 1.7 m at the start of each evaluation. As shown in the figure, in comparison with the conditions in the fan-off condition, the thermal sensation was cooler and the percentage of dissatisfied subjects due to warmth was ~20% lower in the fan-on condition. The relations for thermal sensation and comfort for an occupant density of 0.3 m^2 per person were almost the same as those for an occupant density of 0.2 m^2 per person. The percentage of dissatisfied subjects due to coldness was lower than 20% in all conditions in the experiments.

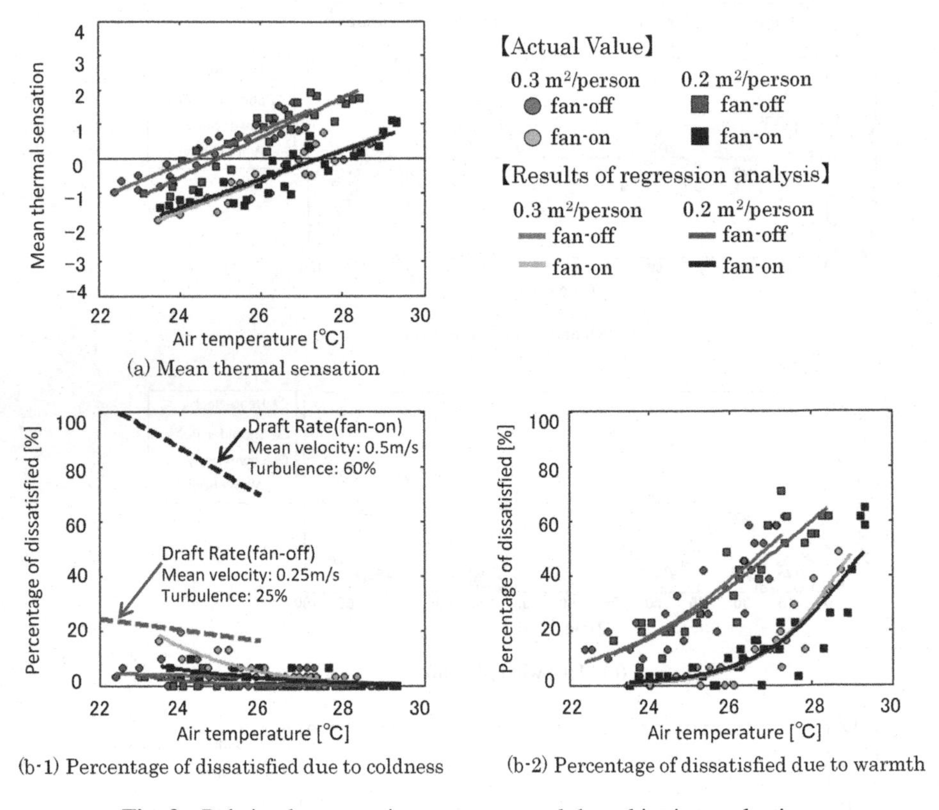

Fig. 9. Relation between air temperature and the subjective evaluations.

Regression lines are also shown in Fig. 9. The mean thermal sensation and the percentage of dissatisfied people are represented as a linear model and a multinomial logistic model, respectively. The results indicate that (1) the air temperature at which the subjects felt a neutral thermal sensation on average was 24.6 °C under the fan-off condition and 27.5 °C under the fan-on condition, (2) the air temperature at which 20% of the subjects felt discomfort due to warmth was 24.3 °C under the fan-off condition and 27.6 °C under the fan-on condition and (3) the relations between air temperature and subjective evaluations under an occupant density of 0.3 m^2 per person were almost the same as those under an occupant density of 0.2 m^2 per person. Since the occupant density of 0.2 m^2 per person corresponds to the upper limit at which subjects are not in contact with each other, the results indicate that the effects of the wind from the cross-flow fans on thermal sensation and comfort are almost the same if the subjects are not in contact with each other.

The percentage of dissatisfied people due to a draft calculated by the draft risk model [1] is shown in Fig. 9(b-1). The air velocity measured at 1.7 m (near the neck) was used to calculate the draft rates, and the prediction curves were drawn at temperatures below 26 °C since the application range of the draft risk model was from 20 °C to 26 °C. As shown in the figure, in the fan-off condition, the draft risk model predicted that ~20% of subjects will feel dissatisfied; however, the actual percentage was less than 10%. In the fan-on condition, the draft risk model predicted that more than 70% of subjects will feel dissatisfied; however, the actual percentage was less than 20%. The results revealed that the draft risk model substantially overestimated cold discomfort for standing passengers in a commuter train at temperatures over 22 °C, which are consistent with the results of previous studies on buildings [2]. It is generally accepted that a draft is perceived as cold for light, mainly a sedentary activity if the thermal sensation for the whole body is close to neutral [1]. In our experiments, the subjects were standing and their exposure to the neutral thermal condition lasted ~15 min, which may be the cause of the large error in the draft risk model.

4 Conclusion

In this study, subjective experiments were performed in a stationary commuter vehicle. In the experiments, the temperature changes and the occupant density simulated typical conditions in commuter trains in Japan in summer. The effects of air flow from cross-flow fans in the ceiling on thermal comfort were examined based on the experimental results. The main results and findings of this study are as follows:

1. From the results of the regression analysis between the temperature and mean thermal sensation of the subjects, the temperature at which the subjects felt a neutral thermal sensation on average was 24.6 °C in the fan-off condition and 27.5 °C in the fan-on condition. The wind from the cross-flow fans can reduce the sensible temperature by about 3 °C.
2. From the results of the regression analysis between the temperature and the percentage of people dissatisfied, the temperature at which 20% of the subjects felt warm discomfort was 24.3 °C in the fan-off condition and 27.6 °C in the fan-on condition. The percentage of people dissatisfied due to warmth was about 20% lower in the fan-on condition than the fan-off condition at temperatures over 24 °C.
3. The relations between air temperature and subjective evaluations for an occupant density of 0.3 m per person were almost the same as those for an occupant density of 0.2 m per person. Since an occupant density of 0.2 m per person corresponds to the upper limit at which the subjects are not in contact with each other, the result

indicates that the effects of the wind from the cross-flow fans on thermal sensation and comfort are almost the same if the subjects are not in contact with each other.
4. At temperatures from 22 °C to 26 °C in the fan-off condition, the draft risk model predicts that about 20% of people will feel dissatisfied; however, the actual percentage was less than 10%. In the fan-on condition, the draft risk model predicts that more than 70% of people will feel dissatisfied; however, the actual percentage was less than 20%. The results indicate that the draft risk model will substantially overestimate the cold discomfort for standing passengers at temperatures over 22 ° C in a commuter train.

From the above findings, the wind from a cross-flow fan can offset the sensation of warmth and improve thermal comfort without causing a draft in a commuter train where the temperature changes more rapidly and the occupant density is higher than those in office buildings.

Applying our findings to commercial trains, for example, it can be considered that the elevated air speed from cross-flow fans may be used to offset the sensation of warmth of passengers instead of increasing the cooling intensity of an air-conditioning system when the temperature increases over the set temperature. This may be a useful way to conserve energy. However, to control the air speed of cross-flow fans appropriately, we need more information on the effects of the wind from cross-flow fans on passengers' thermal comfort under the unsteady and inhomogeneous thermal environment in a commuter train. Moreover, we need to develop a quantitative method to evaluate the effects.

In recent years, methods for evaluating thermal comfort in unsteady and inhomogeneous environments have been proposed [5]. Using such methods, our future work will be to develop a quantitative method for evaluating the effects of wind from cross-flow fans on passengers' thermal comfort in an unsteady and inhomogeneous thermal environment.

In this study, we did not measure the energy consumption of the air-conditioning system. Our future work will also include an examination of the effects on energy conservation of utilising an elevated air speed from cross-flow fans instead of increasing the cooling intensity of the air-conditioning.

References

1. ISO7730 (1995) Ergonomics of the thermal environment—Analytical determination and interpretation of thermal comfort using calculation of the PMV and PPD indices and local thermal comfort criteria. International Standard Organization
2. Arens E, Turner S, Zhang H, Paliaga G (2009) Moving air for comfort. ASHRAE J 51:18–28
3. Endoh H, Izumi Y, Hayashi N (2015) Method for predicting thermal comfort inside the commuter trains in summer. RTRI Rep 29(7):27–32 (in Japanese)
4. Tanabe S, Kimura K (1994) Effects of air temperature, humidity and air movement on thermal comfort under hot and humid conditions. ASHRAE Trans 100:953–969
5. Tanabe S, Nakano J, Kobayashi K (2001) Development of 65-node thermoregulation-model for evaluation of thermal environment. J. Archit. Plann. Environ. Eng. AIJ 541:9–16 (in Japanese)

Naturalistic Observation of Interactions Between Car Drivers and Pedestrians in High Density Urban Settings

Dimitris Nathanael[1], Evangelia Portouli[2(✉)], Vassilis Papakostopoulos[3], Kostas Gkikas[1], and Angelos Amditis[2]

[1] School of Mechanical Engineering, National Technical University of Athens, 9 Iroon Polytechniou Street, 15780 Zografou, Athens, Greece
dnathan@central.ntua.gr

[2] Institute of Communication and Computer Systems, 9 Iroon Polytechniou Street, 15773 Zografou, Athens, Greece
v.portouli@iccs.gr

[3] Department of Product and Systems Engineering Design, University of the Aegean, 2 Konstantinoupoleos Street, 84100 Hermoupolis, Syros, Greece

Abstract. Interactions among drivers and pedestrians especially in heavy urban traffic constitute a key issue that needs to be addressed in the future autonomous vehicles. There is little evidence, however, concerning the signals and cues used by the drivers to infer the future intention of a pedestrian and/or a pedestrian's awareness of the driver's vehicle. The paper reports preliminary findings of an instrumented observational study of naturally occurring vehicle – pedestrian interaction cases at high density un-signalized urban crossings. Specifically, 21 experienced drivers drove their own car in a predefined course while equipped with an eye-tracker. In total 321driver – pedestrian interaction cases were analysed based on driver's eye-gaze analysis and video-assisted retrospective commentary. Several types of signals were identified. These were stratified according to their expressiveness/explicitness. A main finding is that cues with a medium level of expressiveness/explicitness (i.e. eye-gaze from pedestrians) seem to resolve a great number of interaction cases, therefore it is important to explicitly consider this type of cues in future autonomous vehicles. The paper ends with a working model depicting the possible states of mutual attentiveness between driver and pedestrian as identified from the observation data.

Keywords: Eye-contact · Eye-gaze · Crossings · Autonomous vehicles

1 Introduction

Interactions between car drivers and pedestrians constitute a key issue of driving in urban settings. It has long been observed that these interactions are not only directed from legal/formal norms but also from social/informal ones [1]. In addition, drivers and pedestrians alike are often unaware of, or disregard existing regulation to right-of-way. This lack of awareness and/or disregard can lead to conflicts [2, 3]. Moreover in heavy

S. Bagnara et al. (Eds.): IEA 2018, AISC 823, pp. 389–397, 2019.
https://doi.org/10.1007/978-3-319-96074-6_42

urban traffic ambiguous situations often arise that cannot be catered for from a legal standpoint or even from established social norms. These situations are typically resolved through various types of communication oriented behaviour. Especially, in mixed traffic situations with no signalling, both drivers and pedestrians typically seek informal signs so as to anticipate the other road-user's intent and trajectory. Due to road users' proximity and low speeds in such situations, drivers and pedestrians are able - and often do - communicate with each other through hand gestures, head nods and other head movements, eye contact, vehicle signals (e.g. flashing lights or honk) and bodily movements signifying intention.

Communication through these signals is paramount for efficient mixed traffic and a key-aspect to pedestrian safety [4]. A number of studies have recently shown that establishing eye contact with a driver also increases the possibility that the driver will yield to the pedestrian [5, 6] and the pedestrians' perceived safety to cross. Also, a study by Lagström and Lundgren [7] found that pedestrians had an evident desire to know that they were seen. In fact, even when a vehicle had come to a complete stop in front of a pedestrian, but the latter couldn't establish eye contact with the driver, the pedestrian felt reluctant to initiate a crossing. Besides eye-contact, other non-verbal communication cues found to be used by pedestrians are vehicle sound [8], vehicle acceleration and/or lateral displacement, and traffic related cues, e.g. vehicle obstructed by other road user [9]; in addition to the formal and more explicit signals emitted by drivers such as headlight flashing, hand waving and nodding [10].

There is little evidence, however, concerning the signals and cues used by the drivers to infer the future intention of a pedestrian and/or a pedestrian's awareness of the driver's vehicle. For instance, Gueguen et al. [11] showed that a pedestrian's movement trajectory can have a pervasive influence on drivers' behaviour when approaching crosswalks. Little is known about other types of signals and cues coming from pedestrians that possibly enhance a driver-pedestrian interaction. However these may prove of particular importance for enabling future autonomous vehicles to better predict pedestrians' behaviour and intent.

The present paper reports preliminary findings of an instrumented observational study of naturally occurring vehicle – pedestrian interaction cases at high density un-signalized urban crossings. Specifically, the paper elaborates on the types of signals and cues that drivers use to clarify pedestrian intent concentrating on issues of awareness, attentiveness and communicative behaviour, based on driver's eye-gaze analysis and video-assisted retrospective commentary. The paper ends with a working model depicting the possible states of mutual attentiveness between driver and pedestrian as identified from the observation data.

2 Method

A field study was conducted aiming to record naturally occurring driver-pedestrian interactions at high density un-signalized urban crossings from a driver's perspective, using drivers' eye-gaze and scene recording along with drivers' video-assisted retro-spective commentary.

Participants. Twenty-one experienced drivers (10 males and 11 females) with a mean age 39.1 years (Median = 38 years; SD = 11.7) and mean driving experience 18.5 years, took part in this study.

Course. A selected circular route of 0.75 km in an urban area of Athens was driven five times by each participant, resulting to a total course length of 3.75 km. The course included a left turn from a two-way road without a traffic light, a right turn from a smaller to a two-way road, straight segments where pedestrians frequently cross and small one-way roads where pedestrians frequently walk on the road. It was expected that there would be numerous interactions between participant drivers and pedestrians relevant to the left and right turns who would wish to cross or walk on the road.

Equipment. All participants drove their own passenger car while equipped with a wearable eye-tracker, i.e. an eye glass mounted gaze sensor (Tobii Pro Glasses 2). This system records the traffic scene from the driver's point of view and identifies the drivers's eye-fixations points with a 50 Hz sampling frequency and gaze position accuracy of 0.5°. A posteriori verbal data collected during self-confrontation sessions were recorded trough a screen and voice capture software.

Procedure. After arriving at the lab, participants were introduced to the general setup and were calibrated on the eye-tracker, while seated on driver's seat their own passenger car, with a five-point procedure. Then they were instructed to drive at the selected site in their normal style and to repeat the selected course five times in a row. The driving duration was estimated to approximately 15 min. Immediately following the driving session, participants returned to the lab and were asked to watch their eye-gaze video recording while commenting aloud on their behaviour and decision making for each case of interaction with a pedestrian. Such self-confrontation methods have been reported to aid recollection of thought processes [12] and have been used in driving studies [13].

Data Extraction and Processing. Driver's eye gaze and scene video as well as their retrospective commentary were later analysed per typical interaction scenario (i.e. pedestrian crossing one way street at intersection or on straight segment road, pedestrian moving parallel to cars sharing road space). In the present paper, only cases where a pedestrian intended to cross the street in front of the participant's vehicle are examined.

An interaction case was defined when a pedestrian in the vicinity of the participant driver (i) affected the car movement and/or the driver's behaviour in an observable manner and (ii) received at least one eye-fixation from the driver. The starting point for each interaction case was defined by the observers according to the following criteria: either (i) the drivers' first fixation towards to the pedestrian or (ii) the first cue from the pedestrian interpreted as intention to cross. Example of eye gaze video recordings used for the labelling of interaction cases analysis is shown in Fig. 1.

For each interaction case, video data were analysed by labelling the following indices: (i) participant-drivers' eye-fixations on the pedestrians, (ii) eye-contacts between pedestrian and participant-driver, (iii) cues denoting a pedestrian's projected direction (i.e. pedestrian's head orientation, body movement/orientation), (iv) cues denoting pedestrians awareness of the participant's vehicle (i.e. pedestrian's eye-gazes

Fig. 1. Example of eye gaze video recording while a pedestrian intents to cross the street in front of a participant's vehicle.

towards to the participant's vehicle). In addition, based on the video-assisted retrospective commentary (v) participants' expressed confidence about the future intended action of a pedestrian was noted when mentioned.

3 Results

In total 321 driver-pedestrian interaction cases were analysed. An initial analysis of the drivers' eye fixations to pedestrians along with the drivers' video-assisted retrospective commentaries for each interaction case, showed that: (i) in the great majority of cases (82%), a driver fixated to a pedestrian three times or less, and (ii) in a high number of cases (73%), a driver expressed confidence about the future intended action of a pedestrian even though no deliberate signals were emitted from the pedestrian (e.g. hand gesture, head nodding).

The above findings formed the basis for identifying different levels of driver-pedestrian interaction from the drivers' point of view. Specifically, these levels considered on the one hand, the drivers' visual attentiveness to pedestrians, and on the other, the expressiveness of cues and signals emitted by pedestrians.

3.1 Drivers' Attentiveness to Pedestrians

A frequency distribution of the interaction cases according to the number of drivers' eye-fixations on pedestrians is presented in Fig. 2. As mentioned above, in the majority of interaction cases (264 out of 321, i.e. 82%) a driver fixated to a pedestrian three times or less. Taking into account the drivers' video-assisted retrospective commentaries, three fixations was chosen as a criterion to nominally divide the interaction cases into two levels of drivers' attentiveness to pedestrians: (i) the driver is considered to be aware of the pedestrian but does not allocate his full attention to him (hereafter denoted A), and (ii) the driver is considered to be fully attentive to the pedestrian, at least at some point in time during the interaction (hereafter denoted F).

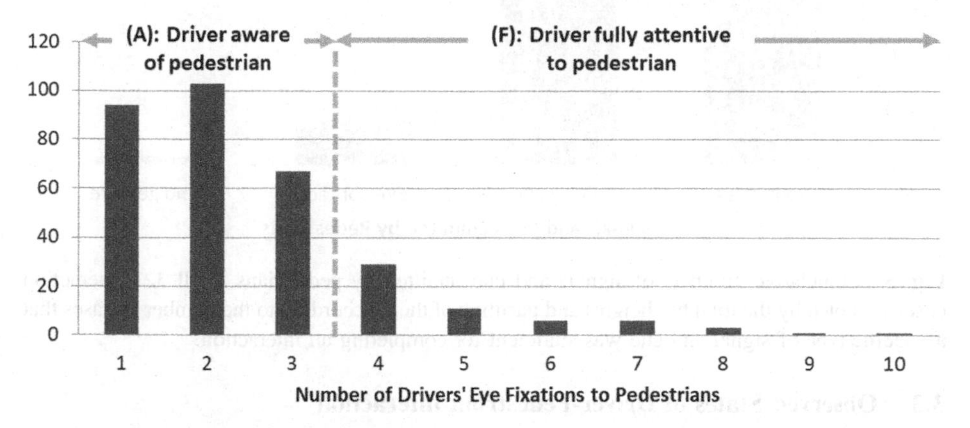

Fig. 2. Distribution of interaction cases according to the number of drivers' eye-fixations on pedestrian. Four fixations or more towards the pedestrian was taken as a nominal criterion signifying that the driver is fully attentive to a pedestrian.

3.2 Pedestrians' Behavioural Expressiveness During Crossing

Observable signals and cues emitted by the pedestrians during an interaction case were stratified according to their expressiveness relative to the pedestrians' crossing intent into three levels: (i) low level of expressiveness/explicitness, namely, body cues denoting a pedestrian's projected direction, e.g. pedestrian's body movement/head orientation, (ii) medium level of expressiveness/explicitness, namely, pedestrians eye-gazes towards the participants' vehicle denoting a pedestrian's awareness/monitoring of the participants' vehicle, and (iii) high level of expressiveness/explicitness, namely, eye-contact between driver-pedestrian, occasionally followed by pedestrian's hand gesture/head nodding, denoting explicit communication.

As it is seen in Fig. 3, body cues emitted from a pedestrian's body movement/head orientation were obviously always conspicuous, i.e. 321 cases (100%), but only a third of the cases (35%) were resolved through these form of cues alone. This is because a pedestrian's eye-gaze towards the participants' vehicle was observed in 65% of interaction cases. In fact, eye-gaze, in conjunction with the above bodily cues, was deemed sufficient for resolving half of interaction cases (52%). Finally, signals with a high level of expressiveness/explicitness as evidenced by eye-contact and hand gestures/head nodding, were observed in 13% of cases. From these cases, most were resolved solely through eye-contact between pedestrian and driver (11%) whereas 2% involved additional signals, e.g. hand gestures/head nodding.

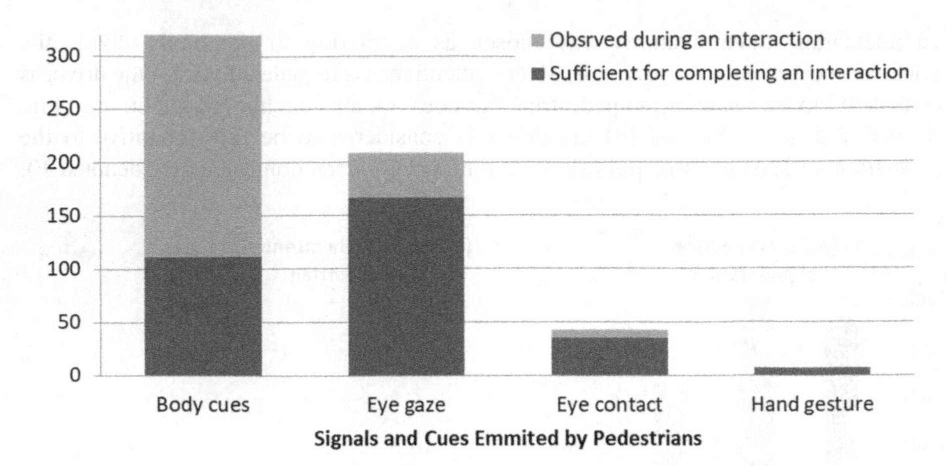

Fig. 3. Cumulative numbers of signals and cues emitted by pedestrians in all 321 interaction cases (denoted by the total bar height) and partition of them according to the number of cases that a specific type of signal and cue was sufficient for completing an interaction.

3.3 Observed States of Driver-Pedestrian Interaction

Considering the two levels of drivers' attentiveness to pedestrians (see Sect. 3.1) and the three levels of pedestrians' expressiveness/explicitness of their crossing behaviour (see Sect. 3.2), a matrix was constructed, encompassing the possible states of driver-pedestrian interaction during crossing. As it is seen in Table 1, a driver-pedestrian interaction may end within all six possible states. Also, cues with a medium level of expressiveness/explicitness seem to resolve a great number of interaction cases.

Table 1. A matrix of all six possible states of driver-pedestrian interaction during crossing and relative observations stemmed from current study.

Drivers' attentiveness	Pedestrians' expressiveness/explicitness			
	Low % (f)	Medium % (f)	High % (f)	Total % (f)
Aware of pedestrian (A)	40% (101)	59% (149)	1% (3)	100% (253)
Fully attentive to pedestrian (F)	16% (11)	25% (17)	59% (40)	100% (68)
Total	35% (112)	52% (166)	13% (43)	100% (321)

4 Discussion

Analysis of naturally occurring vehicle – pedestrian interaction cases indicated a stratified structure of signals and cues emitted by pedestrians according to their expressiveness/explicitness. A main finding of this preliminary analysis is that cues with a medium level of expressiveness/explicitness seem to resolve a great number of interaction cases, therefore it is important to explicitly consider this type of cues besides the ones already recognised in the literature, i.e. body cues on the one hand, and eye-contact/explicit signals on the other.

Based on the observations and analysis above a working model was elaborated depicting driver – pedestrian levels of mutual attentiveness from a driver's point of view (Fig. 4).

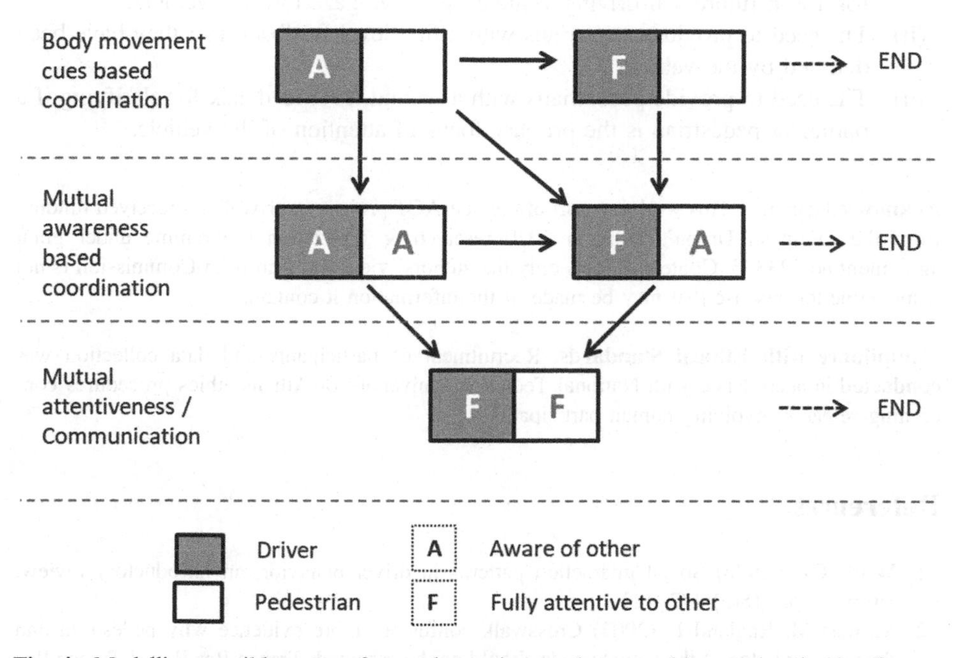

Fig. 4. Modelling possible states of mutual attentiveness between driver and pedestrian from a driver's point of view.

As it is seen in Fig. 4, at the start of an interaction, a driver becomes aware of a pedestrian; this may be followed by the driver fully attending to the pedestrian. In both these two possible states of drivers' attentiveness, an interaction may be effectively accomplished solely through physical movement co-ordination, e.g. relying on pedestrian body cues (i.e. body movement cues based co-ordination). However, in more demanding cases of time/space resource sharing, mutual attentiveness becomes necessary. Results of this study provide evidence that drivers typically use pedestrians' eye gaze towards to their vehicle as a cue to confirm pedestrians' readiness to co-ordinate and share the same resources; this tactic being often sufficient for completing an interaction, saving perceptual resources (i.e. mutual awareness based co-ordination). Only when this mutual co-ordination leads to vagueness of intent or misunderstanding, both road users are forced to devote their full attention (through eye-contact, and hand gestures) and to communicate explicitly to each other their future intended actions (i.e. mutual attentiveness/communication).

It must be noted that the present model needs to be substantiated with more data and to consider more use cases, e.g. road sharing. Most importantly, the model should be complemented with the Pedestrians' point of view, since the methodology of this study was able to observe only the driver's side of the interaction.

Nevertheless, even at this stage of development, the model can be used to derive a number of recommendations for future automated vehicles.

(i) The need to develop technologies for detecting pedestrians' states of awareness for use in future self-driving vehicles, e.g. eye-gaze towards vehicle.
(ii) The need to provide pedestrians with a first level feedback that they have been detected by the vehicle.
(iii) The need to provide pedestrians with a second level feedback for clarifying if a particular pedestrian is the primary focus of attention of the vehicle.

Acknowledgement. This work is a part of the interACT project. interACT has received funding from the European Union's Horizon 2020 research & innovation programme under grant agreement no 723395. Content reflects only the authors' view and European Commission is not responsible for any use that may be made of the information it contains.

Compliance with Ethical Standards. Recruitment of participants and data collection was conducted in accordance with National Technical University of Athens ethics procedures concerning research involving human participants.

References

1. Wilde GJS (1976) Social interaction patterns in driver behavior: an introductory review. Hum Factors 18(5):477–492
2. Mitman M, Ragland D (2007) Crosswalk confusion: more evidence why pedestrian and driver knowledge of the vehicle code should not be assumed. Transp Res Rec: J Transp Res Board 2002:55–63
3. Hatfield J, Fernandes R, Job RS, Smith K (2007) Misunderstanding of right-of-way rules at various pedestrian crossing types: observational study and survey. Acc Anal Prev 39(4): 833–842
4. Yang S (2017) Driver behavior impact on pedestrians' crossing experience in the conditionally autonomous driving context. Master of Science Thesis, KTH Royal Institute of Technology
5. Guéguen N, Meineri S, Eyssartier C (2015) A pedestrian's stare and drivers' stopping behavior: a field experiment at the pedestrian crossing. Saf Sci 75:87–89
6. Ren Z, Jiang X, Wang W (2016) Analysis of the influence of pedestrians' eye contact on drivers' comfort boundary during the crossing conflict. Procedia Eng 137:399–406
7. Lagström T, Lundgren VM (2015) AVIP - Autonomous vehicles' interaction with pedestrians. Master of Science Thesis, Chalmers University of Technology
8. Wu J, Austin R, Chen C-L (2011) Incidence rates of pedestrian and bicyclist crashes by hybrid electric passenger vehicles: An update. DOT HS 811 526. National Highway Traffic Safety Administration, Washington, DC (2011)
9. Faria JJ, Krause S, Krause J (2010) Collective behavior in road crossing pedestrians: the role of social information. Behav Ecol 21(6):1236–1242
10. Šucha M (2014) Road users' strategies and communication: driver-pedestrian interaction. In: 2014 Proceedings of transport research arena (TRA)

11. Guéguen N, Eyssartier C, Meineri S (2016) A pedestrian's smile and drivers' behavior: when a smile increases careful driving. J Saf Res 56:83–88
12. Ericsson KA, Simon HA (1993) Protocol analysis: verbal reports as data. MIT Press, Cambridge, MA
13. Portouli E, Nathanael D, Marmaras N (2014) Drivers' communicative interactions: on-road observations and modelling for integration in future automation systems. Ergonomics 57 (12):1795–1805

A Review of Driver State Monitoring Systems in the Context of Automated Driving

Tobias Hecht[1(✉)], Anna Feldhütter[1], Jonas Radlmayr[1], Yasuhiko Nakano[2], Yoshikuni Miki[2], Corbinian Henle[1], and Klaus Bengler[1]

[1] Chair of Ergonomics, Technical University of Munich, Boltzmannstr. 15, 85748 Garching, Germany
t.hecht@tum.de
[2] Denso Ten Europe GmbH, Tiefenbreucher Weg 15, 40472 Duesseldorf, Germany

Abstract. Conditionally automated driving (CAD) will lead to a paradigm shift in the field of driver state monitoring systems. High underload and the possibility of engaging in non-driving related activities will greatly influence the driver state. Level 3 also requires drivers to act as a fallback level in a take-over situation. Drivers have to get back in the loop and regain control with possible challenges due to their state. Therefore, driver state assessment will gain importance in order to ensure a safe and comfortable hand-over. The purpose of this paper is to provide an overview of driver state models and monitoring systems in the context of automated driving. Based on three driver state models, we focus on the commonly used driver state constructs fatigue, attention and workload. As part of this review, different definitions are summarized and possible metrics to operationalize these constructs were identified and critically reviewed. When reviewing the literature, it became apparent that driver state and the different constructs lack a common definition. Overall, eye-tracking is the technology with the most potential, but it needs further development to increase reliability. EEG lacks practicability and subjective measures are prone to misjudgement and may counteract extreme levels of fatigue.

Keywords: Driver state assessment · Fatigue · Drowsiness · EEG Eye-tracking

1 Introduction

Driver state monitoring has been an important field of research for many years. Especially with the background of fatigue and distraction, researchers have been looking for ways to assess driver state and design the interaction between driver and vehicle to be as safe as possible. The ongoing development of vehicle automation challenges current driver state monitoring systems: Intense research and development made it possible to establish partly automated driving in production vehicles. The subsequent automation level 3, the conditional automation (CA), is announced to be introduced in a few years. CA enables the driver to engage in other activities, as

S. Bagnara et al. (Eds.): IEA 2018, AISC 823, pp. 398–408, 2019.
https://doi.org/10.1007/978-3-319-96074-6_43

monitoring the system is not necessary anymore [1]. In the literature, these activities are usually called non-driving related tasks. Since we are of the opinion that the voluntary engagement makes it an activity instead of a task, we propose the term non-driving related activity (NDRA). In level 3 automation, the driver still serves as a fallback level at system boundaries and must regain vehicle control timely and properly. This take-over performance is composed of quality and timing aspects and is, amongst others, influenced by situational aspects like traffic density and time budget [2]. Beyond that, Gold [2] assumes the driver state to be a further influencing factor. Therefore, driver state assessment could play a bigger role in order to ensure a safe and comfortable hand-over of control. In the past, manufacturers developed fatigue alert systems based on the driver's steering behavior and the vehicle position in lane [3], e.g. the Attention Assist (Daimler AG) or the Driver Alert Control (Volvo Group). However, with automation performing lateral and longitudinal guidance, those systems become irrelevant. The purpose of this paper is to discuss the concept of driver state in the context of automated driving, to provide an overview of driver state constructs, and to examine advantages and disadvantages of several monitoring systems and metrics.

2 Driver State Models

Knowledge about different driver state constructs and their connection is key for the assessment of the driver state. The following models explain the research on human behavior with regard to the task of (automated) driving.

A literature-oriented approach has been chosen by Heikoop et al. [4], who propose a consensus-based psychological model of automated driving. It is based on a model of driving automation developed by Stanton et al. [5] with eight psychological constructs. Seven extra constructs with a focus on long-term effects on driving psychology in automated driving were selected in a subsequent literature research. This was then reduced to the nine most popular ones: Mental Workload, Attention, Feedback, Stress, Situation Awareness, Task Demands, Fatigue, Trust and Mental Model. Even though this model provides a deeper insight into the interrelations of the psychological constructs, it has several limitations, like different terms used for the same construct or the cut-off mark after the nine most relevant constructs. This model can be seen as an overview regarding ongoing research, but without a clear focus on e.g. the take-over process.

Focusing on assisted and automated driving, the model by Rauch et al. [6] is motivated by the limits of the driver's performance capabilities. These limits are mainly set by the driver's current physiological and psychological state. An optimum driver state includes an optimum alertness level (=energetic state) and a task-oriented attentiveness (=attentional state). The self-initiated or system-initiated execution of additional tasks lead to an allocation of attention away from the driving task. According to the proposed model, the alertness level is mainly impaired by vigilance decrements, fatigue and drowsiness/sleepiness. The authors conclude that the output of this driver state assessment module can be used to identify the driver's need for automation and whether the automation has to be up- or downgraded.

With even more of a focus on the transition from automated to manual driving, the model developed by Marberger et al. [7] as part of the KoHAF project introduces a comprehensive model of this transition process. The concept of driver availability is regarded as a quantitative measure that relates the estimated time required to safely take over manual control to the available time budget. The model allows for the integration of additional time or quality metrics if a quantitative relationship between take-over performance and predicted performance is found. Besides situational factors, the model takes central aspects of driver state research into account. The researchers propose that a task switch affords a time-consuming reconfiguration of the driver's sensory, motoric and cognitive state to meet the demands of manually controlling the vehicle. Further influencing factors are the arousal level and motivational conditions. The predicted duration of an individual take-over can be modelled by considering all above-mentioned sub-processes required to reach the target driver state.

These three models highlight the complexity of this topic. Different driver state models include different theoretical constructs, which in turn overlap and influence each other.

3 Driver State Constructs and Assessment Methods

The following constructs were identified as most relevant for driver state assessment: fatigue, attention and vigilance, as well as stress/workload. This was done because they change short-term and are highly researched in the context of automated driving. Assessment methods are structured into subjective, behavior-based and physiological methods. The latter includes direct and indirect measures.

3.1 Fatigue

Definition. There is no universally accepted definition in literature, neither for fatigue, nor for drowsiness or sleepiness. The terms are often used synonymously and without providing a definition, and among studies that define these terms, definitions vary. For example, Croo et al. [8] state that fatigue "concerns the inability or disinclination to continue an activity, generally because the activity has been going on for too long". They furthermore distinguish between local physical, general physical, central nervous and mental fatigue. Rauch et al. [6] suggest defining fatigue as "weariness or exhaustion from labor, exertion or stress" mainly caused by time on task and mention an "impairment of vigilance" as a consequence. However, unlike the very distinguished definition by Croo et al. [8], Rauch et al. [6] conclude that those phenomena are not independent and the result is all the same: they reduce the driver's arousal level and lead to decrements in driving performance. Therefore, the different phenomena are summarized into the concept of drowsiness. Similar, Karrer-Gauß [9] reasons after an extensive literature review that fatigue, sleepiness and drowsiness cannot be distinguished, neither in their subjective nor in their objective effects. A commonly used, but also distinguished definition of fatigue is the one by May and Baldwin [10], who subcategorize fatigue based on the causal factors: A distinction is drawn between active task-related (TR) fatigue, passive TR fatigue and sleep-related (SR) fatigue. Based on the described literature review, the authors have decided to continue working with the term fatigue as it is broader and the classification by May and Baldwin [10] provides a good basis for further work. The term fatigue therefore covers active TR fatigue caused by the driving task or by a NDRA, as well as the impact on the driver caused by an underload situation like a monotonous conditionally automated drive.

Assessing Fatigue. Using subjective methods, participants are asked to rate their fatigue level based on a scale. The most common self-rating scales for fatigue are the nine-level Karolinska Sleepiness Scale (KSS) [11] and the seven-level Stanford Sleepiness Scale (SSS) [12]. KSS and SSS are often used, for example in fatigue studies by Goncalves et al. [13] and Jarosch et al. [14]. As an alternative option, observer ratings can be conducted, either in real-time or post hoc. One of the most common rating scales is the five-descriptor Wierwille-scale [15]. Trained observers rate participants' fatigue level based on specific behavioral indicators like eyelid closure behavior, facial expression and posture from "not drowsy" to "extremely drowsy". Also, recent studies used adapted versions of the Wierwille-scale down to four levels of fatigue [9, 16]. Moreover, measuring the electric brain activity of alpha, beta, gamma and theta waves via EEG has proven to be reasonably accurate [17] and is therefore also used to validate other monitoring systems [18]. For heart activity tracking, there are three different forms: Electrocardiography (ECG), pulse oximetry and photoplethysmogram (PPG). Whereas ECG works with electrodes, PPG uses flashing LEDs placed on extremities like the wrist or the ears to measure light absorption. Measured indicators are heart rate and heart rate variability (HRV) [19]. Eye movements can be tracked either through a camera-based system or through electrooculography (EOG). The latter requires a pair of electrodes being placed either above and below the eye or to the left and right of the eye to record changes in the electrical dipole field of the eye. This can be used to track different eye-related metrics, such as blink duration, lid closure and opening speed, as well as peak closing and opening velocities [17]. Camera-based eye-tracking systems allow for the assessment of eye movements via remote or head-mounted cameras. Valid metrics for fatigue are blink frequency and blink duration [20] and, most importantly, the percentage of eye-closure (PERCLOS) within a specific time interval [21]. Some systems also track the driver's head. According to Murata et al. [22], the head pitch angle tends to increase with increasing fatigue. This method has rarely been used so far, but additional head-tracking typically improves the validity of the eye tracker and allows for a rough estimation of gaze direction without the tracking of the pupils [23].

3.2 Attention and Vigilance

Definition. Attention is one central aspect in the model by Heikoop et al. [4], and Rauch et al. [6] define attention as one of two main constructs, which is negatively influenced by distraction. The difference between inattention and distraction is that distraction involves an activity that competes for the attention of the driver [24]. This definition outlines that distraction is a construct used in the context of manual driving. We assume that in CAD, attention towards the driving scene without a driving task is still beneficial for a successful take-over. It is also worth noting that drivers can pay attention visually but fail to direct cognitive attention towards a task, e.g. mind off road vs. eyes off road. This phenomenon is called mind wandering and is linked to vigilance [25]. Even though definitions for vigilance vary, sustained attention or tonic alertness are the most common ones [26]. Vigilance is supposed to be of interest mainly for partial automation. In CAD, the driver can either decide to relax, thus furthering fatigue, or deal with NDRA and thus adjust to an optimal level of vigilance.

Assessing Attention. Schmidt et al. [27] used a nine-level questionnaire similar to the KSS for self-assessment of attention. Measuring attention in general can also be done by means of a video analysis [28] or physiological methods like eye-tracking, mouth- and body-tracking and EEG [17, 28, 29]. A possible way of subjectively assessing vigilance is the Dundee Stress State Questionnaire [30]. Further assessment of attention and vigilance can be done via reaction tests like the auditory oddball reaction time task [27]. Most often used is tracking the eyes-off-road time with an eye-tracking system [31]. While attention in general and vigilance measurements are based on or typically correlated with eye-tracking metrics, the visual attention can be directly measured using eye-trackers. Since manual driving and take-overs are highly dependent on visual information as a source of information, eye-tracking provides a direct method of assessing visual attention. This is more applicable for quantitatively predicting performance decrements in manual driving and CAD alike.

3.3 Workload/Stress

Definition. Selye [32] defines stress as a reaction from a calm state to an excited state for the purpose of preserving the integrity of the organism. If the (mental) workload of the driver is too high, this will result in elevated levels of stress and a decrease in performance [4]. The model by Heikoop et al. [4] connects high workload with increasing levels of stress and fatigue, while a very low workload has similar effects. Conti-Kufner [33] links cognitive workload imposed by a DRT with negative effects on attention. Another contribution to the construct workload can be found in Wickens' Multiple Resource Theory [34]: four resource dimensions are defined and, wherever overload is imposed by multiple tasks, this theory can make an important contribution to the mental workload by "predicting how much performance will fail once overload has been reached".

Assessing Workload/Stress. Widely used methods to assess the subjective workload are the NASA Taskload Index (NASA TLX) [35] and the Driving Activity Load Index (DALI) [36]. Detection response tasks (DRT) can also help gain insight into the driver's workload [33]. As for other driver state constructs, several physiological measures are applied for workload: pupillary, heart rate and skin resistance [37], nose temperature [38] and the following eye-tracking metrics [39]: blink latency, PERCLOS, fixation duration, pupil, blink duration and blink rate [39]. In addition, horizontal gaze dispersion is most sensitive to changes in workload when it comes to gaze concentration measures [40]. Stress can also be assessed via questionnaires [41, 42] and behavior-based measures [41]. Using physiological measures, heart rate [41], heart rate variability [43], blood pressure measurement [44], skin conductivity [41] and skin temperature [42] are correlated with driver stress level.

3.4 Advantages and Limitations of Driver State Assessment Systems

This section summarizes the various methods that can be used to assess the driver state and allows for a critical overlook. Questionnaires have the advantage of being easy to implement. As described in the previous sections, there are different questionnaires for fatigue, attention and workload. Goncalves et al. [13] even suggest a better correlation of fatigue with the SSS than with eye-tracking. However, Schmidt et al. [45] argue that participants are reactivated and, hence, biased by being asked for the self-rating of their fatigue level. Furthermore, findings by Schmidt et al. [27] indicate a lack of self-assessment abilities. Additionally, often different scales are used in experiments making it difficult to compare results. Observer-based ratings do not interrupt studies and can be done both online and post-hoc. Those ratings are non-intrusive and more objective than subjective measures. The main disadvantage is that observers base their judgement on external indicators of the subjects, which can be misleading. Observable symptoms within a specific driver state can greatly vary. Observer-based ratings allow for an additional assessment and should be compared with objective metrics from the same experiment, since they will not be available in the future use-case. As part of this review, behavior-based ratings could be found for fatigue, attention and stress. EEG is regarded as a strong indicator of the transition between wakefulness and sleep, as well as different stages of sleep. Distraction levels can also be read from an EEG measurement [17]. But besides the fact that attaching electrodes directly to the scalp to measure the electric potential is highly intrusive and difficult to implement in a car, there are also problems regarding data loss or corruption [27]. Talking can alternate results and expert knowledge is required to understand the results [17]. Together with eye blinks, EEG is the most used physiological measure [26, 46]. In regard to driver monitoring, heart activity tracking has been mostly used trying to detect fatigue and stress in the subject. ECG, which works by recording the electrical activity of the heart over a period of time, is considered the most reliable way of measuring one's heart rate. With electrodes placed on the skin, it has the same disadvantages as EEG. Non-intrusive electrodes can also be used, but need continuous direct contact with the skin and are prone to data loss [47]. Concluding, they can hardly be used in the context of automated driving. Regarding eye movements, EOG suffers similar problems as EEG [6]. By contrast, camera-based eye-tracking devices do not have these constraints, but problems were observed with glasses and sunglasses [48], bad light conditions and reflections from outside light [49]. Friedrichs and Yang [50] also report difficulties in distinguishing between gazes to the dashboard and blinks, and [51] found great differences in detected blinks between EOG and camera-based systems. Remote eye-tracking systems have problems when the head moves out of the system's head box and multiple cameras are needed to track the eye when the driver deals with NDRA. The ISO 15007-2 [52] defines criteria for the quality of data available for eye-tracking and states that at least 70% of the valid data must be available in order to make a statement about a driver state. The implementation of the PERCLOS algorithm differs in the literature, making it more complicated to compare results [46]. Nonetheless, remote eye-tracking systems offer great potential for non-intrusive and real-time assessment of different driver state constructs.

In the literature reviewed, the rates for detecting a specific driver state were rarely found. Another frequently observed limitation is that researchers do not mention the quality of the data collected and often fail to mention data losses. This leads to the assumption that the detection systems were working properly at all times, which can be doubted with respect to having worked with several different eye-tracking systems.

4 Conclusion

Throughout the literature review, it became apparent that researchers lack a common definition of driver state and its constructs. Constructs differ in their definition and influence and overlap each other. For example, vigilance is closely linked to both attention and fatigue and differentiating between those constructs with assessment systems can sometimes be impossible. Overall, the most prominent driver state monitoring systems were eye-tracking, EEG and subjective measures. EEG proved to be a valid metric especially in the transition between awake and asleep, but lacks practicality, as it is less reliable when non-intrusive. Subjective metrics demonstrated good correlations with EEG, but are prone to misjudgment and may counteract the development of extreme levels of fatigue. In addition, future systems cannot include expert assessment and neither rely safely on self-reported measurements. Eye-tracking holds the highest potential due to the possibility of non-intrusive measurements and the multitude of information about the driver state. On the other hand, the availability and robustness of this measurement has to be questioned in many situations (e.g. lighting, posture, glasses). Thus, motion tracking, which is not well researched so far, could gain in importance. But each of these systems contains weaknesses to reliably detect a driver state, so they were often used in conjunction with each other. We are of the opinion that driver state monitoring will be important for level 3 automation, but in real road situations, a hybrid approach with multiple assessment systems will be inevitable.

References

1. SAE International (2016) Taxonomy and definitions for terms related to on-road motor vehicle automated driving systems. SAE Standard J 3016_201609
2. Gold CG (2016) Modeling of Take-Over Performance in Highly Automated Vehicle Guidance. Dissertation, Technische Universität München. http://mediatum.ub.tum.de?id=1296132
3. Bhatt PP, Trivedi JA (2017) Various methods for driver drowsiness detection: an overview. Int J Comput Sci Eng (IJCSE) 9(03):70–74
4. Heikoop DD, de Winter JCF, van Arem B, Stanton NA (2015) Psychological constructs in driving automation. A consensus model and critical comment on construct proliferation. Theor Issues Ergon Sci. https://doi.org/10.1080/1463922X.2015.1101507
5. Stanton N, Young M (2000) A proposed psychological model of driving automation. Theor Issues Ergon Sci. https://doi.org/10.1080/14639220052399131

6. Rauch N, Kaussner A, Boverie S, Giralt A (2009) HAVEit Deliverable D32.1 Report on driver assessment methodology. HAVEit - Highly automated vehicles for intelligent transport, Regensburg
7. Marberger C, Mielenz H, Naujoks F, Radlmayr J, Bengler K, Wandtner B (2018) Understanding and applying the concept of "Driver Availability" in automated driving. In: Stanton NA (ed) Advances in human aspects of transportation: proceedings of the AHFE 2017 international conference on human factors in transportation. Springer International Publishing, Cham, pp 595–605
8. Croo HD, Bandmann M, Mackay GM, Rumar K, Vollenhoven P (2001) The role of driver fatigue in commercial road transport crashes. European Transport Safety Council, Brussels, Belgium
9. Karrer-Gauß K (2012) Prospektive Bewertung von Systemen zur Müdigkeitserkennung - Ableitung von Gestaltungsempfehlungen zur Vermeidung von Risikokompensation aus empirischen Untersuchungen. Technische Universität Berlin (2012)
10. May JF, Baldwin CL (2009) Driver fatigue. The importance of identifying causal factors of fatigue when considering detection and countermeasure technologies. Transp Res Part F Traffic Psychol Behav. https://doi.org/10.1016/j.trf.2008.11.005
11. Åkerstedt T, Gillberg M (2009) Subjective and objective sleepiness in the active individual. Int J Neurosci. https://doi.org/10.3109/00207459008994241
12. Hoddes E, Zarcone V, Smythe H, Phillips R, Dement WC (1973) Quantification of sleepiness. A new approach. Psychophysiology. https://doi.org/10.1111/j.1469-8986.1973.tb00801.x
13. Goncalves J, Happee R, Bengler K (2016) Drowsiness in conditional automation. Proneness, diagnosis and driving performance effects. In: Proceedings of the 2016 IEEE 19th international conference on intelligent transportation systems (ITSC), Rio de Janeiro, Brazil, pp 873–878
14. Jarosch O, Kuhnt M, Paradies S, Bengler K (2017) It's out of our hands now! effects of non-driving related tasks during highly automated driving on drivers' fatigue. In: Proceedings of the 9th international driving symposium on human factors in driver assessment, training, and vehicle design: driving assessment 2017. Driving assessment conference, Manchester Village, Vermont, USA. University of Iowa, Iowa City, Iowa, 26–29 June 2017, pp 319–325. https://doi.org/10.17077/drivingassessment.1653
15. Knipling RR, Wierwille WW (1994) Vehicle-based drowsy driver detection. Current status and future prospects. In: IVHS America fourth annual meeting, Atlanta
16. Feldhütter A, Feierle A, Kalb L, Bengler K (2018) A new approach for a real-time non-invasive fatigue assessment system for automated driving. In: Proceedings of the human factors and ergonomics society (HFES) (in Press)
17. Dong Y, Hu Z, Uchimura K, Murayama N (2011) Driver inattention monitoring system for intelligent vehicles. A review. IEEE Trans Intell Transp Syst. https://doi.org/10.1109/tits.2010.2092770
18. Coetzer RC, Hancke GP (2009) Driver fatigue detection. A survey. In: Proceedings of the AFRICON, AFRICON 2009, Nairobi, Kenya. IEEE, pp 1–6. https://doi.org/10.1109/afrcon.2009.5308101
19. Vicente J, Laguna P, Bartra A, Bailón R (2016) Drowsiness detection using heart rate variability. Med Biol Eng Comput. https://doi.org/10.1007/s11517-015-1448-7
20. Hargutt V (2000) Eyelid movements and their predictive value of fatigue stages. In: 3rd international conference of psychophysiology in ergonomics, San Diego, California, 30.07.2000 (2000)

21. Wierwille WW, Wreggit SS, Kirn CL, La Ellsworth, Fairbanks, R.J (1994) Research on vehicle-based driver status/performance monitoring; development, validation, and refinement of algorithms for detection of driver drowsiness. Final report. U.S. Department of Transportation. Springfield, Virginia
22. Murata A, Koriyama T, Hayami T (2012) A basic study on the prevention of drowsy driving using the change of neck bending and the sitting pressure distribution. In: Proceedings of SICE (Society of Instrument and Control Engineers) Annual Conference 2012, Akita, pp 274–279
23. Schmidt J. Braunagel C, Stolzmann W, Karrer-Gauss K (2016) Driver drowsiness and behavior detection in prolonged conditionally automated drives. In: 2016 IEEE intelligent vehicles symposium (IV), Gotenburg, Sweden. IEEE, Piscataway, NJ, 19–22 June 2016, pp 400–405. https://doi.org/10.1109/ivs.2016.7535417
24. Regan MA, Hallett C, Gordon CP (2011) Driver distraction and driver inattention. Definition, relationship and taxonomy. Accident; analysis and prevention. https://doi.org/10.1016/j.aap.2011.04.008
25. Schooler JW, Smallwood J, Christoff K, Handy TC, Reichle ED, Sayette MA (2011) Meta-awareness, perceptual decoupling and the wandering mind. Trends Cogn Sci. https://doi.org/10.1016/j.tics.2011.05.006
26. Oken BS, Salinsky MC, Elsas SM (2006) Vigilance, alertness, or sustained attention. physiological basis and measurement. Clin Neurophysiol. https://doi.org/10.1016/j.clinph.2006.01.017
27. Schmidt EA, Schrauf M, Simon M, Fritzsche M, Buchner A, Kincses WE (2009) Drivers' mis-judgement of vigilance state during prolonged monotonous daytime driving. Accid Anal Prev. https://doi.org/10.1016/j.aap.2009.06.007
28. Sathyanarayana A, Nageswaren S, Ghasemzadeh H, Jafari R, Hansen JHL (2008) Body sensor networks for driver distraction identification. In: Proceedings of the 2008 IEEE international conference on vehicular electronics and safety (ICVES 2008), Columbus, OH, 22.09.2008–24.09.2008, Columbus, USA, pp 120–125. https://doi.org/10.1109/icves.2008.4640876
29. Azman A, Meng Q, Edirisinghe E (2010) Non intrusive physiological measurement for driver cognitive distraction detection. Eye and Mouth Movements. In: 2010 3rd international conference on advanced computer theory and engineering (ICACTE). IEEE, pp 595–599 (2010)
30. Körber M, Cingel A, Zimmermann M, Bengler K (2015) Vigilance decrement and passive fatigue caused by monotony in automated driving. Procedia Manuf. https://doi.org/10.1016/j.promfg.2015.07.499
31. Young K, Regan M (2007) Driver distraction. A review of the literature. In: Faulks IJ, Regan M, Stevenson M, Brown J, Porter A, Irwin JD (ed) Distracted driving, NSW, Sydney, pp 379–405 (2007)
32. Selye H (1980) Selye's guide to stress research. Van Nostrand Reinhold, New York
33. Conti-Kufner A-S (2017) Measuring cognitive task load: an evaluation of the Detection Response Task and its implications for driver distraction assessment. Dissertation, Technische Universität München. http://mediatum.ub.tum.de?id=1340561
34. Wickens CD (2008) multiple resources and mental workload. Hum Factors. https://doi.org/10.1518/001872008X288394

35. Hart SG, Staveland LE (1988) Development of NASA-TLX (Task Load Index). Results of empirical and theoretical research. In: Meshkati N, Hancock PA (eds) Human Mental Workload. Advances in Psychology, vol 52, 1st edn. Elsevier textbooks, s.l., pp 139–183 (1988)
36. Pauzié A (2008) A method to assess the driver mental workload. The driving activity load index (DALI). IET Intell Transp Syst. https://doi.org/10.1049/iet-its:20080023
37. Kahneman D, Tursky B, Shapiro D, Crider A (1969) Pupillary, heart rate, and skin resistance changes during a mental task. J Exp Psychol. https://doi.org/10.1037/h0026952
38. Itoh M (2009) Individual differences in effects of secondary cognitive activity during driving on temperature at the nose tip. In: Proceedings of the 2009 international conference on mechatronics and automation (ICMA), Changchun, China, 09.08.2009–12.08.2009, IEEE, Changchun, China, pp 7–11. https://doi.org/10.1109/icma.2009.5246188
39. Marquart G, Cabrall C, de Winter J (2015) Review of eye-related measures of drivers' mental workload. Procedia Manuf. https://doi.org/10.1016/j.promfg.2015.07.783
40. Wang Y, Reimer B, Dobres J, Mehler B (2014) The sensitivity of different methodologies for characterizing drivers' gaze concentration under increased cognitive demand. Transp Res Part F Traffic Psychol Behav. https://doi.org/10.1016/j.trf.2014.08.003
41. Healey JA, Picard RW (2005) Detecting stress during real-world driving tasks using physiological sensors. IEEE Trans Intell Transp Syst. https://doi.org/10.1109/TITS.2005.848368
42. Yamakoshi T, Yamakoshi K, Tanaka S, Nogawa M, Shibata M, Sawada Y, Rolfe P, Hirose Y (2007) A preliminary study on driver's stress index using a new method based on differential skin temperature measurement. In: Proceedings of the 29th annual international conference of the IEEE EMBS, vol. 2007, Lyon, France, pp 722–725
43. Lee HB, Choi JM, Kim JS, Kim YS, Baek HJ, Ryu MS, Sohn RH, Park KS (2007) Nonintrusive biosignal measurement system in a vehicle. In: Proceedings of the 29th annual international conference of the IEEE EMBS, Lyon, France, pp 2303–2306
44. Jeong IC, Jun Sh, Lee DH, Yoon HR (2007) Development of bio signal measurement system for vehicles. In: Proceedings of the 2007 international conference on convergence, pp 1091–1096
45. Schmidt EA, Schrauf M, Simon M, Buchner A, Kincses WE (2011) The short-term effect of verbally assessing drivers' state on vigilance indices during monotonous daytime driving. Transp Res Part F Traffic Psychol Behav. https://doi.org/10.1016/j.trf.2011.01.005
46. Lenné MG, Jacobs EE (2016) Predicting drowsiness-related driving events. a review of recent research methods and future opportunities. Theor Issues Ergon Sci. https://doi.org/10.1080/1463922X.2016.1155239
47. Heuer S, Chamadiya B, Gharbi A, Kunze C, Wagner M (2010) Unobtrusive in-vehicle biosignal instrumentation for advanced driver assistance and active safety. In: Proceedings of the IEEE EMBS conference on biomedical engineering and sciences (IECBES), Kuala Lumpur, Malaysia. IEEE, Piscataway, pp 252–256. https://doi.org/10.1109/iecbes.2010.5742238
48. Bergasa LM, Nuevo J, Sotelo MA, Barea R, Lopez ME (2006) Real-Time system for monitoring driver vigilance. IEEE Trans Intell Transp Syst. https://doi.org/10.1109/TITS.2006.869598
49. Danisman T, Bilasco IM, Djeraba C, Ihaddadene N (2010) Drowsy driver detection system using eye blink patterns. In: 2010 international conference on machine and web intelligence (ICMWI), Algiers, 03.10.2010–05.10.2010. IEEE, pp 230–233. https://doi.org/10.1109/icmwi.2010.5648121

50. Friedrichs, F., Yang, B.: Camera-based drowsiness reference for driver state classification under real driving conditions. In: Proceedings of the 2010 IEEE intelligent vehicles symposium (IV), La Jolla, CA, USA, 21.06.2010–24.06.2010. IEEE, La Jolla, USA, pp 101–106. https://doi.org/10.1109/ivs.2010.5548039
51. Fors C, Ahlström C, Sörner P, Kovaceva J, Hasselberg E, Krantz M, Grönvall J-F, Kircher K, Anund A (2011) Camera-based sleepiness detection. Final report of the project SleepEYE. ViP publication: ViP - Virtual Prototyping and Assessment by Simulation. Statens vägoch transport-forskningsinstitut, Linköping
52. ISO/TS 15007-2:2014 (2014) Road vehicles - Measurement of driver visual behaviour with respect to transport information and control systems: Part 2: Equipment and procedures. International Organization for Standardization, Switzerland

How Will the Driver Sit in an Automated Vehicle? – The Qualitative and Quantitative Descriptions of Non-Driving Postures (NDPs) When Non-Driving-Related-Tasks (NDRTs) Are Conducted

Yucheng Yang(✉), Jan Niklas Klinkner, and Klaus Bengler

Chair of Ergonomics, Technical University of Munich,
Boltzmannstr. 15, 85747 Garching, Germany
yucheng.yang@tum.de

Abstract. Highly-automated driving (HAD) is currently one of the most discussed innovative topics and likely to become a series product within the next few decades [1]. From the level-3 automation (SAE) on, the driver does not have to constantly monitor the vehicle while driving [2], this enables the driver to carry out different activities and be out of the control loop. By conducting the non-driving related tasks (NDRT) like eating, texting, talking, relaxing and so on [3], the driver may take other sitting positions – defined as 'non-driving postures (NDPs)' – rather than the driving position. In this work, an online survey (n = 122) and an experiment (n = 16) were conducted, which found out that there are 13 activities which would be conducted by significantly ($\alpha = 0.05$) more drivers in HAD, compared with the manual driving. Four basic NDPs are mapped (many-to-many) to the NDRTs. In the experiment, 10 NDPs of each participant are measured, where the descriptive statistics of torso, thigh and knee angles offer a quantitative description of NDPs. Based on the results, 30 new requirements for the interior of automated vehicles are derived.

Keywords: Non-driving related task · Non-driving posture
Interior requirement · Automated vehicle

1 Introduction

Drivers actually already conduct NDRTs in the current manual driving condition [4], e.g. talking to their passengers, using a smartphone. It is expected that drivers would spend more time on the NDRTs as the automation level rises, as the driver does not have to constantly monitor systems starting from the level-3 automation [2]. NDRTs could be conducted for a longer continuous period of time without being interrupted by the driving task. Driving posture may not be optimal anymore, if the majority of time spent in a vehicle does not consist of driving. Instead, the driver may like to text, send an email, eat, etc. [3]. This results in different requirements for the seat and other interior elements [5], which are not necessarily fulfilled by the current driving-task-orientated interior. Current concepts used in SAE Level-0 or 1 are restricted by the

S. Bagnara et al. (Eds.): IEA 2018, AISC 823, pp. 409–420, 2019.
https://doi.org/10.1007/978-3-319-96074-6_44

constant necessity of dynamic control by the driver. It presumes that the driver is able to reach the pedals, steering-wheel and the gear shift in any sitting position [6]. This fact, on the one hand, restricts the possibility of interior concepts (e.g. position, type, range of controls and displays) and, on the other, it restricts the driver's behaviour (e.g. movements, postures). For example, when conducting a visual-manual task (e.g. radio-tuning, navigation-setting), the Alliance of Automobile Manufacturers requires that: (a) single glance durations should generally not exceed two seconds; (b) task completion should require no more than 20 s of total glance time to task displays and controls [7]. NDRTs, as a scenario in the current design process, are considered as a secondary or tertiary priority because of their possible distraction to the primary driving task. However, in automated vehicles, the driver could conduct NDRTs as the primary task, as long as the automation works.

In this work, an online survey (n = 122) was conducted to identify the NDRTs and the NDPs belonging thereto. In addition, a follow-up study (n = 16) was conducted in the Modular Ergonomic Mock-up (MEPS) at the Chair of Ergonomics. This enables a quantitative description of the NDPs with the torso, thigh and knee angle (°) and seat back angle (°). Combining the results of the survey and the experiment, 11 requirements of the driver's seat, in terms of its construction, positioning, adjustment, and material; five requirements of the driver's space; three requirements of storage and 11 other requirements of infotainment, lighting system, etc., have been derived.

2 Hypotheses

Four alternative hypotheses are made as follows: $H_{11:}$ There are more NDRTs in the automation, than in the manual drive. $H_{12:}$ There are NDPs in the automation. $H_{13:}$ There is a many-to-many mapping between NDRTs and NDPs. $H_{14:}$ There are new requirements of the interior for automated vehicles to fit NDPs.

3 Methods

3.1 Online Questionnaire

The online questionnaire, generated on the LimeSurvey from the Chair of Ergonomics, consists of four parts: personal demographic information, NDRTs in manual driving, NDRTs and their associated postures in automated driving and interior concepts. It took about 20 min to finish the online questionnaire (the amount equals to eight A4 pages), which was widely distributed through social media and in the university. After six weeks, 122 participants in total were able to complete all the questions. They were aged (M = 42, SD = 18) between 18 and 79, amongst whom 64.75% were male and 35.25% female. 49% drive between 5000 and 20,000 km a year – over 50% only privately. However, most of them had never used the advanced driving assist systems (ADAS), like the automated parking assistant (83%), active cruise control (70%), and lane-keeping assistant (75%).

3.2 Experiment

Based on the results of the online questionnaire, an experiment was designed in the MEPS. The steering-wheel was fixed at its end positions of x-axis (+) and z-axis (+) to ensure the largest space for the driver. It had no practical function in this experiment. The driver's seat was adjustable in three dimensions: 60 mm adjustment range for the seat height, 220 mm adjustment range for the seat longitudinal positioning, and a 68° adjustment range for the seat back angle (0° to 68°). The seat-height H30 distance [8] was adjustable between 250 mm and 310 mm, which corresponds to a typical limousine according to [6]. The seat inclination angle was fixed to 10°. There were 16 participants (six females and 10 males) aged 31 years old on average (SD = 13). Before the experiment, an introduction to HAD was given. Each participant took 10 individually-preferred postures and seat adjustments following the instruction. The first four postures were pre-defined and without any NDRT (supervised postures), the other six were not specified, but with specified NDRTs (unsupervised postures). Factors of their choices, explanations of their activities and postures, wishes and opinions about the interior were collected by a recorded semi-structured interview. Three photos were taken for each posture, by means of three cameras installed around the driver, for the front, side and top view. The simplified 2-D human model was considered (Fig. 1, left) with torso, thigh and knee angle. Angles were measured by means of the side-view pictures (Fig. 1 right) by identifying the ankle joint, calf, knee, thigh, hip point, torso and the C7 spinal segment respectively.

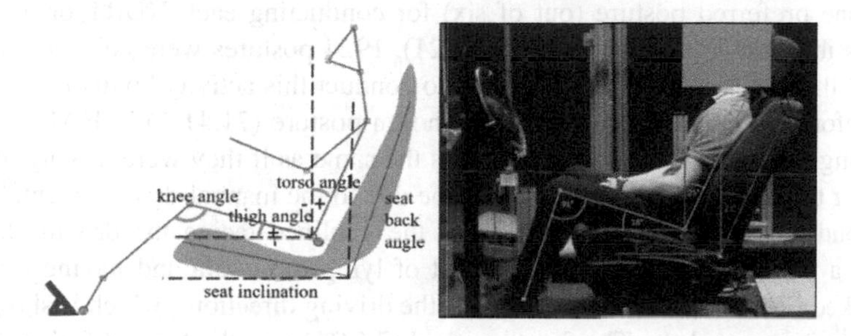

Fig. 1. The simplified human model used and the measurement in the experiment

4 Results of Online Questionnaire (n = 122) and Discussion

4.1 Non-Driving Related Task

Among the 21 listed NDRTs (Fig. 2), there are 13 activities mentioned significantly ($\alpha < 0.05$) more frequently (Chi-square test [9]) in HAD, than in the manual driving condition, which are denoted by the '*' (s) in Fig. 2. This fact would confirm the first hypothesis: there are more NDRTs in the automation, than in the manual drive. More NDRTs lead to not only new functional, but also ergonomic, requirements of the vehicle interior to allow and support drivers with conducting NDRTs more effectively, efficiently, and with greater satisfaction, thus shifting the focus of the current driving-task-orientated driver's workplace towards multi-focus.

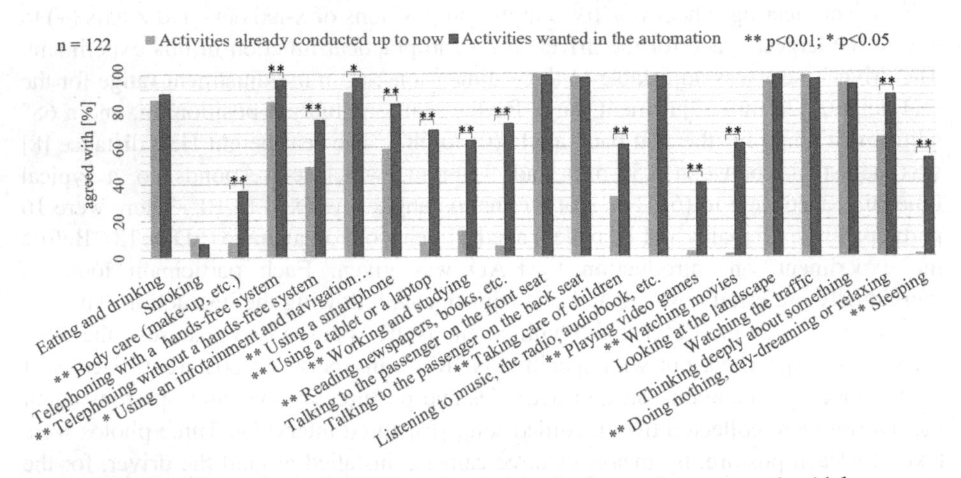

Fig. 2. Comparison of NDRTs today and in the feature in automated vehicles

4.2 Qualitative Results of Non-Driving Postures

There are 21 NDRTs in the questionnaire. From the six postures, participants could choose one preferred posture (out of six) for conducting each NDRT or not. Out of 2562 possible individual postures (122 * 21), 1924 postures were collected, the other 638 were the answer: "No, I do not want to conduct this activity" without any posture and therefore, were excluded. The most chosen posture (71.41%) in HAD, is still the 'seat facing the driving direction', which is the same as if they were driving manually. Except for their preferences, it could also be due to the manual-driving mental models. The second-ranked posture (10.29%) was the 'reclined facing the driving direction', which is a relaxing posture similar to that of lying on a sofa and having a rest. The third-ranked (7.90%) was the 'seat against the driving direction', which is shown in the buses or trains nowadays. The fourth-ranked (7.54%) was the 'seated facing the front-seat passenger', which is lateral to the driving direction. Two other reclined postures were only minimally chosen. These facts could confirm the second hypothesis that there are NDPs in the automation. Apart from the normal seated posture, the reclined posture could have enormous potential in HAD, which requires a wider range of seat back angles and sufficient space behind. In addition, the seat should be rotatable, meaning the seat-belt should be mounted on the seat. There should be enough room around the seat, when rotating, to avoid the crushing between the seats and the driver's legs or other interior elements, e.g. door panels. The seat should be able to be fixed at different angles after rotating, which raises critical safety issues in the crash tests. Furthermore, those rotated postures in a moving cabin could cause motion sickness [6], since the visual input corresponds less well with the acceleration being sensed by the vestibular system.

4.3 Mapping of NDRTs and NDPs

Figure 3 shows the 12 very significantly more ($p < 0.01$) mentioned NDRTs and their associated postures. The 'seated facing the driving direction' is the most preferred posture by 10 activities, except for 'sleeping' where the majority (71%) chose to recline facing the driving direction, and for 'taking care of children' where 40% chose to sit against the driving direction. Figure 3 could confirm the third hypothesis that there is a many-to-many mapping between NDRTs and NDPs, meaning each NDRT would be conducted in more NDPs; one NDP would also be suitable for more NDRTs. This fact means different functions of the interior could be used in various contexts in terms of the different adjustments of interior elements and different driver postures. For example, the display should be seen not only by a seated, but also a reclined driver. The transitions among such contexts shall not derogate or interrupt the activities.

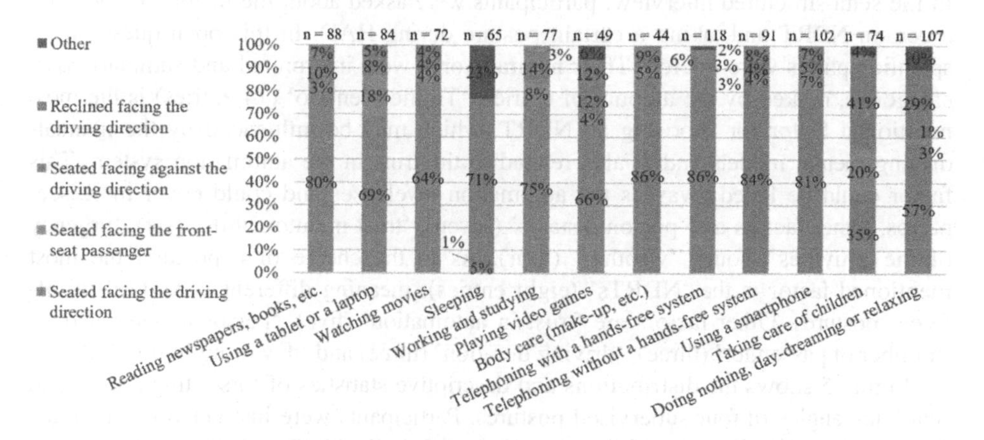

Fig. 3. Many-to-many relationship between NDRTs and NDPs

4.4 Space Requirements

Figure 4 shows that more than 50% of the people are satisfied with the current dimension of the foot-well (dimensions A and D). To ensure that the pedals are reachable, dimension A should not be too large, even when the feet do not have to press the pedals in HAD, as it may convey a feeling of safety. More people prefer the dimensions C and E to be larger to have more space around the driver's torso and knees, which is in line with the requirements of a reclined posture and rotatable seat. Adjusting the steering-wheel forwards, or the seat and seat back backwards enables a larger C, however, these adjustments should be set back to a proper driving condition in case of a system limit.

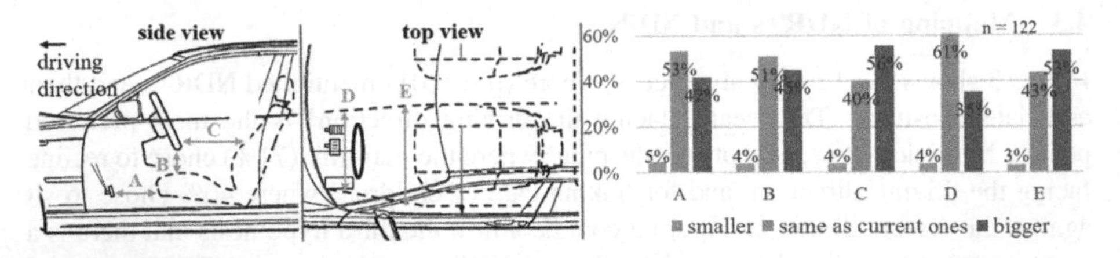

Fig. 4. Expectations of space in HAD in comparison with current vehicles (Modified images from: the-blueprints.com/bmw_e46)

5 Results of Experiment (n = 16) and Discussion

In the semi-structured interview, participants were asked about the factors of choosing a certain NDRT and taking a certain posture during HAD. In this open question, no specific options were offered. Their natural words were interpreted and summarised in categories, ranked by the amount of entries. 'Traffic scenario' (14 entries) is the most mentioned factor for choosing an NDRT, which may be influenced by the manual-driving mental models and is also related to the trust in the automation system. This factor could be faded away as the automation level rises and could cover more scenarios. Other factors are 'personal states' (seven), 'trust in automation' (six), 'urgency of the activities' (four), 'weather' (four). As to the choice of a posture, the most mentioned factor is the 'NDRTs' (eight entries), meaning different tasks lead to different postures. Other factors are 'trust in automation' (five), 'personal states' (four), 'number of passenger' (three), 'driving duration' (three) and 'day or night time' (three).

Figure 5 shows the distributions and descriptive statistics of torso, thigh, knee and seat back angles of four supervised postures. Participants were instructed to sit/recline facing/against/lateral to the driving direction, as described in Fig. 5. Nevertheless, their exact individual postures and seat adjustments were free to choose.

Figure 6 shows the distributions and descriptive statistics of torso, thigh, knee and seat back angles of unsupervised postures with six defined NDRTs. Participants were instructed to read, use a tablet or laptop, watch movies, sleep, work and study and play video games, their individual postures and seat adjustments were free to choose.

Summarising Figs. 5 and 6, Table 1 shows the descriptive statistics of all 16 participants' ten postures, whose medians are close to the reference posture: the RAMSIS neutral driving posture [10]. This indicates that most NDPs are located around the conventional driving posture although they could vary and stay at some point in a very wide range. It also means that the current optimisation, based on the RAMSIS neutral driving posture, could satisfy the major usage, while the adjustment range should be wider and consider that, unlike in the manual drive condition, many more extreme postures, which deviate from the medians, would also be assumed over a long period of time – ergonomics and functionality should therefore be considered at every point within the range.

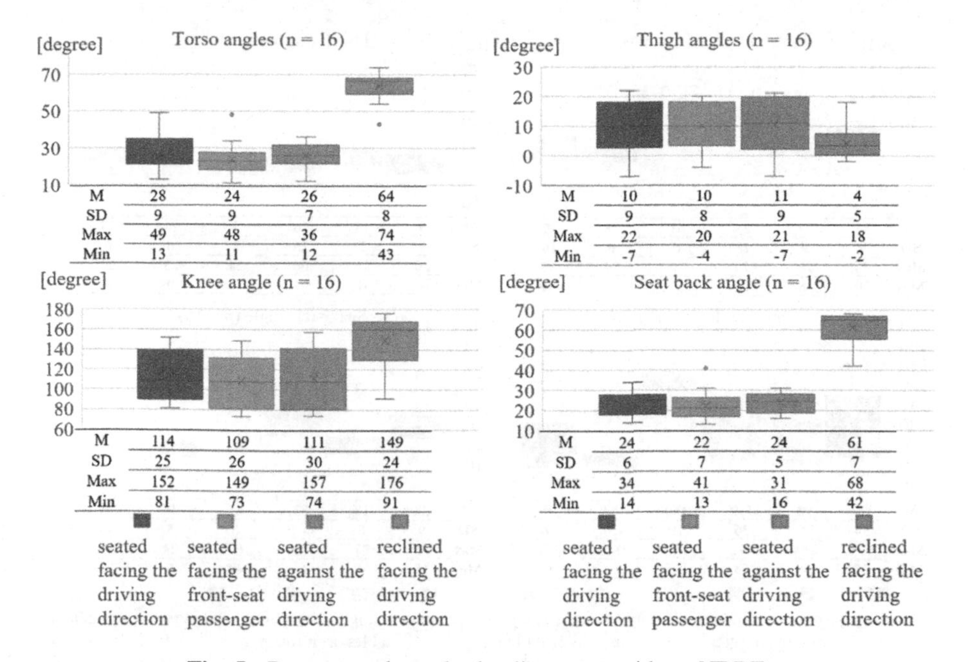

Fig. 5. Postures and seat back adjustments without NDRTs.

6 Requirements of Interior for Automated Vehicles

Table 2 shows 30 preliminary ergonomic requirements which are derived based on the results of the online questionnaire and the experiment.

7 Limitation

The reliability of an online questionnaire is not able to be validated. 122 participants are inexperienced with ADAS, which allowed the results to be strongly influenced by their mental models of manual drive scenarios. In the experiment in the MEPS, n = 16 is too small to have representative results. The driver's seat was not rotatable in the experiment, all postures were measured facing the steering wheel as if they were lateral to, or against the driving direction. The participants are being observed, cameras and audio devices are used, thus their postures may not be natural, due to the Hawthorne effect [11]. Interior concepts are not correspondingly available for every posture, e.g. armrest, which may change the posture if there were actually an armrest there. The MEPS is a static environment, the preference of the NDRTs and NDPs may, in reality, change in a moving cabin due to different brightness, acceleration, etc.

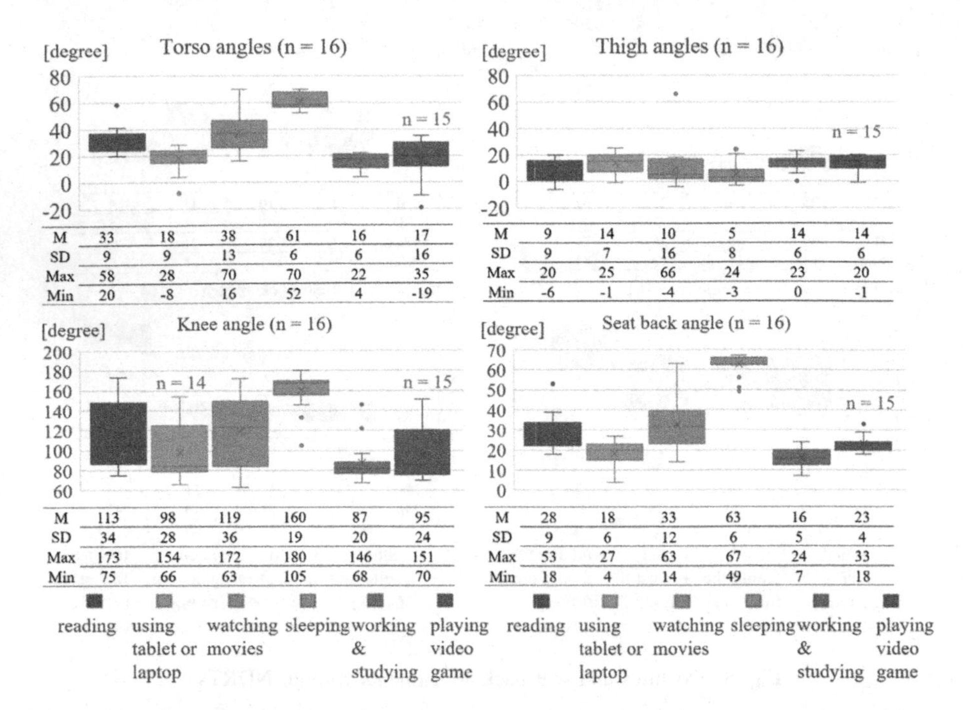

Fig. 6. Postures and seat back adjustments of six NDRTs (The case 'n = 14' for the 'Knee angle', because two cross-legged postures with both legs on the seat are excluded. For the task 'Playing video games' (n = 15), one dataset was damaged.)

Table 1. Ranges of all postures and seat back adjustments

		Torso angle	Thigh angle	Knee angle[a]	Seat back angle[b]
Summarised all postures (n = 159)	Max	74	66	180	68
	Mean	32	10	116	31
	Median	26	10	114	24
	Min	−19	−7	63	4
	SD	19	9	34	18
RAMSIS posture		27	18	119	–

[a]'n = 157' for knee angle, due to two excluded cross-legged postures with both legs on the seat.
[b]If the surfaces of the human's back and thigh have fully close contacts with the seat back and seat surface: torso angle = 0.93 * seat back + 1.66 (in the range: seat back = 26° ± 4°); thigh angle = 1.16 * seat inclination + 8.52 (in the range: seat inclination = 8° ± 2°). These are two linear regressions of the measured data collected from an H-point machine on a car seat.

Table 2. Preliminary requirements of interior for automated vehicles

No.[c]	Requirements
Seat	
1-Q	Rotatable
2-E	In the driving direction: driver's seat supports torso angles: −19° to 90°[d], thigh angles: −7° to 66°, knee angles: 63° to 180°
3-E	Lateral to the driving direction: driver's seat supports torso angles: 11° to 48°, thigh angles: −4° to 20°, knee angles: 73° to 149°
4-E	Against the driving direction: driver's seat supports torso angles: 12° to 36°, thigh angles: −7° to 21°, knee angles: 74° to 157°
5-E	Headrest tiltable around y-axis
6-E	Headrest has a soft pillow and side supports
7-E	Fully-flat seat without side supports in the seat back, seat is wider
8-QE	Left and right armrest adjustable in all positions
9-E	Seat-height adjustable, like an office chair (H30 = around 50 cm)
10-E	Electric seat adjustment with memory
11-E	Longer range of longitude positioning
Space	
12-QE	More space for knees between the door panel and the transmission hump
13-Q	More space between the steering-wheel and the driver's torso
14-E	Sufficient space between the knees and the steering-wheel
15-E	Enough space behind the driver's seat for the longitudinal adjustment
16-E	Retractable steering-wheel
Storage	
17-E	Cup-holder reachable from all postures
18-E	Table for food, laptop, etc.; reachable and retractable from all postures
19-E	Storage for glass, pens and controller

(continued)

Table 2. (*continued*)

No.[c]	Requirements
Other	
20-Q	No middle transmission hump
21-Q	Touch screen and voice-control
22-E	Socket for various electric devices
23-E	Internet access
24-E	Dimmable internal space (very bright for a workplace to dark for sleeping)
25-E	Lighting (e.g. ambient, reading-light)
26-E	Interfaces for different tablets and laptops to the in-vehicle displays
27-E	Displays (possible positions: sun visor, roof, in front of steering-wheel)
28-E	Large Head-up-display for watching movies or working
29-E	Keyboard for the on-board PC
30-E	Games console with wireless controller

[c]The number and source of the requirements, Q: questionnaire; E: experiment
[d]The max. seat back angle in the experiment is 68°, but it is wished to be 90° when ‘sleeping’.

8 Summary

This work presented contributes to giving the first introduction of the driver's NDPs in automated vehicles in terms of their background, motivation and influential factors, as well as the qualitative and quantitative descriptions.

In the online survey (n = 122), 13 NDRTs are identified to be conducted by significantly ($\alpha < 0.05$) more drivers in HAD (Fig. 2). By doing those, the most chosen posture has been shown to still be the current driving posture; nevertheless, other NDPs, e.g. the reclined position, are shown. NDRTs map (many-to-many) to NDPs (Fig. 3), which means there could be transitions among different NDRTs in one NDP and vice-versa. In addition, more space around the torso and the knee is expected (Fig. 4). In the experiment (n = 16) in the MEPS, 'traffic scenario' and 'trust in automation' as the (in)direct factors influence the choices of NDRTs and NDPs. The four supervised postures (Fig. 5), and six unsupervised postures (Fig. 6), are measured with the torso angle, knee angle, thigh angle and their seat back adjustments, whose enlarged ranges are summarised in Table 1. In the end, 30 preliminary requirements (Table 2) for the interior of automated vehicles in terms of seat, space, storage, etc., are derived based on the results of these two empirical studies. Four hypotheses are therewith confirmed.

The results make it clear that users of HAD will conduct more NDRTs and stay in different postures, known as NDPs, which could differ from, and have wider ranges than, the driving posture. These facts bring the future research into two directions. On the one hand, from a static point of view, driving-task/driving-posture orientated interior concepts should be adjusted to be NDRT and NDP-orientated in terms of functionality and ergonomics. These changes of goals and contexts will lead to fundamental changes in the whole development process, starting from requirements, specification, design, and prototyping, to testing and validation. On the other hand, from a dynamic point of view, different NDPs means different drivers' motoric states [12], which all have to be transited to a certain state that is adequate for handling the dynamic control of the vehicle within a very limited period of time, should the automation system reach its boundary. It is therefore meaningful to investigate these transitions among different driver's postures when the take-over request (TOR) happens, and compare their take-over performances in different scenarios, to evaluate if and how NDPs derogate the take-over performance. In addition, at the physical level, transitions among different states of interior elements (e.g. lighting, steering-wheel, seat-belt, seat adjustments) may have an enormous potential to assist drivers in different transition phases by, for example, raising their situation awareness (SA) during HAD [13], helping to perceive the TOR, understand the scenario and recover the driving posture more quickly to achieve a better take-over performance.

References

1. Gold C (2016) Modeling of take-over situations in highly automated vehicle guidance
2. SAE J3016 (2016) SAE J3016 Taxonomy and definitions for terms related to driving automation systems for on-road motor vehicles RATIONALE

3. Pfleging B, Rang M, Broy N (2016) Investigating user needs for non-driving-related activities during automated driving. In: 15th international conference on mob ubiquitous multimedia (MUM 2016), pp 91–99
4. Huemer AK, Vollrath M (2012) Ablenkung durch fahrfremde Tätigkeiten – Machbarkeitsstudie
5. Winner H, Wachenfeld W (2015) Auswirkungen des autonomen Fahrens auf das Fahrzeugkonzept. Autonomes Fahren. Springer, Heidelberg, pp 265–285
6. Bubb H, Grünen RE, Remlinger W (2015) Anthropometrische Fahrzeuggestaltung. In: Automobilergonomie. Springer Fachmedien Wiesbaden, Wiesbaden, pp 345–470
7. Alliance of Automobile Manufacturers (2003) Statement of principles, criteria and verification procedures on driver interactions with advanced in-vehicle information and communication systems, p 23
8. SAE International (2009) SAE J1100: motor vehicle dimensions
9. Campbell I (2007) Chi-squared and Fisher-Irwin tests of two-by-two tables with small sample recommendations. Stat Med 26:3661–3675
10. Lorenz S (2011) Assistenzsystem zur Optimierung des Sitzkomforts im Fahrzeug. (Dissertation). Lehrstuhl für Ergonomie, TUM
11. McCarney R, Warner J, Iliffe S et al (2007) The Hawthorne effect: a randomised, controlled trial. BMC Med Res Methodol 7:30
12. Marberger C, Mielenz H, Naujoks F et al (2017) Understanding and applying the concept of "Driver Availability" in automated driving. Springer, Cham, pp 595–605
13. Yang Y, Götze M, Laqua A, et al (2017) A method to improve driver's situation awareness in automated driving. In: Proceedings of the human factors and ergonomics society Europe chapter 2017 annual conference, pp 29–47

Use of a H∞ Controller on a Half Semi-trailer Truck Model to Reduce Vibrations and Its Implications on Human Factor

A. Oguzhan Ahan[1], Dogan Onur Arisoy[2(✉)], Kenan Muderrisoglu[3], Burcu Arslan[4], and Rahmi Guclu[1]

[1] Mechanical Engineering Department, Yıldız Technical University, Istanbul, Turkey
[2] Control and Automation Engineering Department, Yıldız Technical University, Istanbul, Turkey
arisoy@yildiz.edu.tr
[3] Anadolu Isuzu Automotive Industry and Trading Inc., Kocaeli, Turkey
[4] Psychology Department, Middle East Technical University, Ankara, Turkey

Abstract. Comfort of a large goods vehicle driver has a crucial factor on fatigue one of which being physical (muscular) fatigue, which has common indicators of pain in the back and legs caused by prolonged voluntary and involuntary muscle activities as an attempt by the body to counteract the vibrations induced via the seat. Therefore, drivers working under uncomfortable driving conditions, encounter ergonomic problems, increased fatigue and require additional rest time. Consequently, these rests increase both in time and numbers resulting with lost logistics time and reduced efficiency. Additionally, these drivers suffer from medical issues both physically and mentally. It is shown that muscular health complaints is mostly accompanied by mental health complaints (e.g.: stress, burnout etc.). These muscular and mental health complaints increase drivers need for recovery, however drivers undergo problems to recover from this work-related fatigue at the end of work day. Though it may be seen as an acute issue, in long term, the accumulation of restlessness may induce the development of psychosomatic health problems. Hence, drop in medical status and efficiency of drivers burden the organizations with major financial costs, via sickness absence of worker, compensations of absenteeism, errors or mistakes occurred at work caused by sudden pain, stress, and etc. Therefore an uncomfortable driving experience becomes not only a health issue for the drivers but also an economical inconvenience to the company. To increase comfort and prevent these issues a H∞ controller with a dynamic output feedback was proposed in this study. This controller was designed and modeled with linear matrix inequality (LMI) method, and implemented on a model of a half semi-trailer truck augmented with a human-seat couple model. Both uncontrolled and controlled cases were simulated and compared in terms of the comfort level of the driver with respect to ISO 2631 standard. As a result, controller decreased the root mean square (RMS) acceleration on the human-seat couple and increased the comfort. Originating from these results, a discussion was made in the perspective of human factor such that, reducing the vibration is not only meant to increasing comfort of driver, but also eliminating the negative effects of discomfort on safe driving and reducing fatigue.

S. Bagnara et al. (Eds.): IEA 2018, AISC 823, pp. 421–435, 2019.
https://doi.org/10.1007/978-3-319-96074-6_45

Keywords: Driver ergonomics · Vehicle vibrations · Multi-body analysis Modern control · Output feedback control · H∞ controller · Human factor

1 Introduction

Driver comfort and ergonomics are important cases in overland logistics which is a major area of employment [1]. Therefore research areas such as driver comfort and health are crucial topics of interest. In the research done by Kyung et al. they evaluated their experimental results on 6 different driver's seats in terms of comfort on 27 participants [2]. In the continuation of their study they concluded that the pressure between hip and seat is the primary reason for an uncomfortable driving experience [3]. Tiemessen et al. proposed that whole-body vibrations (WBV) to be the main reason for lumbago (lower back pain) and labor time losses [4]. Coyte et al. gave an extensive literature review on modeling and analysis of WBV. In their paper a detailed methodology for analysis of vibration on ergonomics and health was presented [5]. Blood et al. studied front-end loader operator working conditions. Under different work tasks WBV were measured, then which were analyzed using ISO 2631-1 and -5 standardizations [6]. Thamsuwan et al. researched bus drivers' biodynamic responses under different road conditions. Additionally effects of these conditions on health were investigated using the ISO 2631-1 and -5 standardizations [7]. Ahan et al. studied effects of vibration mitigation on driver comfort utilizing a semi-trailer truck model and an optimal controller [8].

Control theory is widely used to mitigate undesired conditions of various systems. Methods, which may be distinguished by direct and indirect approximations, are researched and applied on many systems for decades. Fuzzy logic controllers, which are indirect methods of control, were applied on vehicle systems by Guclu [9]. Thence developing methodology, Guclu and Gulez studied on artificial neural networks to achieve passenger comfort at full-vehicle model via the control of linear permanent magnet synchronous motors [10]. Direct methods such as optimal control create significant results due to their efficacy on the establishment of desired conditions. So called H∞ controllers can mitigate dominant parts of mechanical vibrations effectively. Li et al. applied an H∞ controller on a quarter car model and reduced vibratory mechanical responses of the system [11]. Sun et al. studied a dynamic output feedback H∞ controller to lessen the vibratory effects on human and achieve comfort [12]. On the other hand, Zhao et al. studied critical application problems such as actuator delays, parameter uncertainties etc. and as a result, they presented a design methodology to mitigate these problems [13].

Extensive review studies done by Ljungberg and Parmentier [14], and Marras [15] showed that vibrations are risk factors for both physical and psychological health of drivers. Exposure of professional drivers to WBV are fundamental issues regarding their safety, resulting in great disadvantages for their employers. Long exposure times to vibrations mostly result in lumbago [16, 17], fatigue [18], which is one of most common cause of fatal accidents on roads [19, 20], and also impairment of driver performance [21]. Muscular health complaints (back pain) along with mental health complaints (elevated stress and fatigue) increase driver's need for recovery. In some cases due to work-related fatigue drivers might have problems while recovering at the end of work day. Though need for recovery is seen as immediate and acute issue, in long term, it can be reason for development of psychosomatic health problems, which cause drivers to be absence due to sickness [22]. Kresal et al., argued that muscular health complaints are major financial cost for organizations due to sickness absence of drivers. Compensation of absenteeism, occurrence of errors or mistakes due to sudden pain, stress and painkillers cause significant burdens both for drivers and organizations [23]. In addition to health complaints, high exposure to vibration impair performance of the drivers. High levels of vibration reported to have adverse effects on tracking performance [21]. Reduced performance, in turn, affects work efficiency and increases the risk of serious accidents [14].

In this study, physiological effects of the vehicular WBV have been researched and their implications on human factor has been discussed. Physiological effects have been studied in according to ISO 2631-1 [24] and ISO 2631-5 [25] specifications. Whole Body Vibration Dose, Static Compressive Stress Value and Adverse Health effects have been calculated in accordance with the procures given in respective standards. Study has been conducted on a half semi-trailer truck model, for uncontrolled and controlled cases and between 0–100 km/h velocities.

2 Materials and Methods

2.1 Modelling

A half semi-trailer truck model (see Fig. 1), which consists of a 3 axle tractor and a 2 axle trailer, has been used as a working platform in this study. Due to the chosen half vehicle approach only vertical and pitch motions have been taken account. Hence roll, yaw and maneuvering coordinates have been neglected. Tractor part has been augmented with a 3 degrees of freedom (DOF) biodynamic model of a driver body-seat couple. Masses and moment of inertias for the tractor and semi-trailer have been represented with m_1, J_1 and m_2, J_2 respectively. Additionally J_3 represents the moment of inertia of the trailer bogie. Wheel-axle set masses have been defined as m_3, m_4, m_5, m_6 and m_7. In the biodynamic model seat chassis mass, combined mass of seat cushion-hip couple and combined mass of human torso-head couple have been represented with m_{s1}, m_{s2} and m_{s3}, respectively. Stiffness and damping confidents between the masses have been denoted with k and c letters, respectively. Lengths have been defined with L letters. In the vehicle part of the model, rotational and translational coordinates have been represented with θ and x letters, respectively. Furthermore, in the biodynamic part translational displacements have been shown with z letters. Road profiles have been defined with X_y letters. Finally actuator, which has been denoted with u, placed between the tractor and the seat frame.

2nd kind Lagrange equations methodology has been used to obtain the equations of motion. Then a system matrix of A which as shown in Eq. 1 have been acquired. In the system matrix mass matrices, stiffness matrix and damping matrix have been denoted with M, K and C respectively.

$$A = \begin{bmatrix} [0]_{n\times n} & I_{n\times n} \\ -M^{-1}K & -M^{-1}C \end{bmatrix}_{2n\times 2n} \tag{1}$$

Moreover to define the external effects and observer dynamics on the whole system B, C and D matrices have been utilized. State-space formulation has been obtained as follows.

$$\dot{x}(t) = Ax(t) + Bu(t) \tag{2}$$

$$y(t) = Cx(t) + Du(t) \tag{3}$$

Model parameters given in Table 1 have been used.

Table 1. Model parameters of semi-trailer truck and human-seat couple.

Masses and moment of inertias	Stiffness coefficients	Damping coefficients	Lengths
m_1 = 3750 kg	k = 5000000 N/m	c = 34811 N • s/m	L_1 = 0.775 m
m_2 = 12500 kg	k_1 = 128820 N/m	c_1 = 1464 N • s/m	L_2 = 1.395 m
m_3 = 487.9 kg	k_2 = 992000 N/m	c_2 = 1380 N • s/m	L_3 = 2.395 m
m_4 = 487.9 kg	k_3 = 149420 N/m	c_3 = 1698 N • s/m	L_4 = 2.294 m
m_5 = 487.9 kg	k_4 = 992000 N/m	c_4 = 1380 N • s/m	L_5 = 1.613 m
m_6 = 487.9 kg	k_5 = 149420 N/m	c_5 = 1698 N • s/m	L_6 = 0 m
m_7 = 487.9 kg	k_6 = 992000 N/m	c_6 = 1380 N • s/m	L_7 = 0.982 m
m_{s1} = 15 kg	k_7 = 149420 N/m	c_7 = 1698 N • s/m	L_8 = 0.388 m
m_{s2} = 8.8 kg	k_8 = 992000 N/m	c_8 = 1380 N • s/m	
m_{s3} = 43.4 kg	k_9 = 149420 N/m	c_9 = 1698 N • s/m	
J_1 = 12221.5 kg • m^2	k_{10} = 992000 N/m	c_{10} = 1380 N • s/m	
J_2 = 14399.5 kg • m^2	k_{s1} = 31000 N/m	c_{s1} = 830 N • s/m	
J_3 = 3300 kg • m^2	k_{s2} = 18000 N/m	c_{s2} = 200 N • s/m	
	k_{s3} = 44130 N/m	c_{s3} = 1485 N • s/m	

2.2 Controller Design

State-space model enables the separation of system dynamics, desired performance and observed conditions. Therefore a separated state-space system can be shown as follows.

$$\dot{x}(t) = Ax(t) + B_1w(t) + B_2u(t) \tag{4}$$

$$z(t) = C_1x(t) + D_{11}w(t) + D_{12}u(t) \tag{5}$$

$$y(t) = C_2x(t) + D_{21}w(t) + D_{22}u(t) \tag{6}$$

Where x(t), w(t), z(t) and y(t) represent system states, exogenous inputs, exogenous outputs and observers, respectively. Hence, a feedback controller may be defined as follows.

$$\dot{x}_k(t) = A_k x_k(t) + B_k y(t) \tag{7}$$

$$u(t) = C_k x_k(t) + D_k y(t) \tag{8}$$

Where $x_k(t)$ represents the actuator dynamics with respect to (w.r.t.) output function y(t). Furthermore, desired performance may be defined in accordance with the exogenous conditions of the system. As a result, a controller that always produces required forces to achieve optimal performance can be designed.

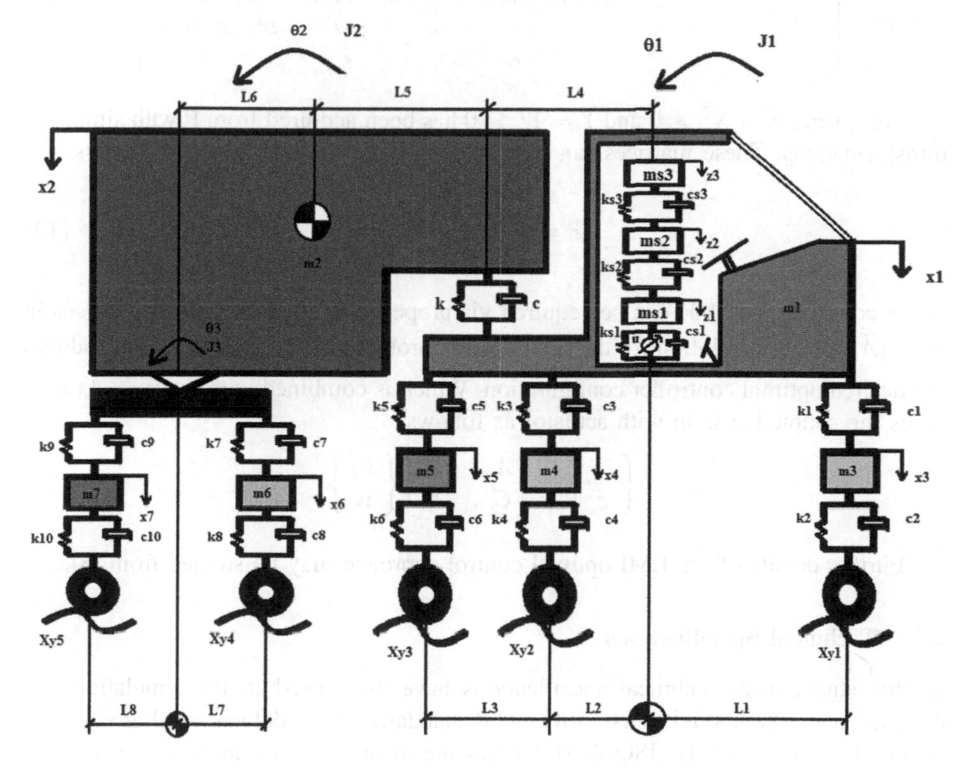

Fig. 1. Semi-trailer truck model with human-seat couple.

Pivoting on this idea, performance can be defined with $\|G\|_\infty = \frac{\|z\|_2}{\|w\|_2} \le \gamma$ and Lyapunov function has chosen as, $P = P^T \succ 0$ and $V(x) = x^T P x$. Therefore a Linear Matrix Inequality (LMI) formulation using Lyapunov function can be constructed as follows.

$$\mathbb{S}_P^\gamma(x, w, u(x_k, y)) = \dot{V}(x) + z^T z - \gamma^2 w^T w \prec 0 \tag{9}$$

Using the definition above, the LMI formulation has been formed. However a goal function had to be defined as $\mathbb{S}_P^\gamma(x, w, u(x_k, y)) \sim = \mathbb{S}_{P,\mathbb{K}}^\gamma(x, w)_{u(t)}$ using a controller characterized with $\mathbb{K} = (A_k, B_k, C_k, D_k)$, before the optimization problem has been introduced as follows.

$$\mathcal{P}_0 : min_{P,\mathbb{K}}(\gamma), const : max_{\forall w}\left\{\mathbb{S}_{P,\mathbb{K}}^\gamma(x, w) \prec 0\right\} \tag{10}$$

After Schur compliment and proper congruence transformations to achieve Linearity, LMI of dynamic output feedback definition has been obtained as follows.

$$\mathbb{S}_{X,Y,\overline{\mathbb{K}}}^\gamma = \begin{bmatrix} XA + A^TX + \bar{B}C_2 + C_2^T\bar{B}^T & \bar{A} + A^T + C_2^T\bar{D}^TB_2^T & XB_1 + \bar{B}D_{21} & C_1^T + C_2^T\bar{D}^TD_{12}^T \\ * & AY + YA^T + B_2\bar{C} + \bar{C}^TB_2^T & B_1 + B_2\bar{D}D_{21} & YC_1^T + \bar{C}^TD_{12}^T \\ * & * & -\gamma I & D_{11}^T + D_{21}^T\bar{D}^TD_{12}^T \\ * & * & * & -\gamma I \end{bmatrix} \prec 0 \tag{11}$$

From here, $X = X^T \succ 0$ and $Y = Y^T \succ 0$ has been acquired from P with similarity transformations. These matrices can be written in their coupled forms as follows.

$$\mathbb{P} = \begin{bmatrix} X & I \\ I & Y \end{bmatrix} \succ 0 \tag{12}$$

A controller set of $\mathbb{K}$ can be acquired via proper simplifications of the matrices in $\overline{\mathbb{K}} = \left(\overline{A}, \overline{B}, \overline{C}, \overline{D}\right)$. Solution for the optimization problem of $\mathcal{P}_0' : \min_{\overline{\mathbb{K}};\mathbb{P}\succ 0}(\gamma)$ produces the desired optimal controller configuration. Which is combined with the $x_{cl} = \{x\,x_k\}^T$ gives the coupled system with actuator as follows.

$$\begin{Bmatrix} \dot{x}_{cl} \\ z \end{Bmatrix} = \begin{bmatrix} A_{cl} & B_{cl} \\ C_{cl} & D_{cl} \end{bmatrix} \begin{Bmatrix} x_{cl} \\ w \end{Bmatrix} \tag{13}$$

Further details of the LMI optimal control approach may be studied from [8].

2.3 Technical Specifications

In this study, three technical specifications have been used in the simulation and evaluation of the model. First, ISO 8608 standardization defines road irregularity standards [26]. Secondly, ISO 2631-1 gives the measurement criteria and evaluation perspective for comfort conditions as well as effects of vibration dosage on health. Finally, ISO 2631-5 describes short and long term health effects of WBV.

In ISO 2631-1 a spectrum between 0.5 Hz and 80 Hz is used for measurements and evaluations. Hence three types of filters are needed for weighted acceleration measurements (a_w) as follows.

$$H_H(s) = \frac{s^2}{s^2 + \frac{2\pi * 100}{0{,}71}s + (2\pi * 100)^2} \tag{14}$$

$$H_L(s) = \frac{(2\pi * 100)^2}{s^2 + \frac{2\pi * 100}{0{,}71}s + (2\pi * 100)^2} \quad (15)$$

$$H_{TV}(s) = \frac{(s + 2\pi * 16)^2}{s^2 + \frac{2\pi * 16}{0{,}63}s + (2\pi * 16)^2} \frac{s^2 + \frac{2\pi * 2{,}5}{0{,}8}s + (2\pi * 2{,}5)^2}{s^2 + \frac{2\pi * 4}{0{,}8}s + (2\pi * 4)^2} * 32{,}768\pi \quad (16)$$

Where H_L, H_H and H_{TV} define a low-pass, a high-pass and a vertical comfort investigation filters, respectively [27]. Therefore a general weight filter may be acquired as follows.

$$H_{vertical}(s) = H_H(s)H_L(s)H_{tv}(s) \quad (17)$$

In accordance with ISO 2631-1 the root-mean square (RMS) value of acceleration is utilized in the evaluation of vibration given as follows.

$$A_w = \sqrt{\frac{1}{T}\int_0^T a_w^2(t)dt} \quad (18)$$

Where, T is the measurement period and has been chosen as 5 s. This approach can be extended to longer exposure times using the correlation given below. In this study, 9 h of exposure have been investigated using the following function.

$$A(8) = \sqrt{\frac{1}{t_d}\sum_{n=1}^{N} A_{wn}^2 t_n} \quad (19)$$

Where, t_d and t_n defined as the duration of exposure time and the period in seconds, respectively. One way of interpreting the effects of the WBV on health is using vibration dose value (VDV), which is calculated as follows.

$$VDV = \left(\int_0^T a_w^4(t)dt\right)^{\frac{1}{4}} \quad (20)$$

In ISO 2631-1 specification, it is advised that an appropriate measurement time should be chosen for the analysis. In accordance 120 s of measurement time has been used in this study. To evaluate the effects of VDV for whole exposure time, the following correlation has been used.

$$VDV(8) = VDV\left(\frac{t_d}{t_m}\right)^{\frac{1}{4}} \quad (21)$$

Where t_m is the measurement time in seconds.

ISO 2631-5 investigates the physiological effects of the acceleration, especially on the lumbar spine. To analyze these points weighted acceleration (a_w) is utilized again, as previously defined in ISO 2631-1, and spine accelerations (a_{li}, i = x, y, z) are obtained as specified in ISO 2631-5 standard [25].

Due to the limitations of the study, only the accelerations in the z-direction have been taken into account. Sampling frequency has been chosen to be 160 Hz, as suggested in the ISO 2631-5 standard. Moreover appropriate measurement time has been advised by this standard. Therefore a measurement time of 120 s has been chosen.

After the calculation of z-directional lumbar spine acceleration, only the positive peak values have been taken into account due to the importance of compressive force on the lumbar spine. Hence the dose have been calculated as follows.

$$D_z = \left[\sum\nolimits_i A_{iz}^6\right]^{\frac{1}{6}} \tag{22}$$

Where A_{iz} is the positive peak values of the z-directional lumbar spine acceleration. To extent this dose to the duration of exposure the following formulation have been used.

$$D_{zd}(8) = \left[\sum\nolimits_{i=1}^{n} D_{zi}^6 \frac{t_{di}}{t_{mi}}\right]^{\frac{1}{6}} \tag{23}$$

Where n defines the number of work periods in a day. To evaluate the health effects of the acceleration, static compressive stress has been calculated as follows.

$$S_{ed}(8) = \left[\sum\nolimits_{k=x,y,z} (m_k D_{kd})^6\right]^{\frac{1}{6}} \tag{24}$$

Where m_z = 0.032 MPa/(m/s^2) is the correlation value to define a static compressive stress in terms of the acceleration dosage.

To evaluate the long term effects of vibration on health a Factor R calculation, which takes into account daily exposure as well as increased age and reduced strength, is given in ISO 2631-5 standard. To find Factor R a preliminary value of S_{ui}, which is correlated with bone density of the vertebrae, has been calculated as follows.

$$S_{ui} = 6,75 - 0,066(b + i) \tag{25}$$

$$R = \left[\sum\nolimits_{i=1}^{n} \frac{S_{ed} N^{\frac{1}{6}}}{S_{ui} - c}\right]^{\frac{1}{6}} \tag{26}$$

Where i and b are defined as the time counter and the exposure starting age, respectively. To represent the static stress of the gravitational force on driving posture a constant of c = 0.25 MPa has been used. Additionally, the exposure days per year and the number of exposure years have been described with N and n, respectively.

3 Results

Simulation studies have been conducted under three different set ups using MATLAB/SIMULINK program. First configuration has been used to investigate the driver biodynamics for uncontrolled and controlled scenarios. Second one has been used to observe the comfort condition of the driver. Finally, the third set up has been used to analyze the vibration dosage and its heath effect on the driver.

Biodynamic model analysis has been conducted at 60 km/h vehicle velocity. Relative displacement and acceleration values of frame, cushion and torso have been obtained in time domain as given in Fig. 2. Due to the actuator being positioned between truck chassis and seat frame, absolute displacements and accelerations on the driver model have been mitigated. This may be observed from Fig. 2 in the top center and top right graphs as the displacement differences have been reduced almost to zero for controlled case. For the relative displacement between seat frame and truck chassis, shown in top left graph in Fig. 2, has been increased for controlled case compared to the uncontrolled one. This has been occurred due to the actuator keeping the seat frame still in absolute coordinates therefore transferred vibrations have been minimized for the human-seat couple as shown in bottom row of Fig. 2.

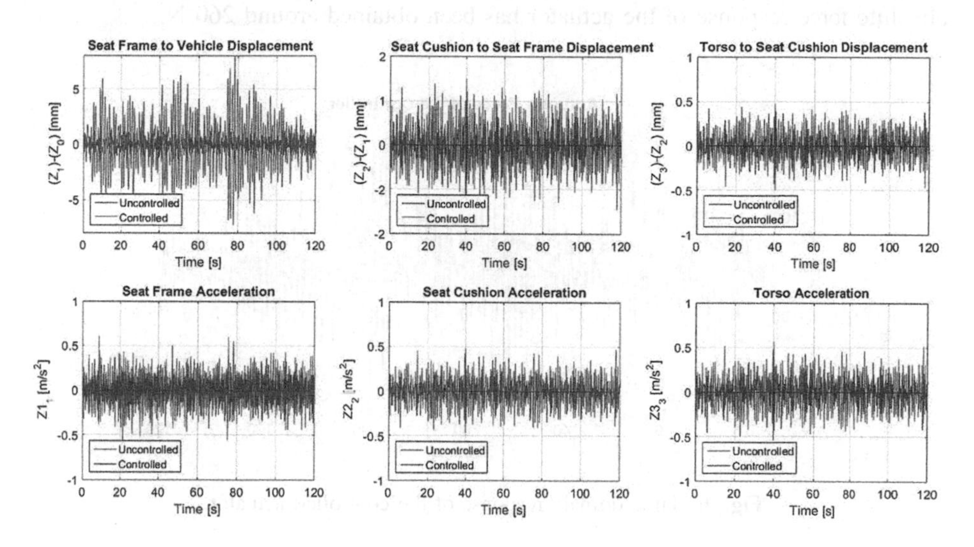

Fig. 2. Time domain responses of the biodynamical human-seat model.

To investigate the frequency response of the system, spectral analysis has been conducted (see Fig. 3). All acceleration frequencies between 0.5 to 80 Hz have been well diminished, except for the relative displacement between truck and seat (see top left graph of Fig. 3). This have been also happened due to the same reason as explained above.

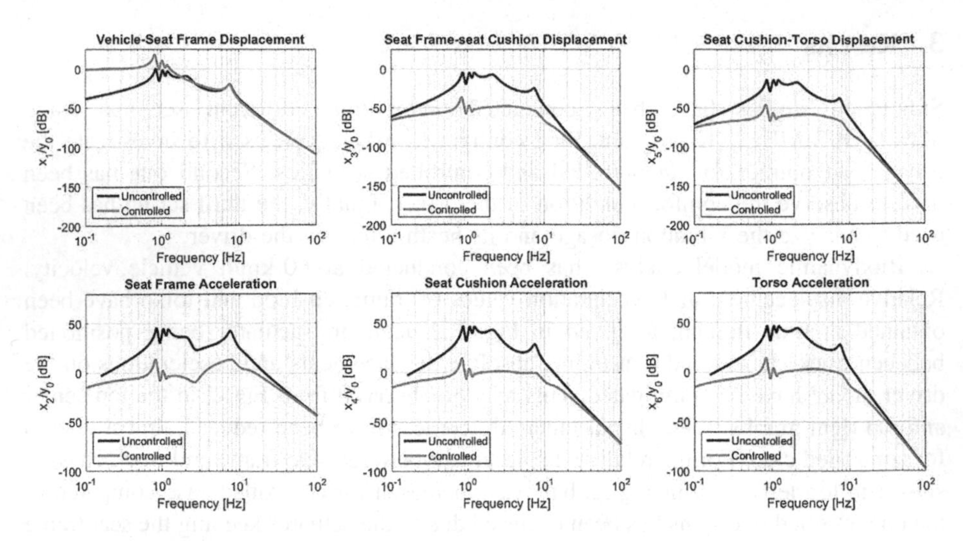

Fig. 3. Frequency domain responses of the biodynamical human-seat model.

Generated actuator forces during this session are presented in Fig. 4. Maximum absolute force response of the actuator has been obtained around 260 N.

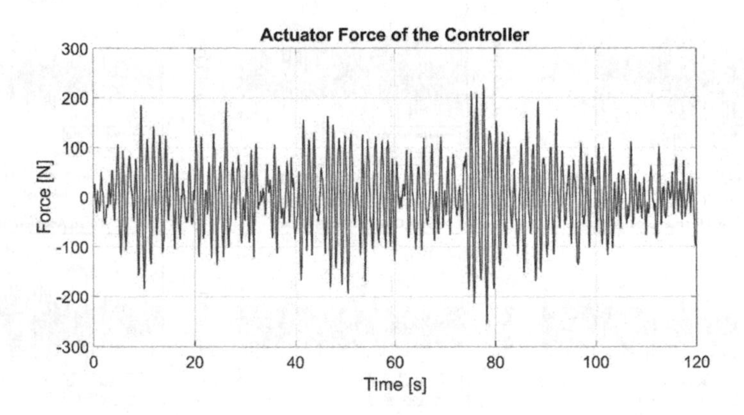

Fig. 4. Time domain response of the controlled actuator.

Comfort, vibration dose value and adverse health effect analyses have been also conducted with varying vehicle velocities. In Fig. 5, comfort parameters and RMS values of actuator forces are given w.r.t. different vehicle velocities. As can be seen from the figure values of the comfort parameter, which is the RMS value of weighted accelerations (a_w), have been dramatically reduced from 0.15–0.25 m/s^2 band in uncontrolled case to practically zero in the controlled one for all velocities. Additionally, a correlation between RMS of the actuator force and the comfort parameters may be observed.

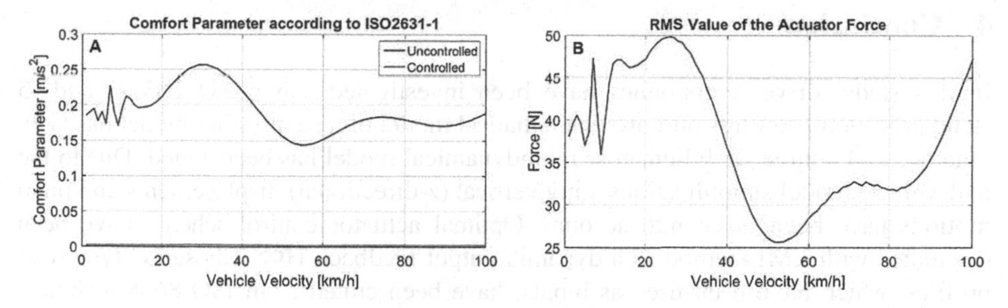

Fig. 5. (A) Comfort parameter and (B) RMS value of the actuator force w.r.t. vehicle velocity.

A driver's condition is not only dependent on the comfort conditions. Dosage of vibrations also play a critical role on it. Therefore in Fig. 6 calculated VDVs, for 9 h of work trend, are shown w.r.t. different vehicle velocities.

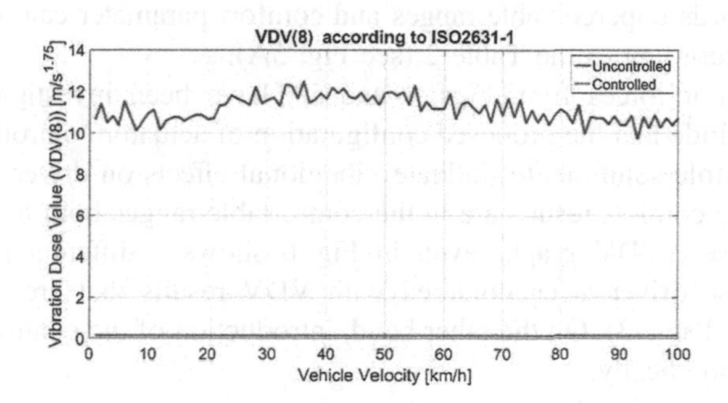

Fig. 6. Vibration dose value w.r.t. vehicle velocity.

Static compressive stress on lumbar spine and adverse health effects have been computed as specified by the ISO 2631-5 standard and presented in Fig. 7. Even though the applied methodology is correct controlled case resulted with zero outputs for both graphs. This is due to the small acceleration values being neglected by the given method in ISO 2631-5 standard.

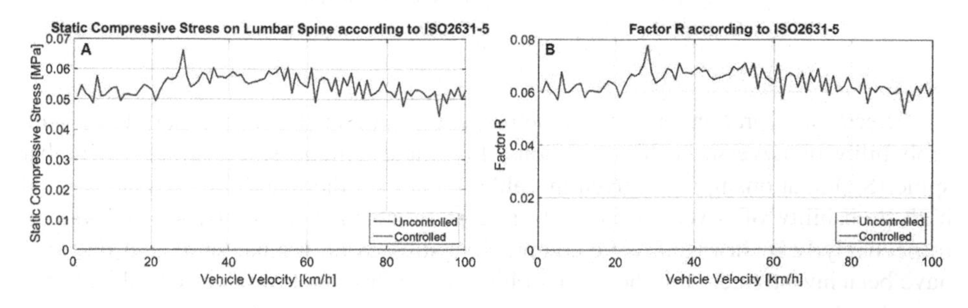

Fig. 7. Effects of vibration on (A) static compressive stress on lumbar spine and (B) Factor R w.r.t. vehicle velocity.

4 Conclusion

In this study, driver ergonomics have been investigated using ISO 2631-1 and -5 standardizations. For this purpose mathematical model of a semi-trailer model has been obtained and a three DOF human-seat biodynamical model has been added. Due to the half-vehicle model simplifications, only vertical (z-directional) displacements and pitch motions have been taken into account. Optimal actuator control scheme have been calculated with LMI method as a dynamic output feedback H∞. Chosen C type road profiles, which have been used as inputs, have been crated from ISO 8608 technical specification. Simulation studies have been done using MATLAB/SIMULINK program. Comfort, vibration dosage and health effect analyses have been conducted, as specified by ISO 2631-1 and -5, both for controlled and uncontrolled cases.

According to ISO 2631-1 standardization, humans cannot be able to sense vibrations that are below 0.01–0.015 m/s^2 [24]. Even though the uncontrolled vehicle is highly comfortable, with the application of proposed controller vibrations can be diminished towards unperceivable ranges and comfort parameter can be improved for all velocities, according to the Table 2 (see Fig. 5A).

When, actuator forces from Figs. 4 and 5A have been investigated, it may be possible to conclude that the proposed configuration of actuator controlled seats can be commercial capable solutions to mitigate vibrational effects on drivers.

Although the comfort results are in the comfortable ranges both for controlled and uncontrolled cases, VDV graph given in Fig. 6 shows a different picture. For the uncontrolled case driver is encountered with VDV results that are above the daily action limit (see Table 3). On the other hand, introduction of the controller reduces the VDV results dramatically.

Table 2. Comfort level w.r.t. comfort parameter [24].

Comfort Parameter (m/s^2)	Comfort Level
Less than 0.315	Not uncomfortable
0.315–0.63	A little uncomfortable
0.5–1	Fairly uncomfortable
0.8–1.6	Uncomfortable
1.25–2.5	Very uncomfortable
Greater than 2	Extremely uncomfortable

Effects of vibration on lumbar spinal stress and Factor R are correlated with probability of adverse health conditions. For static compressive stresses on lumbar spine (S_{ed}) relationship is as given in Table 3. On the other hand, low probability and high probability of adverse health effects correspond to R < 0.8 and R > 1.2 values, respectively [25]. In Fig. 7 static compressive stresses on lumbar spine and Factor R have been investigated. For the uncontrolled case both show low adverse health effects.

However for the controlled case results have been found to be zero. Therefore they may not be interpreted quantitatively but comparatively, based on the indirect evidence provided by Figs. 2 and 5A, as lower possibilities of adverse health effects compared to uncontrolled case.

As expected from the proposed optimal controller increased comfort as well as reduced vibration dosage and adverse health effects have been achieved.

Table 3. Daily (8-h) action and exposure limits for whole body vibration [7].

	ISO 2631-1		ISO 2631-5	
	A_w (m/s^2)	VDV ($m/s^{1.75}$)	S_{ed} (MPa)	Probability of an adverse health effect
Action limit	0.5	9.1	0.5	Moderate
Exposure limit	0.8	14.8	0.8	High

5 Discussion

Implications of the model regarding safety and health of drivers is considerable. Eliminating vibrations during driving may decrease driver's health complaints, back pain, along with reduced psychological impacts such as stress and fatigue. Even though it can be argued that vibrational exposure during driving is not the sole cause of stress or fatigue for a driver, Ljungberg and Neely state that lower exposure to WBV do not jeopardize drivers' performance, even when work is considered as stressful and/or difficult [28]. Therefore, increasing productivity and eliminating health related loses should be most important implications for organizations.

One of the main limitations of the study is the half-vehicle approach which omits the translational movements in x- and y-coordinates as well as rotational movements of roll and yaw. However vibrational effects on these axes is shown to be significant for comfort and health [7]. Additionally, two dimensional structure of the system disregards the road bends that may affect the results. Other important limitations are caused by both the in-silico nature and setup of the study which neglects the road profile changes, vehicle speed fluctuations, braking effects, weather conditions etc. These conditions may cause different levels of change in vibrational behavior of the system. In every system, although unknown exogenous disturbances play important roles on the results, utilization of disturbance rejection methods would guarantee desired performances. Also, standardized ideal postures of biodynamic models impose a conservatism on the system. To overcome this issues a more parameter intensive model with advanced modern control techniques may be utilized in the future studies.

Proposed controller is shown to increase the comfort of the driver in terms of ISO 2631-1 and decrease the adverse health effects in terms of ISO 2631-5. Suggested method provides guaranteed comfort and health conditions for the driver. This approach can also be expanded into different road cases and vehicles, and may be augmented with robust control approaches for better disturbance cancellation in the future studies. That way it would be possible to assure best comfort for the driver in any condition, as well as to decrease vibration induced fatigue and health problems.

In this study, negative effects of long term vibrational exposure on health is studied. Although the results are in the safe zones (except for VDV) for the uncontrolled case, longer exposure to vibrations may still cause health issues due to individual differences, extended shifts or overworking hours. The expected outcome of controlled case is the reduction of adverse health effects of vibrations. However the method recommended in 2631-5 is set to neglect the very small vibrational accelerations (less than 0.015 m/s^2), which is the common occurrence in the controlled case. Therefore the results for adverse health effect analysis returned null (zero) values, as shown in Fig. 7. Hence it is not possible to assess the effects of the controller on health quantitatively. Therefore a comparative indirect approach is used to assess the health effects which utilizes acceleration results (see Figs. 2 and 5A), and said to be controller reduces the adverse health effect on the driver. At this point it may be suggested that, a novel approach which would enable the quantification of adverse health effects for smaller vibrational accelerations should be developed to meet the requirements of today; as the applications on ergonomics tend to employ more modern control approaches with each passing day.

References

1. Bureau of Labor Statistics, Occupational Employment and Wages, 53-3032 Heavy and Tractor-Trailer Truck Drivers. http://www.bls.gov/oes/current/oes533032.htm. 12 Apr 2018
2. Kyung G, Nussbaum MA, Reeves KB (2008) Driver sitting comfort and discomfort (part I): use of subjective ratings in discriminating car seats and correspondence among ratings. Int J Ind Ergon 38(5–6):516–525
3. Kyung G, Nussbaum MA, Reeves KB (2008) Driver sitting comfort and discomfort (part II): relationships with and prediction from interface pressure. Int J Ind Ergon 38(5–6):526–538
4. Tiemessen IJ, Hulsholf CTJ, Frings-Dresen MHW (2007) An overview of strategies to reduce whole-body vibration exposure on drivers: a systematic review. Int J Ind Ergon 37 (3):245–256
5. Coyte JL, Stirling D, Du H, Ros M (2015) Seated whole-body vibration analysis, technologies, and modeling: a survey. IEEE Trans Syst Man Cybern Syst 46(6):725–739
6. Blood RP, Rynell PW, Johnson PW (2012) Whole-body vibration in heavy equipment operators of a front-end loader: role of task exposure and tire configuration with and without traction chains. J Saf Res 43(5–6):357–364
7. Thamsuwan O, Blood RP, Ching RP, Boyle L, Johnson PW (2013) Whole body vibration exposures in bus drivers: a comparison between a high-floor coach and a low-floor city bus. Int J Ind Ergon 43(1):9–17
8. Ahan AO, Arisoy DO, Muderrisoglu K, Yazici H, Guclu R (2016) Vibration control of a semi-trailer truck for comfort with an output feedback H∞ controller. In: II. International conference on engineering and natural sciences (ICENS), Sarajevo, pp 2053–2061
9. Guclu R (2004) The fuzzy-logic control of active suspensions without suspension-gap degeneration. J Mech Eng Strojniski Vestnik 50(10):462–468
10. Guclu R, Gulez K (2008) Neural network control of seat vibrations of a non-linear full vehicle model using PMSM. Math Comput Model 47(11–12):1356–1371

11. Li H, Jing X, Karimi HR (2014) Output-feedback-based H∞ control for vehicle suspension systems with control delay. IEEE Trans Ind Electron 61(1):436–446
12. Sun W, Zhao J, Li Y, Gao H (2011) Vibration control for active seat suspension systems via dynamic output feedback with limited frequency characteristic. Mechatronics 21(1):250–260
13. Zhao Y, Sun W, Gao H (2010) Robust control synthesis for seat suspension systems with actuator saturation and time-varying input delay. J Sound Vib 329(21):4335–4353
14. Ljungberg JK, Parmentier FBR (2010) Psychological effects of combined noise and whole-body vibration: a review and avenues for future research. Proc Institut Mech Eng Part D J Automob Eng 224(10):1289–1302
15. Marras WS (2001) Spine biomechanics, government regulation, and prevention of occupational low back pain. Spine J 1(3):163–165
16. Tamrin SBM, Yokoyama K, Jalaludin J, Aziz NA, Jemoin N, Nordin R, Abdullah M (2007) The association between risk factors and low back pain among commercial vehicle drivers in peninsular Malaysia: a preliminary result. Ind Health 45(2):268–278
17. Robb MJ, Mansfield NJ (2007) Self-reported musculoskeletal problems amongst professional truck drivers. Ergonomics 50(6):814–827
18. Boggs C, Ahmadian M (2007) Field study to evaluate driver fatigue on air-inflated truck seat cushions. Int J Heavy Veh Syst 14(3):227–253
19. Asleep at the Wheel: The prevalence and impact of drowsy driving, american automobile association foundation for traffic safety. https://www.nhtsa.gov/sites/nhtsa.dot.gov/files/documents/12723-drowsy_driving_asleep_at_the_wheel_031917_v4b_tag.pdf. Accessed 19 May 2018
20. Ting P, Hwang J, Doong J, Jeng M (2008) Driver fatigue and highway driving: a simulator study. Physiol Behav 94(3):448–453
21. Sommer HC, Harris CS (1973) Combined effects of noise and vibration on human tracking performance and response time (No. AMRL-TR-72-83), Air Force Aerospace Medical Research Lab Wright-Patterson AFB OH
22. de Croon EM, Sluiter JK, Frings-Dresen MH (2003) Need for recovery after work predicts sickness absence: a 2-year prospective cohort study in truck drivers. J Psychosom Res 55 (4):331–339
23. Kresal F, Roblek V, Jerman A, Meško M (2015) Lower back pain and absenteeism among professional public transport drivers. Int J Occup Saf Ergon 21(2):166–172
24. International Organization for Standardization (ISO), ISO 2631-1 (1997) Mechanical vibration and shock - evaluation of human exposure to whole-body vibration. Part 1: general requirements
25. International Organization for Standardization (ISO), ISO 2631-5 (2004) Mechanical vibration and shock - evaluation of human exposure to whole-body vibration. Part 5: method for evaluation of vibration containing multiple shocks
26. International Organization for Standardization (ISO), ISO 8608 (2016) Mechanical vibration - road surface profiles - reporting of measured data
27. Orvnas A (2011) On active secondary suspension in rail vehicles to improve ride comfort, KTH School of Engineering Science, PhD thesis, Stockholm
28. Ljungberg JK, Neely G (2007) Stress, subjective experience and cognitive performance during exposure to noise and vibration. J Environ Psychol 27(1):44–54

Take-Overs in Level 3 Automated Driving – Proposal of the Take-Over Performance Score (TOPS)

Jonas Radlmayr(✉), Madeleine Ratter, Anna Feldhütter, Moritz Körber, Lorenz Prasch, Jonas Schmidtler, Yucheng Yang, and Klaus Bengler

Chair of Ergonomics, Technical University of Munich, Boltzmannstr. 15, 85748 Garching, Germany
jonas.radlmayr@tum.de

Abstract. Research on take-over performance in conditionally automated driving (Level 3) has led to many publications that analyze human performance and behavior in take-over situations. These are of utmost importance for the safety and comfort of Level 3 systems. Take-over performance is reported using metrics that can be divided into time aspects such as reaction times and quality aspects such as vehicle accelerations. While these metrics provide a very detailed reflection of take-over performance, a comparison between different studies remains challenging. We propose a novel way of combining the relevant metrics to facilitate an easier comparison between different experimental settings. For this purpose, the metrics are aggregated to a vehicle guidance parameter (VGP), a mental processing parameter (MPP) and a subjective rating parameter (SRP). We based the aggregation on published data and included correlation matrices per parameter. An initial standardization by range is necessary to allow the aggregation of the metrics leading to the VGP, MPP and SRP. The three resulting parameters integrate relevant metrics of take-over performance to allow for a quicker post-hoc understanding and a horizontal comparison of results from take-over experiments.

Keywords: Conditionally automated driving · Take-over performance

1 Introduction

The introduction of partially automated (Level 2, [1]) vehicles to the consumer market in addition with announcements concerning the availability of Level 3 vehicles [1] has fueled human factors research on automated driving. Past and present research is heavily focused on take-over situations. In case the automated vehicle reaches a system limit, a request to intervene (RtI) is issued that prompts drivers to continue driving manually. Level 3 systems allow drivers to engage into non-driving related tasks (NDRTs). Thus, the take-over needs to be described with detail to allow an understanding of human reactions and choices after a RtI. Reviewing previous literature, we found a variety of metrics describing drivers' reactions in a take-over situations. Most prominently, these metrics represent reaction times such as gaze reaction time, hands-

S. Bagnara et al. (Eds.): IEA 2018, AISC 823, pp. 436–446, 2019.
https://doi.org/10.1007/978-3-319-96074-6_46

on time, first undeliberate reaction, first deliberate or conscious reaction, duration of lane change or end of driver intervention. In addition, metrics of take-over quality such as maximal lateral and longitudinal accelerations, minimum time to collision, use of indicators, checking the mirrors or crash rate were reported. The take-over performance can be understood to consist of time and quality aspects [2]. Several authors also reported additional subjective ratings such as perceived criticality and complexity of the take-over situation, time pressure or stress.

The described abundance of metrics jeopardizes a sensible comparison of take-over performance between different studies because it increases the chance of a non-overlap of metrics between the studies. While all of these metrics add to a deeper understanding of take-over performance, standardization and aggregation of some of them may allow an easier and more comprehensive understanding of what researchers are interested in: "What happened in that take-over?".

We propose an integrative framework, the *take-over performance score* (TOPS), that includes the most important take-over metrics from the interval between RtI and system limit and aggregates them to three main dimensionless parameters.

Our proposal of the TOPS includes the following individual metrics that are aggregated to:

1. **Vehicle Guidance Parameter (VGP):** crash (yes/no), time to collision, maximal lateral and longitudinal acceleration
2. **Mental Processing Parameter (MPP):** lane check (yes/no), gaze reaction time, eyes on road reaction time, take-over time (TOT)
3. **Subjective Rating Parameter (SRP):** perceived criticality, perceived complexity of the situation, subjective time budget

In order to allow a sensible aggregation, the framework includes a correlation matrix for each parameter including the proposed metrics based on data from [3]. This work looked at the influence of different NDRT's and situations on take-over performance and found significant results between situations. We chose the database due to its availability and because the reported metrics cover the most relevant ones for the three parameters. Prior to calculating the correlation matrices, we preselected the metrics based on an expert review from the authors. Results from the correlation matrix show whether specific metrics offer only small incremental insight into take-over performance (high correlation) or if they offer a high amount of unique information (no/low correlation).

Each parameter is calculated by adding the standardized metrics, weighted based on their correlation with other metrics. The final three parameters representing the TOPS offer a common and comprehensible ground for authors and allow a consistent, objective rating of take-over performance throughout different situations and experiments while they simplify their comparison at the same time.

2 Literature Review and Reported Metrics

The TOPS is based on a comprehensive literature review on take-overs in level 3 automated driving. We identified various metrics of take-over performance. The results can be seen in Table 1.

Table 1. References on take-over performance and reported metrics.

References	Reported metrics
[4]	TOT [s], success rate of lane change [%], braking rate (unnecessary deceleration) [%], eyes-on-road time [s], hands-on time [s], steering and braking reaction times (TOT) [s], lane change time [s], head angle [°], workload (NASA RTLX) [], acceptance [], standard deviation of steering wheel angle [°], absolute lateral position [cm]
[5]	Subjective evaluation of perceived automated driving (19 items) [], reaction time (first glance away from NDRT) [s], TOT [s], max. lateral and longitudinal accelerations [m/s^2], TTC [s]
[2, 6]	Reaction time (first glance away from NDRT) [s], eyes-on-road time [s], time to undeliberate reaction [s], TOT [s], TTC [s], max. lateral and longitudinal accelerations [m/s^2], accident rate [%], lane check []
[7]	Intervention time (TOT) [s], TTC [s], max. lateral and longitudinal accelerations [m/s^2], checking mirrors [%]
[8]	Steering and braking response times (TOT) [s], TTC [s], intervention time [s], max. lateral and longitudinal accelerations [m/s^2], min. and max. velocity [km/h], max. yaw [°], steering wheel angle [°], overshoot lane change [m], TTCL, TTCH, ETTCL, ETTCH [s], obstacle clearance [m], road clearance [m]
[9]	TOT [s], TTC [s], max. lateral and longitudinal accelerations [m/s^2]
[10]	TOT [s], max. lateral and longitudinal accelerations [m/s^2], steering angle and angle-speed [°,°/s], SDLP
[11]	Gaze reaction time [s], Road fixation time [s], Movement time [s], Hands-on time [s], Side mirror time [s], TOT [s], max. lateral and longitudinal accelerations [m/s^2], max. steering wheel angle [°], SDLP [m], rate of lane change error [%]
[12]	TOT [s], TTC [s], max. lateral and longitudinal accelerations [m/s^2], solution of situation (crash rate) []
[13]	Hands-on time [s], TOT [s], max. lateral and longitudinal accelerations [m/s^2], expert rating of take-over quality []
[14]	Gaze reaction time [s], road fixation time [s], hands-on time [s], side mirror time [s], TOT [s], indicator [], distribution of reaction types [%], max. lateral and longitudinal accelerations [m/s^2]
[15]	TOT [s], action time [s], inverse TTC [s], max. derivative of control input (steering and braking)
[16]	Max. lateral and longitudinal accelerations [m/s^2], velocity [m/s], subjective safety and effort rating [], undeliberate reaction time [s], TOT [s], brake pedal pressure [%], TTC [s], criticality rating [], rating of usefulness of RtI []
[17]	Steer touch, steer initiate, steer turn [s], car avoid, lane change [s], TOT [s], rating of usefulness and satisfaction of RtI's [], NASA-TLX [], absolute deviation from lane center [m], gaze and head reaction times [s], acceptance rating []

(continued)

Table 1. *(continued)*

References	Reported metrics
[18]	Eyes-on-road time [s], hands-on time [s], automation off time [s], TOT (braking) [s], mirror time [s], SDLP [m], rating of controllability and mental workload []
[19]	TOT (steering and braking) [s], hands-on time [s], crash rate [%], eyes-on-road time [s], lateral accelerations [m/s^2], SDLP [m], rating of distraction []

The table also includes explanations in case the term or definition of a single certain metric diverged among authors. The calculation of the take-over time (TOT) varies in its definition between different studies regarding thresholds or considered time frame. Since the reviewed studies strongly vary in their denomination and corresponding definition of these metrics, we provide the original name of these metrics in parentheses.

Table 1 indicates that there is no common understanding between researchers on which metrics are essential or sufficient to fully describe a take-over. Different reaction times play an important role to describe the process from RtI until drivers have taken over manual control. Take-over quality is typically reported as accelerations or the criticality of a maneuver expressed as time to collision. The TOPS is based on data and metrics reported by [3] but can be adjusted to integrate more metrics if necessary.

3 The Take-Over Performance Score – TOPS

In order to aggregate the multiple metrics into a single parameter, they first have to be standardized. Since our goal is to simplify horizontal comparisons across different studies, methods like z-score are not suitable, since they represent relative values based on a specific data set. We choose to standardize the metrics by dividing the values by their range of possible values. For example, the range of possible values of negative longitudinal acceleration for typical cars spans between -10 and 0 m/s^2. The physical limit is set by the friction coefficient in combination with the normal force on the ground from the car, resulting in a maximum of -10 m/s^2. The maximum range of the TTC (∞) does not allow a standardization by range, so we use the available time budget in the take-over situation at the RtI. In case drivers initially brake hard, values greater than 1 would be cut down to 1, since a min. TTC equal or greater to the time budget at moment of RtI without braking or executing big lateral accelerations can be understood to represent low vehicle dynamics. In addition, most reported values for min. TTC in the presented literature are smaller than the available time budget.

The TOPS calculation transforms the observed values to create a common direction for all scales. A high metric value represents a high take-over performance for any metric. As a result, a TOPS close to 0 represents high vehicle dynamics, slow reaction times and a low or bad subjective rating. A TOPS close to 1 represents a fast, subjectively well rated take-over with low vehicle dynamics (Table 2).

The following equations for VGP, MPP and SRP can be expanded following the example of this framework.

Table 2. List of all metrics used in the proposal of the TOPS and the means of standardizing them in combination with a brief explanation. The metrics rely to the interval between request to intervene and system limit responsible for the take-over.

Metric []	Range of scale and standardization	Explanation
Vehicle Guidance Parameter		
Crash [] – vgp_crash	Binary, "no crash" = 0, "crash" = 1	No standardization necessary
Lateral and longitudinal acceleration [m/s^2] – vgp_long, vgp_lat	0–10 m/s^2 vgp_long(standardized) = 1 − (vgp_long/10 m/s^2) vgp_lat (standardized) = 1 − (vgp_lat/10 m/s^2)	Standardized values are subtracted from 1, to have values of the VGP closer to 1 represent very little vehicle guidance
Minimum time to collision [s] – vgp _ttc	0 (crash) – ∞ (braking to stop) s vgp_ttc (standardized) = vgp_ttc/TimeBudget	See text for detailed explanation. Values > 1 are cut to 1
Mental Processing Parameter		
Lane Check [] – mpp_lane	Binary metric, "no lane check" = 0, "lane check" = 1	No standardization necessary, lane check will only be integrated into the MPP when a lane change maneuver is necessary in the situation
First saccade reaction time away from NDRT (visual NDRT's only) [s] – mpp_react	0 – time budget (7 s in [3]) mpp_react (standardized) = 1 − (mpp_react/time budget)	–
Gaze reaction time (eyes on road time, visual NDRT's only) [s] – mpp_gaze	0 – time budget (7 s in [3]) mpp_gaze (standardized) = 1 − (mpp_gaze/time budget)	–
Time of undeliberate reaction (minimum between Hands-on reaction time and foot-on-pedal reaction time) [s] – mpp_undelib	0 – time budget (7 s in [3]) mpp_undelib (standardized) = 1 − (mpp_undelib/time budget)	–
Take-over time (TOT, minimum between start of deliberate steering or braking reaction time) [s] – mpp_tot	0 – time budget (7 s in [3]) mpp_tot (standardized) = 1 − (mpp_tot/time budget)	–

(continued)

Table 2. (*continued*)

Metric []	Range of scale and standardization	Explanation
Subjective Rating Parameter		
Subjective Criticality [] – srp_crit	−3–3 srp_crit(standardized) = 1 − ((srp_crit + 4)/7)	A high criticality rating resulted in values close to 0, values close to 1 of the SRP represent a very good subjective rating from participants
Subjective Complexity [] – srp_comp	−3–3 sub_comp(standardized) = 1 − ((srp_comp + 4)/7)	–
Subjective Time Budget [] – srp_time	−3–3 srp_time (standardized) = (srp_time + 4)/7	High values of sub_time represent drivers feeling like they had enough time for the take-over

3.1 Vehicle Guidance Parameter

The vehicle guidance parameter (VGP) integrates metrics related to the resulting vehicle dynamics in the take-over situation. The crash rate is integrated by multiplying the value with the sum of the other metrics to obtain a VGP of 0 in case an accident occurred during the take-over. The correlation matrix in Table 3 shows that the minimum longitudinal acceleration is correlated with the time to collision. This can be explained by people staying longer in the ego lane (lower TTC's) tend to brake harder before executing the lane change in [3]. We also observed a small correlation between the lateral and the longitudinal acceleration with higher lateral accelerations also showing higher longitudinal accelerations. Higher longitudinal accelerations (below zero) represent softer braking. The correlation reflects drivers either braking hard and executing the lane change smoothly or little to no braking in combination with a more dynamic lane change maneuver. This is underlined by results from [2]. Thus, we propose weighing the longitudinal acceleration and TTC equally, while the weight for the lateral acceleration is doubled to adequately represent the lateral dynamic within the VGP. The TTC and the longitudinal acceleration can be understood to represent more of the same information due to their correlation with respect to the lateral acceleration adding substantially more information (no correlation with TTC).

$$VGP = vgp_{crash} * \left(\frac{vgp_{ttc} + vgp_{long} + 2 * vgp_{lat} + \sum_1^n k}{4 + n} \right) \tag{1}$$

with *k*: *n* additional, standardized metrics representing vehicle dynamics in a take-over, e.g. SDLP, yaw rate error.

Table 3. Overview of the correlation results for the vehicle guidance parameter metrics. All results are based on data from [3] because the data was fully available. We based the TOPS proposal only on the available data. A validation with additional data is highly recommend for the outlook. $n(TTC)$ = 173 take-over situations, $n(vgp_{long}, vgp_{lat})$ = 189. Note: the errors of the correlation data are not independent because one participants experienced a total of four take-overs. *: p-value < .05, **: p-value < .01, ***: p-value < .001. Confidence intervals are reported in [], upper and lower 95% CI.

Person's r	Longitudinal acceleration	Lateral acceleration	Time to collision (TTC)
Longitudinal acceleration	–	.157* [.293, .015]	.211** [.349, .064]
Lateral acceleration		–	.096 [.242, −.054]

3.2 Mental Processing Parameter

The correlation matrix in Table 4 shows high correlations between all reactions times except of the correlation between the time until the first saccade away from the visual NDRT and the take-over time, being almost significant. While the first saccade away from a visual NDRT is the initial reaction after the RtI, it is typically succeeded by a glance towards the road. After that, drivers relocate their hands and feet resulting in the reaction time until the first undeliberate reaction. This is succeeded by drivers choosing an appropriate reaction for the prevalent take-over situation resulting in the start of the driver reaction, the time to a deliberate reaction, or take-over time (TOT, minimum between either actively steering or braking/accelerating). The maximum distance between all considered reaction times lies between the initial reaction time and the TOT. We conclude, that the reaction times up to the TOT describe a rather automatic process in direct reaction to the RtI, while the TOT is additionally affected by choosing the appropriate maneuver for the prevalent situation. Therefore, we propose weighing the TOT double in comparison to the other reaction times. The lane check is integrated with less importance compared to the crash rate and the VGP. We propose weighing the lane check half in comparison to the reactions time up to the TOT. Participants could also "look but fail to see" when executing a lane check, adding to the feasibility of the divided lane check weight.

$$\mathrm{MPP} = \frac{.5 * mpp_{lane} + mpp_{react} + mpp_{gaze} + mpp_{undelib} + 2 * mpp_{tot} + \sum_1^n k)}{5.5 + n} \quad (2)$$

with k: n additional, standardized metrics representing mental processing after a RtI in a take-over, e.g. maneuver time.

Integration of additional mental processing metrics would focus most likely on additional reaction times.

Table 4. Overview of the correlation results for the mental processing metrics. Results are based on data from [3]. All reaction times are reported in seconds and start at the request to intervene (RtI). $n(mpp_{react}, mpp_{gaze}) = 63$, $n(mpp_{undelib}) = 126$, $n(mpp_{tot}) = 189$. Note: the errors of the correlation data are not independent because one participants experienced a total of four take-overs. *: p-value < .05, **: p-value < .01, ***: p-value < .001. Confidence intervals are reported in [], upper and lower 95% CI.

Person's r	First reaction time	Eyes-on-road reaction time	Time to undeliberate reaction	Take-over time
First reaction time	–	.587*** [.729, .397]	.588*** [.730, 0.396]	.242 [.464, –.008]
Eyes-on road reaction time		–	.428*** [.612, .199]	.282* [.497, .035]
Time to undeliberate reaction			–	.268** [.423, .097]

3.3 Subjective Rating Parameter

Results from the correlation matrix of the subjective metrics can be seen in Table 5.

Table 5. Overview of the correlation results for the subjective metrics. Results are based on data from [3]. The ratings are based on a seven-item Likert scale from −3 to 3, low/high critically and complexity, and sufficient time for take-over, very little to plenty accordingly. $n(sub_{crit}) = 180$, $n(sub_{comp}, sub_{time})$, = 192. Note: the errors of the correlation data are not independent because one participants experienced a total of four take-overs. *: p-value < .05, **: p-value < .01, ***: p-value < .001. Confidence intervals are reported in [], upper and lower 95% CI.

Person's r	Subjective Criticality – sub_crit	Subjective Complexity – sub_comp	Subjective Time Budget – sub_time
Subjective Criticality – sub_crit	–	.717*** [.781, .638]	−.450*** [−.325, −.560]
Subjective Complexity – sub_comp		–	−.432*** [−.310, −.541]

The correlation matrix shows very strong correlations between all individual subjective metrics. Thus, we propose integrating correlating metrics equally to the subjective rating parameter. The data from [3] is missing a subjective comfort rating. Since all other subjective rating metrics are highly correlated, we hypothesize that an additional comfort rating of the take-over would also correlate with the criticality, complexity and time budget rating. We propose the subjective rating parameter to consist of the individual metrics calculated to:

$$SRP = \frac{\sum_{1}^{n} k}{n} \quad (3)$$

with n standardized, individual subjective rating metrics k weighed equally. Values of k range between 0 and 1, with values of 1 representing the highest possible subjective rating.

For integration of any new metrics to either VGP, MPP or SRP, we propose an expert preselection of the individual metrics to the three parameters including the calculation of new correlation matrices following the example of this framework.

4 Discussion

While the TOPS allows a sensible and transparent way of describing take-over performance with only three remaining parameters, external factors such as traffic density, time of day, weather, additional passengers are not factored into it. Resulting differences between TOP-scores should not be understood to represent better or worse take-over performances but different ones. This is apparent when looking at the inherent trade-off when assessing take-overs: a very fast reaction is very good in situations with limited time budget but could be interpreted to be hasty in uncritical and planned take-over situations such as exiting the interstate. In addition, the TOPS is robust towards how the individual metrics are measured. Varying thresholds or methods are tolerated, so small TOPS differences should not be overinterpreted. For a general guide of practice, we would recommend reporting the TOPS with not more than one decimal place.

Looking at the proposed equation for the MPP, the applicability of the TOPS is limited. Aggregating reaction times in take-overs with a time budget between five and ten seconds, as can be found in most of the reported references, the MPP can significantly reduce the complexity of the results while containing the most important information. In case the time budget would increase to 30 s or more, very fast reactions should not be credited with a high MPP limiting the applicability of the TOPS. Thus, we propose the TOPS for take-over situations with a time budget between five and ten seconds. The MPP also incorporates scenario specific information like the time budget, because it can vary from one experiment or situation to another. The VGP incorporates a standardization based on fixed external references, like the physical limit of accelerations a typical car can exert. Looking at the MPP, results from the correlation matrix could also be understood to mainly look at the take-over time to understand the mental processing of participants during the take-over. [20] investigated the effect of different seat-back angles on take-over performance. While they found no significant results concerning the TOT, the time to hands-on differed significantly. Hence, we propose integrating all available reaction times into the MPP, to allow consideration of all possible effects on reaction times into the MPP.

The SRP shows a high correlation between subjective complexity and criticality, making it easy to aggregate the individual, subjective rating metrics. It can be deducted, that participants did not differ greatly between complexity and criticality in addition with perceiving less time to simply add to the subjective criticality and complexity as well.

Overall, the TOPS represents a first proposal but would benefit from validating the proposed equations for VGP, MPP and SRP respectively with additional data from different experiments. If the TOPS is combined with validation from many experiments, interpreting specific TOPS values to be good or bad take-overs could be added following the examples of the system usability scale (SUS).

References

1. SAE, S.o.A.E.I. (2016) J3016 - Taxonomy and definitions for terms related to on-road motor vehicle automated driving systems. Surface Vehicle Recommended Practive (2016)
2. Gold CG (2016) Modeling of take-over performance in highly automated vehicle guidance. Dissertation, München, Technische Universität München
3. Radlmayr J, Gold C, Lorenz L, Farid M, Bengler K (2014) How traffic situations and non-driving related tasks affect the take-over quality in highly automated driving. Proc Hum Factors Ergonom Soc Annu Meet 58:2063–2067
4. Eriksson A, Banks VA, Stanton NA (2017) Transition to manual: comparing simulator with on-road control transitions. Accid Anal Prev 102:227–234
5. Feldhütter A, Gold C, Hüger A, Bengler K (2016) Trust in automation as a matter of media influence and experience of automated vehicles. Proc Hum Factors Ergonom Soc Annu Meet 60:2024–2028
6. Gold C, Damböck D, Lorenz L, Bengler K (2013) "Take over!" How long does it take to get the driver back into the loop? In: Proceedings of the Human Factors and Ergonomics Society Annual Meeting, SAGE Publications Sage CA: Los Angeles, pp 1938–1942
7. Gonçalves J, Happee R, Bengler K (2016) Drowsiness in conditional automation: proneness, diagnosis and driving performance effects. In: IEEE 19th International Conference on Intelligent Transportation Systems (ITSC), pp 873–878
8. Happee R, Gold C, Radlmayr J, Hergeth S, Bengler K (2017) Take-over performance in evasive manoeuvres. Accid Anal Prev 106:211–222
9. Hergeth S, Lorenz L, Krems JF (2017) Prior familiarization with takeover requests affects drivers' takeover performance and automation trust. Hum Factors 59:457–470
10. Jarosch O, Kuhnt M, Paradies S, Bengler K (2017) Its out of our hands now! effects of non-driving related tasks during highly automated driving on drivers' fatigue. In: International Driving Symposium on Human Factors in Driver Assessment, Training and Vehicle Design
11. Kerschbaum P, Lorenz L, Bengler K (2015) A transforming steering wheel for highly automated cars. In: IEEE Intelligent Vehicles Symposium (IV), pp 1287–1292
12. Körber M, Gold C, Lechner D, Bengler K (2016) The influence of age on the take-over of vehicle control in highly automated driving. Transp Res Part F Traffic Psychol Behav 39:19–32
13. Kreuzmair C, Gold C, Meyer M-L (2017) The influence of driver fatigue on take-over performance in highly automated vehicles. In: 25th International Technical Conference on the Enhanced Safety of Vehicles (ESV) National Highway Traffic Safety Administration
14. Lorenz L, Kerschbaum P, Schumann J (2014) Designing take over scenarios for automated driving. Proc Hum Factors Ergonom Soc Annu Meet 58:1681–1685
15. Louw T, Markkula G, Boer E, Madigan R, Carsten O, Merat N (2017) Coming back into the loop: drivers' perceptual-motor performance in critical events after automated driving. Accid Anal Prev 108:9–18
16. Naujoks F, Purucker C, Neukum A (2016) Secondary task engagement and vehicle automation–comparing the effects of different automation levels in an on-road experiment. Transp Res Part F Traffic Psychol Behav 38:67–82
17. Petermeijer S, Bazilinskyy P, Bengler K, de Winter J (2017) Take-over again: investigating multimodal and directional TORs to get the driver back into the loop. Appl Ergonom 62:204–215

18. Vogelpohl T, Kühn M, Hummel T, Gehlert T, Vollrath M (2018) Transitioning to manual driving requires additional time after automation deactivation. Transp Res Part F Traffic Psychol Behav 55:464–482
19. Zeeb K, Buchner A, Schrauf M (2015) What determines the take-over time? an integrated model approach of driver take-over after automated driving. Accid Anal Prev 78:212–221
20. Yang Y, Gerlicher M, Bengler K (2018) How does relaxing posture influence take-over performance in an automated vehicle? In: Proceedings of the Human Factors and Ergonomics Society Annual Meeting (in Print)

Can Computationally Predicted Internal Loads Be Used to Assess Sitting Discomfort? Preliminary Results

Ilias Theodorakos, Léo Savonnet, Georges Beurier, and Xuguang Wang(✉)

Univ Lyon, Université Lyon 1, IFSTTAR, LBMC UMR_T 9406, 69622 Lyon, France
xuguang.wang@ifsttar.fr

Abstract. Whether internal body loads such as muscle forces and joint forces could be used as objective factors to assess sitting discomfort remains an open research question. The present study investigated the potential relationship between computationally predicted internal loads and subjective discomfort ratings. Volunteers were recruited to provide discomfort ratings on a wide range of sitting configurations resulted in by altering the seat pan angle and the backrest angle of a multi-adjustable experimental seat. Moreover, two preferred seat pan angles were selected by the participants, starting from two different initial seat pan angles, allowing the classification of the trials to preferred and not preferred. Kinematic, force and pressure data served as inputs on a musculoskeletal model that enabled the computation of internal loads. Significant positive correlations were found between the subjective ratings and muscle force, compressive force between L4–L5 and seat pan shear force. Significant reduction in the seat pan shear force was found for the preferred compared to the not preferred trials, but no significant differences were found for the muscle and joint forces. Our results suggest that the seat pan shear force and potentially computationally predicted internal loads could be used to assess sitting discomfort. However, the outcome should be interpreted with caution due to limited number of observations.

Keywords: Sitting discomfort · Musculoskeletal modeling · Inverse dynamics Internal loads

1 Introduction

A possible relationship between subjective feeling of discomfort and objective measures such as pressure distribution, muscle activity and loads could be beneficial for evaluating different seats [1]. In a review study, aiming to determine whether sitting comfort and discomfort ratings are related with objective measurements, it was reported that only pressure distribution has a clear relationship with subjective ratings [1]. On the other hand, the authors reported that more evidence is needed to establish whether posture, internal body loads such as muscle forces and joint forces could be used as objective factors to assess sitting discomfort [1].

S. Bagnara et al. (Eds.): IEA 2018, AISC 823, pp. 447–456, 2019.
https://doi.org/10.1007/978-3-319-96074-6_47

Internal load measurements can be obtained only invasively and thus only a limited number of studies exist in the literature [2–4]. Therefore, computational models are employed to provide estimations of internal loads. In the past, computational studies [5, 6] used a full body musculoskeletal model to assess sitting discomfort based on computed internal loads such as muscles and trunk joint forces. Recently, significant correlations between postural comfort curves and computationally predicted muscle forces were reported [7]. However, no experimental validation of these results has been reported.

The present study aims to investigate whether computed muscle forces and joint reaction forces can be used to assess sitting discomfort. Experimental measurements were obtained using a recently developed multi-adjustable seat [8] and served as inputs to a musculoskeletal model. We investigated the possible relationship between computationally predicted internal forces, based on experimentally collected data, and subjective discomfort ratings given during the experimental procedures. Furthermore, we tested for potential differences between preferred and not preferred trials defined based on participants' selected preferred seat pan angles. We hypothesized that muscles forces, joint forces and seat pan shear forces will be lower for the preferred trials and when the participants experienced less discomfort.

2 Methods

2.1 Experimental Data

Six healthy individuals were recruited for the study (5 males, 1 female; mass: 70.2 (12.1) kg, stature 1745 (100) mm, body mass index (BMI) 22.95 (2.47) kg/m^2). An informed consent form was signed by each participant, and the study was approved by the committee of research involving human participants of IFSTTAR (French Institute of Science and Technology for Transport, Development and Networks).

A reference trial of a standing position was recorded by Vicon motion capture system and also scanned by a 3-D body scanner for each participant. This trial served for scaling a generic musculoskeletal template to each individual's anthropometry.

Next, the participants provided their subjective discomfort ratings on a series of different sitting configurations, resulted in by altering the seat pan and backrest angles of an experimental seat [8] (Fig. 1). A rigid deformed surface obtained from a previous study [9] was used on the seat pan. The backrest angle ranged from 100° to 130° in steps of 15°, while the seat pan angles ranged from 5° forward inclination to 15° backward inclination in steps of 10°. For each backrest angle, two extra trials were performed to determine the preferred seat pan angles selected by the participants. For the first trial, the initial seat pan angle was set to the maximum forward inclination of the seat (−5°, PRLE) while for the second trial, the initial seat pan angle was set to 20° backward inclination (PRHE). Subsequently, the participants were instructed to change the seat pan inclination until they reached a preferred seat pan angle. The trials were randomized with respect to the backrest angle and then the seat pan angle.

For each trial kinematic, force and pressure data were recorded. Fourteen Vicon cameras (1 s, 100 Hz) recorded marker trajectories placed on participants' anatomical

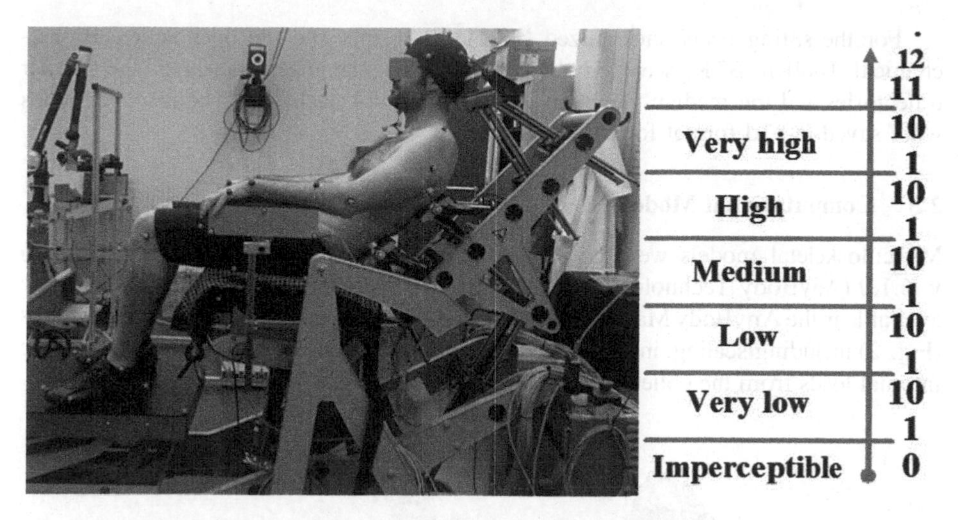

Fig. 1. The experimental seat on the left and the discomfort rating scale on the right.

body landmarks, following a full body marker protocol. Customized force sensors (1.25 s, 20 Hz, LFtechnologie) recorded reaction forces between the experimental seat and participants' feet, thighs, pelvis, arms, head and back in the seat symmetric plane. A PX100.48.48.2 Xsensor pressure mat system recorded the pressure distribution between the participant and the seat pan for 1 s at 25 Hz.

For each seat configuration, after setting the force sensors to zero, participants were instructed to seat on the experimental seat and remain still, until the investigator recorded the kinematic, force and pressure data. Next, the participants were asked to provide their subjective discomfort rating following the CP50 scale [10] (Fig. 1).

2.2 Data Reduction

For the reference trial, the locations of the markers on the 3-D body scans were virtually palpated. Furthermore, the trunk external shape was identified.

Due to experimental limitations, markers could be placed on only two pelvic anatomical landmarks, the right and the left anterior superior iliac spine (RASIS and LASIS). To overcome this limitation, the collected pressure data were used assuming that the pressure peaks would be located under the right and left ischial tuberosity. A MATLAB script was used to automatically detect the two pressure peaks on the seat pan pressure map. A reference trial was employed to identify three points on the pressure mat from both the VICON and the pressure mat systems. Three markers were placed on corners of the seat pan pressure mat and they were manually loaded to allow the simultaneous track from both systems. Then, the positions of those three markers were recorded by the Vicon system for each trial. The two pressure peaks locations were expressed in the global reference system and defined as the ischial tuberosity points (RIT and LIT). These two additional points were appended to the kinematic data. Furthermore, the center of pressure was identified for each trial and expressed to the global reference system following the same procedure as for the RIT and LIT points.

For the sitting trials, customized MATLAB scripts and the open-source Biomechanical ToolKit BTK, were employed to identify the mean values of the marker trajectories and the median values of the force data for each trial. The resulted values were saved in c3d format for further analysis.

2.3 Computational Models

Musculoskeletal models were developed in the AnyBody Modeling System (AMS) v. 6.1.0 (AnyBody Technology A/S, Aalborg, Denmark), based on a generic template available in the AnyBody Managed Model Repository v. 1.6.4. The modeling procedures (Fig. 2) including scaling, inverse kinematics and inverse dynamics followed to provide internal loads from the collected data are shortly described in the next paragraphs.

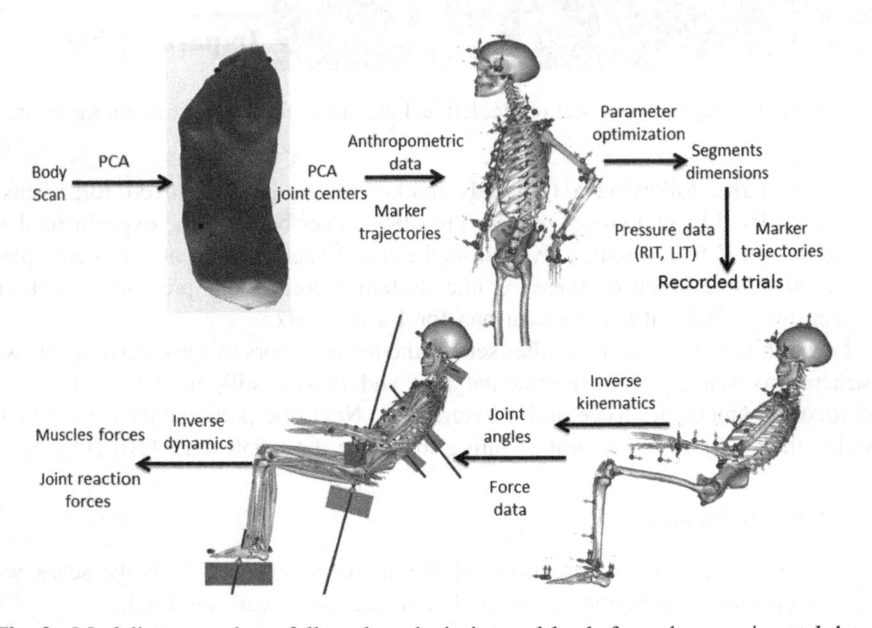

Fig. 2. Modeling procedures followed to obtain internal loads from the experimental data.

Scaling. The reference standing trial was employed to scale the model. Initially, the trunk body shape was used to obtain the thoracic and the lumbar joint centers using a principal component analysis (PCA) based method, as described by Nerot et al. [11]. Subsequently, these joint centers were appended to the external landmarks which were virtually palpated on the body scan. This set of external and internal points (experimental markers) was used as input for scaling the generic template to individuals' geometry.

The model scaling was accomplished by employing an optimization approach [12] that minimizes the least-square differences between the experimental and the model markers. The model markers were virtually palpated on the generic template on locations corresponding to the experimental markers. The output of this procedure was the optimized segments lengths for each participant.

Inverse Kinematics. The optimized segment lengths for each participant, the averaged experimental marker values and the locations of the RIT and the LIT for each trial, served as input for the inverse kinematics analysis for all trials. Joint angles were identified for each trial, using the over-determined kinematic solver [13] which minimizes the least-square differences between the experimental and the model markers.

Inverse Dynamics. The joint angles as computed from the inverse kinematics and the recorded force data served as inputs for the inverse dynamics analysis. Since the employed force sensors do not record reaction moments to allow the computation of the center of pressure for each force platform, assumptions were made regarding the locations of the reaction forces application points. The application points for the feet reaction forces were assumed as the midpoints between the heel and toe markers, for the armrests, the application points were assumed the midpoints of the wrist and the proximal armrest markers. For the back sensors the application points were assumed on the median plane while the center of pressure, as computed from the pressure data, was assumed as the application point of the seat pan reaction force.

The min/max criterion [14] served to solve the muscle recruitment problem. The employed criterion ensures that the maximum relative load of any muscle is as small as possible. Physiologically, the min/max criterion is equivalent to a minimum fatigue criterion which makes it an attractive choice for ergonomic applications such as the one presented in this study.

A constant strength muscle model was used. It requires as only input the maximum muscle force (isometric strength). Although, this is a simple muscle model and does not describe the physiological behavior of muscles, for static models such as the ones in the present study, the simple muscle model works well. The muscles strengths were scaled based on each participant's BMI and gender using the regression equations by Frankenfield et al. [15].

Muscles forces and joint reaction forces were the outputs of the inverse dynamics analysis.

2.4 Data Analysis

Trials that violated the equilibrium of forces were excluded from further analysis. The selected trials satisfied the following criteria:

$$|Fz - BW| < 0.05 * BW \tag{1}$$

$$|Fx| < 11N \tag{2}$$

where Fz: the total vertical force, BW: body weight of each participant, and Fx: the total horizontal force.

The peak value of the muscle activity envelope (MA_max), the normalized to the body weight absolute shear force between the seat pan and the participants, and the normalized to the body weight compression force between L4 and L5 (Fcompr_L4L5) and between L5 and sacrum (Fcompr_L5S) were identified for each trial.

For each backrest angle and participant, a preferred range for the seat pan angle was defined, using the two chosen seat pan angles from the preferred trials (PRHE, PRLE) as extremes. Subsequently, all trials were classified depending whether their seat pan angle was within the preferred range (Preferred vs Not preferred), for the respective participant and backrest angle. Statistical tests were performed to identify possible differences for seat pan shear force, MA_max, Fcompr_L4L5 and Fcompr_L5S between the two groups.

Moreover, Pearson product-moment correlations were performed to determine whether there is a relationship between the CP50 values and computationally computed internal loads (MA_max, Fcompr_L4L5 and Fcompr_L5S) and between the CP50 values and the seat pan shear force. The significance level was set to 0.05 for all analyses.

3 Results

The mean preferred seat pan angle was 16.2° (5.2°) when the initial/starting angle of the seat pan was 20° backward inclined (PRHE trials), while it was 7.6° (8.1°) when the initial seat pan angle was set at 5° forward inclination (PRLE). Table 1 summarizes the selected seat pan angles for the different backrest angles and for the different initial seat pan angles.

Table 1. Mean, standard deviation (SD), minimum and maximum of the selected seat pan angles for the different backrest angles (BA) for the PRHE and PRLE trials.

BA	PRHE				PRLE			
	Mean	(SD)	Min	Max	Mean	(SD)	Min	Max
100	15.4	(4.7)	10.9	21.0	5.0	(8.1)	−3.7	15.8
115	16.9	(5.8)	10.1	22.7	6.8	(5.5)	−1.5	13.4
130	20.2	(4.6)	14.1	24.1	8.3	(7.4)	−0.7	17.5
All	16.2	(5.2)	10.1	24.1	7.6	(8.1)	−3.7	17.5

The absolute seat pan shear force was significantly smaller for the preferred trials compared to the not preferred trials ($p < 0.001$). No significant differences were found between the preferred and not preferred trials for the computationally predicted muscle and joint forces. The configuration angle (A_Conf) defined as the relative angle between the seat pan and the backrest was 104° for the preferred trials, while it was 109° for the not preferred trials (Table 2).

Significant positive correlations were observed between the CP50 values and peak muscle activity ($p < 0.01$), between the CP50 values and the normalized to the body weight compressive force between L5 and sacrum ($p = 0.04$) and between the CP50 values and the normalized to the body weight seat pan shear force ($p < 0.001$). Table 3 summarizes the results of the analysis, while Fig. 3 presents variables boxplots for the different CP50 subgroups.

Table 2. Mean, standard deviation (SD), minimum and maximum of selective variables for the Preferred and Not preferred trials. Significant differences ($p < 0.05$) are denoted with *.

SBA	Variable	Preferred				Not preferred			
		Mean	(SD)	Min	Max	Mean	(SD)	Min	Max
100	Shear force (N/BW)	0.06	(0.04)	0.01	0.13	0.09	(0.06)	0.01	0.18
100	A_Conf (°)	88.9	(8.6)	79.1	103.7	95.8	(9.0)	85.5	105.7
100	MA_max (·100)	17.67	(2.42)	14.63	21.36	16.16	(3.99)	11.18	25.15
100	Fcompr_L4L5 (N/BW)	0.53	(0.10)	0.43	0.73	0.52	(0.11)	0.36	0.71
100	Fcompr_L5S (N/BW)	0.35	(0.06)	0.29	0.46	0.35	(0.09)	0.20	0.49
100	CP50	21.6	(9.0)	11.0	35.0	28.1	(14.0)	5.0	54.0
115	Shear force (N/BW)	0.06	(0.03)	0.02	0.10	0.11	(0.06)	0.01	0.19
115	A_Conf (°)	102.1	(5.7)	92.3	110.4	112.8	(8.4)	100.3	120.7
115	MA_ max (·100)	16.32	(3.15)	12.85	23.01	17.58	(5.75)	11.67	27.55
115	Fcompr_L4L5 (N/BW)	0.51	(0.13)	0.33	0.77	0.59	(0.19)	0.40	0.89
115	Fcompr_L5S (N/BW)	0.31	(0.07)	0.18	0.42	0.33	(0.08)	0.23	0.47
115	CP50	14.8	(8.5)	2.0	27.0	28.1	(11.0)	11.0	49.0
130	Shear force (N/BW)	0.04	(0.02)	0.01	0.08	0.09	(0.05)	0.01	0.15
130	A_Conf (°)	116.3	(6.9)	106.2	126.0	126.7	(8.4)	115.4	135.5
130	MA_ max (·100)	17.31	(4.15)	11.78	24.90	16.42	(3.94)	12.23	24.55
130	Fcompr_L4L5 (N/BW)	0.52	(0.17)	0.30	0.74	0.55	(0.11)	0.39	0.76
130	Fcompr_L5S (N/BW)	0.26	(0.08)	0.17	0.42	0.29	(0.07)	0.16	0.40
130	CP50	17.2	(8.4)	5.0	29.0	20.8	(10.7)	11.0	45.0
All	Shear force (N/BW)*	0.06	(0.03)	0.01	0.13	0.10	(0.05)	0.01	0.19
All	A_Conf (°)	104.0	(13.2)	79.1	127.5	109.3	(16.6)	75.6	140.3
All	MA_ max (·100)	16.62	(3.50)	10.53	24.90	16.56	(4.87)	10.61	27.55
All	Fcompr_L4L5 (N/BW)	0.50	(0.13)	0.30	0.77	0.52	(0.14)	0.32	0.89
All	Fcompr_L5S (N/BW)	0.29	(0.09)	0.10	0.46	0.33	(0.09)	0.12	0.49
All	CP50	15.1	(8.6)	2.0	35.0	25.2	(11.8)	5.0	54.0

Table 3. Summary of the Pearson correlation results for the computationally predicted internal loads and the absolute values of the seat pan shear force. Significant correlations ($p < 0.05$) are denoted with *.

Variable	Correlation coefficient	P-value
MA_max	0.30	0.004*
Fcompr_L4L5	0.17	0.1150
Fcompr_L5S	0.22	0.041*
Absolute shear force	0.37	<0.001*

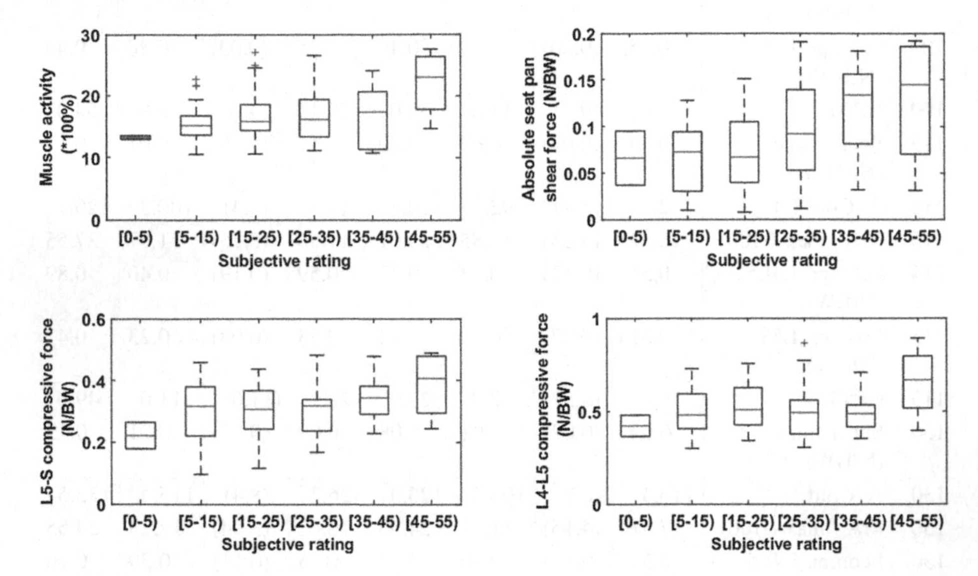

Fig. 3. Boxplots of selective variables for the different CP50 groups. [0–5) very low, …, [45, 55) very high discomfort ratings.

4 Discussion

The present pilot study was inspired from a previous study performed in our lab [9], where sitting discomfort was investigated on a small range of variation for the sitting parameters (seat pan angle [0–5°, preferred], backrest angle [100–110°]). In this study, a higher range for the sitting parameters was employed (seat pan angle [−5–15°, PRHE, PRLE], backrest angle [100–130°]). We investigated whether biomechanical variables could describe sitting discomfort by grouping the experimental trials to preferred and not preferred trials, by searching for potential correlations between biomechanical variables and objective ratings and by searching for potential differences for the trials received extremely low or high discomfort ratings

High differences ($\sim 10°$) were observed between the selected by the participants preferred seat pan angles for the two different initial inclinations of the seat pan. The participants were instructed to choose the seat pan angle for which they would experience the least discomfort, starting from two predefined initial seat pan angles and being free to test the full range of motion of the seat pan. However, the preferred seat pan angles were higher, for all the different backrest angles, for the PRHE trials compared to the PRLE trials. The selected seat pan angles were closer to the initial seat pan extreme than the opposite one, indicating that the selection of a preferred seat pan angle depends on the initial seat pan inclination. This observation supports the notion that trial order may affect the subjective ratings. Furthermore, our data suggest that defining a single preferred sitting configuration is not accurate. Instead, a range of seat pan angles with similar comfort level may exist and thus, the definition of a preferred sitting configuration range could be recommended.

Positive significant correlations were observed between the subjective ratings and the computationally predicted internal loads (maximum muscle forces, compressive force between the L5 and sacrum). Such variables have been used in the past to assess different sitting configurations. Computational studies [5, 6] have discussed the significance of identifying sitting configurations that minimize the muscle activity, while others [7] have reported significant correlations between postural comfort curves and computationally predicted muscle forces however, no experimental verification has been reported. To our knowledge, this is the first study that reports results, which support the existence of a positive correlation between subjective ratings and computationally predicted internal loads.

The trends observed on the variables when classified with respect to the CP50 ratings were consistent with the correlation results. We chose this classification to investigate whether differences on the tested variables would be observed for the extreme CP50 ratings (very low discomfort (0–5) and very high discomfort (>45)). Low muscle activity and joint reaction forces were observed for the very low CP50 values, while high muscle activity, joint reaction forces and seat pan shear forces were observed for the very high CP50 values. Such observations can provide valuable recommendations to researchers and seat designers regarding which variables should be minimized to achieve low discomfort and which variables should be kept in low values to avoid high discomfort. However, our observations based on the current data should be interpreted with caution, since only a small number of trials received very low (11) or very high (4) discomfort ratings.

Our expectations for differences between the preferred and not preferred trials were partially supported from the present data. A significant reduction on the seat pan shear forces for the preferred trials was observed compared to the not preferred trials. However, no statistical differences were found for the computationally predicted internal loads, possibly due to the limited number of the observations.

In conclusion, the presented data suggest that seat pan shear force and potentially computationally predicted internal loads such as muscle and joint forces could be employed to describe sitting discomfort. However, the small number of observations collected does not allow solid conclusions. In the future, studies should employ more participants and investigate sitting discomfort in a wide range of variation for the sitting parameters. Such design would allow a wide range of subjective ratings, which potentially could result in more observations with extreme subjective ratings.

Acknowledgement. The work is partly supported by Direction Générale de l'Aviation Civile (project n°2014 930818).

References

1. de Looze MP, Kuijt-Evers LFM, van Dieen J (2003) Sitting comfort and discomfort and the relationships with objective measures. Ergonomics 46(10):985–997
2. Andersson BJG, Ortengren R, Nachemson A, Elfstrom G (1974) Lumbar disc pressure and myoelectric back muscle activity during sitting. IV. Studies on a car driver's seat. Scand J Rehabil Med 6:128–133
3. Wilke HJ, Neef P, Caimi M, Hoogland T, Claes LE (1999) New in vivo measurements of pressures in the intervertebral disc in daily life. Spine 24:755–762
4. Zenk R, Franz M, Bubb H, Vink P (2012) Technical note: spinal loading in automotive seating. Appl Ergon 43:290–295
5. Grujicic M, Pandurangan B, Xie X, Gramopadhye A, Wagner D, Ozen M (2010) Musculoskeletal computational analysis of the influence of car-seat design/adjustments on long-distance driving fatigue. Int J Ind Ergon 40(3):345–355
6. Rasmussen J, Torholm S, de Zee M (2009) Computational analysis of the influence of seat pan inclination and friction on muscle activity and spinal joint forces. Int J Ind Ergon 39:52–57
7. Cappetti N, Naddeo A, Soldovieri VM, Vitillo I (2017) A study on the correlation between the perceived comfort and the muscular activity, using virtual simulation techniques. In: 1st International comfort congress. Publisher, Salerno, pp 1–2
8. Beurier G, Cardoso M, Wang X (2017) A new multi-adjustable experimental seat for investigating biomechanical factors of sitting discomfort. SAE Technical Paper 2017-01-139
9. Wang X, Beurier X (2018) Determination of the optimal seat profile parameters for an airplane eco-class passenger seat. SAE Technical Paper 2018-01-1324. SAE world congress 2018, 10–12 April 2018, Detroit, USA. https://doi.org/10.4271/2018-01-132
10. Shen W, Parsons KC (1997) Validity and reliability of rating scales for seated pressure discomfort. Int J Ind Ergon 20:441–461
11. Nerot A, Skalli W, Wang X (2016) A principal component analysis of the relationship between the external body shape and internal skeleton for the upper body. J Biomech 49 (14):3415–3422
12. Andersen MS, Damsgaard M, MacWilliams B, Rasmussen J (2010) A computationally efficient optimisation-based method for parameter identification of kinematically determinate and over-determinate biomechanical systems. Comput Methods Biomech Biomed Eng 13 (2):171–183
13. Andersen MS, Damsgaard M, Rasmussen J (2009) Kinematic analysis of over-determinate biomechanical systems. Comput Methods Biomech Biomed Eng 12(4):371–384
14. Damsgaard M, Rasmussen J, Christensen ST, Surma E, de Zee M (2006) Analysis of musculoskeletal systems in the AnyBody Modeling System. Simul Model Pract Theory 14 (8):1100–1111
15. Frankenfield DC, Rowe WA, Cooney RN, Smith JS, Becker D (2001) Limits of body mass index to detect obesity and predict body composition. Nutrition 17(1):26–30

Autostereoscopic Displays for In-Vehicle Applications

Andre Dettmann(✉) and Angelika C. Bullinger

Professorship of Ergonomics and Innovation Management, Chemnitz University of Technology, Erfenschlager Straße 73a, 09125 Chemnitz, Germany
andre.dettmann@mb.tu-chemnitz.de

Abstract. Novel technologies like autostereoscopic 3D displays are providing a perception of depth in a scene towards users. Those added spatial informations allow a better user performance in recognizing and classifying on-screen objects as well as enabling better judgements of positions and distances of displayed objects and on-screen elements. Autostereoscopic 3D displays, if implemented user-friendly into Advanced Driver Assistance Systems (ADAS) or In-Vehicle Information Systems (IVIS), can increase the effectiveness of such systems by providing distinguishable spatial relationships. Possible applications using an autostereoscopic display where users' can benefit from spatial cues are for instance the instrument cluster, the navigation device or an intersection assistant. When implemented correctly, 3D displays will allow a better understanding of complex user interfaces and are overall capable of lowering driver distraction and therefore, benefit directly towards traffic safety. We present a study with 40 participants judging the criticality of an intersection manoeuvre in a simulated traffic environment using an autostereoscopic display. The assumption of the experiment is that autostereoscopic monitors in comparison to 2D monitors allow a better assessment of traffic situations in the context of ADAS/IVIS applications. Results show, that 3D displays enable a better accuracy and judgement of positions in simulated traffic situations. While the technology has an impact on the participants' judgements, perspective does not. Regarding visual fatigue, the usage of autostereoscopic displays seems to be unproblematic despite a long exposure time. Also, regarding the special requirements in content creation we recommend a disparity level with a high perceptual performance and low visual fatigue.

Keywords: Autostereoscopic · 3D · HMI · ADAS · IVIS · Disparity

1 Introduction

Driver assistance and information systems in combination with their user interfaces require mental resources. In complex driving situations, they can add mental load in addition to the driving task on the driver [33]. One challenge is the ever-increasing choice of ADAS/IVIS when buying a new car. Each system can increase the visual load on the driver and therefore, distracting him when operating the system or looking at it [23]. Consequently, if further assistance systems are integrated into vehicles, the problem of visual load is further intensified [20, 26]. The proposed solution of

S. Bagnara et al. (Eds.): IEA 2018, AISC 823, pp. 457–466, 2019.
https://doi.org/10.1007/978-3-319-96074-6_48

optimized driver assistance by autostereoscopic monitors is based on the requirements on the driving task which can be reduced by technical systems, and thereby increasing the driver's performance [8]. Such monitors allow the three-dimensional structuring of information on user interfaces, which can reduce the duration of a visual search of relevant information. Tory and Möller [31] refer to this type of visual assistance as cognitive support. However, the literature indicates that the use of stereoscopic displays can lead to increased visual fatigue and discomfort and needs to be considered when designing ADAS/IVIS based on autostereoscopic monitors [16].

2 State of the Art

General application scenarios of stereoscopic representations are all ADAS/IVIS, which use an optical interface. A major application scenario that has been the focus of several studies on stereoscopic displays has been the instrument cluster. The instrument cluster has a high importance for the primary driving task, since this display communicates information about the vehicle state to the driver [4]. Due to the high frequency drivers use to extract information from the instrument cluster, the instrument cluster has the greatest potential for optimizing the perceptual performance of a driver. Stereoscopic displays can thus help the driver to capture the increasingly complex information designs faster and to extract the relevant information [30]. This benefit can also be used for navigation devices. In the route guidance mode of the navigation display, three-dimensional views can be used depending on the technical design. Another potential scenario are intersection assistants. Designed as an IVIS in combination with emerging Car2X technologies, three-dimensional representations can enable the situation assessment of an intersection and enable the driver to detect non-visible road users [33]. In a stereoscopic view, the user is supported in estimating the relations of objects and distances. Other applications are parking assistance and rear-view camera systems.

The research field of stereoscopic displays for optimizing the interaction with displays is in the constant focus of empirical investigations. McIntre et al. [21] and van Beurden et al. [32] were able to show from the meta-analyses that these advantages occur especially in difficult or complex tasks. Regarding ADAS/IVIS applications, these advantages can be grouped into three main categories: "position and/or distance estimation", "navigation" as well as "finding, identifying, classifying objects" [21]. For the categories "position and/or distance estimation" and "navigation", advantages such as increased speed and accuracy in perception were found [5, 6, 19, 22]. Broy et al. [2] could also prove it in an automotive scenario. In the category "finding, identifying, classifying objects" [28, 29] found advantages in terms of shorter visual searches. Test subjects were able to find the relevant information faster and made fewer mistakes. The results were confirmed by [25] in an automotive context.

Therefore, it can be concluded that stereoscopic displays have a largely positive effect on the perceptual performance of users. In addition, several studies have demonstrated a positive impact on satisfaction, user experience, and intuition [3, 22, 29].

For an on-vehicle application, however, the ergonomic requirements for ADAS/IVIS as well as the ergonomic requirements for stereoscopic displays to generate high perceptual performance without visual fatigue are essential. The goal is to design a user interfaces that allow an error-free and ergonomically stress-free perception.

3 Method

Based on the driving simulator experiments of [9] and studies on hazard perception [13], the influence of the monitor technology on the assessment of the criticality of traffic situations were examined. Also the characteristics of the perspectives "driver's view" and "bird's-eye view" were taken into account. For this purpose, videos containing a left turn manoeuvre with an oncoming vehicle at a crossroad from both perspectives are presented to the subjects. By looking at the situations in a passive way, the subjects were able to concentrate on the left turn manoeuvre without further distraction and rate the situation. This procedure ensures that the effects of the monitor technology and the perspective are isolated from the evaluation behaviour and excludes interferences such as handling a vehicle in a driving simulator and the individual driving behaviour [13]. Furthermore, by applying the method of constant stimuli, a controlled-randomized stimulus presentation in relation to the distances between the two vehicles was realized [10]. The aim was to determine if and when an effect with regard to the perception of danger occurs depending on the used stereoscopic depth and perspective. The detailed course of the experiment will now be described below.

The requirement for participating in the experiment is the ability to perceive 3D content. At the beginning of the experiment, the Lang stereo tests I and II [17] and the visual function questionnaire [18] were used. Furthermore, we collected socio-demographic data and the distance measurement to the monitor as an ergonomic feature.

For the main part of the experiment, several videos were presented in which two vehicles drive to and cross an intersection. The ego vehicle turns to the left, whereas the oncoming vehicle just crosses the intersection, which corresponds to one of the most common accident scenarios in Germany [15]. This was done according to the trajectories shown in Fig. 1. The distances of the vehicles to the intersection were controlled analogously to [9] by the independent variable time-to-intersection (TTI). The TTI is the time taken by a vehicle from its current position to the point of entry into the intersection (see red line, Fig. 1). The distance of the ego vehicle to the intersection was kept constant and only the distance of the oncoming vehicle varied in twelve stages from 0.3 to 2.0 s. Using the smallest value of 0.3 s, equivalent to about 4.2 m distance of the vehicles to each other, it is clear that the vehicles will not collide at the intersection, but a sufficiently critical situation was created for the evaluation. With increasing TTI the distances become larger and the situation more uncritical. This left turn manoeuvre was randomly presented to the subjects 48 times in twelve variations (four repetitions per measurement) and rated by the participants using a 4-point rating scale. To achieve a standardized rating, the video was paused at the point where the ego vehicle crossed the centreline. The image of the situation could be observed by the subjects for another ten seconds.

The used scale was developed by [12] and was revised by [9] to evaluate the subjective criticality in intersection situations. The measurement of criticality shown in Table 1 is suitable for assessing traffic scenarios but needs to be controlled to allow a

Fig. 1. Intersection situation in both perspectives (Color figure online)

consistent interpretation. Driving experience and willingness to take risks are possible moderating variables of a criticality assessment [11, 13, 27]. Hence, questionnaires on the duration of driving license ownership and the mean annual kilometrage as well as the subjectively assessed driving style were asked. As a personality trait, the risk behavior was analyzed by the Brief Sensation Seeking Scale (BSSS) according to [14].

Table 1. The 4-point criticality scale.

Rating	Meaning	Example
1	No danger	Hardly distinguishable to passing the intersection without other road users
2	Increased level of danger	Attentive observation of the position and speed of the other participants, but no doubt from the safe completion of the manoeuvre
3	Higher level of danger	At least at one point in the manoeuvre execution first doubts about the safe completion
4	High danger	A collision with the other participant appeared unavoidable

In a second experimental section, a five-minute city trip was presented to both groups on the autostereoscopic monitor. The aim was to investigate whether a subjective perceived visual fatigue occurs when the autostereoscopic monitor is used for a longer time or when switching from a conventional monitor to a 3D monitor. For this purpose, the Visual Fatigue Questionnaire (VFQ) according to [1] was issued to all subjects. The total duration of the experiment was about 50 min.

For the experiment, a laboratory was chosen, in which constant light conditions could be created. Before each test, the test manager closed all blinds and turned on all ceiling lights. The subject positioned himself according to the group and task in front of the relevant monitor. For the study, the SF3D-133CR monitor from SeeFront GmbH based on a lenticular lens system was used. The monitor has a 13.3" screen size with a native resolution of 2560 × 1440 pixels. For viewing three-dimensional content, the monitor accepts 1920 × 1080 pixel materials in side-by-side mode. The 2D monitor used was an EIZO FlexScan S2402 W with a native resolution of 1980 × 1080 pixels. The 2D monitor was masked to the size of the autostereoscopic 3D display.

The videos were cut to identical starting points and lengths. The created material was designed using the "puppet theater" analogy. This means that a stereoscopic image of the reality with mostly uncrossed depth was created. The uncrossed disparity of the oncoming vehicle at the time of evaluation with the respective TTI was based on the recommended 1 degree rule according to [16]. The values were approximately 0.2 to 1.1°. As point with no disparity, the entry point of the intersection was chosen. Regarding the bird's-eye view, the virtual camera was raised to a height of four meters and tilted 20° forward in comparison to the driver's perspective to give the observer a better overview of the crossing situation.

4 Sample

The study counted 40 participants with a mean age of 30.9 (SD_{age} = 9.4), whereby 27.5% were female. All participants were allocated randomly in four groups according to the experimental conditions "2D" and "3D" with the subgroups "driver's view" and "bird's-eye view". Based on an ANOVA, no differences could be found in the groups regarding age and viewing distance from the monitor. All subjects were able to perceive stereoscopic contents. In terms of self-reported vision capabilities, the results were good to very good and no significant differences between the groups were found. For the driving experience, the average annual mileage and the BSSS also no differences were found. In general, it can be stated that homogeneous groups have been generated.

5 Results

During the experiment, 1920 samples were recorded (40 subjects with four repeated measurements and twelve variations of the TTI). First of all, the general response behaviour was examined. The variations of the TTI were well chosen with respect to the measuring range of the criticality scale, since neither floor nor ceiling effects were recorded, and the scale was largely exploited. A high correlation of the response behaviour and the TTI ($r = -.770$, $p < .001$) was found. This shows a generally good understanding of the scale by the subjects.

Analysis of the responses on the criticality scale was based on a three-factorial analysis of variance (see Table 2). For all factors a significant effect on the ratings could be found. The factor "TTI variations" has the main effect on the criticality evaluation and explains 62% of the variance, which is due to the experimental setup. Regarding the "monitor technology", a small effect could be found, and a negligible effect could be found regarding the factor "perspective". For the interactions TTI × technology and technology × perspective statistically significant interaction effects were found. Only for TTI × perspective, no significant interaction effect was found. For all interaction effects on the factor perspective, the effect is negligible, but a small effect has been found for the interaction between TTI × technology. The interaction effect between all factors is not significant.

The interaction effect between the TTI variations and the perspective thus has no influence on the subjective ratings, but the monitor technology generates differences in relation to the TTI and the perspective. Figure 2 reflects the results of the criticality scale.

Table 2. Between-group comparison on the criticality scale

	Between-group comparison	Statistical significance (p)	Partial η^2
TTI variation	F (11, 1872) = 277.21	<.001	.620
Technology (2D/3D)	F (1, 1872) = 39.01	<.001	.020
Perspective	F (1, 1872) = 39.01	.002	.005
TTI × technology	F (11, 1872) = 5.29	<.001	.030
TTI × perspective	F (11, 1872) = 0.43	.941	.003
Technology × perspective	F (1, 1872) = 4.34	<.001	.002
TTI × technology × perspective	F (11, 1872) = 0.59	.839	.003

Due to the negligible effect of the perspective, the ratings of the TTI were clustered into the groups "2D" and "3D" for further analysis. Unpaired t-tests were used to analyse at which specific TTI a permanent difference in the rating behaviour occurs. A reliable distinctness was found for a TTI $\geq$ 1.0 s (see Figs. 2 and 3).

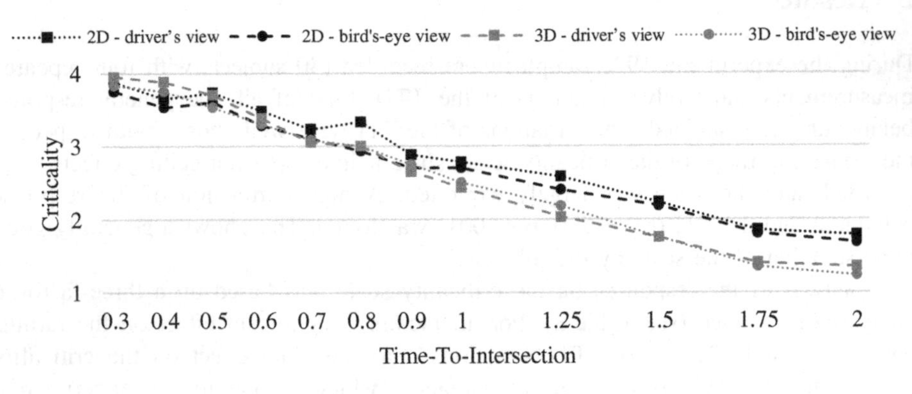

Fig. 2. Rating behavior on the criticality scale

For the analysis of visual fatigue and the subjectively perceived image quality, the four groups were combined into the main groups "2D" and "3D". For all items, the values are at a very low level. Only the items "dry eyes" and "mental fatigue" were slightly increased. In sum, there are only minor effects on visual fatigue, which have no effect when comparing the two groups. Given the duration of the experiment, also no "noticeable" effect on the scale was achieved at any time. Concerning the subjectively perceived image quality, the negative perceived effects are low, and the overall quality was rated "okay" for both screens.

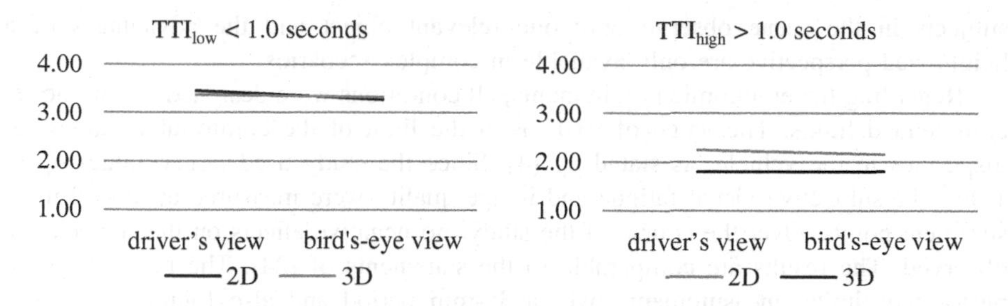

Fig. 3. Main effect of the monitor technology on the criticality rating

A detailed comparison of the subjective image quality items over the course of the experiment shows no difference between the groups "2D" and "3D". When the group "2D" switched to the autostereoscopic display, the subjects experienced a slight deterioration compared to the conventional monitor. For the items "insufficient contrast" ($t\ (19) = 2.43$, $p = .025$), "graphics looked incorrectly sized or distorted" ($t\ (38) = -2.227$, $p = .032$) and "screen reflections were bothersome" ($t\ (38) = -2.999$, $p = .005$), significant differences were found. Lack of contrast was perceived as less annoying on the 3D monitor, but the graphics and the reflections were rated worse. No significant differences were found for the first and second measurement of the 3D group (no changes in monitor technology).

6 Discussion

The study investigated the effect of monitor technology on the assessment of critical traffic situations in the context of an ADAS/IVIS application. The results show a clear effect of the monitor technology on the subjective ratings. Subjects who observed the traffic situations with the autostereoscopic display were able to anticipate the traffic situation better and estimate the sufficient distance to the oncoming vehicle. This rating behaviour was constant at a $TTI \geq 1.0$ s with a disparity greater than 0.61°. These results generally confirm the improved estimation of position and/or distances on autostereoscopic monitors and thus stands in the context of [5, 6, 22]. The fact that no distinctness could be determined below this value cannot be attributed to the threshold values of the perception of depth stimuli, since the lowest value of 0.20° in the present experiment was five times higher than that of [7] empirically raised a threshold for 97.3% of the population. Rather, there was no purely stereoscopic condition in the experiment but a variety of monocular cues for the estimation of distances were available. It can be concluded that the advantage of perceived depth at low levels of disparity did not benefit the observers and was masked by monocular clues (see [21]). In practical terms this means, despite the high human ability to perceive the smallest differences in stereoscopic vision, greater discernible differences due to monoscopic clues in the scene must be applied. Furthermore, the condition "Perspective" within the groups "2D" and "3D" shows no differences. The rating behaviour essentially follows the progression of the technology conditions. This can be explained by the fact that the subjects in the scene observe only one relevant object and the advantages of a heightened perspective are only available in complex scenarios.

Regarding the ergonomic requirements, all conditions were designed within the recommended limits. The value of 0.61° is at the limit of the comfortable values for disparities in the vehicle, as stated by [4]. Since the study used larger values up to 1.09°, the subjective visual fatigue and image quality were measured at several measurement points. Over the course of the study, no negative effects on the subjects are observed. The results are comparable to the statements of [24]. The research group performed similar measurements over a 36-min period and also found only minor effects on visual fatigue independent of the monitor technology. Regarding image quality, only slight differences at a low level were found for the autostereoscopic monitor used. Those were classified as unproblematic, which argues for a good quality of the 3D monitor in comparison to the 2D monitor.

From the methodological point of view, the criticality scale and the material used demonstrate good suitability for a time-independent evaluation without floor or ceiling effects. The scale was exploited across the bandwidth along the chosen TTI values and the transmitted values of the cross disparity of the subjects.

7 Conclusion

It has been proven that stereoscopic images enable an improved perception in an automotive context. Autostereoscopic 3D monitors provide better situational assessment of traffic situations in the context of ADAS/IVIS applications compared to 2D monitors. Thus, a transfer into a real-world application seems to be possible. With regard to visual fatigue and perceived image quality, no differences could be found between the technologies, which further argues for a general applicability of autostereoscopic monitors in an automotive context. However, there are limitations of the present experiment: the insights gained cannot be applied directly to time-critical applications within a driving task. The focus of the experiment lies on the effects of monitor technology and perspective isolated from a driving task. A typical user behaviour within a normal driving task has not been depicted. Accordingly, further research is needed on the evaluation of an intersection assistant while performing a driving task. However, when a non-critical assessment horizon of a situation exists, such as a parking manoeuvre, then the results may provide implications for further applications. Regarding the research domain of stereoscopic displays, existing insights into the applicability of ergonomic limits of disparities and their effects on visual fatigue could be confirmed. Furthermore, a contribution was made to practical relevant disparities which provide a high perceptual performance and low visual fatigue in stereoscopic presentations.

Acknowledgements. The authors acknowledge the financial support by the Federal Ministry of Education and Research of Germany in the framework of IVIS-3D (project number 03ZZ0406).

References

1. Bangor AW (2000) Display technology and ambient illuminat ion influences on visual fatigue at VDT workstations. Dissertation, Virginia Polytechnic Institute and State University
2. Broy N, Alt F, Schneegass S et al (2014) 3D displays in cars. Exploring the user performance for a stereoscopic instrument cluster. In: Boyle LN (ed) Automotive UI 2014, proceedings of the 6th international conference on automotive user interfaces and interactive vehicular applications, pp 1–9
3. Broy N, Guo M, Schneegass S et al (2015) Introducing novel technologies in the car - conducting a real-world study to Test 3D dashboards. In: Burnett G (ed) Automotive UI 2015, proceedings of the 7th international conference on automotive user interfaces and interactive vehicular applications. ACM, New York, pp 179–186
4. Broy N (2016) Stereoscopic 3D user interfaces. Exploring the potentials and risks of 3D displays in cars. Dissertation, Universität Stuttgart
5. Chen J, Oden R, Kenny C et al (2010) Stereoscopic displays for robot teleoperation and simulated driving. Proc Hum Factors Ergon Soc Ann Meet 54(19):1488–1492
6. Chen J, Oden R, Merritt JO (2014) Utility of stereoscopic displays for indirect-vision driving and robot teleoperation. Ergonomics 57(1):12–22
7. Coutant BE, Westheimer G (1993) Population distribution of stereoscopic ability. Oph Phys Optics 13(1):3–7
8. Fricke N (2009) Gestaltung zeit- und sicherheitskritischer Warnungen im Fahrzeug. Dissertation, Technische Universität Berlin
9. Geyer S (2013) Entwicklung und Evaluierung eines kooperativen Interkationskonzepts an Entschiedungspunkten für die teilautomatisierte, manöverbasierte Fahrzeugführung. VDI Verlag GmbH, Düsseldorf, Fahrzeugtechnik TU Darmstadt
10. Goldstein EB (ed) (2015) Wahrnehmungspsychologie. Der Grundkurs, 9th edn. Springer Lehrbuch, Berlin
11. Heino A, van der Molen HH, Wilde GJ (1996) Differences in risk experience between sensation avoiders and sensation seekers. Pers Individ Differ 20(1):71–79
12. Hohm A (2010) Umfeldklassifikation und Identifikation von Überholzielen für ein Überholassistenzsystem. Fortschrittberichte VDI, vol 727. VDI Verlag GmbH, Düsseldorf
13. Horswill MS (2016) Hazard perception in driving. Curr Dir Psychol Sci 25(6):425–430
14. Hoyle RH, Stephenson MT, Palmgreen P et al (2002) Reliability and validity of a brief measure of sensation seeking. Pers Individ Differ 32(3):401–414
15. Kühn M, Hannawald L (2015) Verkehrssicherheit und Potenziale von Fahrerassistenzsystemen. In: Winner H, Hakuli S, Lotz F et al (eds) Handbuch Fahrerassistenzsysteme. Grundlagen, Komponenten und Systeme für aktive Sicherheit und Komfort, 3. Auflage. Springer Vieweg, pp 55–70
16. Lambooij M, Ijsselsteijn WA, Fortuin M et al (2009) Visual discomfort and visual fatigue of stereoscopic displays: a review. J Imaging Sci Technol 53(3):1–14
17. Lang J (1982) Mikrostrabismus. Die Bedeutung der Mikrotropie für die Amblyopie, für die Pathogenese des grossen Schielwinkels und für die Heredität des Strabismus, 2. Auflage. Bücherei des Augenarztes, Heft 62. Enke, Stuttgart
18. Mangione CM, Lee PP, Gutierrez PR et al (2001) Development of the 25-item national eye institute Visual Function Questionnaire (VFQ-25). Arch Ophthalmol 119:1050–1058
19. Martinez Escobar M, Junke B, Holub J et al (2015) Evaluation of monoscopic and stereoscopic displays for visual-spatial tasks in medical contexts. Comput Biol Med 61:138–143. https://doi.org/10.1016/j.compbiomed.2015.03.026

20. Mattes S, Hallén A (2009) Surrogate distraction measurement techniques: The Lane Change test. In: Regan MA, Lee JD, Young KL (eds) Driver distraction. Theory, effects, and mitigation. CRC Press, Boca Ratón, pp 107–122
21. McIntire JP, Havig PR, Geiselman EE (2014) Stereoscopic 3D displays and human performance: a comprehensive review. Displays 35(1):18–26
22. Mikkola M, Boev A, Gotchev A (2010) Relative importance of depth cues on portable autostereoscopic display. In: Proceedings of the 3rd workshop on mobile video delivery. ACM, New York, pp 63–68
23. NHTSA (2010) Overview of the National Highway Traffic Safety Administration's Driver Distraction Program, Washington, USA
24. Ntuen CA, Goings M, Reddin M et al (2009) Comparison between 2-D & 3-D using an autostereoscopic display: the effects of viewing field and illumination on performance and visual fatigue. Int J Ind Ergon 39(2):388–395. https://doi.org/10.1016/j.ergon.2008.07.001
25. Pitts MJ, Hasedžić E, Skrypchuk L et al (2015) Adding depth: establishing 3D display fundamentals for automotive applications. SAE Technical Paper 2015-01-0147
26. Regan MA, Hallett C, Gordon CP (2011) Driver distraction and driver inattention: definition, relationship and taxonomy. Accid Anal Prev 43(5):1771–1781
27. Rudin-Brown CM, Edquist J, Lenné MG (2014) Effects of driving experience and sensation-seeking on drivers' adaptation to road environment complexity. Saf Sci 62:121–129
28. Sandbrink J, Rhede J, Vollrath M et al (2017) 3D-Displays - Das ungenutzte Potential? Die Wahrnehmung von stereoskopischen Informationen im Fahrzeug. Der Fahrer im 21. Jahrhundert. Der Mensch im Fokus technischer Innovationen. VDI Verlag GmbH, Düsseldorf, pp 153–164
29. Sassi A, Pöyhönen P, Jakonen S et al (2014) Enhanced user performance in an image gallery application with a mobile autostereoscopic touch display. Displays 35(3):152–158
30. Szczerba J, Hersberger R (2014) The use of stereoscopic depth in an automotive instrument display. Proc Hum Factors Ergon Soc Annu Meet 58(1):1184–1188
31. Tory M, Möller T (2004) Human factors in visualization research. IEEE Trans Vis Comput Graph 10(1):72–84
32. van Beurden M, van Hoey G, Hatzakis H et al (2009) Stereoscopic displays in medical domains: a review of perception and performance effects. In: Rogowitz BE, Pappas TN (eds) Human vision and electronic imaging XIV. SPIE, Bellingham, pp 1–15
33. Winner H, Hakuli S, Lotz F et al (eds) (2015) Handbuch Fahrerassistenzsysteme. Grundlagen, 3. Auflage. ATZ/MTZ-Fachbuch. Springer Vieweg

Optimization of Occupational Whole Body Vibration Exposure for Rotavation Operation

Amandeep Singh[1(✉)], Lakhwinder Pal Singh[1], Sarbjit Singh[1], and Harwinder Singh[2]

[1] Department of Industrial and Production Engineering, Dr. B. R. Ambedkar National Institute of Technology, Jalandhar, India
ip.nitj@gmail.com, {singhl, balss}@nitj.ac.in
[2] Department of Mechanical Engineering, Guru Nanak Dev Engineering College, Ludhiana, India
harwin75@rediffmail.com

Abstract. The study focused to investigate the relative importance of forward speed, pulling force and tilling depth on tractor ride comfort in terms of overall vibration total value (OVTV). Taguchi's L_{27} orthogonal array used to formulate a systematic design of experiments. It provided twenty-seven (27) experiments with five (5) replications for each set to get an average of root mean square (RMS) weighted acceleration value (a_w). Point vibration total value (PVTV) was calculated at three different locations (i.e. tractor platform, seat-pan and seat backrest) in order to obtain OVTV. The magnitude of overall vibration total value (0.637–0.843 m/s^2) represents the ride little to fairly uncomfortable as per ISO 2631-1 (1997). The fast fourier transform (FFT) analysis resulted dominant frequencies around 10 Hz and 12 Hz, which is observed to be very sensitive frequencies for human body due to existence of natural frequencies of various body parts. The optimum values of forward speed, pulling force and tilling depth are 1.3 m/s, 6 kN, and 0.16 m to obtain reduced OVTV.

Keywords: Whole Body Vibration (WBV) · Agricultural tractor
Ride comfort · Point Vibration Total Value (PVTV)
Overall Vibration Total Value (OVTV) · Weighted acceleration (a_w)
Fast Fourier Transform (FFT)

1 Introduction

In current technological era, the agricultural sector has influenced manual labor as well as characteristics of workload [1]. Most of the rural population (around 3 million) are using tractors with an average growth of 0.25 million tractors per year in the current Indian scenario. Population growth has raised the demand for agricultural products to fulfill their requirements. This is the key factor to adopt mechanized machinery such as tractor and its operated implements. Farmers are devoting much effort to reduce the window period between two successive crops. Therefore, the traditional soil tillage operations are being replaced by tractor operated implements such as plougher, cultivator, harrow, rotary tiller, planker etc. Tillage operations are the prime requirement to

S. Bagnara et al. (Eds.): IEA 2018, AISC 823, pp. 467–473, 2019.
https://doi.org/10.1007/978-3-319-96074-6_49

prepare and manipulate the field soil as optimum for the crop. Tillage operations are essential in every agricultural field and can be categorized into two types namely, primary and secondary tillage. Rotavation is known as secondary soil tillage operation used to prepare field after the crop harvesting in order to manipulate the soil as desired. This operation has been carried by using a rotary tiller mounted to the tractor with the application of three-point hitch driven through power take off (PTO). However, tractor drivers are being exposed to whole body vibration (WBV) during working hours. This is due to operating tractor and its mounted machinery over uneven terrain with respect to varying ride conditions [2]. It can be observed that vibration transmitted through tractor platform, seat, backrest and steering wheel enters into the driver's body significantly affects the ride comfort [3]. Moreover, tractor drivers have to work for long working hours in fields that may cause low back pain issues [4]. It is normally due to the high vibration amplitudes with low frequencies during off-road activities [5]. Most of the past research studies have been carried out to investigate ride comfort with varying frequencies, magnitudes, amplitudes, seating conditions, postures, foot locations etc. [6–10]. However, these studies are performed in controlled laboratory conditions on simulators. A few researchers attempted to study the WBV response in real field conditions [12–14]. As far as the literature reviewed, it has been found that occupational investigations into real field rotary soil tillage operation are still unexplored. Therefore, present study attempted to investigate the relative effect of forward speed, pulling force, tilling depth on ride comfort during real field rotary tillage operation.

2 General Details

Twenty-five years old male subject was chosen with stature 1.531 m, weight 79 kg and body mass index (BMI) 33.09 kg/m^2. The selected subject had an experience of five (5) years in tractor driving. A consent form has been signed by the subject to carry out experiments in order to get ethical approval. The study was conducted on a post-harvested paddy agricultural field located at Punjab Agricultural University (PAU), Ludhiana, Punjab (India). A 2015 model tractor 'T' of 50 hp was used for the experimentation. The tire inflation pressure was maintained as per standard provided in tractor catalogue (Fig. 1).

Fig. 1. Tractor mounted with cultivator

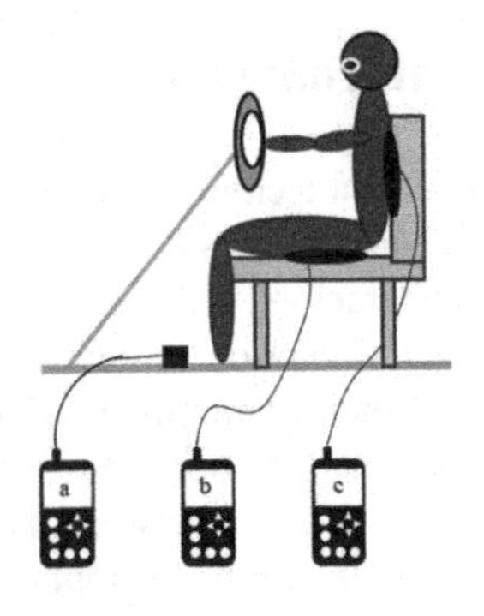

Fig. 2. Apparatus mounting locations

A six (6) feet rotavator with weight 450 kg was mounted to the tractor for performing rotavation operation. There were forty eight (48) C-shape cutting blades welded on eight flanges so as to provide the rotary motion for soil cutting purpose. The cutting depth and width capacity of the rotavator was 2.18 m and 0.15 m, respectively. Ride comfort response was evaluated in terms of overall vibration total value (OVTV) with respect to three different locations, namely tractor platform, seat pan and backrest [16]. The whole body vibration levels were recorded in terms of root mean squared frequency weighted acceleration (a_w) values at each location along the three translational axis (i.e. x, y and z axis). The weighting filters for each location and axis were used as per ISO 2631-1 (1997).

3 Measurement Devices/Equipments

The whole-body vibration levels were recorded in terms of acceleration weighted root mean squared (rms) values along fore-and-aft (x), lateral (y), and vertical (z) axes, respectively. The floor vibrations were recorded by using 4-channel SV 958 vibration analyzer (a). Seat and backrest vibration were recorded by using two SVAN 106 human vibration monitors (b & c) mounted at individual locations as shown in Fig. 2. The measurement and calculations were done as per ISO 2631-1 (1997) standard.

4 Design of Experiments

The design of experiments has been formulated by using Taguchi's L_{27} orthogonal array with respect to three input factors as shown in Table 1. This orthogonal array provided twenty seven (27) combinations of the input parameter to form systematic set of experiments. Tractor was driven as per each set to record a_w values at respective locations. Further, the data was analyzed in the form of signal-to-noise (S/N) ratios.

Table 1. Input factors and their levels

Levels	Input factors		
	A: Forward speed (m/s)	B: Pulling force (kN)	C: Tilling depth (m)
1	1.3	2	0.10
2	1.5	4	0.13
3	1.7	6	0.16

This method aimed to obtain optimum ride conditions in order to get reduced overall vibration total value. Therefore, the S/N ratios were calculated by using mathematical relation corresponds to "smaller-the-better" option as shown below:

$$(S/N) = -10\log\left[1/R\left(y_1^\wedge 2 + y_2^\wedge 2 + \ldots y_n^\wedge 2\right)\right] \quad (1)$$

The maximum value of S/N ratio will represent the optimum level for each input factor. The experimental design and corresponding results of each set has been tabulated in Table 2. This includes five number of replication represented as R1…R5 and further, the mean OVTV was calculated. Correspondingly, the computed S/N ratio for each set of experiments was reported as shown in Table 2. The mean OVTV with respect to conducted experiments was ranges 0.637–0.843 m/s^2. Similarly, the computed S/N ratios were found between 1.663–3.888 (dB). These S/N ratios were used to represent the main effect plot with respect to forward speed, pulling force and tilling depth as shown in Fig. 3. The maximum S/N ratio value represents the condition at which minimum OVTV was obtained, e.g. experiment 7–9. On the other hand, the minimum S/N ratio exhibits worse ride conditions results into increased OVTV, e.g. experiment 19–21.

Table 2. Experimental design and results of L_{27} orthogonal array

Trial no.	Input factors			Output (OVTV) m/s^2						S/N (dB)
	A	B	C	R_1	R_2	R_3	R_4	R_5	Mean	
	Trial conditions			Replications						
1	1	1	1	0.731	0.746	0.721	0.752	0.742	0.738	2.766
2	1	1	1	0.749	0.729	0.757	0.742	0.747	0.745	
3	1	1	1	0.736	0.712	0.748	0.719	0.758	0.735	
4	1	2	2	0.68	0.696	0.718	0.677	0.672	0.689	3.423
5	1	2	2	0.665	0.688	0.698	0.662	0.664	0.675	
6	1	2	2	0.688	0.672	0.705	0.697	0.712	0.695	
7	1	3	3	0.646	0.66	0.681	0.641	0.664	0.658	3.888
8	**1**	**3**	**3**	**0.634**	**0.626**	**0.641**	**0.648**	**0.635**	**0.637**	
9	1	3	3	0.652	0.648	0.654	0.665	0.671	0.658	
10	2	1	2	0.821	0.792	0.820	0.801	0.819	0.811	2.087
11	2	1	2	0.799	0.782	0.809	0.791	0.809	0.798	
12	2	1	2	0.778	0.798	0.784	0.804	0.769	0.787	
13	2	2	3	0.752	0.731	0.768	0.757	0.761	0.754	2.392
14	2	2	3	0.768	0.785	0.778	0.749	0.760	0.768	
15	2	2	3	0.792	0.779	0.802	0.777	0.808	0.792	
16	2	3	1	0.736	0.749	0.764	0.742	0.737	0.746	2.491
17	2	3	1	0.768	0.788	0.768	0.767	0.797	0.778	
18	2	3	1	0.754	0.764	0.777	0.759	0.768	0.764	
19	**3**	**1**	**3**	**0.841**	**0.844**	**0.855**	**0.831**	**0.845**	**0.843**	**1.663**
20	3	1	3	0.845	0.846	0.826	0.835	0.825	0.835	
21	3	1	3	0.824	0.835	0.828	0.859	0.826	0.834	
22	3	2	1	0.819	0.799	0.835	0.825	0.818	0.819	2.009
23	3	2	1	0.775	0.772	0.802	0.767	0.807	0.785	
24	3	2	1	0.812	0.795	0.818	0.795	0.842	0.812	
25	3	3	2	0.802	0.801	0.792	0.768	0.777	0.788	2.335
26	3	3	2	0.736	0.754	0.743	0.755	0.739	0.745	
27	3	3	2	0.786	0.777	0.815	0.799	0.797	0.795	

A, B and C: Input factors; R1, R2, R3, R4 and R5: Number of replications for each respective set of trial.

In Fig. 3, optimum ride conditions with respect to forward speed, pulling force and tilling depth were found 1.3 m/s, 6 kN and 0.16 m, respectively. It means that the tractor operated at these conditions could result into reduced OVTV, thereby enhance the ride comfort.

It can be seen that OVTV increases with the increase in forward speed. This trend was due to operating over uneven field surface, which results into increased vibration levels (i.e. acceleration values) with increasing the forward speeds [17]. However, the OVTV trend found to decreases with increase in pulling force. Similarly, this trend gets decreases up to a tilling depth of 0.13 m and further, slightly increases while increasing depth up to 0.14 m. It means that the vibrations may get damped or absorbed with the increase in tilling depth. Afterwards, Analysis of variance, percentage contribution of each input factor was tabulated in Table 3 and their respective ranking was shown in Table 4.

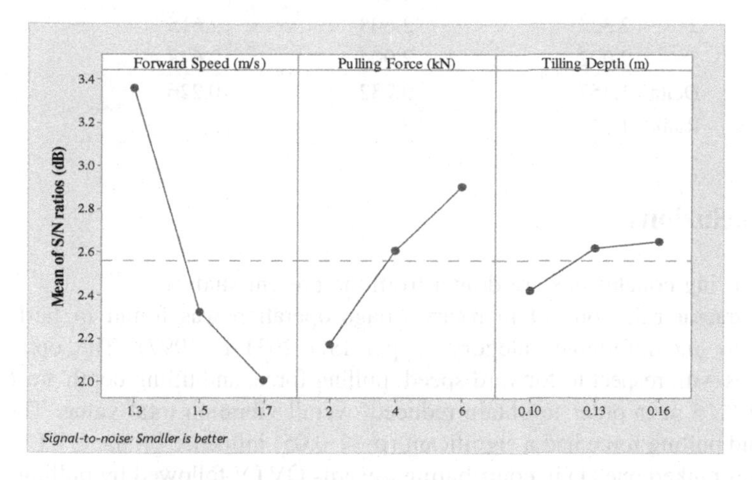

Fig. 3. Main effects plot for Signal to Noise (S/N) ratios

The effect of forward speed and pulling force on overall vibration total value was found significant at 95%. However, the effect of tilling depth was insignificant on the output response. Furthermore, the contributing effect (P%) of each factor was calculated by dividing individual sequential sum of square by total sequential sum of square value. The forward speed was found to have maximum effect (76.36%) on the OVTV followed by pulling force (21.07%) and tilling depth (1.77%), respectively. In Table 4, the ranking of each factor has been calculated by means of delta values. It has been computed by the difference between maximum and minimum signal-to noise ratio. It was seen that the forward speed has maximum delta value, thereby ranked 1 followed by pulling force and tilling depth with respect to their influence on OVTV.

Table 3. Analysis of Variance for overall vibration total value

Source	dof	Seq SS	Adj MS	F	p	P %
Forward speed	2	0.021135	0.010567	95.23	0.01*	76.36
Pulling force	2	0.005831	0.002916	26.27	0.04*	21.07
Tillage depth	2	0.000490	0.000245	2.21	0.31	1.77
Residual error	2	0.000222	0.000111			0.80
Total	8	0.027678				100.00

Table 4. Response Table for Signal to Noise (S/N) Ratios (Smaller-the-better)

Level	Input factors		
	Forward speed (m/s)	Pulling force (kN)	Tilling depth (m)
1	3.359	2.172	2.422
2	2.323	2.608	2.615
3	2.002	2.904	2.648
Delta	1.357	0.732	0.226
Rank	1	2	3

5 Conclusions

The following conclusions are drawn from the present study:-

The tractor ride comfort in rotary tillage operation was found in fairly uncomfortable to uncomfortable category as per ISO 2631-1 (1997). The optimum ride conditions with respect to forward speed, pulling force, and tilling depth are 1.3 m/s, 6 kN, and 0.16 m in order to obtain reduced overall vibration total value. The forward speed and pulling force had a significant ($p \leq 0.05$) influence on the OVTV. Forward speed was ranked one (1) in contributing towards OVTV followed by pulling force and tilling depth, respectively.

Acknowledgements. The authors would acknowledge The Institution of Engineers (IEI), India for assisting this research work with financial support [Grant Code: RDDR2016067]. Authors would also like to thank The Department of Farm Machinery and Power Engineering, Punjab Agricultural University, Ludhiana, Punjab, India for providing experimental facilities.

References

1. Salokhe VM, Majumder B, Islam MS (1995) Vibration characteristics of power tiller. J Terramech 32:181–197
2. Mehta CR, Tiwari PS, Varshney AC (1997) Ride vibrations on a 7·5 kW rotary power tiller. J Agr Eng Res 66:169–176
3. Village J, Trask C, Chow Y, Morrison JB, Koehoorn M, Teschke K (2012) Assessing whole body vibration exposure for use in epidemiological studies of back injuries: measurements, observations and self-reports. Ergon 55:415–424

4. Tiemessen IJ, Hulshof CT, Frings MHW (2008) Low back pain in drivers exposed to whole body vibration: analysis of a dose–response pattern. Occup Environ Med 65:667–675
5. Griffin MJ (2012) Handbook of human vibration. Academic press, London
6. Ciloglu H, Alziadeh M, Mohany A, Kishawy H (2015) Assessment of the whole-body vibration exposure and the dynamic seat comfort in passenger aircraft. Int J Ind Ergon 45:116–123
7. Rakheja S, Dong RG, Patra S, Boileau PE, Marcotte P, Warren C (2010) Biodynamics of the human body under whole-body vibration: synthesis of the reported data. Int J Ind Ergon 40:710–732
8. Nawayseh N, Griffin MJ (2009) A model of the vertical apparent mass and the fore-and-aft cross-axis apparent mass of the human body during vertical whole-body vibration. J Sound Vib 319:719–730
9. Patra SK, Rakheja S, Nelisse H, Boileau PE, Boutin J (2008) Determination of reference values of apparent mass responses of seated occupants of different body masses under vertical vibration with and without a back support. Int J Ind Ergon 38:483–498
10. Rakheja S, Haru I, Boileau PE (2002) Seated occupant apparent mass characteristics under automotive postures and vertical vibration. J Sound Vib 253:57–75
11. Scarlett AJ, Price JS, Stayner RM (2007) Whole-body vibration: evaluation of emission and exposure levels arising from agricultural tractors. J Terra 44:65–73
12. Servadio P, Marsili A, Belfiore NP (2007) Analysis of driving seat vibrations in high forward speed tractors. Bio Eng. 97:171–180
13. Loutridis S, Gialamas T, Gravalos I, Moshou D, Kateris D, Xyradakis P, Tsiropoulos Z (2011) A study on the effect of electronic engine speed regulator on agricultural tractor ride vibration behavior. J Terra 48:139–147
14. Huang ML, Hung YH, Yang ZS (2016) Validation of a method using Taguchi, response surface, neural network, and genetic algorithm. Meas 94:284–294
15. Wyllie IH, Griffin MJ (2007) Discomfort from sinusoidal oscillation in the roll and lateral axes at frequencies between 0.2 and 1.6 Hz. J Acoust Soc Am 121:2644–2654
16. Roy RK (1990) A primer on the Taguchi method. Competitive Manufacturing Series, New York
17. Vrielink HHO (2009) Exposure to whole-body vibration and effectiveness of chair damping in high-power agricultural tractors Report (2012-0601). ErgoLab Research BV

The Mobility in Belo Horizonte Through the Macroergonomics and Service Design

Róber Dias Botelho(✉), Jairo José Drummond Câmara, Ivam César Silva Costa, and Erick Tadeu Teixeira Costa Maia

Universidade do Estado de Minas Gerais/Escola de Design, Av. Antônio Carlos, 7545, São Luiz, Belo Horizonte, MG 31270-010, Brazil
roberrubim@yahoo.com, camara.jairo@gmail.com, ivamcesar@gmail.com, ericktmaia@gmail.com

Abstract. The urban organization passed through successive revolutions over the years, with regards to styles, social-political strategies and impacts in human daily life. These conditions have forced the implementation of new mobility concepts, in which incipient complex structures feature in urban centers. Brazil relies on, primarily, the individual transport through the road/car pair, contradicting the strategic world thinking, in which prevails the intermodal system of transportation. The purpose is analyzing the capital's transport system, under the macro ergonomic and service design perspective (its organization and physical/psychological impacts on the population). Thus, considering the present situation of Belo Horizonte city, we noted that just implementing new vehicles (MICRO approach) won't solve the problem, because it doesn't integrate the urban mobility global system (MACRO approach) – interconnect, in an efficient and safe way, the point of departure to the destiny. The preliminary results point to two directions: A - urban space should be thought as a "system of systems integration" and not just the association/approximation of products; B - the previous concept demands a systemic reading of the MACRO/MICRO, since the intrinsic values must be attacked to the individual and not less to the collective in relation to the organized society.

Keywords: Ergonomics applied · Urban transport
Occupational health and safety

1 Introduction

Cities have undergone profound transformations throughout history, regarding not only its forms but also its political and social organizations. In this sense, the street occupies a prominent place in the city, it is here that we can see the changes, the wrinkles, the demonstrations. It will reveal the social life that exists in the city.

Transports interact primarily with the development of economy as a whole, as their availability has implications both with changes in inventories and relative combinations of production factors, and with the changes in the structure of intermediate and final demand [1].

S. Bagnara et al. (Eds.): IEA 2018, AISC 823, pp. 474–479, 2019.
https://doi.org/10.1007/978-3-319-96074-6_50

Following the development of urban areas, means of transport started to exert important influence on the location, size, characteristics of cities and especially in the population habits which will be, sometimes molded, sometimes led by the manner the local transport system constitutes. The almost simultaneous appearance of public transport in several cities was a result of Industrial Revolution.

With the advent of the electric tram, a permanent revolution about the city's growth, its structure and interrelationships were possible. The tram became a success, remaining for many years as the main mean of urban transport in the world. Belo Horizonte was the fifth Brazilian city to have electric traction tram system, administered by the company General Electric and opened in 1902. Although ephemeral, the success of the first tram line, stimulated similar initiatives in the local business community and, in subsequent decades, several concessions to companies seeking to explore different areas of the city were issued [2]. The tram symbolized a picture of modernist spirit, which the capital of Minas Gerais going through.

The modern city appears as a place where the new outstands, showing elements that break with parameters which correspond to the past. [...] The distinctive architecture, public services installation as telegraphs, parks and squares, electricity and trams, take considerable part of the socially shared reports, when it is mentioned "modernity" lived in the state capital [3].

Until around 1920, public transport was virtually the only alternative way of passenger locomotion in cities. There was a reduction in the coefficient of manufactured goods imports, due to the global crisis, which generated additional flows of intermediate goods for the internal market in the 1930s, providing good conditions for the beginning of the industrial development in the country. Few countries had a boom in road construction as occurred in Brazil in the 1945–1980 period. Road transport was then seen as a necessary way to displace these flows [4]. The increasing manufactured production and consumption of petroleum products justified the supplementation of railways, establishing road transport as a development priority at that time.

Thus, the country went into an industrialization cycle associated with the construction of road infrastructure and the low cost of oil, reaching in 1980 a position among the top ten world producers of vehicles. Such context consolidated, then, urban development patterns centered on road vehicles without real development focused on integrating the infrastructure to other modes of transport, as shown in "The National Policy for Sustainable Urban Mobility" [6].

The funds were directed to create an infrastructure that served the demands related to the car. Thus, the car was treated as an object capable of giving the optimal speed to the new society [6]. The car has overcome, in Brazil, all other means of transport - the automotive euphoria is settled in society's imagination through the speed, power, autonomy and status ideas. This symbol introduced in urban areas caused a number of social, spatial and environmental impacts within the city. Cities would not be what they are today if the car did not exist [7].

The fleet of cars has increased 175.0% since 1998. It is essential for us to reverse this logic, favoring the public, systemic and integrated transport. The quality of public transport service and the choice of technology and operational solutions suitable for each demand are very important factors in shaping the urban development of large cities, which directly impacts the population's quality of life [8].

2 Objectives and Methodology

This work results from the study of mobility in Brazil, is supported by several other researches and aims to study the mobility problems (centered in the urban context) based on service design and macro ergonomics approach.

3 The Urban Transport as a Service

Design has many different definitions, but its basic foundation can be translated as the process of transposing ideas into reality, making abstract thoughts become tangible and concrete [9].

The user's in the design process has changed considerably. The design underwent a design centered design transformation - based on observation, to user-centered design - based on immersion [10]. This new design model, with the presence of participative methods and tools make the user part of the project, involving it in a collaborative design process [11].

"Design must continually redesign its speech and itself" [13] and in this "re-designing" a design approach is very important: the service design, centered in user and contact points that promote interaction between tangible and intangible, can join the objects that a design project must supply on these recent days.

Thus, when the transport of Belo Horizonte city is approached focusing on design perspective, it is possible to analyze the transport system under the macro ergonomic plans and at the same time, the design services (their demands, their organization and their physical and psychological impacts on the community).

The discrepancy between the mobility concept and circulation idea created a number of problems in urban centers. Defining the daily lives of people and relate it to the social, economic and environmental aspects are of fundamental importance to think mobility as a service that should be centered on human beings [14]. Technological aspects should be thought later. Analyze the scenario is significantly more important when you want to really meet the needs of urban mobility. The concept of transporting people in large urban centers goes far beyond simply connecting a point "A" to point "B". People have different needs that are related to the concepts of use, rationality and objectivity of the system, while demanding symbolic visions and related to sympathy and its relation to emotional values.

3.1 Each Case Is Different!

3.1.1 An Always Complex System - The City

Belo Horizonte city and the metropolitan area of Belo Horizonte (MRBH) are facing challenges arising from the lack of proper integration between the municipal systems and the metropolitan passenger transport system. "Failures as overlapping lines and poor services cause delays and loss of quality of the transport system of the region" [2].

The state capital has a population of 2.5 million people and a fleet of 1.6 million vehicles [15]. The public transport fleet has 3,000 buses and 28.2 km of subway.

In the conventional bus system 26,000 trips per working day are performed with the above-mentioned fleet, covering 14 million kilometers and carrying 36.0 million people, according to the Management Control and Tariff Studies [16] BHTrans, public company established in 1992 and responsible for the management of public transport in Belo Horizonte city. The metropolitan management of transportation systems, including buses and underground train, is under the coordination of Setop, Secretary of State for Transport and Public Works [8].

Traffic in Belo Horizonte is becoming more chaotic every day. The travel time increased from 32.4 min in 1992 to 36.6 min in 2012. More than 16.0% of the population spends more than an hour to get to work [17]. These data indicate a strong increase in mobility problems and the renovations done so far were not enough to improve the population displacement conditions. To this situation, still add up the movement of 19.000 daily truck trips in the city and MRBH.

The state capital traffic will only begin to improve in 2020 and the difficulty of expanding the municipal subway system is one of the factors that slow the signs of improvement in Belo Horizonte's traffic [18]. The subway, which opened 28 years ago, is only 28 km and the plan was to reach 120. In the meantime, there is still the problem of air quality. With the fleet rising and fast-growing city without an urban planning suited to collective needs, the air which people from Belo Horizonte breathe lose quality every year. In 2011 the air quality in the city reached its worst mark since 2002. Among the 353 days when there was air monitoring, in 62.99% of them air quality in the city was considered "good". However, about 250 new vehicles are added per day on the streets. So, the trend is that this percentage drops further and takes the city authorities to adopt emergency actions to public health and urban planning [18].

Currently, 45.0% of citizens from Belo Horizonte use public transportation in their daily movements [18]. 2001–2002 data show that in this period, only 25.87% of citizens used motorized individual vehicles to travel in the city, a percentage that in the last years has increased due to government incentives for purchasing vehicles and City Hall projects that resulted in duplication of routes and construction of streets, aiming 2014 World Cup, which took this city as one of its host cities.

The average speed of buses is far below the speed of private vehicles: bus drivers, in 2012, drove around 18,0 km/h on average, while the cars were moving at an average speed of 25.0 km/h. These data indicate how urban mobility is discrepant in this city. The numbers are regressing: in 2001, 35.0% of the passengers considered the public transport in the city excellent/good and in 2011 this figure had its lowest rate, only 12.0% of users [18].

The MRBH has also introduced distortions in the metropolitan transport system that affect not only the mobility of people and goods, but also generates economic losses for the region. Much of the problems stems from the lack of proper integration between the local and metropolitan passenger transport systems [2].

Because of this combination of factors, the need of creating an Urban Mobility Plan emerged focused on the feasibility of a multimodal structural network of public transport consisting of subway lines and corridors of BRT (Bus Rapid Transit). This plan has as guidelines the densification of urban occupation around the mass transport corridors, providing attractive elements to increase the number of trips by non-motorized modes. Making a more attractive public transport facing the individual transport is also one of the objectives of PlanMob [18].

4 Final Considerations

Until the half of twentieth century transport was seen as the most efficient way to move loads and people between two points. However, when we extrapolate the concept of mobility and deepen the human and urban character, there is a considerable expansion of the system complexity. If the simple move had as limits measurable, rational and sometimes absolute factors (I - the distance - directly related to the geography and geology of the region, etc.; II - speed - in addition to the previous points, we have the quality of the route, the amount of traffic flow, directions, etc. and III - the time - this is presented as a result of points I and II), currently there are factors, that even being partially measurable, many present higher level of subjectivity (pollution, symbolism of the goods, the importance of the journey, landscape quality, socio-cultural references of vehicles and stations, etc.).

With such perspective transport should be seen as a service oriented to collective mobility, since this is a distinguished service in the provision of an ability to offer an immaterial and not storable use for a group of users. Offering a mobility service represents the availability of economic and material resources that are intangible to a group of people. When we think mobility in isolation (in the case of a person or limited group of people), vehicles and support buildings it currently appears as a solution that is likely to evolve. However, thinking mobility of a community (city, state, country, continent, planet), there is in vehicles and their surroundings a problem that demands, in a sense, a revolution. A thinking, systems designing revolution that can be adapted to local adversities without neglecting global impacts. In this local/global duality, there is also the individual/collective relationship. So, think mobility of tomorrow is a strategic challenge in making it increasingly collective without removing the personal character of individuals that are part of the local system or localities that constitute the global system.

References

1. Araújo MP (2006) Infraestrutura de transporte e desenvolvimento regional: uma abordagem de equilíbrio geral inter-regional. In: Universidade de São Paulo – USP, São Paulo, Brasil. http://www.teses.usp.br/teses/disponiveis/11/11132/tde-07062006-162615/publico/MariaAraujo.pdf. Accessed 13 Apr 2016
2. BHTrans (2013) Plano de Mobilidade Urbana de Belo Horizonte - PlanMob-BH. In: Prefeitura de Belo Horizonte – PBH (report). http://www.bhtrans.pbh.gov.br/portal/page/portal/portalpublicodl/Temas/Observatorio/observatorio-da-mobilidade-publicacoes-2013/PlanMob-BH_-_apresentacao.ppt.pdf. Accessed 25 Mar 2015
3. Rodrigues DUA (2016) Belo Horizonte no início do século XX: uma cidade entre rupturas e continuidades. In: Revista Três Pontos, Universidade Federal de Minas Gerais – UFMG, Belo Horizonte. https://seer.ufmg.br/index.php/revistatrespontos/article/download/1561/1120. Accessed 13 Apr 2016

4. Barat J (2007) Logística, Transporte e Desenvolvimento Econômico: A Visão Macroeconômica. Cla Cultural Ltda, 8585454296, São Paulo, pp 36–41
5. Portal PBH (2016) Estatísticas e Indicadores: Síntese de Indicadores de Belo Horizonte, Prefeitura de Belo Horizonte, - PBH, Belo Horizonte. http://portalpbh.pbh.gov.br/pbh/ecp/comunidade.do?app=estatisticaseindicadores. Accessed 12 Apr 2016
6. Ministério das Cidades (2005) Mobilidade Urbana é Desenvolvimento Urbano. Instituto Pólis, Brasil, 39 p
7. Santos FANV dos, Ferroli PCM (2002) Experiências Pedagógicas no Curso de Design Industrial da Univali. In: Anais do P&D Design, AEnD-BR: Rio de Janeiro, Brasil, CD-ROM
8. Hermont LD (2016) Oferta e demanda de transportes integrados. https://www.ufmg.br/pos/geotrans/images/stories/diss034.pdf. Accessed 13 Apr 2016
9. Fraser H (2013) Design para negócios na prática. Elsevier, 978-85-352-6415-9, São Paulo
10. Moritz S (2013) Service design: pratical access to an involving field. http://stefanmoritz.com/_files/Practical%20Access%20to%20Service%20Design.pdf. Accessed 22 Oct 2013
11. Junior J da C (2010) Codesign no design de serviço: o uso de métodos participativos para o melhor atendimento das necessidades dos usuários. In: 1st Brazilian Symposium on Service Science, Universidade Federal do Paraná - UFPR, Curitiba. http://www.redlas.net/materiali/priloge/slo/77268.pdf. Accessed 05 Apr 2018
12. G1 (2016) Brasil perde para México e agora é 8º em ranking de produção de veículos. In: G1. Portal de Notícias. http://g1.globo.com/carros/noticia/2015/03/brasil-perde-para-mexico-e-cai-para-8-em-ranking-de-producao-de-veiculos.html. Accessed 10 Apr 2016
13. Krippendorf K (2000) Design centrado no ser humano: uma necessidade cultural. In: Estudos em Design – Design e Ser Humano, Rio de Janeiro, vol 8(3), Set. http://periodicos.anhembi.br/arquivos/Hemeroteca/Periodicos_MO/Estudos_em_Design/107170.pdf. Accessed 10 Apr 2018
14. Néspoli LCM (2016) Encontro debate saídas para a cultura transporte individual. http://www.cartacapital.com.br/dialogos-capitais/dialogos-capitais-9762.html. Accessed 10 Apr 2016
15. IBGE: Instituto Brasileiro de Geografia e Estatística, Cidades. http://cidades.ibge.gov.br/xtras/home.php. Accessed 10 Apr 2016
16. GECET: Gerência de Controle e Estudos Tarifários (2018) Relatório de dados gerenciais - Belo Horizonte: BHTRANS (report). http://www.bhtrans.pbh.gov.br/portal/page/portal/portalpublicodl/Temas/Onibus/gestao-transporte-onibus-2013/Dados%20Gerenciais%20do%20Sistema%20de%20Transporte%20P%C3%BAblico%20por%20%C3%94nibus%20do%20Munic%C3%ADpio%20de%20BH.pdf. Accessed 10 Apr 2016
17. IPEA: Instituto de Pesquisa Econômica Aplicada (2012) Indicadores de mobilidade urbana da PNAD. http://www.ipea.gov.br/agencia/images/stories/PDFs/comunicado/131024_comunicadoipea16_graficos.pdf. Accessed 10 Apr 2016
18. Tampieri GA (2015) Mobilidade Urbana em Belo Horizonte. http://www.nossabh.org.br/noticias.php?q=123/Artigo:_A_Mobilidade_Urbana_em_Belo_Horizonte. Accessed 28 Mar 2015

The Quality of Roads in Brazil: The Interrelation of Its Multiple Stressors and Their Impact on Society

Róber Dias Botelho(✉), Jairo José Drummond Câmara, Ivam César Silva Costa, Júlia Silveira Pereira Guimarães, and Humberto Carvalho Dias

Universidade do Estado de Minas Gerais, Escola de Design, Av. Antônio Carlos, 7545, São Luiz, Belo Horizonte, MG 31270-010, Brazil
roberrubim@yahoo.com, camara.jairo@gmail.com, ivamcesar@gmail.com, juliaguimaraes997@gmail.com, hcd@gmail.com

Abstract. The road transport in Brazil has evolved in a disorderly way, so today the modal is placed as the main articulator of the economic circuit of goods distribution and passenger transport, representing 61.1% and 48.16%, respectively of the total transported. This condition represents a limiting factor in terms of national competitiveness. The purpose of this article is to study the variables related to the quality of roads in Brazil, the interrelation of its multiple stressors and their impact on society. The act of driving is a task of monitoring and multiple occupational stressors for professional drivers. The diversity of these worries has direct implications on the work conditions, also affecting health and the security of the system – system fails, or the called "human error". The cognitive overload required in national roads, due to the diversity of incognitos caused by the lack of structure and conservation, significantly contributes to the occurrence of accidents. Old trucks, poorly maintained highways, desperate drivers and the lack of supervision are the ingredients of a chaotic situation. Although only 12.0% of the national highways are paved, a significant portion of this amount (Federal, Transit State, State and Municipal highways) presents serious problems of quality and safety. The accidents registered only in 2016, on the federal highways audited, resulted in a cost of US $ 3.02 billion for the country. The poor quality of the roads burdens the cost of operation and compromise the safety and health of society as a whole.

Keywords: Applied ergonomics · Highway quality · Truck driver
Social security · Sustainable development

1 Introduction

The cargo transportation in Brazil has evolved in a disorderly manner so that today the road transportation represents nearly 60.0% of the total volume transported, having already reached more than 70.0% in the decades of 1990–2000. This is due to a government policy strategy grounded in concepts even from the 1920s, when the

S. Bagnara et al. (Eds.): IEA 2018, AISC 823, pp. 480–488, 2019.
https://doi.org/10.1007/978-3-319-96074-6_51

notion of governing used to fell under the factor "open roads", as stated at the time by Mr. Washington Luiz – then Governor of the State of São Paulo [1].

By making a detailed evaluation, it is made explicit that the quality and growth of the road network does not accompany the demand for infrastructure to the flow of production or to the displacement of people. Giving an idea, the vehicle fleet increased by 194.1% from 2001 to 2016, but the roads continue having serious quality problems (with regard to design and correction), compromising safety once that the total mesh, 1.7 million kilometers, only 12.2% (slightly more than 210.0 thousand kilometers) have pavement [2].

1.1 Objectives and Methodology

The purpose of this article is to study the variables related to the quality of roads in Brazil, the interrelation of its multiple stressors and their impact on society. For this, the research followed a methodological reasoning of basic nature, of qualitative approach and exploratory character in the following phases: literature review and information cross-check with the market/context reality; evaluation and interpretation of indices and transport linked figures in Brazil with emphasis on road conditions; discussions among the co-authors, representatives and industry experts, besides market professionals. Finally, there is the preparation of a report from which this article was generated.

2 Literature Review

2.1 Scenario

With about 1.7 million kilometers of highways cutting Brazil, this modality moves 58.0% of the national cargo volume. However, 80.3%, or 1,364 million kilometers, are unpaved (78.63% of the total), leaving only 213.5 thousand kilometers of paved roads (12.3% of the total). Another 9.06% are planned routes, that is, they have not left the project yet [3].

The lack of paving is just one of the Brazilian problems of logistics infrastructure. In 2013, 300 professionals from 250 companies with the largest merchandise movement participated in a survey and 99.0% said they believe that logistics infrastructure causes a loss of competitiveness for the country. In all, 97.0% indicated that poor roads are the main problem. A 2013 National Transportation Confederation (CNT) Survey went even further: it saw that the average increase in operating costs due to pavement conditions on Brazilian highways is 25.0%. According to the Federal Highway Police - PRF, the fleet allowed to circulate - only of trucks and trailers-, in June of this year of 2014, was 5,035 million vehicles. As a result, Brazil loses US $ 2.2 billion per year, 50.0% in cargo handling and transportation [4].

In a survey carried out in 2017 by CNT, the overall quality of Brazilian highways fell in 2017 compared to 2016. The survey encompasses 103,259 km, equivalent to all federal paved roads (total of 65,530 km) and major state responsibility highways (148,061 km paved in total). In 2017, 61.8% of the extension of the roads surveyed had the general condition considered "regular", "bad" or "very bad" and in 2016 58.2%

were classified in these same conditions. The main reason for the worsening, according to the CNT, was the reduction of investment in road infrastructure. The most marked deterioration was observed in signaling, in which the classification as "good" or "great" fell from 48.3% to 40.8%. The quality of the pavement had worsened from 48.3% to 50.0% with "regular", "bad" or "awful" evaluation. Regarding geometry, the low-quality index repeated last year's numbers, at 77.9% under "regular", "bad" or "awful" conditions [3, 5]. From this total, we have the following classification, according to Table 1.

Table 1. Evaluation of federal and state paved roads [6].

Rating of general conditions of Roads in km – 2016					
Great	Good	Regular	Bad	Awful	Total
11,936	31,158	**35,840**	17,838	**6,487**	103,259
Rating of Paving in km					
45,876	7,485	36,968	10,227	**2,703**	103,259
Rating of Signaling in km					
14,453	**35,371**	30,040	13,128	**10,267**	103,259
Rating of Geometry in km					
5,158	17,705	28,783	18,819	**32,794**	103,259

Even the percentage of kilometers evaluated (103,259) representing 48.34% compared to the total of 213,591 paved kilometers (Federal roads, Transit Roads, State and Municipal), important limitations are observed in all the questions. This represents 61.1% of the cargo transport share (with 485.62 million TKU) and approximately 48.16% of the passengers (92.529.213 out of a total of 192.137.422 passengers) [3].

It is noteworthy that between 2007 and 2017 the number of vehicles passing through these highways practically doubled, jumping from 46.0 million to 95.0 million. According to the president of the cargo transportation section of the CNT, Flávio Benatti, "Brazil has no infrastructure". For the poor quality of the roads costs the average cost of operating transport companies by 27.0%. This percentage can exceed 90.0% of the operating cost on very poor roads. The precariousness of the roads is due to the decrease of federal investments and also other facts, such as the economic crisis. The CNT Transport and Logistics Plan point out that US$ 81.61 billion of investments in road infrastructure would be necessary to adapt it to national demand. Only for the maintenance, restoration, and reconstruction of the nearly 83.0 thousand kilometers worn it would be necessary to disburse US $ 14.3 [5, 6].

Such pathway conditions have direct and indirect implications on the number of accidents, in addition to the number of deaths and injuries. In 2016 only 96,363 accidents were recorded, 6,398 fatalities and 86,672 injured. Between 2007 and 2016, the 2011 year stands out with the highest rates of accidents (188,925) and the number injured (104,448). In 2012, in turn, it had the highest number of deaths (8,655). When analyzing the economic costs related to road accidents in 2016, we have the following situation, Table 2 [7].

Table 2. Economic cost of road accidents – 2016 [7].

Type of accident	Average cost US$*	Number of accidents	Total cost of accidents (US$* and billions)
Fatal	211,331.84	5,355	1,13
With victims	29,467.48	54,873	1,62
Victimless	7,535.9	36,135	0,27

*Values in Reais – R$. 1US$ is equivalent to R$3,6 in May 30, 2018.

Accidents, deaths, and injuries recorded in 2016 on federal police roads resulted in a cost of US $ 3.02 billion for the country. This figure is higher than the investment made in highways in 2016, which was US $ 2.39 billion. In addition, it is estimated that only in 2017, the transportation sector will have an unnecessary consumption of 832.30 million liters of diesel. This waste will cost $ 705.55 million to haulers. The calculation is based on the inadequacies found on the pavement. Another critical point, as a consequence of roads conservation conditions, refers to the costs of freight and passenger transportation. The overall quality of the roads and, especially, the condition of the pavement have a direct impact on the operational transporters cost. In 2017, with the deterioration of road quality, the estimated operating cost rose to 27.0%, compared to the 24.9% calculated in 2016. In order to have a better idea, when the road maintenance status is "good". The operating cost reaches 18.8%, jumping to 41.0% when the state of the road is "regular ", 65.6% when "bad" and 91.5% when in awful conditions. In another comparison, while the average operating cost related to public roads reaches 30.5%, it is restricted to 12.0% in the roads granted to concessionaires [7 and 5].

2.2 Ergonomic Prerogatives

There are several factors that contribute to the occurrence of an incident. According to studies by the Volvo Trucks Accident Investigation Group in Sweden, the following causes can be attributed as contributors to the occurrence of a road incident: 90.0% human factor; 10.0% vehicle factor and 30.0% road factor. Regarding the human factor. The error can be as much of the driver as of another track user and that resulted in the incident. The most common problems in these cases are: inattention; error of speed judgment, causing instability of the vehicle and resulting in overturning in curves, jackknifing in the event of breaking or sliding of the truck on a slippery track due to bad weather and, finally, the poor perception of the risk in certain situations of traffic [8].

According to the National Highway Traffic Safety Administration – NHTSA, the human factors (or Human Factors Engineering) consist of the application of knowledge about human abilities, limitations, and other human characteristics to the design of equipment, tasks, and jobs. The role of human factors research is to provide an understanding of how drivers perform as a system component in the safe operation of vehicles. This role recognizes that driver performance is influenced by many environmental, psychological, and vehicle design factors. The focus of the research is to determine which aspects of vehicle design should be modified to improve driver performance and reduce unsafe behaviors. An additional focus is to evaluate driver's capabilities to benefit from existing or new in-vehicle technologies. The research supports Federal Motor Vehicle Safety Standards, safety defects investigations, consumer information, and advancement of knowledge about driver behaviors and performance that can be applied to the development of vehicle technologies that are compatible with driver capabilities and limitations [9].

Thus, road transport should be approached as a "human/machine/system" set, as it presents relevant design flaws and requires a comprehensive systemic treatment. In the search for tools related to the optimization of roadway and roadway design and implementation processes, William Haddon, an American researcher, developed a matrix focused on identifying risk factors before, during and after a collision, in relation to human, the vehicle and the track. For the phase before the collision, it is necessary to select all the measures that prevent its occurrence. The collision phase is associated with measures that prevent the injury occurrence or reduce its severity if it occurs. Finally, the post-collision phase involves all actions that reduce the adverse outcome of the event after its occurrence. Such a tool can be best understood from Table 3 [8].

Table 3. Matrix for risk factors identification before, during and after a collision, in relation to human, vehicle, and road [8].

-		Factors		
-	Phase	Human	Vehicles/Equipment	Environment
Pre-shock	Shock prevention	• Information/education • Attitudes • Physical disability (alcohol/drugs) • Police effort	• Mechanical conditions • Lighting • Brakes • Maneuverability • Speed control	• Road Project and layout • Velocity limits • Pedestrian Protection
Shock	Prevention of Injury during Shock	• Use of Retention Devices (seatbelt) • Physical disability	• Protected occupants/seatbelt • Other safety accessories • Shock protection design	• External impact protection objects
Post-shock	Preservation of life	• Notions of first aid • Access to doctors	• Internal access facilities • Fire risk	• Facilities for rescuing victims • Congestionation

Most accidents are caused by human error. But a well-designed, well-signaled and operated road can greatly reduce the error rate of drivers who are often led to committing it for lack of proper guidance. A safe road can still reduce the severity of accidents. That is if that highway still has adequate protection devices, not only will the number of accidents fall, but its consequences will certainly be minimized. They are responsible for accidents, besides the human factor, the vehicle, the road, the environment and institutional and social factors [10]. The accident locations on the highways can be classified into four distinct types, each with typical characteristics that require different solutions, namely: urban crossings; intersections; curves; bridges and viaducts [11].

The act of driving is a task of monitoring and multiple occupational stressors for professional drivers. The cognitive overload required in national roads, due to the diversity of incognitos caused by the lack of structure and conservation, significantly contributes to occurrence of accidents. Old trucks, poorly maintained highways, desperate drivers and the lack of supervision are ingredients of a chaotic situation. Apart from contributing to the occurrence of accidents, these variables have increased the appalling rates of accidents and deaths on Brazilian roads. On this account, such variables present themselves not only single and/or separately, but combined. The diversity of these worries has direct implications on the work conditions, also affecting health and the security of the system – system fails, or the called "human error". Human error in human/system interaction was a major influence in establishing the area of human factors [12].

It should be noted that an approximation between the human error, the three main criteria elucidated by CNT, is perfectly plausible, as regards the assessment of the quality of national roads (pavement, signaling and road geometry), and the inherent connotations of design ergonomics, correction, awareness, and participation. With regard to design ergonomics, this is the best situation to mitigate accidents and failures. The ergonomic contribution is made during product design, machine, environment or system. Complementing this ergonomic contribution exists the ergonomics of correction. This is applied in real situations, already existing, to solve problems that are reflected in the safety, excessive fatigue, worker illness or quantity and quality of production. Frequently, the solution adopted at the conception moment does not prove satisfactory, requiring changes and adaptations to optimize the operation and suppression of commercial demand and usability. In the present study such an approach could lead to the substitution or reformulation of inadequate parts or materials, and in some cases, certain improvements, such as the adoption of conduits, placement of safety devices, etc., may be implemented with relative facility [13].

Pavement, holes, signs, protection devices, the slope of a curve, over or insufficient lighting, vegetation and climatic conditions. In this item, we consider highway design factors, such as closed curves, excessive ramps, poles and trees near the runway, etc.; maintenance factors and natural factors (rain, snow, ice, etc.). In addition, institutional factors are essential, since they are those that induce users to behave appropriately: timely and correct information, adequate signaling, consistent attitudes on the part of highway operators, policing, etc. [10].

Associated with the two contributions already cited is the ergonomics of awareness. This is directly related to the training and awareness in ergonomics of the users that compose and use the system. Frequently ergonomic problems are not completely solved in the design and correction phases, since new parameters may arise with the very dynamics and interaction between the various processes inherent in the system. Finally, the ergonomics of participation seeks to involve the user of the system in the solution of ergonomic problems. This can be the worker, in case of road maintenance teams, highway police, rescue teams, etc., or the driver, cyclist, pedestrian, etc. This principle is based on the hypothesis that those involved have a practical knowledge and that the systems are not operated in the "correct" form, that is, as it was idealized at the moments of conception and/or correction [13].

In the case of roads, there may be, for example, natural wear and tear of the roads, facilities and support equipment, changes introduced by managers and maintenance staff, changes in schedule, demands for new maintenance procedures, diversity of drivers and vehicles, among many other factors. Contingencies can arise at any time and those involved in the system must be prepared to deal with them. The social factors directly interfere in the process of awareness of those involved in the process, because these factors arise from the cultural context and are manifested from the environment in which the citizen lives and grows until the formal education that he received, including for the transit [10].

Of the eleven presumed causes that appear in the PRF form, seven can be classified as imprudent (Table 4). Of the 6,862 accidents registered by the PRF on federal highways involving trucks and resulting in deaths, 3,051 were identified as "recklessness" of drivers. It should be noted that 1,947 accidents were registered as "other causes" because it was not possible for the police to determine the reason for the incident. Of course, in many of these "other causes" will also be other rashes of drivers [8].

Table 4. Fatal accidents on federal highways -- presumed causes [8].

-	2016	-
Animals on the runway	80	-
Mechanical defect in vehicle	114	-
Defect on track	**63**	-
Signaling disobedience	301	Imprudence
Sleeping	223	
Lack of attention	1,076	
Alcohol ingestion	370	
Do not keep safety distance	90	
Wrong overdrive	383	
Incompatible speed	708	
Others	1,947	-
Grand Total	5,355	-

Of the total deaths, 3.151 are related to recklessness (that is, 58.84%). Defects in the pathways were responsible for 63 deaths (only 1.18%). Nevertheless, we can not disregard the fact that the quality of the roads does not indirectly influence those of another category. The degradation of pavements and accessories does not explain all aspects of road accidents. In reality, even if it is not degraded, the road may have lost the possibility of meeting basic safety criteria, owing to the constant development of the adjoining regions and the evolution of traffic. This, by itself, already constitutes a risk or expose the different groups of users to the imminent risks.

Risk exposure is defined as the frequency of traffic events that create a risk of accidents. The purpose of risk assessment is to estimate the probability and severity of a particular event to occur. When analyzing the safety problems, it is convenient to relate the occurrence of traffic accidents with the movement of individuals in the traffic system, looking for parameters that express or measure how drivers, pedestrians and passengers are exposed to risk when driving along the tracks. The judgment of risk depends on an individual perception of results or expected values and the socio-cultural context. Risk perception is a complex skill involving a varied number of components, a process whose steps, dependent on one another, can be summarized in such a way: drivers must record the existence of events that constitute potential risks; drivers should then judge whether the trajectory or nature of any of the events and that of their own car have the potential to cause a conflict, and then drivers must then check that the event requires an answer proper [14].

Such exposure to a number and diversity of risks favors the occurrence of errors-accidents, also entails a double rate of work which, therefore, favoring, even more, the occurrence of errors and the triggering of more errors-accidents.

3 Discussion and Conclusion

Annually, Brazil counts more than 45.0 thousand dead and more than 500.0 thousand injured with some severity in the traffic. It is estimated at well over a million incidents in the streets and roads happen, killing, injuring, destroying assets across the country each year. Whatever the correct number of accidents attributed to recklessness, the fact is that it is too high, with an unbearable weight both for the companies, for the self-employed and for the whole society [8].

The cognitive overload required in national roads, due to the diversity of incognitos caused by the lack of structure and conservation, significantly contributes to the occurrence of accidents. Old trucks, poorly maintained highways, desperate drivers and the lack of supervision are ingredients of a chaotic situation.

There is a consensus on the liability of driver behavior in most road accidents. Although the human factor is directly related to the occurrence of accidents, this solution does not always have the best cost-benefit ratio. The most efficient solution may not be directly related to the main "cause" of the accident and may even fall on another component of the one that gave rise to the accident. The whole road may create situations where it induces drivers to make mistakes. The set of stressors present on national roads has an important influence (both individual and associated) on the generation of risk factors and accidents. It is believed that the interrelationship of different variables present in the road system should allow the various users to use and interact in a safe, comfortable, simple and clear, including mitigation of consequences and, no less, corrections of any errors caused by the set of factors.

Said this, we realize that the solution of the problem requires a mobilization of the entire society and, in particular, of the government. In addition, education and inspection campaigns are essential for the overall quality of traffic routes. The investments are at the same time the net profit for the country with a reduction of the accidents. The gains, as presented earlier, are higher in the new samples initially observed. The lack of a cultural continuity of security in Brazil leaves a delicate position in the ranking of countries of low competition, with high rates of road accidents.

References

1. CNT[1] – Confederação Nacional do Transporte (2006) Atlas do Transporte – 1ª CNT, Edição, Brasília. http://www.cnt.org.br/informacoes/pesquisas/atlas/2006/arquivos/pdf/Atlas_Transporte_2006.pdf. Accessed 05 May 2017
2. CNT[6] – Confederação Nacional do Transporte (2017) Anuário CNT do transporte: estatísticas consolidadas. CNT, Brasília. http://anuariodotransporte.cnt.org.br/2017/File/MaterialImprensa.pdf. Accessed 28 May 2018
3. CNT[2] – Confederação Nacional do Transporte (2018) Boletim estatístico: CNT – Janeiro. CNT, Brasília. http://cms.cnt.org.br/Imagens%20CNT/BOLETIM%20ESTAT%C3%8DSTICO/BOLETIM%20ESTAT%C3%8DSTICO%202018/Boletim%20Estat%C3%ADstico%20-%2001%20-%202018.pdf. Accessed 28 May 2018
4. Benevides C (2014) No Brasil, 80% das estradas não contam com pavimentação. In: Journal O Globo. Globo, São Paulo, Brazil, 23 August 2014. http://oglobo.globo.com/brasil/no-brasil-80-das-estradas-nao-contam-com-pavimentacao-13710994. Accessed 12 Nov 2017
5. CNT[5] – Confederação Nacional do Transporte (2017) Pesquisa CNT de rodovias 2017. CNT, Brasília. http://pesquisarodoviascms.cnt.org.br//PDFs/Resumo_Principais_Dados_Pesquisa_CNT_2017_FINAL.pdf. Accessed 28 May 2018
6. CNT[3] – Confederação Nacional do Transporte (2017) Anuário CNT do transporte – estatísticas consolidadas 2017. CNT, Brasília. http://anuariodotransporte.cnt.org.br/2017/Apresentacao. Accessed 28 May 2018
7. CNT[4] – Confederação Nacional do Transporte (2017). Informativo: acidentes – 2007–2016. CNT, Brasília. http://cms.cnt.org.br/Imagens%20CNT/BOLETIM%20INFORMATIVO%20ACIDENTES%20BRASIL/Boletim%20Informativo%20Acidentes%20Brasil%20-%202016.pdf. Accessed 28 May 2018
8. Oliveira A et al (2017) Guia zero acidentes para transportadores: orientações para baixar a acidentalidade no transporte rodoviário de cargas e de passageiros. Programa Volvo de Segurança no Trânsito, Curitiba, Brasil, Agosto 2017. https://pvst.com.br/wp-content/uploads/2017/10/Guia-Zero-Acidentes-para-Transportadores.pdf. Accessed 28 May 2018
9. NHTSA – National Highway Traffic Safety Administration (2018) Human factors, Washington, DC. https://www.nhtsa.gov/research-data/human-factors. Accessed 30 May 2018
10. Branco A.M (1999) Segurança Rodoviária. Editora CL-A, São Paulo
11. DNIT – Departamento Nacional de Infraestrutura de Transportes (2006) Metodologia para tratamento de acidentes de tráfego em rodovias. DNIT, Ministry of Transport, Brasília/DF, Brazil, 57p. http://www.dnit.gov.br/download/rodovias/operacoes-rodoviarias/convenios-com-a-ufsc/convenio-242006-produto-complementar-2.pdf. Accessed 28 May 2018
12. Sharit J (2006) Part 5: Design for Healt, Safety, and Comfort. Chapter 21: Human Error. In: Salvendy, G (eds.) Handbook of human factors and ergonomics. Gavriel Salvendy, New Jersey, United States, pp 708–760
13. Iida I (2005). Ergonomia Projeto e Produção. Edgard Blucher, São Paulo, Brazil, 614p
14. LEAL, Bruno Alexandre Brandimarte (2013) Análise da relação das características das rodovias e vias urbanas com as causas de acidentes. UFRJ/Escola Politécnica, Rio de Janeiro, 109 p. http://monografias.poli.ufrj.br/monografias/monopoli10009111.pdf. Accessed 30 May 2018

Getting the Right Culture to Make Safety Systems Work in a Complex Rail Industry

Michelle Nolan-McSweeney[1,2(✉)], Brendan Ryan[1,2], and Sue Cobb[1,2]

[1] Network Rail Infrastructure Limited,
1 Eversholt Street, London NW1 2DN, UK
michelle.nolan-mcsweeney@networkrail.co.uk
[2] University of Nottingham, University Park, Nottingham NG72RD, UK

Abstract. Network Rail is a large rail engineering company sitting within a complex industrial landscape. That said, Britain's railways are the safest they have ever been. However, it is acknowledged that there is still room for improvement in safety and operational performance, and technological transformation is needed to meet the huge increase in passenger demand.

As the challenges have grown more complex Network Rail has sought to develop a high-performance culture, where people work collaboratively, strive for continuous improvement, and where everyone feels able to perform at the limit of their own potential. To achieve such ambitions two major change programmes have been developed to support the transformation of safety and operational performance in the maintenance part of the organisation.

Using in-depth interviews with 10 senior managers in the rail industry (involved with, or impacted by, the two major change programmes), it has been identified that there are several contemporary challenges. These relate to: the need for greater emphasis on participative initiatives that involve the frontline staff in decision making, so that changes are well designed and implemented; and the need to focus on the effective management of change. Possible solutions and best practices are included, outlining the shift in organisational culture needed for both leaders and employees alike.

It is also proposed that the study be extended to include surveys of frontline staff and supervisors to assess the change practices in Network Rail. Follow up interviews will also be undertaken with Senior Managers regarding the alignment with identified 'best practice'.

Keywords: Culture · Safety systems · Complexity · Change Decision-making

1 Introduction

Rasmussen's (1997) well known risk management framework has been used by many to explain the layers of complex systems. In the railway, as Wilson (2014) was careful to describe, at the top level is Government policy and strategies, especially in terms of the organisation of the industry (how it is divided up between infrastructure owners, rail and freight operating companies) and the consequent commercial and operating contracts, which have a profound effect on meeting human factors requirements.

S. Bagnara et al. (Eds.): IEA 2018, AISC 823, pp. 489–503, 2019.
https://doi.org/10.1007/978-3-319-96074-6_52

At another level we have the regulators, those concerned with health and safety as are found in any safety critical industry, and those concerned with commercial contracts and public value for money. Then there are the layers of the organisation: its management, supervisors, all direct and indirect staff, and the hardware and software systems they work with. The UK rail network is certainly a very large complex system of systems.

It is apparent there are diverse areas, traditions and concepts within the overall field of human factors and ergonomics (HFE) that bear on the question of designing, planning and managing human interactions within complex work systems like rail. There are also other traditions beyond HFE dealing with similar concerns within the management sciences where the focus is on organisational and employee behaviour, motivations and the like. Areas reviewed in detail include, but were not limited to, safety climate (Salas et al. 2014; and Patterson et al. 2004); resilience engineering (Hollnagel 2012), and safety culture and risks in organisations (Hopkins 2005).

In safety critical organisations rules and procedures are often considered to be a way of improving the reliability of the activities of humans and organisations (Reiman 2010). Also, some suggest, rule designers often think of procedures as tools for controlling the worker, not as tools for the worker to control their work (Dien 1998, p.181). How safety and operational performance improvement can be achieved with increasing complexity, and how to trade off issues such as safety, reliability, costs and productivity against one another, is a question previously posed (Wilson et al. 2009), but without a definitive answer.

The work by Reiman (2010) and others also shows that research on the maintenance function of an organisation has focused mainly on human errors and individual-level issues, though social and organisational factors do receive attention on occasion because of high profile incidents/accidents. Studies of normal work, practices and cultures of maintenance functions appear scarce in the literature, particularly for rail, and very few appear to have examined how the implementation of major change programmes can affect the frontline staff and their immediate supervisors, particularly in influencing behaviours and decision-making.

Network Rail is a large rail engineering company and, given the geographical and functional breadth of its operations, is a complicated organisation with many different departments – including a maintenance organisation. That said, Britain's railways are currently the safest they have ever been. The progress of the last decade has been built on a shared commitment by industry leaders, managers, workers, trade unions, government and regulators to improve risk management.

It is acknowledged, however, that there is still room for improvement, from a position where safety and performance (i.e. productivity and operational performance) still falls short of Network Rail's ambitions. This is particularly true of workforce safety and is the reason why there are two major change programmes that have been developed to support the transformation of the organisational culture, and to make safety systems more effective.

The Business-Critical Rules programme was initiated in 2012 and seeks to replace the very complex suite of Standards that Network Rail has in place with a simpler, risk-based, rules framework which is underpinned by the bow-tie methodology of risk management[1]. The long-term plan is to provide a clear line of sight from Network Rail's risks to how they control them. The aim is to improve the way risks to people, assets and the success of the business are managed, by providing a clear understanding of the controls necessary, including individual role accountabilities, responsibilities and capabilities.

The planning and delivering safe work programme is also focused on improving workforce safety, by delivering a change in the way Network Rail approaches the management and planning of work to reduce harm to its people. Roll out of the new processes and introduction of technology began in 2015, and made one person accountable for managing task, site and operational safety risk, with that person involved in the planning of their worksite. Learning from the initial phase led to programme changes, and a re-launch of the initiative in November 2016, with further revisions to the process in July 2017.

Both of Network Rail's two major change programmes have a focus on workforce safety aimed at improving how work is planned and undertaken, removing potential for error, increasing understanding of the way tasks are to be executed, and making it easier to share best practice through standardisation. However, they have generally evolved as separate initiatives and the approaches to implementation have been different, including the level of end-user involvement, and how change is managed and measured.

Network Rail is clear that it needs to develop strategies that make the changing landscape seem less complex, rather than more so for its employees and its stakeholders. However, the challenge is whether, in the present environment – where Government and organisations like Network Rail face huge pressure to reduce costs and deliver more, for less – there is a shared sense of purpose and a clear understanding of the interconnections between human actions, decisions and technological factors.

This paper describes research work in rail human factors to understand the perceptions of those personnel considered best placed to have an appreciation for the demands of implementing change within a complex work system, seeking to explore the gaps between the 'work as imagined' in the formal rules and procedures, and the 'work as done' (Hollnagel et al. 2006). The paper explains the method used to gather stakeholder views and to integrate these with best practice ideas identified in the array of literature available.

High-quality engineering and operations management has been described as "*.... key to meeting all the requirements of a successful railway – quality of service, reliable and safe performance, and maximum possible use of capacity*" (Wilson et al. 2007). Wilson and his co-authors describe the railway as a socio-technical system and that human factors are at its core, thus requiring a strong integrated ergonomics contribution. The discussion in this paper, therefore, addresses organisational needs with respect to managing transformational change in a complex rail industry, and the role of

[1] Bow Ties have been developed under the Business-Critical Rules programme to do two things: (1) give a visual summary of all plausible accident scenarios that could exist around a hazard, and (2) identify ways in which control measures fail.

human factors within this. Findings from the research are summarised on the perceptions of the challenges faced in implementing the two major change programmes to transform safety and operational performance, including the simplification of rules (known as the Business-Critical Rules framework), and implementing new/revised processes for planning and delivering safe work (PDSW).

2 Method of Investigation

2.1 Preliminary Activities

The study has involved observations of programme boards over the past 2 years – attended by the researcher in their capacity as a PhD part-time student – noting reported progress, achievement against key milestones, programme risks and issues. There has also been a desk-top review of 'lessons learnt' reports commissioned by Network Rail, that the researcher was given access to, related to the two major change programmes. This part of the research established that the focus of the two change programmes is primarily workforce safety, aimed at:

- improving how work is planned and undertaken;
- removing potential for error;
- identifying key risk exposures, and having controls to prevent these from materialising;
- increasing understanding of the way tasks are executed; and
- making it easier to share best practice through standardised tasks.

The programme boards, themselves, have evolved since the two major change programmes began, and their constitution has changed due to changes in personnel, programme sponsors, and required governance arrangements. In-depth interviews have been undertaken with individuals connected to the two change programmes to better understand the success (or otherwise) of the changes to-date.

2.2 Interviews with Rail Executives and Senior Managers

In-depth, interviews were undertaken with 10 selected individuals – all are senior managers in the rail industry - involved with, or impacted by, the two major change programmes.

The participants were approached directly via email, with the researcher able to contact individuals through an industry 'directory' and giving them full details of what they were expected to do as part of the study. The aim of the interviews was to understand the two change programmes being implemented, with the participants having knowledge and experience of these programmes and the effectiveness of changes made to-date.

The benefits of the qualitative research with a select group of 10 meant that the interviews have afforded the time to delve more deeply into the individuals, settings, and sub-cultures, to generate an understanding of how and why participants perceive, reflect, interpret, interact in the way that they do. This style of qualitative research was open-ended and followed emergent empirical and conceptual findings, often in unexpected ways. What this will also allow in practice is further interviews (in future) with other managers, and surveys with frontline staff, to explore the significant features of the socio-technical system and their effectiveness in getting the right culture to make safety systems work in a complex industry such as rail.

2.3 Method of Conducting and Analysing the Interviews

The 10 interviews were exploratory in nature, though were structured around several common topic areas to guide the discussion, i.e. the structure of the industry and interfaces, safety leadership, change management processes, decision making, rules and procedures, and change programme goals. Participants were asked to give informed consent for participation in the interviews.

Each interview took between 60 and 90 min, and these were arranged at a time that was convenient to the participant to maximise the opportunity of the allotted time. The interviews were audio recorded and notes were taken; these were later transcribed, input to an interview record form (for each interview) and sent to interviewees for checking, enabling later categorisation and comparative analysis across the interviews.

The different functions and responsibilities that were included in the interview programme gave an opportunity to develop a picture of the complexity of the rail industry, from a range of perspectives. The interviewees were also the people who are best placed to answer, i.e. having the necessary experience and authority on the subject matter(s) at hand associated with the two major change programmes.

Five themes (see Fig. 1) were used to classify the interview responses (based on Rasmussen's (1997) breakdown of the information available to decision makers and their capability of control). The themes were selected having also been previously used in a study regarding the challenges and strategies for an effective organisational structure in a complex rail socio-technical system (Nolan-McSweeney et al. 2017). The five themes have resonance in the way that Network Rail operates its business and is regulated; helping to understand work system constraints, boundaries of acceptable performance, and criteria for making/adapting to change.

Objectives: are objectives and values with respect to operational as well as safety issues properly communicated within the system?
Status information: are the individual decision makers (staff, management, regulators) properly informed about the system status in terms comparable to the objectives? Are the boundaries of acceptable performance around the target state 'visible' to them?
Capability: are these decision makers competent with respect to the functional properties of the organisation, of the technical core and the basic safety design philosophy? Do they know the parameters sensitive to control of performance in a changing environment?
Awareness: are decision makers prompted to consider risk in the dynamic flow of work? Are they - continuously during normal work - made aware of the safety implications of their every-day work business decisions?
Priorities: are decision makers committed to safety? Is management, for instance, prepared to allocate adequate resources to maintenance of defences? Does regulatory effort serve to control management priorities?

Fig. 1. Themes related to the information available to decision makers and their capability of control (Rasmussen 1997)

A series of passes at analyses were conducted to collate interview responses related to the two major change programmes, using the 5 themes initially (referred to in Fig. 1). Coding was then undertaken to align the 5 themes to Network Rail's critical success factors for managing change (Network Rail 2013), which also include dimension(s) of such change, e.g. sponsor behaviour in the context of effective change leadership. This approach led to the systematic organisation of information, enabling the extraction of recurrent issues. Relevant supporting or contradictory evidence was also identified, based on a particular theme, or a participant's opinion – see Table 1 – to aid further analyses. This approach was useful in being able to highlight the relative importance of issues arising from the interviews, using the dimensions as an effective way to review responses – especially where the use of the five themes lacked the granularity to be able to critically analyse what senior managers truly thought about the two major change programmes.

Table 1. Critical success factors in managing successful (change) programmes (Network Rail 2013)

Network Rail (2013)		Rasmussen (1997) themes
Critical success factors	Dimensions	
Effective change leadership: occurs when change leaders provide direction, guidance and support for the people implementing change and the people transitioning through the change	**Effective change leadership:** 1. Sponsor behaviour 2. Confidence in change agents 3. Informal influence	**Objectives**
Committed local sponsors: role models, supporting people through local managers	**Committed local sponsors:** 4. Role modelling 5. Local manager support	
Shared change purpose: When people involved in a change understand - why change is a necessity, why things cannot remain the way they are, where the organisations needs to get to, and how it will get there	**Shared change purpose:** 6. Organisational imperative 7. Future state vision 8. Solution visibility	**Status information**
Personal connection: personal imperative, solution viability, and being successful	**Personal connection:** 9. Personal imperative 10. Solution viability 11. Being successful	**Capability**

(*continued*)

Table 1. (*continued*)

Network Rail (2013)		Rasmussen (1997) themes
Critical success factors	Dimensions	
Engagement processes: scope of involvement, learning, rewards & incentives, and communication	**Engagement processes:** 12. Scope of involvement 13. Learning 14. Communicating the plan	**Awareness**
Sustained performance: future security, financial impact, work relationships, levels of responsibility, and learning curve	**Sustained performance:** 15. Financial impact 16. Work relationships 17. Level of responsibility 18. Learning curve	**Priorities**

3 Results

As Fig. 2 shows, those interviewed cited the 'future state vision' and 'solution visibility' as important to the successful implementation of the change programmes. Many were clear that the 'vision' for the future needs to be aligned among many of the other, different, dimensions, such as organisational learning, working relationships, and the drivers that afford people to commit to change (e.g. personal imperatives such as motivation, engagement, clarity of roles and responsibilities, a sense of purpose, trust).

A recurring part of the interviews was often a belief that people resist change, or at least are fearful of the unknown. Many considered that the way to tackle this was to address the matter head on through more regular communication and reminders of the 'future state vision' for the change programmes. However, some did not always appreciate that employees require different information and training targeted at specific challenges relevant to their organisational level, such as front-line staff implementing new technology for the first time. This resonates with some of the findings from the lessons learnt review undertaken specifically for the PDSW change programme (Network Rail 2017), where it was identified that a more-tailored approach to change is needed at a local level.

Analysis of the interview responses, from which the key dimensions were derived, and from the observations undertaken of programme boards, highlighted some barriers to success which have affected the success of the two major change programmes to-date. These include:

- Overly ambitious plans initially, but now with new timelines, and tranches of work identified;

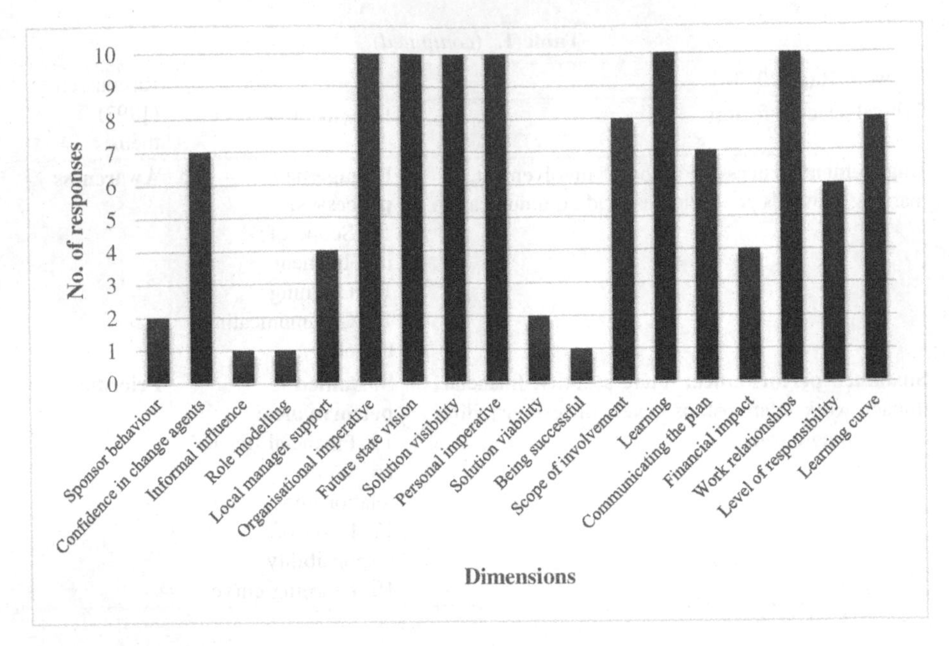

Fig. 2. Key dimensions of implementing successful change programmes, and their relative importance to the Senior Managers interviewed

- Insufficient attention to the people element regarding what people do/need to do;
- Lack of readiness and capacity to manage change within a complex system of systems;
- An absence of visible leadership across the two change programmes; and
- Lack of experience in programme governance and the 'managing successful programmes' process.

3.1 Views on the Change Programme's Timelines

When asked for their views on the two major change programmes, particularly the effectiveness of these to-date in terms of their implementation, the interviewees' responses were almost unanimous. Most felt that although there was a genuine intent to improve safety, there was also a desire to realise the safety benefits as quickly as possible, leading to a pace of change that the organisation was not prepared for. A number of the interviewees also felt that the scopes of the two programmes were vast, and required significant changes to underlying processes, organisation, and technologies, but without the necessary resources or change capability in place.

Overall, it seems that to achieve successful change many of the interviewees felt that each programme should have been developed into specific tranches of work, trialled in specific areas of the business, and then – following a 'lessons learnt' review – rolled out further into the organisation. The term 'big bang approach' was used by several senior managers to highlight their concern at the scale and scope of the programmes; this phrase has also previously been used during programme boards that were observed as part of the researcher's preliminary study activity. Some senior managers said they felt more time should have been spent at the outset in agreeing the approach to implementation and the 'organisational imperative', including roll out.

3.2 Views on the Preparedness of the People Affected by Change

Comments, related to questioning around goals, demonstrate to a large degree a real desire for improvement from a position where performance, in terms of safety, and operationally, still falls short of Network Rail's ambitions; particularly true of workforce safety. There is clearly an emphasis that Network Rail's transformation plans require a focus on taking decisions faster, thereby requiring as much effort on developing the right kind of culture and behaviours in the organisation as the technological and process factors that have previously held sway. Interviewees, when talking about transformation in the context of the two change programmes, mentioned more inclusive engagement and communications with end users, and industry stakeholders. Several said they believed a proactive approach to include trade unions is key to the change programmes' success, and they thought better communication of the benefits of the changes was now bringing dividends through 'solution visibility'.

3.3 Views on Readiness and Capacity for Change, and Leadership of the Change

The comments captured during the interviews seem to suggest that whilst the 'case for change' has been spelt out for the two programmes, many interviewees believe the challenges ahead will depend on the abilities of the individuals and the organisation to change effectively in a dynamic, complex, industry – paying attention to both the external and internal factors that impact individual, team, and organisational performance.

It is apparent from the interviews that the way key messages are delivered, and the company's goals articulated, may differ depending on what is being said and to who. Most interviewees stated that leadership starts from the very top, and some thought the change goals and expectations are explicit, delivered through positive/proactive messaging.

Those more familiar with the frontline staff suggested that what is conveyed in terms of change, versus what is received and understood, may be very different (e.g. a distrust from staff of the benefits of change despite management outlining the reasons for the transformation and what this might mean in terms of safety, simplicity, and more effective processes). This could sometimes result in a disconnect in the importance or value placed on the change.

Other interviewees added that more should have been done to consider the people who are at the centre of the programme changes, building in some criteria to determine their readiness for change, but also recognising that their actual 'as is' operating model – the way that they carry out tasks and are motivated to work – may be very different in practice. As a consequence, individuals may feel marginalised, or have no real engagement or desire for involvement in the change.

Pertinent to 'readiness for change' are two quotes from two different senior managers.

"Looking at the change programmes in isolation makes them appear achievable; put them all together and it exposes the enormity of the task and that the same resources (usually the frontline) are impacted over and over again". Senior Manager No. 1
"It is inevitable in an organisation the size of Network Rail that we will have a number of change programmes running in parallel. Change cannot be managed as simply as deciding to deliver them in 'series', but we need to help 'condition' people for this". Senior Manager No. 9

Some interviewees, following questioning around the role of rules and procedures, discussed the culture of the organisation and the rationale for implementing the change programmes such that *"risk-awareness among employees would be further encouraged"*. This has resonance with the work of Hopkins (2005), and the view that it is impossible to devise a set of safety rules which adequately covers every situation, but that rules are still essential and so must be *managed* in a dynamic industry.

Both the Business-Critical Rules programme, and the Planning and Delivering Safe Work (PDSW) programme have safety improvement at the heart of their successful implementation, and within this a 'rule regime' that is never fixed and is able to evolve as circumstances dictate (e.g. a manual intervention in the case of disturbances or incidents). The ambition is that there are clearly defined parameters (and rules) around risk-based decision making, and the way work is planned and delivered, allowing for flexible routines (Grote 2015), but also where people are drilled to apply the appropriate rules and procedures until they are second nature.

What the interviews have, however, highlighted is that whilst some managers believe the changes have empowered the workforce to make risk-based decisions at the lower level, and that the change programmes have provided tools for this, there are contrary views too. These suggest crucial decisions are often delayed because of the hierarchical nature of the organisation – waiting for a more senior level input – and that the decision-making and planning processes are not quite as empowering as portrayed.

As one manager said: *"We say one thing but mean another. We set the tone of the organisation, so if we question a decision it, surely, cannot be a surprise to us that those we say we have empowered then distrust us"*. Senior Manager No. 3

3.4 Views on the Organisational Culture and Programme Governance

A previous study, undertaken by the authors of this paper, has considered the challenges and strategies for an effective organisational structure in a complex socio-technical system in GB railways, and this highlighted among other things that leaders and employees needed to be better equipped to manage and/or make change.

Those interviewed in this more recent study have acknowledged that transforming the business will not always be smooth and that regular engagement with those affected by the two major change programmes will hopefully improve ownership and develop competence, so those best placed to implement the changes do so confident in the knowledge that they will be supported.

Interview participants recognised the part they must play in supporting and cultivating a culture of cooperation, communication, openness, and/or a greater tolerance of small mistakes through greater employee engagement and less seeking to apportion blame when things do not quite go according to plan. However, some said that the ultimate source for driving the organisational (and safety) culture may lie outside the organisation. Public pressure on Government may focus the minds of rail executives on 'on time' running, rail fares, and passenger safety; similar pressure might be necessary from the Regulator to generate equivalent attention to workforce safety and risk management. The inference here is that adequate planning, training, resources and system testing are necessary to engage the workforce in the change programmes – this will take time and money *("not the threat of enforcement action")* that the Regulator could fund to help gain the confidence of end users, and for the change programme benefits to be realised.

Some Senior Managers, during their interviews, questioned whether the organisation should restrict the span of control and decision latitude of the frontline workforce to help avoid confusion, or whether they want them to be far more autonomous (recognising that the system constantly changes and modifies its state and the interactions within it in light of circumstances, events and decisions made). Most of the interviewees, however, believed that autonomy is an inevitable outcome of the change programmes but accept it may take time to embed as a culture, for leaders and employees alike, and so the disconnect between the future state vision, organisational goals/imperatives, individual roles and responsibilities, and learning, could continue for a period.

Almost two-thirds of Senior Managers also expressed concerns regarding silo-focused employees, despite an organisation re-design to move to a matrix structure in 2014. That said, participants report positively that there is emerging trust between employees in different parts of the organisation as lessons learnt from the earlier phases of the change programmes have been reviewed, reported and acted upon. 'Communicating the plan' was less of a priority now that more focused briefings have been undertaken – particularly for the PDSW programme – and 'sponsor behaviour' has been addressed through changes in programme personnel and governance bodies being established.

Finally, many interviewees believe Network Rail have *"turned the corner"* on the change programmes in more recent months, and that the collaborative approach in developing improved processes is building credibility and increasing the chance of successful delivery. They acknowledge that there have been 'bumps' along the way, but not to the detriment of safety, and in fact several were keen to emphasise that when it was recognised that the two major change programmes were not delivering the intended outcomes that the organisation was prepared to pause them, critically evaluate the lessons learnt, and re-design processes as necessary.

4 Discussion and Conclusion

From the interviews, related to the two major change programmes, it is evident there is:

- the need for greater emphasis on participative initiatives that involve the frontline staff in decision making, so that changes are well designed and implemented (aligned to Rasmussen's (1997) view of 'capability' and the parameters sensitive to control of performance in a changing environment).

- the need to focus on the effective management of change if programmes are to be successful, such as communicating plans and reasons for change, and developing new processes that have clear accountabilities. This is analogous to Rasmussen's earlier work (1997) and the importance of information being available to decision makers, i.e. 'status information' and whether the boundaries of acceptable performance around the programme's 'future state' are visible.

The results also show that Network Rail has sought to implement two major change programmes whilst considering a host of complex systems and processes – how people and the things they do affect others, the tools and equipment needed to undertake work, job design, and decision-making – but not necessarily in a way that afforded those likely to be most affected with the capacity or capability to change at pace.

Gudela Grote concluded that when uncertainties are managed well, a basic prerequisite for good risk management is established; and she argues for "….the importance of making deliberate operational and strategic choices between reducing, maintaining, and increasing uncertainty in order to establish a balance between stability and flexibility in high-risk systems while also matching control and accountability for the actors involved" (Grote 2015). In practical terms, this means the rail socio-technical system having to continuously adjust the balance between stability and flexibility to secure successful performance. Part of the interview study revealed that Network Rail did not undertake a systematic evaluation of reducing, maintaining, or increasing uncertainty during the various phases of the change programmes, even though the railway tends to operate in a dynamic, reactive state.

What has emerged from the study is that no change programme is exactly the same; comparing just two major change programmes has highlighted similarities but also some differences in approach, usually because of the various interactions involved in integrating human, technical, information, social, political, economic and organisational components. Of the two programmes, PDSW relied very much on planning tools being introduced involving third parties to develop new technologies and for external providers to deliver training, whilst the BCR programme was more process driven initially and was dependent on internal 'know how' to identify where Standards could be rationalised.

Reliable performance occurs in complex organisations like Network Rail *despite* often conflicting constraints imposed by divergent goals or perceived mixed messages with respect to safety, profitability and performance. As Rasmussen (2000) noted, the importance given to activity in situations and to user intelligence has constantly grown. Within imperfect systems, humans often are the ones to mitigate risk. The results show that there was insufficient attention to the human factors and what people do/need to do in their roles which, in part, explains the difficulties in delivering successful change across the two programmes.

Taking a systems-oriented view in the interviews has highlighted that the rail industry is likely to remain in a state of dynamic change and, as Network Rail seeks to transform its business, safety and performance can be viewed as emergent products of the complex rail socio-technical system. These emergent properties of rail engineering and operations means that things are going to happen on occasion in unexpected ways, that work arounds will be found, and/or new and better ways to do things will be identified. Learning lessons even as change programmes are implemented will be critical to success; the results clearly highlight the need for more experience in programme governance and the 'managing successful programmes' process going forward.

As Wilson *et al.* (2009) have previously described, with emergent properties there are also human components that are affected, often requiring people's jobs to change. Practically, for Network Rail, with increasing technological demands and for the major change programmes to become embedded, comes an emergence of new roles, communication channels, relationships, power structures, sources of decision making and collaborations to consider.

The attribute view of complexity (Walker et al. 2010) – where complexity has a number of fairly distinct features – defines the ergonomics problem space as containing, among other things, multiplicity (multiple interacting factors), dynamism (system state changes over time) and uncertainty (difficulty and vagueness in determining the final system state). Therefore, to meet the demands of an increasingly complex industry, Network Rail knows it must change and at pace. The two major change programmes were intended to deliver safety benefits quickly, but criticism of their early phases around *"being too ambitious, too quickly"* has resulted in a review of their scopes and led to redesign and re-baselining of key deliverables.

There are clear parallels between the two major change programmes and their intended outcomes that the interviews and programme observations identified (e.g. simplifying standards, improving workforce safety, maturing the safety culture of the organisation), but there are other programmes too within Network Rail, and so the challenge remains of:

- whether there may be conflicts and contradictions in the intended outcomes of change programmes;
- a lack of a clear strategy and detail regarding implementation of change programmes;
- limited understanding of the impact on Network Rail's maintenance teams, and what support might be required and available pre- and post- implementation to those affected by the changes;
- continuity of senior sponsorship (where senior leaders move on/out of the business);
- building operational competence (e.g. what are the gaps in current competence needed to achieve compliance with the Business-Critical Rules? How training needs to be re-framed around a role rather than for a specific/task-based competence, and the development of behavioural competences for PDSW necessary to enable more local risk-based decision making).

The list below reflects what is considered to be 'best practice' identified through the Nicols Group (2013a; 2013b) comparator studies and structured interviews of organisations[2] that have successfully implemented programme/system change(s) related to their management of risk (and who generally operate as large, safety critical organisations, with asset management businesses).

[2] The comparator organisations included: Ministry of Defence, Royal Dutch Shell, National Grid, Exxon, NASA, EDF, and BHP Billiton.

- Focusing on the effective management of change, including appropriate training (linked to risk and accountability);
- Using risk to prioritise and shape programme change decisions;
- Educating the whole company in the importance of the systems, and reasons for change;
- Articulating what 'good' looks like, with performance measures to monitor progress; and
- Have teams that 'own' the respective programmes and who are able to provide guidance post-implementation when the system(s) are 'Business As Usual'.

Given the significance of the changes being brought about, and the pace of change apparent in the timescales for implementing the two major change programmes, it is important that further research is undertaken to provide them with a timely perspective on the state of implementation, as part of building assurance that the actions Network Rail is taking are appropriate and likely to deliver the intended benefits.

It is, therefore, proposed that this study is extended to include surveys of frontline staff and supervisors to assess the change practices in Network Rail. This will mean questioning around the impact of the Business-Critical Rules programme, and planning and delivering safe work programme, on safety and operational performance.

Follow up interviews will also be undertaken with Senior Managers regarding the alignment with the 'best practice' outlined above. The aim is that the early learning from the two major change programmes can be considered when setting up and implementing further changes so that the intended culture is achieved, and safety systems are effective.

Outputs from this programme of work will also form part of wider PhD research leading to a description of the rail engineering socio-technical system that aids understanding of the interconnections between human actions, decisions and technological factors, assisting Network Rail to develop strategies that make the changing landscape seem less complex, rather than more so for its employees.

References

Adler PA, Adler P (1987) Membership roles in field research. Sage, Newbury Park

Dien Y (1998) Safety and application of procedures, or 'how do 'they' have to use operating procedures in nuclear power plants? Saf Sci 29:179–187

Grote G (2015) Promoting safety by increasing uncertainty – Implications for risk management. Saf Sci 71:71–79

Hollnagel E, Woods DD, Leveson N (2006) Resilience engineering. Ashgate Publishing Ltd., Farnham

Hollnagel E (2012) FRAM: the functional resonance analysis method: modelling complex socio-technical systems. Ashgate Publishing Ltd., Farnham

Hopkins A (2005) Safety, culture and risk. The organisational causes of disasters. CCH Australia Ltd., Sydney

Network Rail (2013) Managing Successful Programmes for Network Rail (MSP4NR)

Network Rail (2016) Business-Critical Rules Programme Brief

Network Rail (2017) PDSW Tranche 1 Stakeholder Lessons

Nicols (2013(a)) Standards Efficiency Study - Independent Reporter (Part C) Mandate CN/024. Office of Rail Regulation and Network Rail. Final Report 7 June 2013
Nicols (2013(b)) Standards Efficiency Study - Independent Reporter (Part C) Mandate CN/024. Office of Rail Regulation and Network Rail. Report Summary 5 July 2013
Nolan-McSweeney M, Ryan B, Cobb S (2017) The challenges and strategies for an effective organisational structure in a complex rail socio-technical system. In: Human factors rail conference, 6–9 November 2017, London, UK
Patterson M, Warr P, West MA (2004) Organisational climate and company productivity: the role of employee affect and employee level. J Occup Organ Psychol 77(2):193–216
Rasmussen J (1997) Risk management in a dynamic society: a modelling problem. Saf Sci 27:183–213
Rasmussen J, Svendung I (2000) Proactive risk management in a dynamic society. Risk & Environmental Department, Swedish Rescue Services Agency, Karlstad
Reiman T (2010) Understanding maintenance work in safety-critical organisations – managing the performance variability. Theor Issues Ergon Sci 12(4):339–366
Salas E, Shuffler ML, Thayer AL, Bedwell WL, Lazzara EH (2014) Understanding and improving teamwork in organisations: a scientifically based practical guide. Hum Resour Manag 54(4):599–622
Walker GH, Stanton NA, Salmon PM, Jenkins DP, Raffety L (2010) Translating concepts of complexity to the field of ergonomics. Ergonomics 53(10):1175–1186
Wilson JR, Norris BJ (2005) Rail human factors: Past, present and future. Applied Ergonomics 36(6):649–660
Wilson JR, Farrington-Darby T, Cox G, Bye R, Hockey GRJ (2007) The railway as a socio-technical system: human factors at the heart of successful rail engineering. Proc Inst Mech Eng Part F J Rail Rapid Transit 221(1):101–115
Wilson JR, Ryan B, Schock A, Ferreira P, Smith S, Pitsopoulos J (2009) Understanding safety and production risks in rail engineering planning and protection. Ergonomics 52(7):774–790
Wilson JR (2014) Fundamentals of systems ergonomics/human factors. Appl Ergon 45(1):5–13

The Relationship of Space Experience and Human Anthropometric Sizes in Aircraft Seat Pitch

Shabila Anjani[1](✉), Wenhua Li[1,2], Peter Vink[1], and Iemkje Ruiter[1]

[1] Faculty of Industrial Design Engineering, Delft University of Technology, Landbergstraat 15, 2628 CE Delft, The Netherlands
S.Anjani@tudelft.nl

[2] Shaanxi Engineering Laboratory for Industrial Design, Northwestern Polytechnical University, No. 127, Youyi Road (West), Beilin District, Xi'an 710072, People's Republic of China

Abstract. This study explores the relationship between space experience and human anthropometric sizes in different aircraft seat pitch. 294 participants experienced economy class seats in a Boeing 737 with 28 in, 30 in, 32 in and 34 in pitches for 10 min each. The sizes taken were: stature, sitting height, eye height seated, buttock-knee length and popliteal height sitting with shoes. A space experience questionnaire was completed by the 294 participants while sitting in the seat after the 10-min period given to explore the seat. The results show that passengers with a higher popliteal height, a longer buttock-knee depth, a higher eye height sitting and a higher sitting height show more discomfort with reduced pitch then shorter passengers. Eye height did not correlate as good with space perception as was expected.

Keywords: Seat pitch · Comfort · Discomfort · Space experience Anthropometric measurements

1 Introduction

People travel in many different ways. The choice of transportation modes differs among individuals, depending on their wishes and needs. Comfort is one of the important considerations in choosing a certain transportation mode [1] and it also has a strong correlation with repetitive choices [2].

Vink, Hallbeck [3] defined comfort as a pleasant state or relaxed feeling of a human being in reaction to its environment, while discomfort is an unpleasant state of the human body in reaction to its physical environment. This comfort perception is found in a slightly different form in Zhang et al. [4]'s model, where comfort is driven by well-being and plushness, while discomfort is due to poor biomechanics and tiredness (Fig. 1). This study will use this definition. Vink, Hallbeck [3] further developed the model which is able to predict discomfort (Fig. 2). The model presents how comfort is perceived as well as the factors which may contribute to comfort perception.

S. Bagnara et al. (Eds.): IEA 2018, AISC 823, pp. 504–511, 2019.
https://doi.org/10.1007/978-3-319-96074-6_53

This experiment focuses on the link of the environment to (I) interaction with environment (see Fig. 2) which at the end of the model will result in either comfort, nothing, or discomfort [4]. Furthermore, this comfort and discomfort feeling could also be driven by psychological [5], physiological [6] or emotional influences [7]. This study only focuses on the psychological and physiological relationship with (dis)comfort.

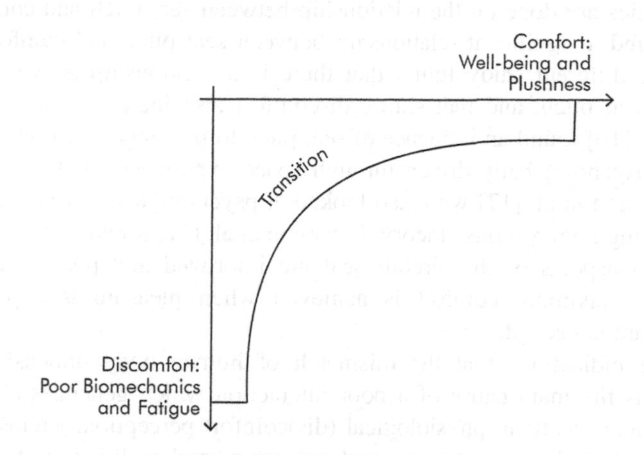

Fig. 1. Comfort model by Zhang et al. [4]

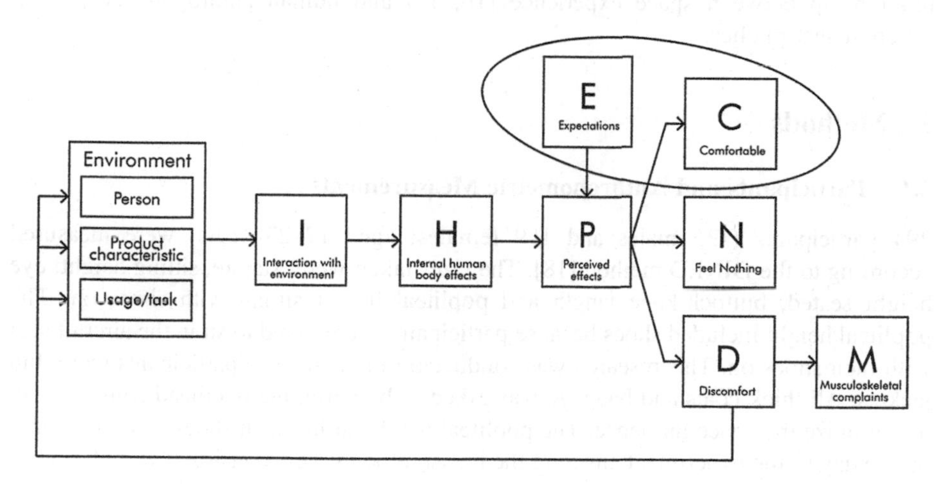

Fig. 2. Comfort Prediction Model by Vink, Hallbeck [3]

Legroom is found to be a major element contributing to comfort when sitting in an aircraft. Vink [8] found that providing legroom enables passengers to stretch legs which results in a changing body posture as a way to prevent discomfort. Curtis et al. [9] also found that legroom is an important factor for frequent flyers' level of satisfaction. Vink et al. [10] showed that legroom has a high correlation with comfort.

In contrary, Blok et al. [11] found that the knee space which is also related to legroom was the lowest rated item. In the last 30 years this legroom decreased as the distance of the rows have decreased 2 to 5 in [12].

Legroom is influenced by the distance between rows of the seats which is known as seat pitch, measured from a point in a seat to the exact same point of the seat in-front or behind. Today's seat pitch sizes, varies from 28 in to 38 in for economy class flights [13].

Some studies are done on the relationship between seat pitch and comfort. Anjani et al. [14] found a significant relationship between seat pitch and comfort as well as discomfort. A different study found that there is a relationship between sitting discomfort and seat pitch, and that sitting discomfort also increases through time [15]. Kremser et al. [16] found an influence of seat pitch to passenger well-being which was found to be psychologically driven through space experience. A different study was done by Menegon et al. [17] who also looked at psychological aspects of aircraft seat comfort by using item response theory. Menegon et al. [17] found that comfort tends to increase when aspects of the aircraft seat are improved and positive emotions are elicited. This maximum comfort is achieved when pleasure is experienced and expectations are exceeded.

There are indications that the mismatch of human body dimensions with the environment is the main cause of a poor interaction which generates changes in the human body and results in physiological (dis)comfort perception. Kremser et al. [16] found relationships between human anthropometry and well-being at different seat pitches, indicating the existence of this relationship as well. This study explores the relationship between space experience [16, 17] and human anthropometry [18] at different seat pitches.

2 Methods

2.1 Participants and Anthropometric Measurements

294 participants (135 males and 159 females; aged 17–23 years) were measured according to the DINED method [18]. The sizes taken were: stature, sitting height, eye height seated, buttock-knee length and popliteal height sitting with shoes on. The popliteal height included shoes because participants were asked to sit in the aircraft seat with their shoes on. This research was conducted in fall so some participants wore thin jackets. All thick coats and baggage was asked to be put in the overhead compartment to minimize the space influence. The popliteal height sitting with shoes were measured on the day of the experiment ensuring the measurements were consistent with the shoes worn that day. The other measurements were done in sessions after the experiment.

2.2 Experiment Setup

To study the relationship between experience, seat pitch and anthropometrics eight rows of economy class seats (see Fig. 3; Table 1) in a Boeing 737 fuselage were used with 28 in, 30 in, 32 in and 34 in seat pitch. The pitch sizes used in this experiment were based on the sizes currently often seen for economy class flights (28 in to 38 in) [13].

The even numbers were selected enabling to match results to other references [14–16, 19]. Half of the setting was arranged to have the participants experience the pitch size from small to large, the other half from large to small. This was done to eliminate the order effect. The changes were small, but participants could refer it to the previous experience and in theory this would be recognisable since human sensors record differences better than absolute values [20]. Participants were not allowed to recline their seats since this might influence the situation of the participants behind them.

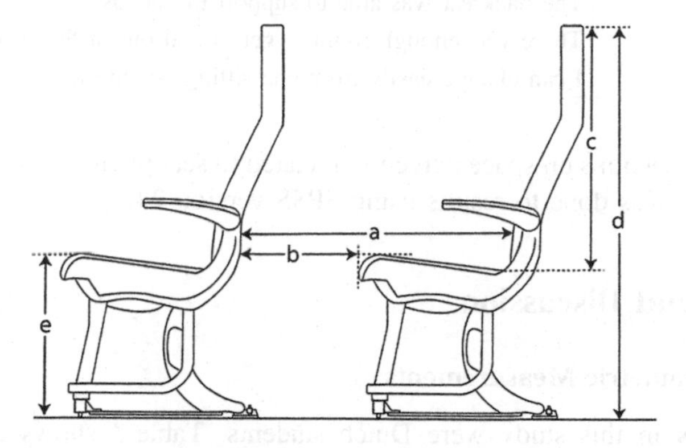

Fig. 3. Seats used in this study

Table 1. Seat dimensions

Dimensions	28"	30"	32"	34"
a	64 cm	69 cm	74 cm	79 cm
b	20 cm	25 cm	30 cm	35 cm
c	71 cm			
d	111 cm			
e	44 cm			

Participants were asked to sit in each seat for 10 min each without any instructions what to do while sitting. All seats were occupied so all participants had a neighbour. As it were students from the same faculty and year most were acquainted with each other. After 10 min of sitting participants were asked to complete an online questionnaire while remain seated. After completing the questionnaire participants were allowed to stand-up and move to a different pitch size.

2.3 Space Experience Evaluation

Eight questions (see Table 2) were used to evaluate the space experience using a 9-scale Likert with half using positive descriptors leading to comfort and the other half using negative descriptors leading to discomfort [21]. Q1 until Q4 are made with negative descriptors and Q5 until Q8 with positive descriptors. These questions were

Table 2. Space experience questions

Question number	Question
Q1	I feel restricted by the distance of the seating rows
Q2	I feel like sitting in front of a wall
Q3	I feel lost because the distance of the seating rows
Q4	I feel stressed out because of the distance of the seating rows
Q5	I was able to stretch my legs without difficulty
Q6	The backrest was able to support my needs
Q7	There was enough room to get in and out of the seat
Q8	I can change easily from one sitting posture to another

psychological questions on space perception related to seat pitch [16, 17]. A Spearman-rank correlation was done to results using SPSS version 24.

3 Results and Discussion

3.1 Anthropometric Measurements

All participants in this study were Dutch students. Table 3 shows that most measurements were close to Dutch reference data. The anthropometric measurements that were not correctly measured were excluded from the results, for example when the eye height seated was longer than the sitting height.

Table 3. Average anthropometric measurements and reference data (mm).

Mean	Observations in this study	Male database [18]	Female database [18]
Stature	1762	1821	1698
Sitting height	906	949	898
Eye height seated	801	840	787
Buttock to knee	596	634	600
Popliteal height with shoes	466	–	–

3.2 Space Experience Evaluation in Different Pitch Sizes

The results of the space experience questionnaires are shown in Table 4. It was found that the positive descriptors of space experience (Q5–Q8) increase with the increased pitch size, with the exception of Q6, while the negative (Q1–Q4) did decrease with the increase of pitch size. Question 6 which was "The backrest was able to support my needs" was found to be higher in the 30" seat pitch than 32". This result could be influenced because the participants were asked to not recline the seat during the test. All results of these questions were found significant (Spearman-rank correlation, $p < 0.01$).

Table 4. Results of space experience questions.

Question	28"		30"		32"		34"		r_s
	Mean	SD	Mean	SD	Mean	SD	Mean	SD	
Q1	6.92	2.00	4.57	1.86	3.18	1.53	1.97	1.51	−.718**
Q2	6.50	2.08	4.49	1.79	3.51	1.66	2.23	1.41	−.652**
Q3	3.48	2.32	2.74	1.53	2.37	1.44	2.15	1.46	−.233**
Q4	5.52	2.15	3.68	1.81	2.80	1.54	1.91	1.23	−.594**
Q5	3.74	2.56	5.37	2.25	6.52	2.33	7.87	2.06	.568**
Q6	5.53	1.90	5.79	1.58	5.55	1.87	6.43	1.71	.162**
Q7	3.60	1.75	5.25	1.71	5.90	1.78	7.48	1.62	.632**
Q8	3.69	2.06	5.37	1.77	6.16	1.80	7.42	1.54	.597**

**. Correlation is significant at the 0.01 level (2-tailed).

3.3 Relationship of Anthropometric Measurement and Space Experience

The relationship between the anthropometric measurements and space experience is shown in Table 5. It was found that the popliteal height with shoes and buttock to knee length were found significant to all space experience questions. Q1, Q4, Q5, and Q8 were also found correlating strongly with all anthropometric measurements. So, the anthropometric measurements indicating physiological comfort highly correlated with space experience, which could indicate that these measurements are a good predictor of space experience leading to psychological comfort.

Table 5. Relationship of anthropometric measurements and space experience

	Sitting height	Eye height seated	Buttock to knee	Popliteal height with shoes
Q1	−.192**	−.196**	−.205**	−.263**
Q2	−.051	−.126*	−.184**	−.167**
Q3	−.055	−.122*	−.153**	−.139**
Q4	−.159**	−.185**	−.177**	−.168**
Q5	−.192**	−.196**	−.205**	−.263**
Q6	−.051	−.126*	−.184**	−.167**
Q7	−.055	−.122*	−.153**	−.139**
Q8	−.159**	−.185**	−.177**	−.168**

**. Correlation is significant at the 0.01 level (2-tailed).
*. Correlation is significant at the 0.05 level (2-tailed).

Kremser et al. [16] also found and Moerland [19] also presented that the buttock to knee measurement strongly correlates with seat pitch and (dis)comfort in this case using space experience questions. In our study the back rests were equal. Seat pitch is not always directly related to legroom as a thicker back rest creates less legroom (Vink and Brauer, 2011). New economy class usually have a thinner backrest for this reason. All questions were also found significant for eye height seated, though some were not strong, which is also in-line with other findings [16]. Sitting height which was assumed to have influence psychologically in space experience only had correlations in Q1, Q4, Q5 and Q8 while all other measurements were also correlated. This might indicate that sitting height is not a predictor for space experience. The two highest correlations were found between Q5 'I was able to stretch my legs without difficulty' and popliteal height with shoes.' (This makes sense as the longer the lower leg the more difficult it is to stretch the legs) and Q1 'I feel restricted by the distance of the seating rows' and popliteal height with shoes. As the back of the seat in front of you comes closer to you at a higher level, it also makes sense that occupants with longer lower legs feel more restricted. Some values were not significant: sitting height did not have a strong relationship with Q2, Q3, Q6 and Q7, which also makes sense as buttock-knee distance has more influence on space experience than the height of the head as there are no physical restrictions above the head.

Limitation of this study is that the population is young and only from one area of the globe and the duration of the seating was 10 min, while in a real flight the duration is longer. It is expected that the effects will be larger for longer flights as other studies show that discomfort increases over time (e.g. Smulders et al. 2016). The anthropometric data are not completely independent as larger persons could have larger lower and upper legs. This means that we are not sure whether the effect can be only contributed to buttock to knee length or popliteal height.

4 Conclusion

A relationship between space experience and human anthropometric sizes in aircraft has been established. Passengers with a higher popliteal height and a longer buttock-knee depth show more negative results in space experience with reduced pitch compared to shorter passengers. Therefore, the taller the passenger, the larger the problems could be expected with low seat pitches, physiological as well as psychological.

Acknowledgement. Lembaga Pengelola Dana Pendidikan Republik Indonesia as the scholarship provider for the PhD of Shabila Anjani.

References

1. Nijholt N, Tuinhof T, Bouwens JM, Schultheis U, Vink P (2016) An estimation of the human head, neck and back contour in an aircraft seat. Work (Read Mass) 54(4):913–923. https://doi.org/10.3233/wor-162355
2. De Lille C, Bouwens JM, Santema S, Schultheis U, Vink P (2013) Designing the cabin interior knowing high and low peaks in a passenger flight. Paper presented at the AEGATS 2016, Paris, France
3. Vink P, Hallbeck S (2012) Editorial: comfort and discomfort studies demonstrate the need for a new model. Appl Ergon 43(2):271–276. https://doi.org/10.1016/j.apergo.2011.06.001
4. Zhang L, Helander MG, Drury CG (1996) Identifying factors of comfort and discomfort in sitting. Hum Factors 38(3):377–389. https://doi.org/10.1518/001872096778701962

5. Ahmadpour N, Kühne M, Robert J-M, Vink P (2016) Attitudes towards personal and shared space during the flight. Work (Read Mass 54:981–987. https://doi.org/10.3233/wor-162346
6. De Looze MP, Kuijt-Evers LFM, Van Dieén J (2003) Sitting comfort and discomfort and the relationships with objective measures. Ergonomics 46(10):985–997. https://doi.org/10.1080/0014013031000121977
7. Bazley C (2015) Beyond comfort in built environments. TU Delft
8. Vink P (2016) Vehicle seat comfort and design. Delft
9. Curtis T, Rhoades DL, Waguespack BP: Satisfaction with airline service quality: familiarity breeds contempt. Int J Aviat Manag 1(4) (2012). https://doi.org/10.1504/ijam.2012.050472
10. Vink P, Bazley C, Kamp I, Blok M (2012) Possibilities to improve the aircraft interior comfort experience. Appl Ergon 43(2):354–359. https://doi.org/10.1016/j.apergo.2011.06.011
11. Blok M, Vink P, Kamp I (2007) Comfortabel vliegen: comfort van het vliegtuiginterieur door de ogen van de gebruiker. Tijdschrift voor Ergonomie 32:4–11
12. McGee B (2014) Think airline seats have gotten smaller? They have. In: USA today. McLean
13. TripAdvisor SG (2017) Airline seat comparison charts. https://www.seatguru.com/charts/generalcharts.php. Accessed 12 Oct 2017
14. Anjani S, Li W, Vink P, Ruiter I (2018, in press) The effect of aircraft seat pitch on comfort. Appl Ergon
15. Li W, Yu S, Yang H, Pei H, Zhao C (2017) Effects of long-duration sitting with limited space on discomfort, body flexibility, and surface pressure. Int J Ind Ergon 58:12–24
16. Kremser F, Guenzkofer F, Sedlmeier C, Sabbah O, Bengler K (2012) Aircraft seating comfort: the influence of seat pitch on passengers' well-being. Work (Read Mass) 41(Suppl 1):4936–4942. https://doi.org/10.3233/wor-2012-0789-4936
17. Menegon LdS, Vincenzi SL, de Andrade DF, Barbetta PA, Merino EAD, Vink P (2017) Design and validation of an aircraft seat comfort scale using item response theory. Appl Ergon 62:216–226. https://doi.org/10.1016/j.apergo.2017.03.005
18. Molenbroek JFM, Albin TJ, Vink P (2017) Thirty years of anthropometric changes relevant to the width and depth of transportation seating spaces, present and future. Appl Ergon 65 (Suppl C):130–138. https://doi.org/10.1016/j.apergo.2017.06.003
19. Moerland RG (2015) Aircraft passenger comfort enhancement by utilization of a wide-body lower deck compartment. TU Delft
20. Vink P, Anjani S, Smulders, M, Hiemstra-van Mastrigt S (2017) Comfort and discomfort effects over time: the sweetness of discomfort and the pleasure towards of the end. In: 1st International Comfort Congress, Salerno
21. Helander MG, Zhang L (1997) Field studies of comfort and discomfort in sitting. Ergonomics 40(9):895–915. https://doi.org/10.1080/001401397187739

Is It the Duration of the Ride or the Non-driving Related Task? What Affects Take-Over Performance in Conditional Automated Driving?

Oliver Jarosch[1(✉)] and Klaus Bengler[2]

[1] BMW Group Research, New Technologies, Innovations, Parkring 19, 85748 Garching, Germany
oliver.oj.jarosch@bmw.de

[2] Chair of Ergonomics, Technical University Munich, Boltzmannstr. 15, 85748 Garching, Germany

Abstract. Conditional Automated Driving (SAE Level 3) is expected to be introduced to the consumer market within the next few years. In this level of automation, the dynamic driving task is executed by the system and the human driver can engage in non-driving related tasks and just has to intervene if requested by the system. As it is assumed, that the human driver may have problems regaining control of the vehicle, studies on take-over performance are in the focus of human factors research right now. It is examined whether the human driver is capable of regaining vehicle control and what factors influence take-over performance of the human driver. In this paper, take-over situations of two studies that just differed in the duration of the automated driving are compared. The studies were both conducted at BMW laboratories in a motion based driving simulator. Take-over performance of the participants was rated using the video-based TOC expert rating tool by three trained raters. Results suggest, that take-over performance strongly differs among individuals and that such take-over situations can cause problems for a majority of the participants. Especially in prolonged conditional automated driving the human driver needs to be supported when it comes to a take-over situation. The influence of the duration of the ride seems to be stronger than that of the non-driving related task.

Keywords: Conditional automated driving · Non-driving related task
Take-over performance

1 Introduction

The different levels of automated driving are regulated in the SAE information report "Taxonomy and Definitions for terms related to On-road Motor Vehicle Automated Driving Systems" (SAE 2016). Referring to this taxonomy, in SAE Level 2, which is state of the art right now, the human driver is still responsible for the driving task and has to react immediately in case of malfunctions. In this level, the vehicle executes longitudinal and lateral control.

S. Bagnara et al. (Eds.): IEA 2018, AISC 823, pp. 512–523, 2019.
https://doi.org/10.1007/978-3-319-96074-6_54

The next Level, which will be available for consumers in just a few years is SAE Level 3. In this level the responsibility for the driving task switches for the first time: When the system executes the dynamic driving task, the human driver is allowed to completely turn away from the actual driving task and to engage in non-driving related tasks (NDRTs). However, in SAE Level 3, the human driver still has to be able to drive and regain control of the vehicle, if requested by the system as the human driver represents the fallback performance. Assuming the system detects a situation it cannot handle, a request to intervene (RtI) is issued and the driver has to regain control of the vehicle within a short period of time.

Referring to Gold (2016), there are two types of such take-over situations: For type 1, the imminent takeover situation is known to the system long before the actual RtI. Thus, the driver can be prepared long before the required intervention. For type 2, things look different: the take-over situation gets detected by onboard sensors and the human driver has to react immediately to regain control of the vehicle and to prevent an uncontrollable situation. It is assumed that short-term take-over situations are much more difficult to handle for the human driver and therefore in the focus of research. Eriksson and Stanton (2017) provide an overview of previous empirical studies on take-over performance in CAD.

In CAD the role of the driver changes from active operating to system supervising. From human factors research of the past it is well known, that the human is not a good system supervisor and that automation effects may occur. With increasing time of automated driving, these effects may even get bigger and take-over performance can be impaired. However, in previous studies advantageous effects of NDRTs while driving automated could be shown. So far, it is not known how long these positive effects will persist and how different NDRTs affect the take-over performance.

Therefore two experiments were conducted in BMW laboratories to examine (i) the influence of an activating vs. an fatiguing task and (ii) the influence of the duration of automated driving on take-over performance. Take-over performance was rated with the subjective take-over controllability rating.

2 Theoretical Issues

2.1 Fatigue and the Influence of NDRTs

Fatigue in manual driving is one of the most relevant crash causations and thus it can be assumed, that fatigue may be a relevant causation for impaired take-over performance in CAD. In the report Drowsy Driving 2015 (National Center for Statistics (2015) 3662 fatal crashes occurred due to drowsy driving between 2011–2015.

Fatigue in manual driving can occur due to different reasons. In the model provided by May and Baldwin (2009) the different causations for fatigue are summarized. According to this model, there are two main causations for fatigue. On the one hand, fatigue may arise due to the driving task itself (*task related fatigue*), on the other hand fatigue may be caused through sleep-related reasons (*sleep related fatigue*).

Task related fatigue in manual driving is either connected to underload (*passive task related fatigue*) or overload (*active task related fatigue*) conditions. Underload conditions in manual driving can be caused through monotony and increasing automation. Overload conditions instead through high traffic density, bad sight conditions or the engagement in a secondary task. The consequences of all forms of fatigue are always the same: *Increased crash risk* and *bad driving performance*.

When CAD will be introduced to the consumer market, this will also affect fatigue and emergence of fatigue when driving automated. On the one hand, causations for active task-related fatigue provoked through the driving task may be reduced in CAD. On the other hand, passive task related fatigue caused through monotony and automation in CAD can be increased as the driving task is then executed by the system (Neubauer et al. 2012).

How NDRTs may affect fatigue during CAD was examined in different studies before. In a study conducted by Schömig et al. (2015) a positive effect of a quiz task on the drivers' state could be demonstrated. When participants engaged in an interesting and motivating secondary task (quiz task) while driving with CAD, drowsiness stayed on a low level. Without a secondary task engagement, drowsiness increased. The conditionally automated ride and the task engagement lasted for 15 min. Jarosch et al. (2017) tested effects of two different NDRTs on the drivers' state in CAD and resulting take-over performance. In a within subject design experiment drivers' experienced an activating quiz task and a monotonous monitoring task. When dealing with the monotonous task, fatigue emerged within 25 min of the task compared to the activating quiz task. However, the generated fatigue did not affect take-over performance.

2.2 The Assessment of Take-Over Performance

When the system for CAD detects a situation it cannot handle (e.g. human being on the road, missing lane markings) the human driver gets requested by the system to take-over control of the vehicle. It is expected, that such a take-over situation may cause problems for the human driver as a fast reaction between 5–10 s (Petermann-Stock et al. 2013) can be necessary. Especially it is assumed, that this control transition from the automated vehicle towards the human driver is dependent on the state of the driver, which next to the energetic state (see Sect. 2.1) comprises of the motoric state, cognitive state, sensory state and motivational factors (Marberger et al. 2017). As this paper focuses on the energetic state, the other factors are not described in detail.

The human drivers' performance in such a take-over situation was assessed in many studies before. Most of these studies used methods to evaluate the drivers' reaction after a RtI. Typically reaction times and parameters concerning the input of the driver upon the RtI were evaluated.

As the take-over process is a sequence of actions the human driver has to fulfill, the timing of these actions upon the RtI can be used for an assessment. Frequently time-based measures are *eyes-on-road time*, *hands-on time* (first steering wheel contact), *first braking reaction*, *steering wheel angle* > 2° and a *braking pedal input* > *10%*.

Next to the reaction times, the quality of the drivers' input can be used for the assessment of the take-over. Therefore measures which are directly connected to the quality of the drivers' input after the RtI can be used. Depending on the situation, other metrics can be considered. If the situation forces the driver to brake, *longitudinal acceleration* can be considered, whereas *lateral acceleration* should be used when a steering maneuver is necessary. An overview of these different metrics is given by Gold (2016).

In this classic analysis of take-over performance only one measure is considered at a time. As a result of this, the meaningfulness of these evaluated parameters is severely limited. A human driver can respond very fast upon an RtI which would imply that this subject reacted well when considering the reaction times. However, a quick response does not immediate mean, that the human driver reacts appropriate. Referring to Peterman-Stock et al. (2013), a fast take-over maneuver is often accompanied by a bad reaction when considering the measures connected with the quality of the drivers' input.

For a more holistic consideration on take-over performance ratings, a new measure was developed: The TOC expert rating uses different parameters, which are directly connected to the human drivers' reaction upon the RtI.

2.3 The TOC Rating

To evaluate reactions of the human driver upon an RtI, the expert based take-over controllability rating (Naujoks et al. 2018) was used for this paper.

The TOC rating was developed by the Wuerzburg Institute for Traffic Sciences (wivw) as part of the Ko-HAF project, funded by the German Federal Ministry of Economics and Technology.

This expert based rating tool requires at least three trained raters and videos of all take-over reactions of the human drivers. The videos should contain a bird eye view of the vehicle, the human driver, the interior with the controls and the footwell. With the help of a rating scheme, different aspects of the take-over reaction can be assessed (e.g., quality of lateral and longitudinal control, adequateness of signaling to other road users, etc.) See Fig. 1 for the TOC rating scheme.

Coding sheet for take-over situations

	Perfect	Imprecisions	Driving Errors	Endangerment	Not controllable Event
Braking response			☐ too strong ☐ too weak ☐ too late ☐ missing	☐ endangerment of others ☐ self-endangerment	☐ collision ☐ lane departure/ leaving road ☐ loss of vehicle control
Longitudinal vehicle control			☐ safety-distance too low ☐ inadequate speed		
Lateral vehicle control		☐ jerky steering movement ☐ imprecise lane keeping	☐ safety-distance too low ☐ strong oscillation ☐ crossing lane markings		
Lane change/ lane choice			☐ hesistant/ interrupted ☐ too late ☐ missing ☐ wrong lane		
Securing/communication		☐ unnecessary/wrong use of indicator	☐ missing/too late use of indicator ☐ missing/too late control glance		
Vehicle operation		☐ imprecisions	☐ problems		
Facial expression of driver		☐ visible emotions			
	1	2 3	4 5 6	7 8 9	10

Comment:

Fig. 1. Rating scheme of the TOC rating tool

The TOC rating results in a total score indicating the quality of the take-over reaction per subject. A "1" indicates a "perfect" reaction upon the RtI and a score of "10" indicates a "not controllable event". Next to this detailed classification, there is a division into five categories: "Perfect", "Imprecions", "Driving Errors", "Endangerment" and "Not controllable event".

For further information about the method TOC expert rating please visit www.toc-rating.de.

3 Methode

Focus of this paper is whether the engagement with a monotonous monitoring task or the duration of the automated journey stronger affects a human drivers' take-over performance upon a RtI. For this purpose, two different studies, both conducted at BMW laboratories in 2016 and 2017 were used for the assessment of the take-over. In both studies the same apparatus including the mock-up and HMI concept (see Sects. 3.2 & 3.3), the same NDRTs (see Sect. 3.4) and the same take-over situation (see Sect. 3.6) was used. The studies were identical and only differed by the length of the automated ride.

3.1 Participants

In both experiments all participants were BMW employees. The experiments took place during regular working time and participation was voluntary.

Study 1. Fifty-six subjects (9 female and 47 male) participated in the study. Mean age was 30.10 years (*SD* = 9.00, min = 20, max = 56). The subjects had a mean driving experience of 12.29 years (*SD* = 9.36). The majority of the sample had experienced at least one driving assistance system (79.3%). Adaptive cruise control was the most experienced one (75%).

Study 2. Seventy-three participants (16 female and 57 male) with mean age of 31.5 years (*SD* = 9.8, min = 20, max = 60) participated in the experiment. Mean driving experience of the subjects was 13.9 years (*SD* = 9.6). The sample quoted their driving experience as "experienced" (83.56%) or "very experienced" (15.07%). Just one participant indicated to be "unexperienced" (1.37%) in driving. Most participants had experienced at least one driver assistance system like adaptive cruise control or active lane keeping assistant (83.56%).

3.2 Apparatus

In both experiments the same simulator and the same mock-up were used. The studies were conducted in the motion-based driving simulator at BMW group laboratories in Munich. A horizontal field of view of 240° x 45° gets displayed by 7 projectors with a resolution of 1920 x 1200. The two rear channels, providing the same resolution, can be seen through the side mirrors.

As mock-up a BMW 520i touring with automatic transmission was used. All necessary instrumentation was identically to a production type car. Realistic steering torque was simulated by using a servo motor and a steering model.

The NDRTs were presented on a Windows Surface tablet (12.3 in.) mounted in front of the CID.

3.3 Conditional Automated Driving and HMI Concept

When CAD was activated, the system executed lateral and longitudinal guidance. The system adapted the speed of the vehicle in front and automatically overtook slower driving vehicles. The target speed of the CAD system was 130 km/h. The system could be activated by pressing a marked button on the steering wheel. A deactivation could be carried out by pressing the button again, by a braking- or steering-input. When the system is deactivated, the vehicle would drift in the current direction of the steering wheel and decelerate with drag torque. All drivers' assistance systems were deactivated at this time to just measure the human performance in a take-over situation.

The current status of the system was displayed in the main instrument cluster of the vehicle. Three possible states could be displayed (see Fig. 2):

- CAD available: grey steering wheel icon, "CAD available"
- CAD activated: blue steering wheel icon, "CAD activated"
- RtI: red steering wheel icon, moving hands grabbing the steering wheel, "please drive manually", advisory (earcon).

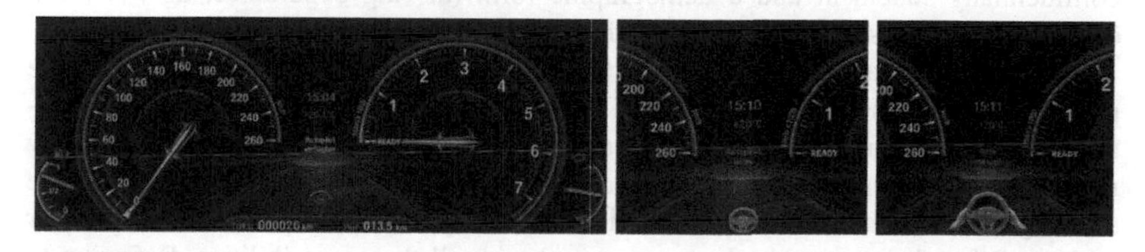

Fig. 2. HMI concept (Color figure online)

3.4 Non-driving Related Tasks

Focus of this study was whether different NDRTs or the duration of the automated ride more affect the take-over performance. Therefore two different NDRTs were used to either keep the driver in an activated state (Quiz-task) or to fatigue the driver (Pqpd-task). In both experiments the same two tasks were used.

In the Quiz-task condition, participants had to engage in a multiple-choice quiz. After choosing one out of four answer options, the right answer was highlighted green. When the question was answered wrong, the chosen answer was additionally highlighted red. The next question followed directly.

In the Pqpd-task condition, the participants had to monitor a screen and always respond by touching the screen whenever a “p” was displayed. Four different letters (“P”, “q”, “p”, “d”) were presented on the screen in randomized order. One letter was always displayed between 10–15 s.

See Fig. 3 for the two NDRTs.

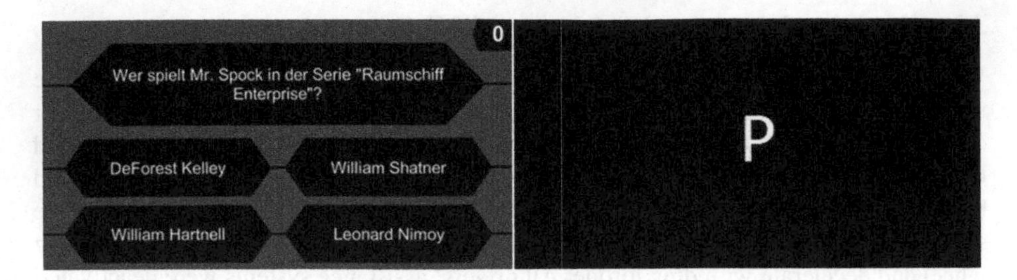

Fig. 3. NDRTs the participants had to deal with

3.5 Experimental Setup

The experimental setup looked similar in both experiments and will therefore only be explained once. After the participants arrived at the simulator, an instruction in the experiment followed. Therefore, a written description of the CAD with its characteristics, system boundaries and the possibility of RtIs was handed out. After signing a confidentially statement and a demographic form (driving experience, age, sex and experience with assistance systems) had to be filled out.

Before the actual experimental ride, a familiarization ride followed to accustom the participants with the driving simulator. Here, the participants should first drive manually to get familiar with the driving simulator and it´s driving behavior. After a few minutes the participants were told to activate CAD by pressing the button on the steering wheel. In this familiarization ride, the participants also experienced a take-over with an RtI and they could take-over control with all three possibilities. Referring to Hergeth et al. (2017) a prior familiarization with CAD and take-over situations strongly influences the take-over performance.

After this familiarization the actual experimental ride followed. In this ride, participants experienced a conditionally automated ride and meanwhile had to engage in the assigned NDRT. After entering the highway, CAD was available. With activation of the system the NDRT was displayed and participants were told to deal with the task for the entire ride. After a certain time a take-over occurred. In experiment 1 the take-over occurred after 25 min and in experiment 2 after 50 min of the conditionally automated ride.

3.6 Scenario

The scenario was identically in the two experiments. The ride was conducted on a three-lane highway with a hard shoulder. Traffic density was low and guidance was mostly straight. Weather conditions were set to a clouded sky. The take-over situation was an accident on the lane of the ego vehicle. To the right of the ego vehicle there was a hard shoulder, on the lane left, there are two further cars with a distance of about 100 m to each other. The velocity of the ego vehicle at the time of the RtI was 130 km/h and time-to-collision (ttc) was 7 s.

3.7 TOC Rating

In order to assess the take-over performance after the automated ride and engaging in the NDRTs the take-over situations of all participants were cut to 30 s videos. The videos were defaced, coded and randomized and handed out to the three trained raters. With the help of the TOC-rating tool (see Sect. 2.3), the take-overs of the two experiments were evaluated.

4 Results

Due to technical problems (missing video data), four participants of the first study had to be excluded from analysis, resulting in $N = 52$. In the second study, five participants could not be included in the analysis, resulting in $N = 68$.

Before analysis, the three individual evaluations from the three raters were combined and average scores were formed.

4.1 Reactions of the Drivers

Upon the RtI the drivers reacted with different driving maneuvers. A distinction was made between stopping on the own lane, doing a lane change maneuver or loss of control. In Fig. 4 you can see the different driver's reactions dependent on the condition. In the two Pqpd-conditions more participants reacted with a less complex stopping maneuver compared to the quiz-task conditions. In the 50 min rides more accidents occurred compared to the 25 min rides. The drivers' reactions differed significantly ($p = .004$, Fisher's exact test).

4.2 TOC Categories According to the Studies

In Fig. 5, the distribution of the TOC categories of the take-over reactions are shown. Not one single participant in both studies has reacted with a perfect take-over performance. The best TOC rating result in both studies were "imprecions". The majority of the take-overs in both studies were rated in the category "driving errors". "Endangerments" and "not controllable events" occurred more frequently in the second study with 50 min of automated driving.

4.3 Mean TOC Rating Scores

In a further analysis, it was examined whether the mean TOC values of the four conditions differ significantly. A Kruskal-Wallis H test showed that there was no statistically significant difference in the TOC rating scores between the different

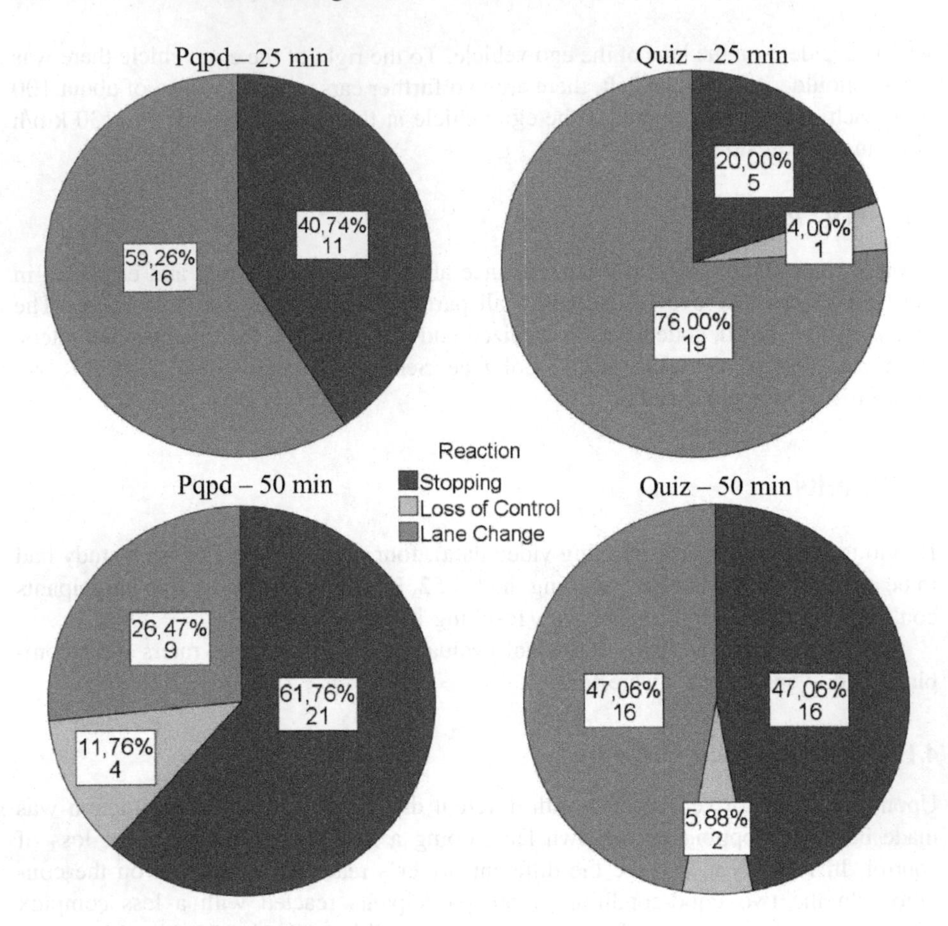

Fig. 4. Different drivers' reactions according to the condition

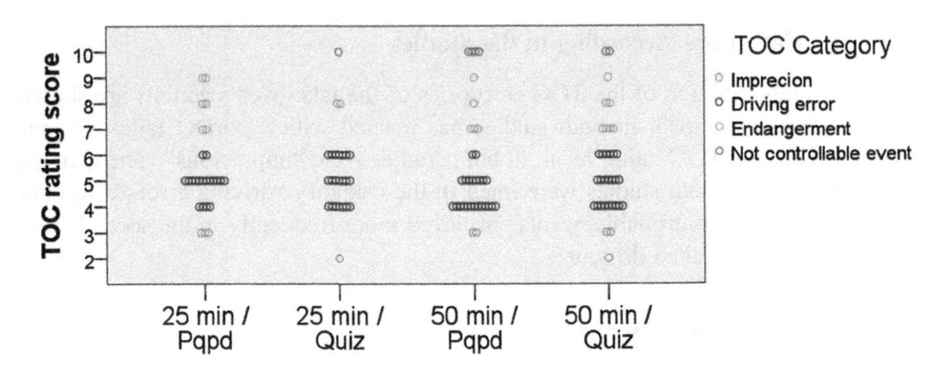

Fig. 5. TOC ratings according to the four conditions

conditions, $\chi2(3) = .473$, $p = .925$, with a mean rank TOC rating score of 61.00 for the 25 min and engaging in the Pqpd task group, 61.08 for the 25 min ride and engaging in a quiz task group, 57.26 for the 50 min ride and engaging in the Pqpd task group and 62.91 for the 50 min ride and engaging in a quiz task group. See Fig. 5 for an overview of the mean TOC scores.

A Mann-Whitney test indicated that NDRT did not affect the TOC rating significantly, with a mean rank score of 58.92 for the Pqpd task and a mean rank score of 62.14 for the quiz task, $U = 1703.00$, $p = .612$.

A Mann-Whitney test indicated that time of the ride did not affect the TOC rating significantly, with a mean rank score of 61.04 for the 25 min ride and a mean rank score of 60.09 for the 50 min ride, $U = 1749.00$, $p = .882$.

5 Discussion

The main purpose of this paper was to examine, whether the time of the conditional automated driving or the type of the NDRT more affect the take-over performance, assessed with the TOC expert rating, of the human driver when it comes to a take-over situation.

Therefore, take-over reactions of two studies were rated by three trained raters with the TOC expert rating tool. The two studies were identical in structure, using the same mock-up, the same scenario and the same NDRTs and differed only in the duration of the automated ride.

The results suggest, that the drivers` reactions upon an RtI differ according to the duration of the automated driving as well as among the type of the NDRT. After a 25 min automated ride and engaging in an activating NDRT the majority of the human drivers reacted by doing a lane change. When engaging in an activating task (quiz-task), even more participants reacted by doing a lane change compared to monotonous monitoring task (Pqpd-task). After a 50 min conditional automated ride and engaging in the same two NDRTs things look different. In these conditions, many more participants reacted by stopping on their own lane. Again, more subjects did a lane change maneuver when engaging in the activating quiz task compared to the monotonous Pqpd-task. In the 50 min condition six subjects could not control the vehicle compared to one single participant in the 25 min condition. These differences in the take-over reaction are significant, suggesting that human drivers can have major take-over issues if they are driving conditionally automated for longer periods of time and cannot engage in any activating activity.

However, when looking at the TOC mean rating scores, these differences will no longer be visible. The mean scores, calculated from three independent expert ratings, barely differ between the four different conditions. In the two conditions with the prolonged driving time (50 min) the TOC rating scores just slightly worse compared to the shorter driving time (25 min).

When considering the different TOC categories, it becomes visible that in the 25 min quiz condition less take-overs were rated as “endangerment” or “not controllable event” category as in the other conditions. Regardless of the duration of the automated ride and the NDRT, most of the subjects showed at least driving errors during the takeover situation. Not one participant in both studies reacted was rated with a "perfect" take-over. According to the toc rating, take-over performances in the four conditions did not differ significantly.

These results show that the TOC rating is barely meaningful in terms of take-over performance if there are multiple options for a driver's response. A stopping on the highway, which is the less complex maneuver is often rated as well as or even better than a lane-change maneuver, which is the more complex maneuver. A stopping maneuver on the highway can lead to a dangerous situation for the driver, especially when the human driver responds with an emergency brake. When considering the TOC rating scheme, it also becomes apparent that if a driver reacts with a complex lane change maneuver when it comes to a RtI more factors can be assessed as a "driving error" or an inadequate reaction than in a possible braking maneuver.

For future studies, it is recommended to use the TOC expert rating as an additional assessment. A consideration of driving data and reaction times is still strongly recommended. Presumably the TOC rating provides the best results if the situation is configured in a way that there is only one possible reaction for the driver.

Nonetheless, the two studies show that it can be quite difficult for people to regain control from CAD. Hardly anyone has shown a correct take-over. Furthermore it seems like take-over performance decreases with increasing time of automated driving and monotonous and fatiguing NDRTs. One possible explanation for this phenomenon can be reduced situation awareness. How the human driver can be supported when it comes to a take-over situation and how situation awareness can be maintained even when driving conditionally automated should be investigated in upcoming experiments.

References

Eriksson A, Stanton NA (2017) Takeover time in highly automated vehicles: noncritical transitions to and from manual control. Hum Factors 59(4):689–705

Gold C (2016) Modeling of Take-Over Performance in Highly Automated Vehicle Guidance (Doctoral dissertation, Universität München)

Hergeth S, Lorenz L, Krems JF (2017) Prior familiarization with takeover requests affects drivers' takeover performance and automation trust. Hum Factors 59(3):457–470

Jarosch O, Kuhnt M, Paradies S, Bengler K (2017) It's Out of Our Hands Now! Effects of Non-Driving Related Tasks During Highly Automated Driving on Drivers' Fatigue

Marberger C, Mielenz H, Naujoks F, Radlmayr J, Bengler K, Wandtner B (eds) (2017) Understanding and Applying the Concept of "Driver Availability" in Automated Driving. Springer

May JF, Baldwin CL (2009) Driver fatigue: the importance of identifying causal factors of fatigue when considering detection and countermeasure technologies. Transp Res Part F Traffic Psychol Behav 12(3):218–224

National Center for Statistics and Analysis (Oktober 2017) Drowsy Driving (2015) (Crash•Stats Brief Statistical Summary. Report No. DOT HS 812 446). Washington, DC: National Highway Traffic

Naujoks F, Wiedemann K, Schömig N, Jarosch O, Gold C (2018) Expert-based controllability assessment of control transitions from automated to manual driving. Manuscript submitted for publication

Neubauer C, Matthews G, Langheim L, Saxby D (2012) Fatigue and voluntary utilization of automation in simulated driving. Hum Factors J Hum Factors Ergon Soc 54(5):734–746

Petermann-Stock I, Hackenberg L, Muhr T, Mergl C. (2013) Wie lange braucht der Fahrer? Eine Analyse zu Übernahmezeiten aus verschiedenen Nebentätigkeiten während einer hochautomatisierten Staufahrt. 6. Tagung Fahrerassistenzsysteme. Der Weg zum automatischen Fahren

SAE International (2016 September 30) Taxonomy and Definitions for Terms Related to Driving Automation Systems for On-Road Motor Vehicles (J3016)

Schömig N, Hargutt V, Neukum A, Petermann-Stock I, Othersen I (2015) The interaction between highly automated driving and the development of drowsiness. Procedia Manuf. 3:6652–6659

Application of Attribution Theory in Crane Incident Analysis in a South African Port

Jessie Mashapa[1(✉)], Jessica Hutchings[2], and Andrew Thatcher[1]

[1] Psychology Department, University of the Witwatersrand, WITS, Johannesburg 2050, South Africa
jessie.mashapa@gmail.com

[2] Transnet Centre of Systems Engineering (TCSE), University of the Witwatersrand, Johannesburg, South Africa

Abstract. Crane incidents occur in the ports of South Africa and around the world in the maritime industry. Crane operators' and crane supervisors' accounts of such events are used in investigations and therefore the accuracy of the findings from the incident investigation is important because these findings inform the remedial actions that must be taken to address the cause(s) of the incident. The aim of this paper is to understand and compare crane operators' and supervisors' perceptions of why incidents happen using attribution theory (Heider 1958). In this qualitative study, 16 participants from a port in South Africa were used. Specifically, 8 crane operators and 8 crane supervisors. The two groups were interviewed using semi-structured interviews. Thematic content analysis was conducted on the interview data to identify the emerging themes between the two groups utilizing a systems thinking approach (Wilson 2014). The results support Gyekye (2010), demonstrating that supervisors were more inclined to attribute crane incidents to internal factors of the crane operators while the crane operators were more inclined to attribute crane incidents to external, systemic factors. Although both groups identified external factors such as mechanical and maintenance issues, weather conditions, communication, training and skills, and safety culture as factors contributing towards crane incidents, the supervisors were more likely to identifying personal factors attributed to crane operators such as a lack of concentration, fatigue, skills, and eyesight as causal factors. Conversely, crane operators mostly felt that these factors were insignificant in causing any incidents.

Keywords: Attribution theory · Crane incidents · Incident investigations

1 Introduction

The development and maintenance of an efficient and competitive transport system is a key objective for developing economies such as South Africa (National Development Plan 2030 2013). Therefore, it is important for researchers to find ways of increasing the efficiency and safety of transport systems by addressing issues that cause delays, inefficiencies, accidents and slow-turnaround times across different disciplines such as psychology, engineering, and systems thinking (Rasmussen 1997; Leveson 2011; Salmon et al. 2014). This study aims to address the above-mentioned challenges,

S. Bagnara et al. (Eds.): IEA 2018, AISC 823, pp. 524–536, 2019.
https://doi.org/10.1007/978-3-319-96074-6_55

specifically in the maritime industry by using attribution theory to understand and compare crane operators' and supervisors' perceptions of why incidents happen. The application of attribution theory to crane incident analysis can explain how crane operators and crane supervisors arrive at their causal attributions in the event of an incident. Crane operators and crane supervisors' accounts of crane incidents are used in incident investigations and the accuracy of the findings from these investigations is important as it informs the remedial actions that must be taken. Accurate remedial actions would decrease delays caused by machine break-downs and time lost during incident investigations leading to increased productivity and a greater contribution to the economic growth of the country.

1.1 Crane Incidents in the Maritime Industry

The Maritime Injury Centre described port terminals as one of the most dangerous places to work because of the crane operations that take place with the loading and offloading of cargo from vessels (Maritime Injury Centre 2017). Employees working in the maritime industry are exposed to a number of risks, but those working with equipment such as cranes either on vessels or on the dock are found to be exponentially exposed to more risks due to the nature of the equipment that is used (Maritime Injury Centre 2017). In recent years, crane incidents in the maritime industry have increased resulting in delays in operations, time lost during investigations and disciplinary hearings, and a loss of income for ports (Lam et al. 2007). This challenge has been addressed through incident investigations that take into account crane operators', crane supervisors', and the technical team's account of the event. Despite incident investigations being carried out after every reported incident, incidents are still highly prevalent. The question that arises is whether these investigations are effective in truly identifying what actually happened. This research aims to investigate how attribution theory may play a role in determining the different perceptions of accident causation in a port in South Africa.

1.2 Attribution Theory

Psychology researchers have been exploring the effect of attribution processes on people for decades (Heider 1958, Weiner 1979; Forsterling 2001). Their interest in the topic stems from the fact that attribution processes affect the manner in which past, present, and future events are interpreted; the manner in which individuals find meaning and make sense of the environment, and individuals' levels of motivation (Heider 1958, Weiner 1976 and Wilcox 2015). Attribution processes also play a pivotal role in establishing and maintaining consistency between the beliefs and thoughts of people and in preventing the adverse impact that may originate from internal inconsistencies. The internal consistencies that are developed through attribution processes allow people to understand, predict and control their daily situations (Forsterling 2001). The essential premise of attribution theory is for people to "understand and predict future events" (Wilcox 2015, p. 1).

According to attribution theory, individual employees will have different factors affecting the way they perceive events and how they arrive at causal attributions, therefore this can result in discrepancies between the views of employees in organisations. In the current research, the employees included the crane operators and their supervisors, and the objective of the research was to determine if there were differing perceptions between the two groups regarding the causes of crane incidents. The differences in work related experiences between crane operators and supervisors can also result in discrepancies in their perceptions of a specific incident. The concept of the interpretation of causes is fundamental in attribution theory with researchers such as Gronhaug and Falkenberg (1994) referring to attribution theory as the study of perceived causation. The following are different factors that can influence or affect the attributions of crane operators and supervisors.

1.2.1 Differences in Observers and Actors' Attributions

Observers can be defined as individuals that attempt to understand other people's behaviour and actors as individuals who attempt to understand their own behaviour (Malle et al. 2007). Two main groups of factors that have been recognised as possibly influencing actor-observer differences are: (1) cognitive factors and (2) motivational factors (Kelley and Michela 1980; Jones and Nisbett 1987). With regards to cognitive factors, the individual observing the event/behaviour usually has a limited amount of information about the actor's behaviour in certain situations, while the actor has a vast amount of information about their own behaviour in different situations.

The second aspect that was recognised to influence the differences amongst observers and actors is motivational factors. Actors and observers have varying interests in the manner which events/behaviours are inferred. The two factors give a limited explanation for the differences between actors and observers. Some researchers propose that the observed differences between actors and observers may be the result of the accurateness of the attributions made. The actors would have more information on the event/behaviour and therefore their attributions should be more accurate (Monson and Snyder 1977).

1.2.2 Hierarchical Level: Supervisors and Subordinate Workers

The hierarchical level in the organisation has been shown to influence individuals' attributions concerning workplace incidents (Turbiaux 1971). Supervisors are inclined to use internal explanations that place the blame on the subordinate in question (Dejoy 1987). Numerous studies by Mitchell and Wood (1980), Hamilton (1986), Dejoy (1987), and Salminen (1992) have demonstrated this notion. Research has also shown that when supervisors assigned internal attributions to incidents, this often resulted in punitive consequences for the subordinate workers, especially in the case of serious incidents and organisations with punitive cultures, therefore subordinate workers and supervisors make defensive attributions to avoid responsibility and blame for the incident (Shaver 1970; Gyekye 2010).

1.2.3 Age

The age of workers has been shown to have a significant effect on their causal attributions. Studies by Gyekye and Salminen (2007) have revealed that 'older subordinate workers' are prone to attributing workplace incidents to external factors more than their younger subordinate worker counterparts. In the current study, 'older subordinate workers would be referred to as 'older crane operators' and younger subordinate workers as 'younger crane operators'.

1.2.4 Worker Experience

Worker experience was also found to make a difference in how workers make causal attributions in the workplace. Gyekye (2010) found that more experienced workers were more prone to make external attributions such as environmental factors for poor performance by their subordinates (Gyekye 2010). This was also observed in supervisors who were categorised as more experienced workers. Less experienced workers were found to make internal attributions for poor performance (Gyekye 2010).

1.2.5 Cultural Background

Research has shown that cultural orientation can also have an effect on the attributions of workers in workplace incidents. In his study on individualistic Finnish workers, Gyekye (2001) found that the individualistic workers were more likely to make internal attributions for incidents than external environmental attributions (Gyekye 2001). Individualistic workers often come from individualistic cultures, such as western cultures, where the people strive to be different and separate from others and their self-esteem is affected by their ability to be autonomous and distinct (Markus and Kitayama 1991). On the other hand, collectivistic workers from Ghana were more likely to attribute accident responsibility to situational or contextual factors (Gyekye 2001). According to Markus and Kitayama (1991), there are two types of self-construal; the interdependent which can be found in collectivistic cultures and the independent which is found in individualistic cultures. Self-construal refers to the degree to which an individual, 'the self', describes themselves interdependently or dependently with others (Cross 2015). People from collectivistic cultures as commonly found in Africa, with an interdependent self-construal, emphasise the importance of having good interpersonal relationships, in-group harmony, and conformity to social norms (Kwan et al. 1997).

1.2.6 Safety Culture and Just Culture

Safety culture is a central part of an organisation's culture, specifically in high risk industries such as crane operations. Four sub-components of safety culture have been proposed in workplace safety literature; a just culture, a reporting culture, a flexible culture, and a learning culture (Reason 1997; GAIN 2004; Jeffcott et al. 2006). Workers and supervisors' accounts of incidents are an essential part of the process of determining blame-worthy behaviour and assigning responsibility for incidents. Unfortunately, due to the blame culture that exists in organisations, workers and supervisors do not give accurate accounts of incidents because of a lack of trust and fear of punishment (Dekker 2007; Laursen 2007; Khatri et al. 2009 and Clarke 2011). A just culture can help to eliminate these challenges by establishing an environment of trust and therefore improving the accuracy of reporting (GAIN 2004). When an organisation is able to effectively and continuously learn from accidents and errors, it can be described as having a learning culture. A just culture lays the foundation for a reporting and learning culture. A just culture facilitates a supportive environment, promotes trust and open communication in an organisation. In addition, it clearly outlines what is unacceptable and acceptable behaviour (GAIN 2004; Khatri et al. 2009). In an organisation with a just culture, employees can openly express their viewsand concerns, and report on accidents without fear of being blamed or punished because of the established trust and ethical nature of the organisation (Khatri et al. 2009; Pepe and Catalado 2011). In contrast to a just culture, an organisation with a blame culture prefers to deal with accidents through a disciplinary process where employees can be terminated or even prosecuted (Clarke 2011). Blame culture can be defined as "a set of norms and attitudes in an organisation characterised by an unwillingness to take risks or accept responsibility for mistakes because of fear of criticism or management astonishment" (Khatri et al. 2009, p. 314). The constant punishment of employees breeds an environment of fear and mistrust in the organisation, resulting in incomplete or no information from employees with regards to accidents. These measures are used by the employees as an attempt to avoid punishment (Dekker 2007; Khatri et al. 2009). An organisation with a blame culture is concerned with finding the guilty party and often assigns blame for accidents to human error and negligence.

2 Research Question

It is proposed that crane operators will have different perceptions of crane incidents to crane supervisors, because according to attribution theory, the differences in work related experiences between crane operators and supervisors as well as other factors such as cultural backgrounds, work experience and gender to name a few, can result in discrepancies in their perceptions of a specific incident. Therefore, this research study aims to analyse and compare crane operators' and supervisors' perceptions relating to crane incident causation by the application of attribution theory. The research question primarily focuses on comparing the task/job factors, environmental factors, and personal factors of both crane operators and supervisors and how these factors can be attributed to the causes of crane incidents in a South African port.

3 Methods

The data was collected at a large transport logistics company in SA as part of a larger study. Data collection focused on interviews with crane operators and crane supervisors to give their accounts of the causes of crane incidents.

3.1 Participants

The sample consisted of 16 employees at the port involved in working with cranes. They participated in the study as interviewees. The sample was divided into two groups consisting of 8 crane operators and 8 supervisors. The participants were selected based on the roster available at the time of this research so as to cause minimal disturbances to crane operations. Participants were interviewed during their rest breaks. The age ranged from 26–52 years old with crane operators' ages ranging from 26–52 years old and the supervisors' ages ranging from 30–40 years old. The sample only consisted of 3 females in the overall group (2 crane supervisors and 1 crane operator). The sample of 16 participants was ethnically diverse and representative of the geographical region.

3.2 Research Design

A qualitative research approach was used in the current research study to allow the researcher to observe the participants in their natural setting and to allow for an in-depth analysis. This type of analysis is the most preferred form of analysis for qualitative studies because it is compatible with both constructionist and essentialist paradigms and thus allows the researcher the freedom to be flexible in analysing the collected data (Braun and Clark 2008). The interviews were scheduled according to the participants' availability so as to prevent any disturbance of crane operations. The interviews took place at the different berths at the port during working hours. The researcher conducted one-on-one, face-to-face interviews in a private room to ensure confidentiality and to allow the participants to speak freely in a 'safe space'. The interviews were recorded so as to allow the researcher to fully engage with the interviewee. The researcher used two interview schedules, one for the crane operators and another for the crane supervisors to elicit appropriate information from the two groups. The use of the schedule was flexible and allowed for probing depending on the responses of the participants. The duration of the interviews was 45 min on average. The researcher kept a journal for documenting field notes which were focused on non-verbal behaviors such as body language and facial expressions. The field notes also included researcher's reflections after each interview. After the interviews, the audio recordings were transcribed and analysed together with the field notes. The audio recordings were then transcribed in preparation for anaylsis.

3.3 Data Analysis

Thematic content analysis was used to analyse the data as adapted from Braun and Clarke (2008) and Ritchie and Lewis (2003). Once the researcher was well acquainted with the content of the transcripts, the process of sorting the data and labeling of the themes and concepts commenced. A spreadsheet was used to record the examples of themes from the text and to record the frequency of themes that emerged from each interview question. The statements from the transcripts were coded according to the 'perceived causes of incidents'. Examples included communication and wind. The themes were then categorized as internal or external attributions using Weiner's (1972) two-dimensional attribution model. Further work is still required, however the relationship between the themes will then be coded by the researcher and also be analysed using Rasmussen's framework as part of a systems thinking approach. The results will be displayed on an Accimap to show the interactions between the different themes.

4 Preliminary Results and Discussion

The preliminary results demonstrate that there are differences in the attributions of crane operators and supervisors. Supervisors were inclined to attribute the causes of incidents to the dispositional factors of the crane operators, with 6 of the 8 supervisors interviewed citing factors such as fatigue and stress of crane operators as significant causes of crane incidents (Tables 1 and 2).

Table 1. Examples of supervisor themes: fatigue and stress

Factor	Themes	Quotes
Personal Factors	Fatigue	"Like I said, fatigue does cause 90% of accidents"
		"It's like people complain but they still work those shifts because the money is good, we need the weekends for overtime so it's a tricky situation because people don't want days off cause they are losing money but they are tired so…"
	Stress	"Some accidents happened because that person, his mind is not here when they are in the crane"

Similarly, crane operators acknowledged the effects of fatigue and stress on their performance. However 7 of the 8 crane operators didn't perceive the above-mentioned factors as being the cause of crane incidents.

Instead, the results in the study demonstrate similar findings to those of Dejoy (1987), Kwan et al. (1997), Kouabenan et al. (2001), Gyekye and Salminen (2007), Gyekye (2010), and Clarke (2011), who demonstrated that factors such as hierarchical level, age, cultural background, and blame have an influence on the attributions of supervisors in organisations. The differences between supervisors and crane operators could be due to a number of factors which are discussed in Sects. 4.1 to 4.4 and with the data presented in Tables 3 and 4.

4.1 Gender and Hierarchical Position

Koubenan et al. (2001) observed an interaction between sex and hierarchical position in their study on attributions. Their findings show that males are more inclined to make internal causal attributions more than females, however the findings in the current study show that female supervisors were more inclined to make internal attributions than their male counterparts. The female supervisors cited the most internal causal factors for incidents and they were the only two supervisors to cite both *negligence* and *fatigue* as causal factors for crane incidents. This finding could be influenced by the fact that the maritime industry is male dominated and male supervisors share personal characteristics with the crane operators. This is because many of them are male and the female supervisors may have less in common with the male operators, causing them to be seen as non-relevant and thus affecting the way they are perceived by the female supervisors.

Table 2. Examples of crane operator themes: fatigue and stress

Factor	Themes	Quotes
Personal Factors	Fatigue	"It's night, your brain wants to shut down. But there's no example of tiredness or fatigue that really cause anything, well not that I know of" "On whether fatigue affects concentration- No, not really hey"

Table 3. Summary of themes that emerged from crane operator's responses.

Factor	Themes	Quotes
Job factors	Communication	"A lot of incidents or accidents here happen because of lack of communication" "that's also something that happens quite often around here we battle a lot because there's a lot of hatch men here they can't communicate in English properly"
	Maintenance of equipment	"So, if they can do the inspection properly and make sure that they give us proper cranes to work with, a reliable crane, then the incidents will be less."
Environmental factors: Physical	Wind	"They don't care about wind gust, they just care about limit 70 if its 70 its ok to stop" "The only thing that affects the crane is wind and maintenance"
Organisational factors	Safety	"The company advertises a lot of safety but personally I don't think there's a lot of safety happening in this place" "I think they've got a good understanding of each other, that's why maybe safety gets pushed aside they follow their own working culture"
	Training	"I passed actually not really knowing much about the crane to be honest" It's not enough, they don't give you time on the crane, any operator can only get better if he spends time on the crane
	Transparency	"No one tells you anything, half the time you don't even hear anything, you're just here to work and get your money. It's sad but that's how it is"
	Punitive culture	"Yah, they are charging us hey, they are serious." "There are a lot of accidents happening and they are swept under the carpet because obviously no one wants to be that person who put someone in trouble"
Personal factors	Fatigue	"Its night, your brain wants to shut down. But there's no example of tiredness or fatigue that really cause anything, well not that I know of"
	Stress	"That would be dangerous with all the stress you're adding more stress, you're going to make incidents or accidents"

Table 4. Summary of the themes that emerged from the crane supervisor's responses

Factor	Themes	Quotes
Job factors	Communication	"I think the major cause of these incidents was poor communication on the part of the operators and stevedoring companies" "One of the hatch lids is the one that keeps on happening because like I said no proper communication"
	Maintenance of equipment	"People report they even fill the log book if there's something wrong in the crane but you find out there is nothing done about it"
Environmental factors: Physical	Wind	"I think if you are a supervisor and you've been an operator before, you can't exactly force someone to work knowing exactly what they are talking about. He can't work so it becomes a challenge" "You get pressure from the management, cause if the management says its ok for the crane driver to operate and on the other hand he reported that it's not safe yah that's the problem"
Organisational factors	Safety	"Yo, its bad, very bad. I can say its bad cause we report things here and nothing is done about it, we report things that can cause an incident we report those things there's nothing maybe if you report it now, they wait until the accident happens then fix it, that's how it works here" "You see that no, this is not the way, but our bosses made it like that, what can we do?"
	Punitive culture	"Sometimes they keep quiet about things like if you damage something you keep quiet. They don't wanna be charged"
Personal factors	Fatigue	"Its like people complain but they still work those shifts because the money is good, we need the weekends for overtime so it's a tricky situation because people don't want days off cause they are losing money but they are tired so…"
	Stress	"Some accidents happened because that person his mind is not here when they are in the crane"
	Negligence	"Some people are reckless drivers, they cause accidents all the time because they are reckless drivers, there are some people that only have their drivers for a month but they are able to drive more carefully than those, so it also comes with the personality

4.2 Age

The current study's findings are in accordance with Gyekye and Salminen (2007) in that older workers were more inclined to make external attributions than their younger counterparts. All crane operators over the age of 40 didn't attribute any internal causes to crane incidents whilst the younger crane operators all mentioned internal factors such as *recklessness, skill,* and *fatigue* as having some effect on performance.

4.3 Cultural Background

The crane operators from collectivistic cultural groups (Black and Indian) didn't cite any internal factors as being the causes of crane incidents whilst the crane operators from individualistic cultural groups (White) cited internal factors as having an effect on their performance. The results could also be influenced by the fact that the same (White) individuals were also from a younger age group. As mentioned previously, younger workers were more inclined to making internal attributions more than their older counterparts (Gyekye and Salminen 2007). The crane supervisors' responses were found to be inconclusive as to whether cultural background influenced causal attributions. This could be affected by the culture of the teams within the organisation. The crane operators and crane supervisors see their colleagues as *'family'* and see each other as a unit. These findings are in contradiction with previous studies by Gyekye (2001, 2006, 2010) in that cultural backgrounds have an effect on the responsibility and causality attributions of workers. In the current study, it seems that the 'organisational culture' or 'team culture' was more of on influence on the attributions of supervisors. On the other hand, these findings are in accordance with Kwan et al. (1997) notion that proposes that people from collectivistic cultures, as commonly found in Africa, emphasise the importance of having good interpersonal relationships, in-group harmony, and conformity to social norms. This observation was demonstrated in one of the crane operator's responses:

> "So basically, even if it's your fault they don't do that, they would be on your side even if it's your fault they will give you the benefit of the doubt because basically we are all here as a family trying to look after our families…"

4.4 Blame Culture

The current study revealed that the crane operators and crane supervisors perceive their organisation as having a punitive culture. This could cause the crane operators and crane supervisors to take a defensive stance when perceiving incidents as a way of protecting themselves from 'punishment'.

> "We don't have enough power as I said they shift the blame to the employees now we find ourselves protecting the employee instead of fixing the problem and dealing with the person that was supposed to fix the problem that was mentioned."

Tables 3 and 4 illustrate a summary of the themes that emerged from the current study. The differences between the crane operators and crane supervisors can be observed in the organisational factors where crane operators emphasised the influence of organisational factors and personal factors as the causes of incidents whereas supervisors emphasised the influence of personal factors in the cause of incidents by crane operators.

5 Conclusion

By applying attribution theory to crane operators' and crane supervisors' perceptions on incident causation, we are able to see that there are differences in their attributions; in other words, the way they perceive events. The attributions of crane operators and crane supervisors are influenced by their role in the incident - whether they are an 'actor' or 'observer', and supervisors are influenced by their hierarchical level in the organisation. Factors such as demographics (age, gender, and cultural background), interpersonal relationships and more especially organisational factors such as 'blame culture' have been found to have an influence on the attributions of crane operators and supervisors. This demonstrates that the current method of conducting incident investigations in the port is not necessarily accurate because it is based on the crane operators and crane supervisors' perceptions of the event. These attributions are further influenced by the organisational culture. Organisations with 'punitive' or 'blame' cultures create an environment of distrust, causing the crane operators and crane supervisors to become defensive in their analysis of events as well as their reporting.

The researcher was present during the one-on-one interviews therefore there are concerns that the crane operators and crane supervisors may have responded in a socially desirable way creating a social desirability bias (Moorman and Podsakoff 1992). The interview questions were focused on general crane incidents and not on a specific incident in which the crane operator and crane supervisor were involved, therefore the results may not be representative of the status-quo because the questions did not arouse a sense of defensiveness from the participants. This may have resulted in the participants recalling recent events rather than all events. Future research should focus on specific crane incidents to allow the researcher to compare crane operators and crane supervisors' perceptions of the same incident in accordance with incident investigation processes. In addition, future research should include technical teams in the sample because their accounts of the crane incidents also need to be taken into consideration during incident investigations and the technical team operates as a separate unit, so this would give the researcher greater range to compare the differences between the 'actors' and 'observers'.

References

Braun V, Clarke V (2013) Successful qualitative research: a practical guide for beginners. Sage

DeJoy DM (1990) Toward a comprehensive human factors model of workplace accident causation. Prof Saf 35(5):11

DeJoy DM (1994) Managing safety in the workplace: an attribution theory analysis and model. J Saf Res 25(1):3–17

Försterling F (2001) Attribution. An introduction to theories, research and applications. Psychology, Hove
Gronhaug K, Falkenberg JS (1994) Success attributions within and across organizations. J Eur Ind Training 18(11):22–29
Gyekye SA (2001) The self-defensive attribution theory revisited: a culture-comparative analysis between Finland and Ghana in the work environment. University of Helsinki, Department of Social Psychology
Gyekye SA (2010) Occupational safety management: the role of causal attribution. Int J Psychol 45(6):405–416
Gyekye S, Salminen S (2007) Workplace safety perceptions and perceived organizational support: do supportive perceptions influence safety perceptions? Int J Occup Saf Ergon 13 (2):189–200
Hamilton V (1986) Chains of command: responsibility attribution in hierarchies. J Appl Soc Psychol 16:118–138
Heider F (1958) The psychology of interpersonal relations. Lawrence Erlbaum, Hillside
Jeffcott S, Pidgeon N, Weyman A, Walls J (2006) Risk, trust, and safety culture in UK train operating companies. Risk Anal 26(5):1105–1121
Jones, EE, Nisbett RE (1987) The actor and the observer: divergent perceptions of the causes of behavior. In: Preparation of this paper grew out of a workshop on attribution theory held at University of California, Los Angeles, August 1969, Lawrence Erlbaum Associates, Inc.
Kelley HH, Michela JL (1980) Attribution theory and research. Ann Rev Psychol 31(1):457–501
Khatri N, Brown GD, Hicks LL (2009) From a blame culture to a just culture in health care. Health Care Manag Rev 34(4):312–322
Kouabenan DR, Medina M, Gilibert D, Bouzon F (2001) Hierarchical position, gender, accident severity, and causal attribution. J Appl Soc Psychol 31(3):553–575
Kwan VS, Bond MH, Singelis TM (1997) Pancultural explanations for life satisfaction: adding relationship harmony to self-esteem. J Pers Soc Psychol 73(5):1038
Lam L, Tok S, Darley P (2007) Crane accidents and emergencies- causes, repairs and prevention. In: Proceedings of TOC Asia 2007, Hong Kong
Leveson NG (2011) Applying systems thinking to analyze and learn from events. Saf Sci 49 (1):55–64
Malle BF, Knobe JM, Nelson SE (2007) Actor-observer asymmetries in explanations of behavior: new answers to an old question. J Pers Soc Psychol 93(4):491
Maritime Injury Centre (2017) Maritime accidents and injuries overview. https://www.maritimeinjurycenter.com/accidents-and-injuries/cargo-and-crane/. Accessed 1 Mar 2017
Markus HR, Kitayama S (1991) Culture and the self: implications for cognition, emotion, and motivation. Psychol Rev 98(2):224
Mitchell TR, Wood RE (1980) Supervisor's responses to subordinate poor performance: a test of an attributional model. Organ Behav Hum Perform 25(1):123–138
Monson TC, Snyder M (1977) Actors, observers, and the attribution process: toward a reconceptualization. J Exp Soc Psychol 13(1):89–111
Moorman RH, Podsakoff PM (1992) A meta-analytic review and empirical test of the potential confounding effects of social desirability response sets in organizational behaviour research. J Occup Organ Psychol 65(2):131–149
National Planning Commission (2013) National development plan vision 2030
Rasmussen J (1997) Risk management in a dynamic society: a modelling problem. Saf Sci 27 (2/3):183–213
Reason J (1997) Managing the risks of organizational accidents. Ashgate, Aldenhot
Salminen S (1992) Defensive attribution hypothesis and serious occupational accidents. Psychol Reports 70(3l):1195–1199

Ritchie J, Lewis J, Nicholls CM, Ormston R (eds) (2013) Qualitative research practice: a guide for social science students and researchers. Sage

Salmon PM, Goode N, Lenné MG, Finch CF, Cassell E (2014) Injury causation in the great outdoors: a systems analysis of led outdoor activity injury incidents. Accid Anal Prev 63: 111–120

Shaver KG (1970) Defensive attribution: effects of severity and relevance on the responsibility assigned for an accident. J Pers Soc Psychol 14(2):101

Turbiaux M (1971) Human factors in occupational accidents. Bulletin de Psychologie 24: 952–960

Weiner B (1976) 5: an attributional approach for educational psychology. Rev Res Educ 4 (1):179–209

Weiner B (1979) A theory of motivation for some classroom experiences. J Educ Psychol 71(1):3

Wilcox AK (2015) Attribution case studies with elite junior australian footballers and their coach (Doctoral dissertation, Victoria University)

Wilson JR (2014) Fundamentals of systems ergonomics/human factors. Appl Ergon 45(1):5–13

Observing Traffic – Utilizing a Ground Based LiDAR and Observation Protocols at a T-Junction in Germany

André Dietrich[1(✉)] and Johannes Ruenz[2]

[1] Chair of Ergonomics, Technical University of Munich, Boltzmannstr. 15, 85748 Garching, Germany
andre.dietrich@tum.de

[2] Chassis Systems Control, Robert Bosch GmbH, 74232 Abstatt, Germany

Abstract. Introducing automated vehicles onto current urban traffic conditions is a demanding task. While trajectory and maneuver planning algorithms needs to cope with complex road structures, the automation needs to deal with another volatile element – human road users. Understanding how traffic participants behave nowadays is indispensable to ensure that automated vehicles can react appropriately to other road users. This paper aims to provide an overview of observation methodologies capable of quantifying different aspects of road user behavior. An observation study of a T-junction in German rush-hour traffic using a ground based LiDAR and an HTML app for manual observations is presented. In normal traffic conditions drivers follows their road prioritization quite strictly. Once congestions emerge on the main road this behavior changes: prioritized drivers yield their right of way to let other vehicles turn onto the main road by increasing gap sizes, sometimes accompanied by flashing their headlights to inform the turning driver of the yielding behavior. Overall, the method of using a ground based LiDAR in combination with human observers has a high potential to quantitatively describe perceivable traffic occurrences while complying with data privacy laws.

Keywords: Traffic observation · LiDAR · Traffic interaction
Automated vehicle interaction · Observation app

1 Introduction

Urban traffic consists of complex situations: drivers have to pay attention to different types of road users but also have to adapt to volatile traffic conditions. In an ideal world, traffic rules would suffice to regulate individual transport in every situation, but the chaotic nature of dense traffic situations requires drivers to adapt to reach their goal safely. The first rule of the German road traffic act [1] stipulates, “The participation in road traffic demands constant attention and mutual regard”. As mutual regard is almost impossible to operationalize, it is indispensable to understand how drivers react to varying traffic situations and obstacles but also how they communicate with other road users in these situations. This will enable future automated vehicles to understand the

S. Bagnara et al. (Eds.): IEA 2018, AISC 823, pp. 537–542, 2019.
https://doi.org/10.1007/978-3-319-96074-6_56

intention behind other drivers' maneuvering and explicit communication including headlights, honks or indicator lights.

Interactions in traffic are defined within the EU Horizon 2020 Project interACT as the "complex process where multiple traffic participants perceive one another and react towards the continuously changing conditions of the situation resulting from actions of the other traffic participant (TP), to achieve a cooperative solution. These actions and reactions involve various means of communication" [2].

Analyzing interaction in urban traffic is a complicated task, though – numerous factors influence the behavior of traffic participants. Furthermore, several aspects of a road user's behavior and the underlying traffic scenario need to be considered. This paper aims to compare several methods to observe interactions in urban traffic and presents insights into an observation study conducted in Germany.

2 Observing Urban Traffic

Understanding the behavior of traffic participants is required for different research areas, ranging from measuring traffic flow to optimize road layouts to controlled experiments focusing on the behavior of drivers in certain situations. To observe interaction in urban traffic, we propose to operationalize the behavior of human road users in a simplified way using the following three aspects:

- Kinematic movements of individual traffic participants,
- Perceivable actions and interactions used for communication (such as head rotation or gestures), and
- Individual psychological behavior of individual road users.

Methodologies to observe these individual aspects depend on the underlying research questions. Preceding studies reveal several methods that could be utilized to acquire data from urban traffic.

Analyzing and modelling kinematic movements can be accomplished by deploying stationary cameras capturing a wide angle from an elevated altitude in combination with tracking algorithms [3] or LiDAR systems [4]. If estimates of velocity ranges and the approximate positions of traffic participants are sufficient, observation protocols could describe an underlying scenario [5], but for accurate quantitative measures, methods that are more sophisticated are required.

Observing actions and communication methods of road users usually requires more information from a ground level, as drivers can rarely be deciphered from bird's-eye perspective. Ground based videos [6] and observation protocols [7–9] enable capturing implicit cues (such as head rotations and gaze directions) and explicit signals (gestures or flashing of headlights) of different traffic participants.

Studies that aim to observe thought processes or motives need to address individual traffic participants. Road users could answer questionnaires to reflect how they perceived a specific situation and provide insights on designing automated vehicles [10]. Controlled field experiments with drivers or pedestrians are another way to understand the thought processes of road users [11, 12].

Using multiple of the aforementioned methodologies enables to combine the observed data creating a holistic understanding of interaction in urban traffic. However, the increasing demand on data privacy (not least because of the newly operative General Data Protection Regulation (GDRP) [13]) limits the methods observing traffic participants without acquiring personal data. While questionnaires could be designed in a way that no personal information is logged, ground based videos might need to distort faces and number plates in runtime. Section 3 describes an approach to acquire kinematic movement data and perceivable road user actions, without generating personal data.

3 Observation of a T-Junction in Germany

A traffic observation was conducted at a T-junction close to Munich, Germany using a stationary LiDAR and an HTML-based protocol to observe interactions between drivers.

The T-Junction consists of one prioritized two-lane road with a speed limitation of 50 km/h and a side-road with a speed limit of 30 km/h merging into the main road. The observation was conducted between 6 and 8 a.m. on three days in December 2017. Overall, 4 h of LiDAR data was acquired, leading to about 4000 observed vehicles with 60 manual observations of explicitly communicated interactions, such as flashing lights or honking. The buildup and dissipation of congested traffic, leading to increasingly cooperative behavior between drivers, was observed (Fig. 1).

Fig. 1. Depiction of the observed traffic scenario. Because of congested traffic in front, the green vehicle yields its right of way to the blue one. (Color figure online)

The study aimed to analyze how traffic participants' kinematic and perceivable behavior changed in interactive situations. This could be achieved by utilizing a stationary LiDAR and observation protocols, which were synchronized in time.

3.1 Methodology – Stationary LiDAR

To generate quantitative measurements of the position, velocity and type of observed road users, a stationary ground based LiDAR system was deployed.

An ibeo LUX LiDAR sensor was connected to a Raspberry Pi and placed in a wooden box. A power supply and solid-state drive (SSD) enabled the box to collect data for up to 12 h without recharging. Observers could connect to the Raspberry Pi using a Wi-Fi connection to control the LiDAR functionality. Furthermore, an integrated webcam allowed to position and align the LiDAR box. The video feed of the webcam was not recorded.

A GNSS receiver was used to synchronize the time on the log files. This way manually observed interactions could be associated with the LiDAR measurements (Fig. 2).

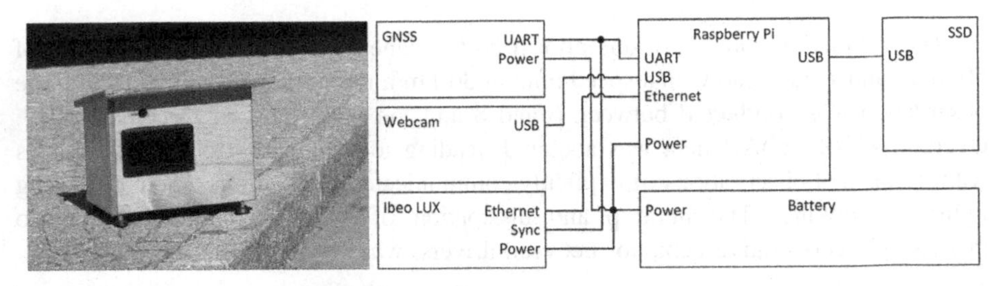

Fig. 2. Functional depiction of the stationary LiDAR

3.2 Methodology – HTML Based Observation App

Within the EU project interACT an HTML based app was developed enabling observers to manually protocol perceived occurrences in traffic. The app ran on a 10′ tablet PC (ASUS Transformer Book T100HA) and each button press was associated with a sequence number to recreate the order of inputs. Furthermore, the device's time was logged to match the manual observations with the LiDAR data.

To reduce input mistakes and enhance user experience, the buttons were designed to be at least 7 mm high. The app was divided into three tabs, which the observers could switch by pressing the left or right on-screen arrow. The three tabs protocolled the following observations:

- Analysis of two interacting vehicles and their drivers, providing buttons to describe the movements and used signals of both vehicles as well as the head and hand movements of the two respective drivers.
- General information, such as weather, the vehicle types and an input box to describe further observations.
- Sketch of the observed encounter, to easier match the manual observations with the LiDAR data.

The app saved all inputs as a .csv file locally, reducing the time to prepare the data for analysis. Observers were trained to use the application prior to the study, reducing the time to fill in the protocols while increasing comparability. All interactions, where explicit communication by the drivers occurred, were protocolled if possible.

3.3 Insights into Results

In normal, uncongested traffic, road users on the main road were not decelerating to let a turning vehicle merge. Therefore, the driver on the side road initiated the merging process, whenever an inter-vehicle gap was deemed large enough. In many cases, trailing drivers on the main road had to decelerate slightly to match the velocity of the turning vehicle, but the observers have perceived no warning signals.

The increased traffic density of the morning rush hour led to congestions on the main road. As the mean velocity on the main road was gradually decreasing to values below 20 km/h, drivers became more cooperative and let vehicles from the side road turn. Cooperative drivers increased their gap size to the leading vehicle by coasting, in some cases accompanied by flashing the headlights signaling their yielding behavior to the driver on the side road. Once one driver let another vehicle merge, following drivers on the main road repeated that behavior leading to a zipping procedure comparable to lane merges on highways. In few cases, a driver on the side road did not immediately accept a created gap. In these encounters the driver on the main road usually flashed his headlights (again) with his vehicle slowly coasting towards the intersection. A repeated omission of the merging driver to turn onto the main road lead to the other driver accelerating, thus terminating the possibility to turn.

The observed strategies – increasing a gap and doing so while flashing headlights – both lead to the turning of vehicles on the side road. This indicates that drivers already anticipate the intent of other drivers in these situations without requiring addition information. However, there might be positive effects on the time drivers on the side road take to merge.

4 Conclusion and Outlook

This paper gave an overview of which observation method to use to obtain specific results when observing interactions in urban traffic. Furthermore, an observation study was presented combining a LiDAR with manual observations using a tablet PC. While the data analysis of the observation is incomplete at the time of writing, we expect that velocity thresholds for cooperative driving behavior when letting someone merge from a side road can be formulated. Moreover, comparing the velocity profile of cooperating drivers who flashed their lights to those who did not, will show, whether explicit communication has any impact on the merging process. Overall, the described method is a good way to observe traffic in fixed locations without generating personal data.

Understanding how human road users behave and interact in complex traffic scenarios will help designing automated vehicles. Expectation conforming kinematic behavior will likely reduce ambiguity increasing traffic flow and safety.

Acknowledgements. This work is a part of the interACT project. interACT has received funding from the European Union's Horizon 2020 research & innovation programme under grant agreement no 723395. Content reflects only the authors' view and European Commission is not responsible for any use that may be made of the information it contains.

References

1. §1 I StVO. Original text in German: "Die Teilnahme am Straßenverkehr erfordert ständige Vorsicht und gegenseitige Rücksicht". https://www.gesetze-im-internet.de/stvo_2013/__1.html. Accessed 28 May 2018
2. interACT deliverable D1.1. https://www.interact-roadautomation.eu/wp-content/uploads/. Accessed 28 May 2018
3. Saunier N, Sayed T, Ismail K (2010) Large scale automated analysis of vehicle interactions and collisions. Transp Res Rec J Transp Res Board 2147:42–50. https://doi.org/10.3141/2147-06
4. Tarko AP, Ariyur KB, Romero MA, Bandaru VK, Lizarazo CG (2016) TScan: stationary LiDAR for traffic and safety studies—object detection and tracking (Joint Transportation Research Program Publication No. FHWA/IN/JTRP-2016/24). West Lafayette, Purdue University, IN. http://dx.doi.org/10.5703/1288284316347
5. Fuest T, Sorokin L, Bellem H, Bengler K (2017) Taxonomy of traffic situations for the interaction between automated vehicles and human road users. In: Stanton N (eds) Advances in human aspects of transportation, AHFE. Advances in intelligent systems and computing, vol 597. Springer, Cham
6. Rasouli A, Kotseruba Y, Tsotsos J (2017) Agreeing to cross: how drivers and pedestrians communicate. In: 2017 IEEE intelligent vehicles symposium (IV), Los Angeles, CA, pp 264–269. https://doi.org/10.1109/ivs.2017.7995730
7. Imbsweiler J, Palyafári R, Puente León F, Deml B (2017) Untersuchung des entscheidungsverhaltens in kooperativen verkehrssituationen am beispiel einer engstelle. Automatisierungstechnik 65:477–488. https://doi.org/10.1515/auto-2016-0127
8. Vollrath M, Huemer AK, Teller C, Likhacheva A, Fricke J (2016) Do German drivers use their smartphones safely? - Not really! Accid Anal Prev 96:29–38. https://doi.org/10.1016/j.aap.2016.06.003
9. Šucha M (2014) Road users' strategies and communication: driver pedestrian interaction. In: Proceedings 5th conference transport solutions from research to deployment, transport research Arena, Paris
10. Merat N, Louw T, Madigan R, Wilbrink M, Schieben A (2018) What externally presented in-formation do VRUs require when interacting with fully automated road transport systems in shared space? Accid Anal Prev. https://doi.org/10.1016/j.aap.2018.03.018
11. Portouli E, Nathanael D, Marmaras N (2014) Drivers' communicative interactions: on-road observations and modelling for integration in future automation systems. Ergonomics 57(12):1795–1805. https://doi.org/10.1080/00140139.2014.952349
12. Connelly ML, Conaglen MH, Parsonson BS, Isler RB (1998) Child pedestrians' crossing gap thresholds. Accid Anal Prev 30(4):443–453. https://doi.org/10.1016/S0001-4575(97)00109-7
13. General data protection regulation. https://eur-lex.europa.eu/legal-content/EN/TXT/PDF/?uri=CELEX:32016R0679&from=EN. Accessed 28 May 2018

Aerospace Human Factors and Ergonomics

Seat Comfort Evaluation Using Face Recognition Technology

Flavia Renata Dantas Alves Silva Ciaccia[1](✉), Jerusa Barbosa Guarda de Souza[1], Alicia Mora[2], Maria Pocovi[2], Guillermo Dorado[2], and Alejandro Lavale[2]

[1] Embraer SA, São José dos Campos, Brazil
fla_silva@hotmail.com
[2] Emotion Research Lab, Valencia, Spain

Abstract. One of the difficulties inherent to comfort assessment is to translate comfort perception into quantifiable variables in order to measure it and use this result to improve seat comfort. This study describes the opportunities of using facial expressions recognition technology to compare comfort perception of two aircraft seats installed in a representative environment. Facial expressions are one of the most apparent ways to capture emotions and it is known that there are six basic emotions which are universal throughout human cultures: fear, disgust, anger, surprise, happiness and sadness. Twenty-one subjects (18 males and 3 females) participated in this experiment and have their faces recorded while using the seats and being asked some questions. The recordings obtained were posteriorly analyzed by Emotion Research Lab facial recognition technology to obtain an emotional analysis of the facial expressions displayed by the participants during the experiment. The facial expressions recognition software of Emotion Research Lab captures the facial micro expressions and uses them to predict the behavior of the participants through the calculation of different metrics such as activation, engagement, satisfaction, valence, relevance and enjoyment. The results showed that seat 1 was better rated by participants and had emotional congruence with their answers. The most important finding was that even subtle differences in seats could be perceived in participants' emotions, suggesting that the use of facial expressions recognition technology to compare comfort perception of aircraft seats is viable and should be better explored during seat development process.

Keywords: Seat comfort · Face recognition · Emotions

1 Introduction

1.1 Seat Comfort Evaluation

Over the past 40 years, passenger comfort has been well studied in different means of transportation [1]. Considering that passengers are seated most of the time on a flight, the passenger seat has an important role to play in meeting the passenger's comfort expectations [2]. Zhang, Helander, Drury (1996) believe that comfort and discomfort are different entities and the absent of discomfort does not mean comfort. In their study,

S. Bagnara et al. (Eds.): IEA 2018, AISC 823, pp. 545–554, 2019.
https://doi.org/10.1007/978-3-319-96074-6_57

discomfort is associated with physical constraints and can be measured through joint angles, tissue pressures, muscular contractions and block of blood circulation. On the other hand, comfort is associated with well-being and aesthetic aspects. Some tools and methods have been investigated to reduce uncertainties of subjective evaluations such as pressure mapping, electromyography, activity analysis and some physiological measures. However, one of the difficulties inherent to comfort assessment is to translate comfort perception into quantifiable variables in order to measure it and use this result to improve seat comfort.

Previous studies have provided findings regarding an individual's affective reaction when exposed to colors, shapes, facial expression and movie scenes [4]. More recently, automotive industry has explored application of face recognition to recognize emotions and mood from drivers in order to prevent car accidents [5]. However, any study that explores the use of facial recognition to evaluate comfort perception. Therefore, this study describes the opportunities of using facial expressions recognition technology to compare comfort perception of two aircraft seats installed in a representative environment.

1.2 Emotions and Face Recognition

Facial emotions are one of the most useful, natural and no conditioned human expression and communication. Emotions play important role in every situation of our lives, from product test to human-computer interaction. An emotion is a complex psychological state that involves three distinct components: a subjective experience, a physiological response, and a behavioral or expressive response. Some researchers have tried to identify and classify different types of emotions and discovered that almost everyone can produce and recognize the associated facial expressions of these emotions, what led them to the assumption that emotions are universal [6]. Psychologist Paul Ekman pointed out that there are six basic emotions which are universal throughout human cultures: fear, disgust, anger, surprise, happiness and sadness [7]. Later, he expanded the list to add a number of other basic emotions including embarrassment, excitement, contempt, shame, pride, satisfaction, and amusement. Many other psychologists have proposed different classifications and claimed that certain emotions are more basic than others for very different reasons, however this manuscript relied on Paul Ekman's definition.

A new generation of facial expression technologies has emerged being fully automated and computer-based. The automatic facial coding technologies use cameras embedded in laptops, tablets, and mobile phones or standalone webcams mounted to computer screens to capture videos of respondents as they are exposed to content of various categories [8].

Emotion Research Lab developed high accuracy machine learning algorithms that detect multi-faces and translate them numerically into emotions. The process starts with face detection in a video or image. The second step is the extraction of face features such as image characteristics, geometric feature-based and appearance-based, using local binary pattern operator (LBP) to compound a main feature vector. The third step is facial coding classification, where the main feature vector will be compared with an internal library. Emotion Research Lab library research both feature extraction in different scenarios in real-time reaching high accuracy of emotions recognition applying different algorithms and techniques to get high real-time performance and accuracy [9].

The emotional facial recognition library of Emotion Research Lab translates facial micro-expression into six basic emotions established by Ekman (2003) and neutrality, considered the absence of emotions. Using the same techniques to detect emotions, this technology has also the capability to recognize age, gender and ethnicity with low performance impact [9] (Figs. 1 and 2).

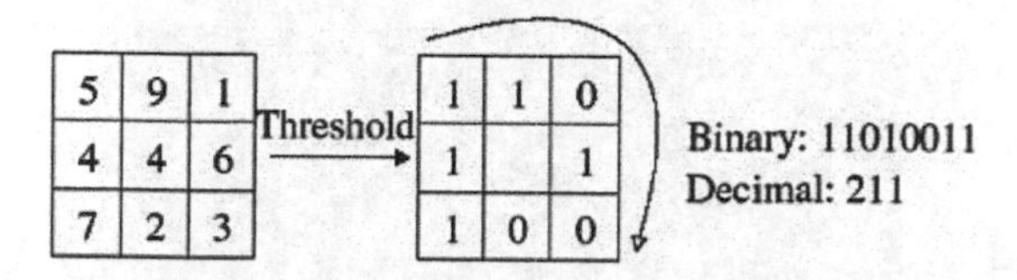

Fig. 1. Local binary pattern operator [9]

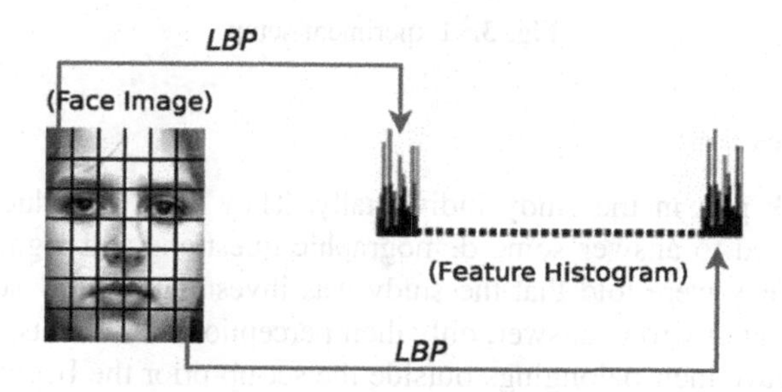

Fig. 2. Local binary features vector [9]

2 Methodology

2.1 Experimental Setup

Twenty one subjects (18 males and 3 females) participated in this experiment. They were aged between 23 and 35 years old.

A physical aircraft cabin mock-up was used in this study (Fig. 3). The mock-up comprised of two rows of seat installed back to back. In order to eliminate any bias that may appear by seat design, color and shape, the seats were covered with a black sheet and stayed like this the entire experiment. There were also 200 cm of space between the front part of the seat and the side walls of the mockup.

An iPad mini was installed in the wall in front of each seat and was used to video record participants face while testing the seat and while being asked questions about different seat characteristics.

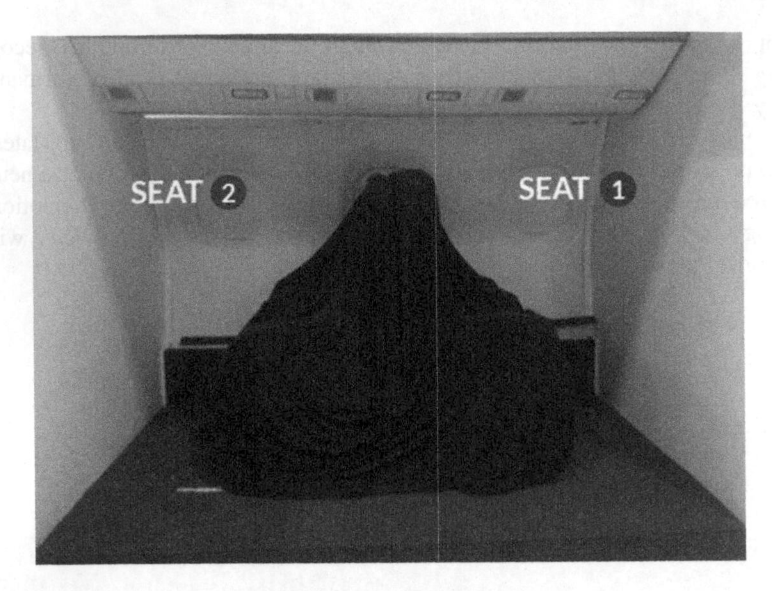

Fig. 3. Experiment setup

2.2 Procedures

Participants took part in the study individually. They were introduced to the study before being asked to answer some demographic questions and signing an informed consent form. They were told that the study was investigating new software and that there was any right or wrong answer, only their perception of the seats. The participants were asked to leave their belongings outside the set-up prior the beginning of the test.

The experiment was divided into 2 main phases. The first phase consisted on first sight comfort test. The participants were invited to seat in a pre-selected seat and the research wait about 20 s before starting talking with them. The idea was to give participants time enough for them sat on the seat and spontaneous testing the overall comfort and characteristics while being recorded.

The second phase consisted on a survey with eight total questions about different characteristics of the seat. The questions were:

1. What do you think of the seat? (Spontaneous response)
2. What do you dislike the most? (Spontaneous response)
3. We would like you to rate some of the features of the seat according to the scale that we are going to show you (1 to 5, being 1 poor and 5 great):
   a. Seat depth
   b. Seat height
   c. Distance between armrests
   d. Headrest height
   e. Body height
   f. Backrest angle
4. Would you travel in this seat for 3 h? (Spontaneous response)

After evaluating one seat, the participants were invited to evaluate the other seat, following the same procedures stated above. The sequence of seat evaluation was randomized in order to minimize bias.

3 Results

The results are divided into first and second phase of data collection. The analysis aimed to find which of the seats was the preferred at an emotional level and which characteristics of the seats were the best according to the participants and their emotional responses. The Fig. 4 shows an example of participant´s face with the virtual mesh and the results of emotions and subemotions provided by the software of Emotion Research lab.

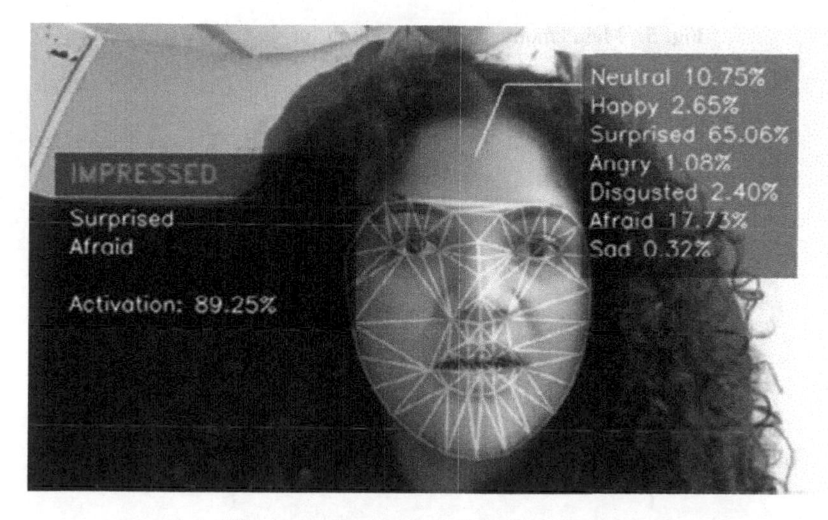

Fig. 4. Example of result of face recognition

3.1 First Phase

The objective was to compare the spontaneous emotions produced by each seat. For the analysis it was used main sub emotions and the metrics obtained. The results showed that the overall impression about both seats was better for the seat 1 during this first evaluation. Seat 1 was also the preferred by the participants when giving spontaneous responses about the seats, as can be seen at Fig. 5. The participants gave better overall opinions about seat 1, showing also emotional congruence with their answers.

Seat 1 is shown to be the preferred seat by the participants at an emotional level during the first 20 s. During the test time, the seat produced significantly higher levels of Satisfaction, more positive levels of Activation and higher levels of Enjoyment, while producing also higher levels of Engagement (see Fig. 6).

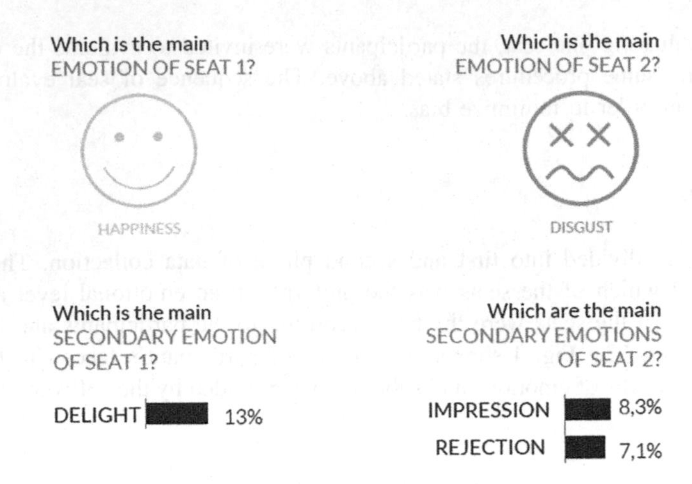

Fig. 5. Main emotion and subemotion for first phase

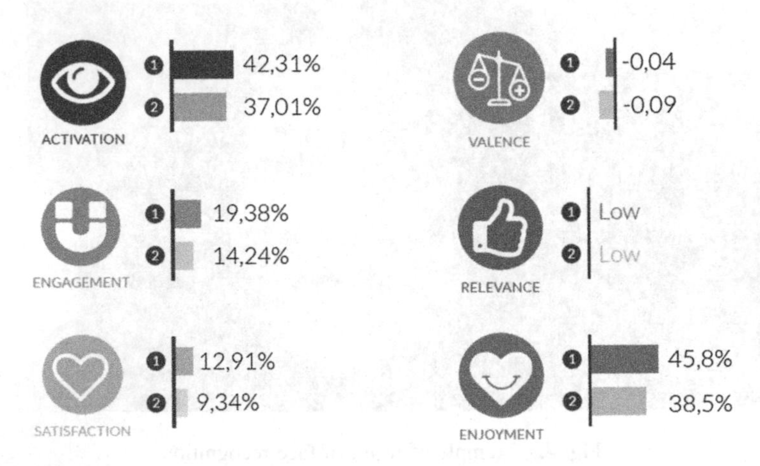

Fig. 6. Summary of results for first phase

The higher levels of Activation indicate better levels of emotional arousal for the seat, while the Valence and the Enjoyment levels indicate that this emotional arousal was more positive. The Relevance of the stimulus is considered "low" due to the neutral Valence levels and the relatively low levels of Activation in both seats, which are perfectly normal taking in account the "calmed" nature of the stimulus analyzed.

Seat 2 is shown to be the least preferred seat by the participants at an emotional level during the first 20 s. During the test time, the seat produced lower levels of Satisfaction, Activation, Enjoyment and Engagement. A lower level of Activation with lower levels of Enjoyment and a slightly worst Valence can be seen for the seat 2. This indicates us that the testing experience during the first 20 s was significantly worse for this seat than for the first seat.

Also, we can see that the two main subemotions related are Impression and Rejection, being Impression a subemotion that enhances the effect of the second subemotion, which in this case is Rejection.

3.2 Second Phase

In this part of the study participants answered several questions about the different parts of the seat and their personal opinion about them. The first question was broad and asked about the general perception of the seat and the second focused on negative aspects of the seat. The Fig. 7 shows the results for these two questions.

WHAT DO YOU THINK ABOUT THIS SEAT?

Seat	Answer	Count
SEAT 1 ECSTASY	- COMFORTABLE	12
	- IT'S OK	6
	- BETTER THAN THE OTHER	3
	- WORSE THAN THE OTHER	2
SEAT 2 ECSTASY	- COMFORTABLE	3
	- IT'S OK	8
	- BETTER THAN THE OTHER	4
	- WORSE THAN THE OTHER	3
	-TOO HARD	2

WHAT DO YOU DISLIKE THE MOST?

Seat	Answer	Count
SEAT 1 DISGUST	- HEADREST	3
	- BACKREST	7
	- ARMREST	4
	- OVERALL SEAT	6
SEAT 2 REJECTION	- HEADREST	1
	- ARMREST	7
	- TOO HARD	4
	- BACKREST	1
	- OVERALL SEAT	7

Fig. 7. Results for the questions 1 and 2

The emotions associated with each seat represent the mean of the subemotions captured by Emotion Research lab software for all participants. On the right side of each seat number, it can be seen the participant´s subjective answers and the number of occurrence of each answer.

The seat 1 was better qualified than the seat 2 when the participants were asked about their overall opinion, moreover, the participants where sure about the opinion they were giving, as shown by the positive subemotions. The seat 2 was the most disliked seat of the two. Also, the participants show an emotional congruence while giving their opinions of what they dislike the most, as showed by their sub-emotional reactions.

The question 3 evaluated the participants' satisfaction regarding different parts of the seats. The Fig. 8 summarizes the mean of answers for this question and shows the emotion and the subjective response for each seat in each seat part. The percentage of T2B is the percentage of satisfaction on subjective evaluation and represents the sum of two higher answers of the scale, i.e. good and great. The percentage of satisfaction was higher for the seat height, the armrest distance and the headrest height of seat 1, while it was higher for the seat depth, body height and backrest angle of seat 2. Participants also showed more emotional congruence when evaluating the characteristics of seat 2, meaning that they were surer about their opinions of this seat.

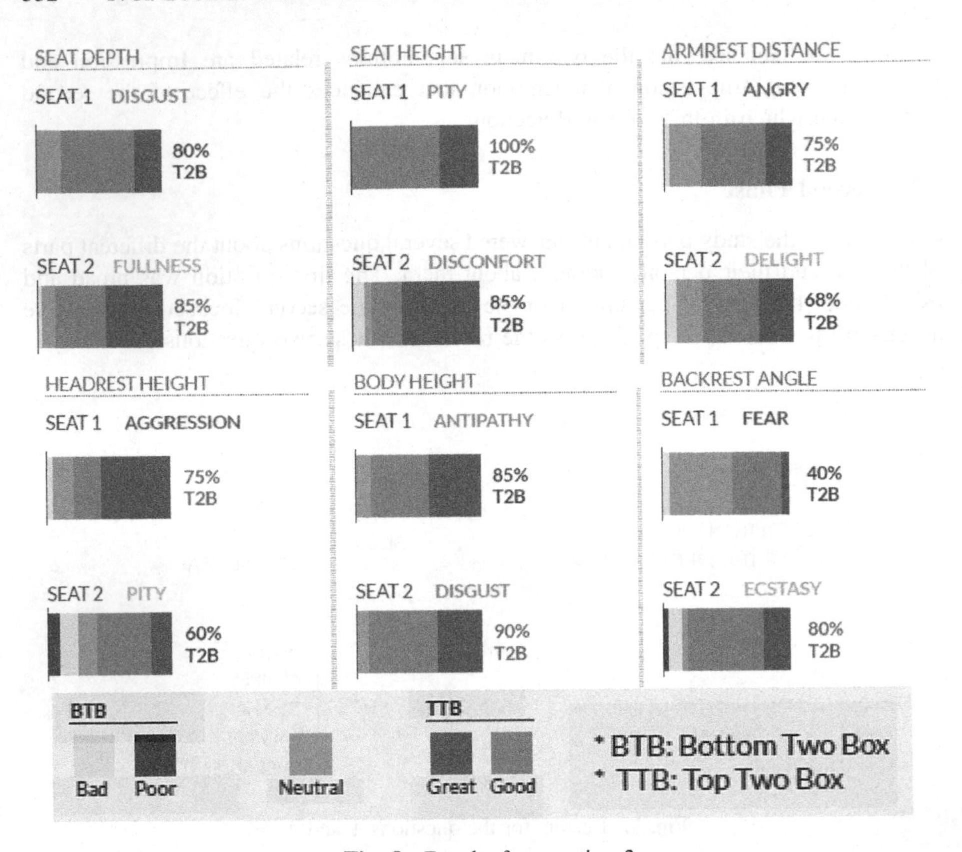

Fig. 8. Results for question 3

The seat 2 depth was slightly better evaluated than the seat's 1 depth. Also the main subemotion related with the seat 2 was fullness, which shows congruence between the verbal opinion given by the participants and their emotions.

Seat's 1 height was the preferred among the participants. The participants showed negative subemotions with both seats, meaning that they were not 100% sure about the opinion they were giving.

Seat's 1 armrest distance was the preferred by the participants. The participants showed negative subemotions while testing the armrest distance of seat 1, while showed a more positive subemotion while testing the armrest distance of seat 2, which means that they were surer when giving the score of seat's 2 armrest distance.

Seat's 1 headrest height was the preferred among the participants. The subemotions related with both seats are negative, indicating that the participants were not sure about the opinions they were giving.

Seat's 2 is shown to be slightly preferred when the participants were asked about the body height of the seats. The emotional analysis of both seats show negative subemotions related to this feature, meaning that the participants were not sure about the opinion they were giving.

Seat's 2 backrest angle was the preferred among the participants. The subemotions related with the analysis of the participants while giving their opinions of seat 2 showed great levels of emotional congruence, meaning that they were sure that they preferred seat's 2 backrest angle.

The question 4 tried to capture the participant´s willing of using this seat in a future flight. The participants liked more the seat 1 when they were asked about travelling for 3 h in the seat. The level of positive subemotions related while answering this question is higher for seat 1 than for seat 2 also, indicating that the emotional congruence with the verbal responses was slightly higher for seat 1, as can be seen at Fig. 9.

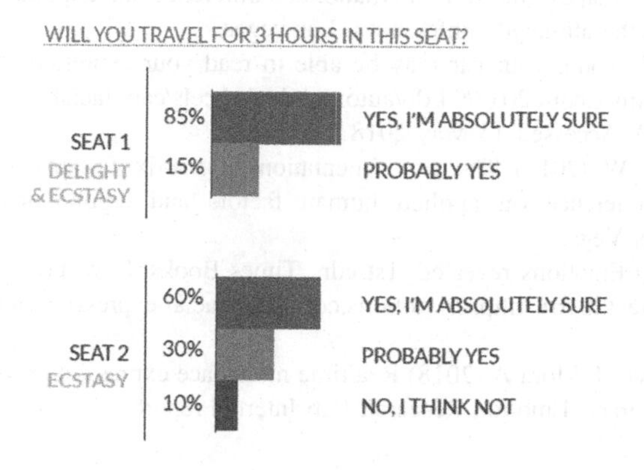

Fig. 9. Result for question 4

4 Comments and Conclusions

The paper presented a pilot study of the use of face recognition to compare the comfort perception of two different seats.

The results showed that the seat 1 was better rated by participants and also had emotional congruence with their answers. The most important finding was that even subtle differences in seats could be perceived in participants' emotions, suggesting that the use of facial expressions recognition technology to compare comfort perception of aircraft seats is viable and should be better explored during seat development process.

It was also possible to verify that the technology allow the identification of perception of a seat without asking a direct question. This is a huge advantage when the subject is comfort, a subjective construct that is difficult to measure. However, the pilot test suggested that the experiment shall be designed carefully in order to not provide false results. It is important to elaborate a consistent questionnaire and take care to not influence the participant during the tests.

More studies should be performed in order to understand better the limitations of the technology. This type of evaluation could also be used in different contexts, for example, to evaluate the client´s perception when entering in an aircraft in a trade show or demonstration.

References

1. Oborne DJ (1978) Techniques available for the assessment of passenger comfort. Appl Ergon 9(1):45–49
2. Tan CF, Chen W, Rauterberg M (2009) Development of adaptive aircraft passenger seat system for comfort improvement. In: International conference for technical postgraduates, Kuala Lumpur, Malaysia, paper no. CC02-05, 14–15 December 2009
3. Zhang L, Helander MG, Drury CG (1996) Identifying factors of comfort and discomfort in sitting. Hum Factors 38(3):377–389
4. Abegaz T, Dillon E, Gilbert JE (2015) Exploring affective reaction during user interaction with colors and shapes. In: 6th international conference on applied human factors and ergonomics and the affiliated conferences, Las vegas
5. Quain JR (2017) Soon, your car may be able to read your expressions. New York Times. https://www.nytimes.com/2017/04/06/automobiles/wheels/cars-facial-recognition-expressions.html. Accessed 15 May 2018
6. Shin D, Zheng W (2015) The experimentation of matrix for product emotion. In: 6th international conference on applied human factors and ergonomics and the affiliated conferences, Las Vegas
7. Ekman P (2003) Emotions revealed, 1st edn. Times Books, New York
8. Imotions website (2016). https://imotions.com/blog/facial-expression-analysis. Accessed 10 May 2017
9. Millan D, Pocovi M, Mora A (2018) Realtime multi-face expression recognition on Emotion Research Lab library. Emotion Research Lab Internal report

Human-Centered Design of a 3D-Augmented Strategic Weather Management System: First Design Loops

Sebastien Boulnois(✉) and Lucas Stephane(✉)

Florida Institute of Technology, Melbourne, FL 32901, USA
sboulnois@my.fit.edu, lstephane@fit.edu

Abstract. Convective weather is one of the main causes of accidents in the National Airspace System according to the Federal Aviation Administration. This paper describes how the use of Human-Centered Design principles, along with Design Thinking, led to the design and development of a 3D-Augmented Strategic Weather Management System aiming to determine how strategic weather information presented both in 2D and 3D could impact pilots' weather situation awareness and decision-making capabilities. This paper explains the five design iterations that were carried out over four years, including several knowledge elicitation, participatory design and evaluation sessions. The results are positive overall; pilots' feedback is very rich and meaningful and will be implemented in the next prototypes. Further human-in-the-loop simulation evaluations will be conducted for consolidating usability and for evaluating pilots' weather situation awareness and decision-making capabilities.

Keywords: Human-Centered Design · Weather · 3D visualization

1 Introduction

While flying in the atmosphere, pilots have to constantly deal with weather. In most situations, weather allows them to conduct their flights safely and efficiently. However, weather can be an indirect or direct cause of delays, incidents or even accidents. Federal Aviation Administration (FAA) statistics between 2002 and 2017 show that 67.4% of delays in the National Airspace System (NAS) involve weather (FAA Operations & Performance Data, n.d.). In addition, 23% of accidents in the NAS also involve weather (National Transportation Safety Board, n.d.). One of the primary causes appears to be convective weather along with visibility and winds, especially during the summer (FAA, n.d.).

Convective weather is characterized by the vertical transport of dry or moist air. Clouds form because the moist air condenses as it rises and cools down; thunderstorms (i.e. rain showers associated with thunder) can often result from this process. Table 1 provides an overview of the main categories of thunderstorms along with their life cycle and associated hazards. Single-cell thunderstorms are not usually a threat for pilots; however, the association of several cells forming more complex systems can be critical due to visibility issues, hail, strong winds (updrafts, downdrafts) and even

S. Bagnara et al. (Eds.): IEA 2018, AISC 823, pp. 555–575, 2019.
https://doi.org/10.1007/978-3-319-96074-6_58

tornadoes, and the forces can exceed the structural limitations of the aircraft (Aircraft Owners and Pilots Association, n.d.). The National Severe Storms Laboratory (NSSL) indicates that "there are about 100,000 thunderstorms each year in the U.S. alone" and that "about 10% of these reach severe levels" (National Severe Storm Laboratory, n.d.).

Table 1. Types of thunderstorms, lifetime and associated hazards.

Type	Lifetime	Associated hazards
Single-cell	20–30 min	Brief heavy rain and lightning
Multi-cell	Up to many hours	Hail, strong winds, brief tornadoes, and/or flooding
Squall line	As long as new cells form	Heavy rain, hail, frequent lightning, strong, straight line winds, and possibly tornadoes and waterspouts
Supercell	More than one hour	Large and violent tornadoes

So far, a tremendous amount of efforts was carried out toward weather instrumentation and visual representations, both in the cockpit and for air traffic control centers. The current research applied from its start an integrated human-centered and design thinking approach. Human-centered design (HCD), part of the multidisciplinary human systems integration (HSI), involves domain experts and operators from the very early stages of the design cycle. HCD is performed iteratively and incrementally, using agile cycles of design and evaluation. As part of HSI, HCD efforts are also in charge of identifying and selecting the best candidate technologies that are aligned with operators' needs and requirements, including critical situations (Booher 2003). Design thinking (DT) reinforces the human-centered paradigm, emphasizing the need to understand domain experts and operators for making sure that the best needs and requirements are gathered at the very beginning of the design cycle. The DT iterative cycle is composed of five major stages that are: (1) (re)defining the problem, (2) need-finding and benchmarking (e.g. understanding the users and the design space), (3) ideating through participatory design, (4) prototyping, and (5) evaluating and testing (Meinel and Leifer 2011).

The current paper shows how both HCD and DT were performed in practice for refining the design of the proposed 3D-augmented strategic weather management system.

2 Related Work

The authors first reviewed in-progress and operational work related to weather representation systems that are presented in this paper based on their content, presentation, and interaction capabilities (Stephane 2013, 2014).

2.1 Strategic Weather Information Systems

Current systems propose strategic weather information that are essentially in 2D. For example, the National Center for Atmospheric Research (NCAR) designed and implemented such a system (Kessinger et al. 2015). It includes different kinds of weather information such as Cloud Top Height (CTOP) and Convective Diagnosis Oceanic (CDO) technologies with flight path. This information is presented on a 2D map via a bird's eye view using polygons. Pilots may zoom in/out and pan the map to navigate in the environment via buttons.

In 2015, Honeywell released their tablet-based application Weather Information Service that seems to be the state-of-the-art commercialized system for flight management (Honeywell, n.d.). This system provides pilots with strategic weather information such as NEXRAD observations and cloud top. A bird's eye view and a profile view are used to present the weather information in 2D. Most of touch-based tablet interaction controls (e.g. zoom in/out, drag, tap) are available so pilots can easily use system.

2.2 Tactical Weather Information Systems

Other onboard weather systems provide tactical weather information and allow for 2D and 3D visualizations. University of Toulouse (Ecole Nationale de l'Aviation Civile) developed two prototypes of an onboard visualization system for integration within navigation displays (Letondal et al. 2015). Both prototypes provide pilots with weather information coming from satellites. The former presents the information via a 2D bird's eye view and a 2D profile view. The latter uses a 2D bird's eye view and two different 3D views: an "axonometric view" and a "perspective view". In the first prototype, pilots can change the heading of the aircraft model with a virtual slider. The selected virtual heading highlights weather cells it crosses so pilots can figure out an ideal trajectory. With the second prototype, pilots are still able to select a virtual heading by rotating a Griffin Powermate controller but can also project themselves into near future by pressing and rotating it at the same time.

NASA also worked on a Cockpit Situation Display similar to a navigation display (Wu et al. 2013). Their approach consists in providing pilots with "historical weather information along the flight path" and allowing "pilots to visualize this information in a way that will support them in generating their own predictions of future weather development". The system mainly provides weather observations coming from the NEXRAD network and includes a few forecast algorithms in order to give an overview of near future weather to pilots. The system supports both 2D (bird's eye view) and 3D (exocentric view) visualization capabilities. Interaction is provided via 3 sliders for changing the time (from −120 min to +120 min), weather visualization (i.e., possibility to see only red, red and yellow), and the altitude (possibility to choose what altitude you want to visualize potential weather events at).

Both Letondal et al. (2015), and Wu et al. (2013) highlight the impact of 3D representations when it comes to "assess the height and spread of weather and the ability to plan alternative trajectories", "temporal navigation", "follow convective cells [...] during a dense meteorological activity where cells levels and positions vary rapidly". However, these systems have a common limitation: the interaction media and associated interaction controls used limit the full potential of 3D weather representations. Wu et al. (2013) acknowledge that "because of the scale issue, it is probably important that in the NextGen Trajectory-Based Operation environment, pilots have access to features like panning and zooming so that they have a better indication of when their routes will take them within some unacceptable distance to the storms. The ability to pan over to weather impacted portions of their trajectory will allow pilots to examine them at a much lower scale".

3 Progress of the Proposed Weather System Since 2014

Table 2 below traces the five main stages of the design thinking process over time illustrated also in Fig. 1. The name "Onboard Weather Situation Awareness System" (OWSAS) is used to refer to this weather system. In this paper, the different versions of the systems will use the term "OWSAS" followed by a version number (e.g. OWSAS-1, OWSAS-2.1). "OWSAS" by itself will be used when referring to the system in general.

Table 2. Stages of the design thinking process for OWSAS.

Term	Step	Activity
Spring 2014	Ideation 1	Collection of ideas for OWSAS-1
Summer 2014	Prototype 1	Development of OWSAS-1 (Microsoft Surface Pro)
	Evaluation 1	Seven expert pilots – Inside and outside of cockpit simulator
Spring 2015	Prototype 2	Development of OWSAS-2 (Objective-C/iPad)
	Evaluation 2	Three Expert pilots (Participated to evaluation 1)
Fall 2015	Prototype 3	Development of OWSAS-3.1 (Change of the architecture)
Spring 2016	Understanding 1	Knowledge Elicitation (interview with four experts)
	Ideation 2	Research of weather information sources that allows for 3D visualization
	Ideation 3	Knowledge elicitation (Focus group with two expert pilots and the design team on weather representation options)
	Prototype 4	Development of OWSAS-3.2
Summer 2016	Evaluation 3	Five pilots – outside of cockpit simulator
	Prototype 5	OWSAS-3.3 (Swift/iPad)
Spring 2017	Understanding 2	Knowledge Elicitation (interview with one expert pilot)
Fall 2017	Understanding 3	Knowledge Elicitation (interview with two expert pilots)
Spring 2018	Understanding 4	Knowledge Elicitation (interview with eight expert pilots)

The design of OWSAS started in 2014 and is based on both HCD principles and DT approach, i.e. identifying problems (i.e. state the problem(s)), understanding (i.e. empathy toward the people who will use the designed system), ideating (i.e. generating design ideas and selecting design solutions), prototyping (i.e. turning selected design solutions into low/high-fidelity prototypes), and evaluating (i.e. testing the prototypes with end users to gather design feedback) (Meinel and Leifer 2011).

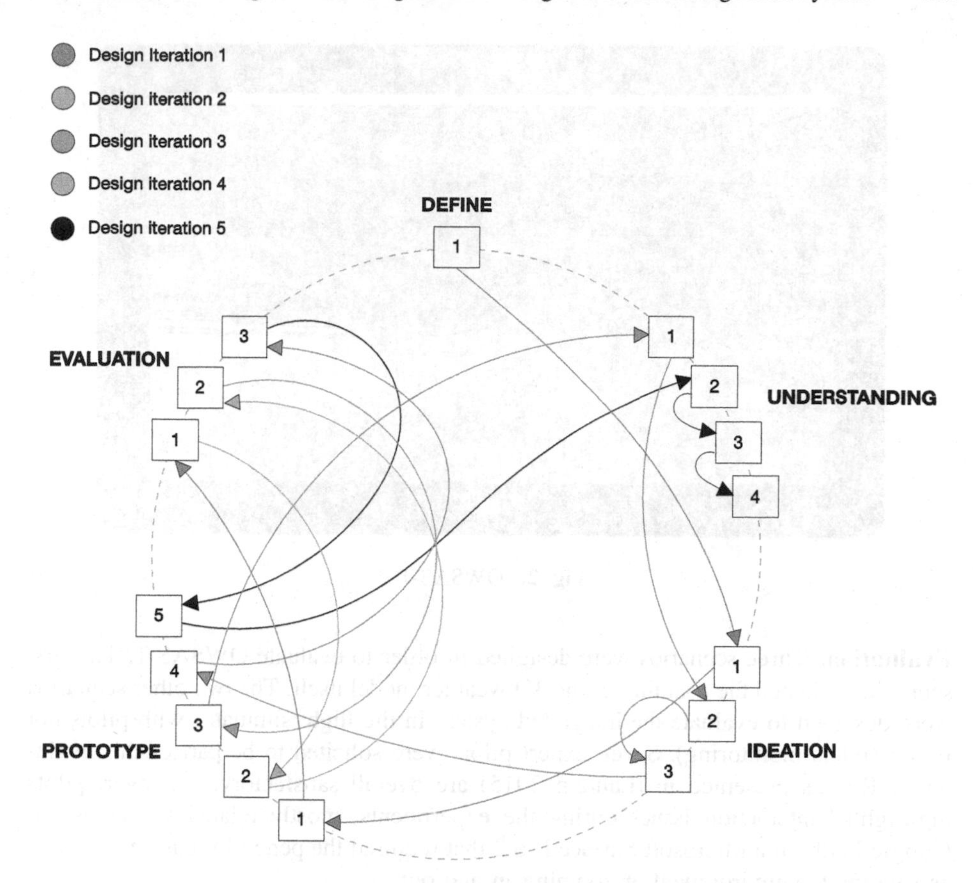

Fig. 1. OWSAS design thinking steps performed based on Meinel and Leifer (2011).

3.1 Prototyping and Evaluation of OWSAS-1

The design of OWSAS was first introduced in 2014 at the Human-Centered Design institute (HCDi), Florida Institute of Technology, as an internship project. The first steps conducted are presented by Laurain et al. (2015) and mainly relate to prototyping and evaluation.

Prototyping. The authors concentrated their work on the construction of the 3D weather model by representing unsafe weather areas based on archives layers from the NEXRAD network as 3D cylinders (Fig. 2). This information along with flight-path information was integrated altogether in Google Earth. A horizontal flight-path modification feature was also added and consisted of providing the pilots with alternative routes around weather perturbations. In order to test the system presentation and interaction features, several devices were tested (i.e. Samsung Galaxy Note, iPad Air, Microsoft Surface Pro 2) and the Microsoft Surface Pro 2 was finally chosen for the evaluation.

Fig. 2. OWSAS-1.

Evaluation. Three scenarios were designed in order to evaluate OWSAS-1. The first scenario evaluated the usability of the 3D weather model itself. The two other scenarios were designed to evaluate the integrated system in the flight simulator with pilots not flying (pilots monitoring). Seven expert pilots were solicited to be part of this evaluation. Results presented in (Laurain 2015) are overall satisfactory. However, pilots highlighted interaction issues during the experiments, mostly related to the use of Google Earth on a Microsoft Surface Pro 2 that required the pencil for basic tasks such as rotating the environment or zooming in and out.

3.2 Prototyping and Evaluation of OWSAS-2

In February 2015, HCDi researchers enhanced OWSAS-1 based on the previous evaluation feedback by focusing on interaction issues highlighted by expert pilots. This led to the implementation of a new prototype (i.e. OWSAS-2) and respective evaluations. This work is presented in (Lang 2015).

Prototyping. Content-wise, two elements were modified. First, the NEXRAD network information layers overlay was removed to lighten the visual scene (Fig. 3). In addition, the horizontal flight-path modification feature was enhanced and the possibility to avoid weather vertically was also added. Nothing was changed presentation-wise. Finally, the system was transferred on an iPad Air to fix interaction issues experienced by pilots on the Microsoft Surface with OWSAS-1.

Evaluation. The main purpose of the second evaluation was to test the new interaction capabilities of the system. Only pilots solicited for OWSAS-1 evaluations participated for consistency purposes. The visual scene was also evaluated with the protocol used in OWSAS-1 evaluations to allow result comparisons. The results are presented in (Lang 2015). The author highlights the fact that "all pilots agreed to prefer the second version

Fig. 3. OWSAS-2.

to the first one, finding it much easier to use." (i.e. OWSAS-2 versus OWSAS-1). In addition, the removal of the NEXRAD network information layers provided more importance to the 3D cylinders and the flight-path as relevant cues for decision-making, although "some pilots requested to have the ability to switch between the two models (layers and cylinders, and cylinders only).".

3.3 Prototyping of OWSAS-3.1

Starting August 2015, the main contribution was the improvement of the OWSAS architecture (i.e. used in OWSAS-1 and OWSAS-2) for enhancing the interactivity with the various scene elements (Boulnois et al. 2016).

OWSAS Architecture. The OWSAS architecture (Fig. 4) used for OWSAS-1 and OWSAS-2 can be found in (Laurain 2014). The presentation of the visual scene requires several actions. Regarding the integration of the weather information, a request has to be sent to NOAA to get the NEXRAD network data. A reference number (i.e. HAS number) is sent by email. The Weather Climate Tool (Ansari, n.d.) is then used to load the visual content of NEXARD network data corresponding to the reference number. This content has then to be saved as a Keyhole Markup language Zipped (kmz) file. This kmz file is then loaded through Google SketchUp (SketchUp, n.d.) 2D/3D modeling tool. The 3D cylinders are manually built based on the NEXRAD weather information layers. The whole content can then be saved as a 3D model. Regarding the integration of the flightpath, a database with the flight waypoints that are needed for the flight is created as an EXCEL file. Then, Earth Point (EarthPoint, n.d.) is used to generate the flightpath according to these flight waypoints. A kmz file of this content is generated. Both 3D models and kmz file are finally loaded in Google Earth.

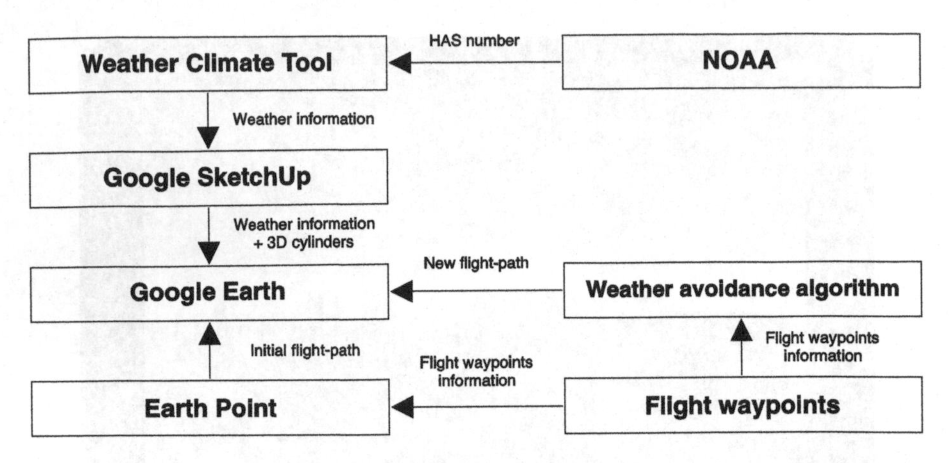

Fig. 4. OWSAS-1 and OWSAS-2 architecture, adapted from Laurain (2014).

Several tools and actions are required in this architecture. In addition, the 3D weather model is built manually based on the NEXRAD network weather archives and 3D cylinders that are not interactive (i.e. information can only be shown are removed). Also, Google announced back in 2015 that Google Earth for tablets would be deprecated, although it is now available again. In other words, this architecture has a significant lack of flexibility. Based on these limitations, a research on available virtual geographic environments was performed to determine what technology would be best to fix the above issues.

Virtual Geographic Environments. Table 3 lists some virtual geographic environments (VGEs) that the OWSAS architecture can be based on for allowing more flexibility. Lots of VGEs are available but are most of the time applications that cannot be customized with proprietary content. Besides, they are not free. Google Earth was kept in the list since it is not deprecated anymore. WhirlyGlobe is a Software Development Kit (SDK) developed by MouseBird (n.d.) running on mobile operating systems (i.e., Android, iOS) providing the possibility of customizing 2D and 3D interactive maps and globes with several features such as shapes (e.g. cylinders, spheres), object models, annotations and so on. Globe3Mobile (n.d.) is supported on Android, iOS and on computer operating systems (HTML5-based applications) and allows for features similar to those offered by WhirlyGlobe. EarthView is an SDK developed by Anderson (n.d.). only available on iOS. It also proposes a 3D interactive globe but seems limited in terms of features (i.e., maps textures that you can use, camera control limitations, and so on).

Table 3. Available VGEs.

	Google earth	WhirlyGlobe	Globe3Mobile	Earth View
Type	Application/SDK	SDK	SDK	SDK
Runs on	Windows/OSX/Linux/IOS/Android	IOS/Android	IOS/Android/HTML5 application	IOS
Environment	2D/3D	2D/3D	2D/3D	3D
Customizable	Yes	Yes	Yes	Yes

WhirlyGlobe and Globe3Mobile seem to be the best 3D environments according to what they propose in terms of content, features, and customization possibilities. Both of them developed a good community network that can be useful in case of problems. Moreover, both of them provide commercial annual support, which means that they can develop new features according to user needs or answer user questions efficiently. Globe3Mobile doesn't provide any documentation to help the user. Therefore, WhirlyGlobe was ultimately selected.

Prototyping. A small prototype (i.e. OWSAS-3.1) was developed on an iPhone in order to test the content, presentation and interaction features offered by WhirlyGlobe that may be used in OWSAS. Figure 5 shows two features that were explored at that time: basic 3D cylinders that represent some weather events, and a round symbol that represents an airport. The difference with OWSAS-1 and OWSAS-2 is that elements are now interactive and dynamic as they can be built programmatically and manipulated as needed.

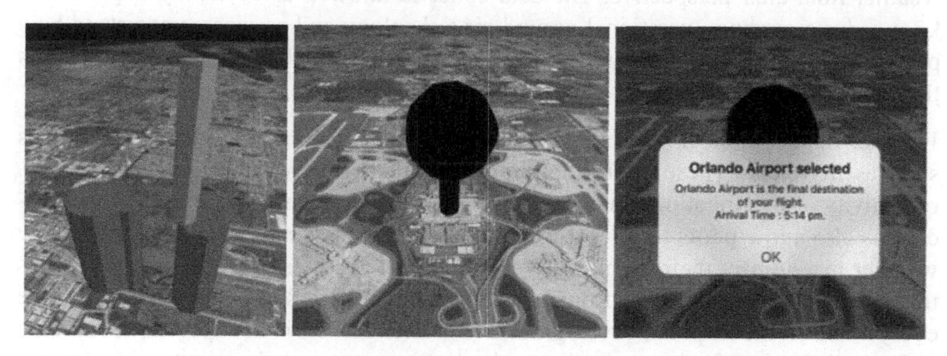

Fig. 5. Weather cylinders (left), airport representation (middle), airport representation selected (right).

3.4 Understanding How Pilots Deal with Weather and Ideation of OWSAS-3.2

In Spring 2016, the design of OWSAS-3.2 was started. First, a phase of knowledge elicitation with expert pilots was conducted to better understand how pilots build weather situation awareness to make weather-related decisions. Next, a brainstorming with the design team was conducted to generate weather representation solutions. Finally, a brainstorming with two expert pilots was conducted to select the best design solution to implement.

Understanding. Interviews were conducted initially with four expert pilots (i.e. airline pilots with a minimum of 2000 h of flight). Currently, eleven additional expert pilots have been interviewed and a grounded theory study is ongoing. The data collected is very rich and can be presented in several ways. Therefore, it will be entirely discussed in another paper. In this paper, we present a summary of the raw critical results. The main research question was stated as follows:

How do pilots get weather situation awareness to make weather-related decisions?

Sub-questions for structuring the answers to the main research question were developed and were stated as follows:

1. What operational ground-air technology and services are available and used by pilots to get weather situation awareness and make decisions?
2. What are the advantages and limitations of these technologies and services?
3. What are the scenarios where these technologies and services limit their weather situation awareness and decision-making capabilities?
4. What weather are pilots the most concerned of, and why?

The main goal of these interviews was to better understand how pilots deal with weather from their perspective. The data collected allowed us to mainly improve our knowledge regarding operational ground-air technology and services, understand how pilots use these technologies and services, and discover challenging weather scenarios that could be used for demonstrating OWSAS capabilities.

Weather Applications. Pilots can obtain weather information via weather applications such as MyRadar. These applications provide pilots with weather information (e.g. reflectivity of precipitation intensity, winds, convective areas) on a map. Some applications also allow pilots to import their flight plan so they can better understand how weather may impact their flight. The main advantage of these weather applications is that the information is coming from ground radars (i.e. NEXRAD) and is then available everywhere on the map. However, ground radar information is not updated in real-time and usually has a refreshing rate that varies between two and six minutes. In addition, the weather information is presented in 2D, therefore pilots are not provided with vertical information that is key for building a 3D mental model of the weather situation.

Onboard Weather Radar. Airline pilots can use an onboard weather radar during a flight. This radar is basically an antenna that sends pulses ahead of the airplane to detect precipitation. The pulses returns are displayed on the navigation display in the cockpit via a color code based on the precipitation intensity (i.e. green, yellow, red and magenta). Pilots can change what they see on the navigation display by adjusting the vertical tilt (i.e. the antenna can sense weather information above and below the airplane flight level), the range (i.e. how much weather information is displayed) and the gain (i.e. sensitivity of the antenna). The main benefit of onboard radars is that they provide pilots with tactical weather information in real-time. However, they have several limitations. First, the weather picture displayed on the navigation display is reliable for weather situation understanding up to a certain range (i.e. pilots cite a maximum range of use between 80 and 150 nautical miles, although the range capabilities can go up to 640 nautical miles). Secondly, the onboard weather radar can experience the attenuation phenomenon. In other words, the radar antenna is sometimes not capable of going through the first layer of precipitation that pilots encounter in case of severe weather. Consequently, the radar does not provide pilots with weather information that may potentially be behind this first layer of precipitation. This phenomenon impacts pilots' weather perception. Thirdly, the navigation display provides pilots with 2D representations of weather information, although some radars are capable of scanning 3D weather information (e.g. Honeywell IntuVue RDR-4000). Hence, building a 3D mental model of the weather situation via this display is challenging and requires a lot of skills and experience (e.g. correct use of manual tilt and gain). This impacts pilots' weather perception and understanding.

Automatic Terminal Information Service (ATIS). Pilots can get weather information at airports and their surroundings in the United States thanks to ATIS. This service mainly provides meteorological terminal aviation routine weather reports (METARs) that pilots can listen to via the radio or retrieve digitally via the aircraft communication addressing and reporting system (ACARS). ATIS information is generally updated every hour and can be updated immediately after weather changes. Sometimes, pilots use ATIS information from several airports to build more weather situation awareness. The main limitation is that the information is either spoken or typed, making it very challenging for pilots to build a mental model of the weather situation.

Communication with Dispatchers and Air Traffic Controllers. Pilots can ask dispatchers and air traffic controllers for direct or indirect updates regarding weather or can receive them when they are proactive. Dispatchers inform the pilots how the weather situation looks like in an area of interest based on the global weather picture they have access to on their weather information displays. However, this information is often not relevant to pilots as it describes strategic weather information in big areas while they are interested in local weather situations. Air traffic controllers (ATCs) can use other pilot reports to inform pilots on local weather situations. They also have a weather radar but most of the time do not use it as their primary task is to track, seperate airplanes and deviate them from severe weather. Unless pilots request weather information in an area of common interest (i.e. ATCs dealing with other flights in the area close to where pilots are), it can be challenging to get helped by ATCs due to their constant overload.

Pilots' Main Concern. Pilots' main concern is definitely convective weather (thunderstorm systems) and how it evolves in time because it can lead to associated hazards such as severe winds, turbulences, microbursts, hail, tornadoes and so on. In addition, pilots mentioned a few situations where convective weather can be very challenging such as flight phases where an altitude change is required (i.e. climbing, descent), at night when eye visibility is reduced, or in cruise when pilots encounter complex thunderstorm systems that are several hundred miles wide.

Summary. The current operational ground-air technology and services limit pilots' weather situation awareness and decision-making capabilities in some ways. Based on the results of these interviews with 15 expert pilots, four concepts were so far generated and taken into account for the next phase of design to improve this issue: (1) strategic weather information that relates to the availability of weather information all along the flight path and its surroundings; (2) 3D weather representation that refers to the representation of weather information in 3D, associated with relevant interaction capabilities based on the visualization technology that is used; (3) evolution of weather information that relates to the evolution of convective weather with time; and (4) shared situation awareness that relates to presenting the same weather information to pilots, dispatchers and any other NAS agents involved in dealing with aviation weather. The research question that emerged is:

How can strategic weather information, presented both in 2D and 3D, impact pilots' weather situation awareness and decision-making capabilities?

Ideation. The first focus was the representation of 3D weather information. Based on the results from the knowledge elicitation with expert pilots, 3D representations of the reflectivity of precipitation intensity was the first item to take care of (e.g. aligned with previous prototypes) along with winds and turbulence information. Participatory design sessions were carried out with multidisciplinary team members (e.g. human machine systems, aviation, environmental psychology) for identifying representation ideas. Four representation solutions were designed: point cloud, cluster of point clouds, polygons and advanced polygons (Fig. 6). A focus group was then created by adding two expert pilots to the initial team in order to discuss these representations and choose the best design option. Out of the two candidate design options, i.e. point cloud and polygons, the latter was retained because the pilots were not sure if the point cloud option would represent weather situation as it really is. They thought polygons would be a better option since it is very close to the shapes they can see on their onboard weather radars.

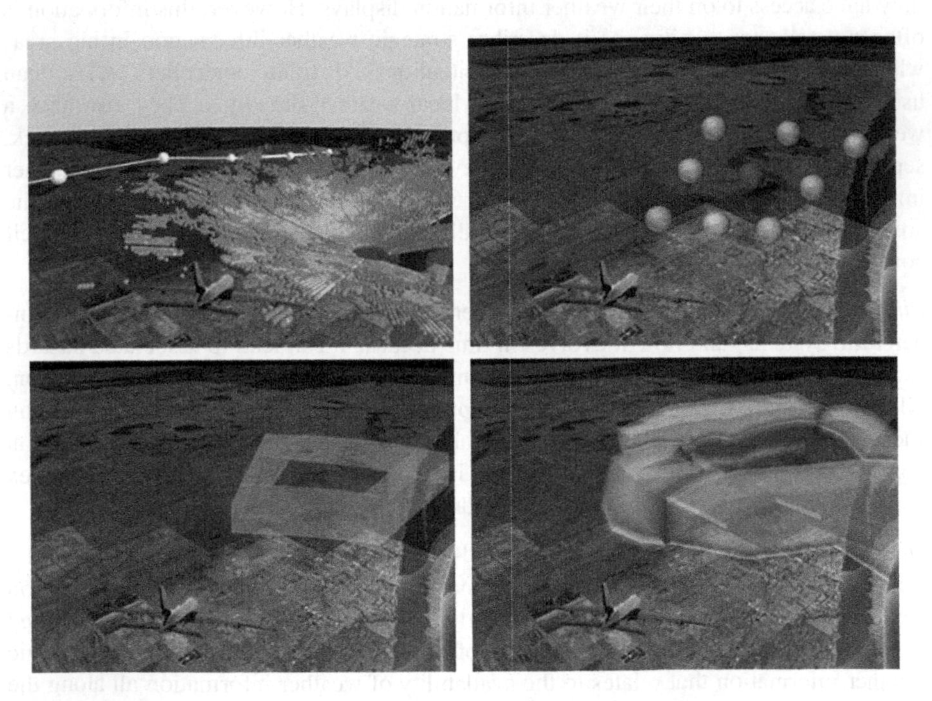

Fig. 6. Weather representation options.

3.5 Prototyping and Evaluation of OWSAS-3.2

In summer 2016, OWSAS-3.2 was developed. All the technical details are available in (Grandin 2016). This prototype was based on an ideal identified system architecture. Most of the efforts were concentrated in the integration of the polygon-based weather representation with a flight plan in the new VGE that was chosen (i.e. WhirlyGlobe). The key points that were generated from the knowledge elicitation, as well as related work on 2D and 3D information visualization were also taken into account to design the prototype. Human-in-the-loop evaluations were then conducted to mainly assess the usability of the system along with pilots' situation awareness and workload.

System Architecture. The ideal OWSAS-3.2 architecture (Fig. 7) is intended to integrate maps with near real-time (e.g. two to six minutes from NEXRAD network) precipitation intensity reflectivity, flight plan (i.e. information coming from PROM-SIM737, a software that can generate flight routes) and airplane information (e.g. 3D location, attitude, speed coming from our B737 cockpit simulator). While near real-time precipitation intensity reflectivity data provided by NOAA was freely available, it was decided to simulate the weather information for feasibility reasons (e.g. scheduled human-in-the-loop evaluations).

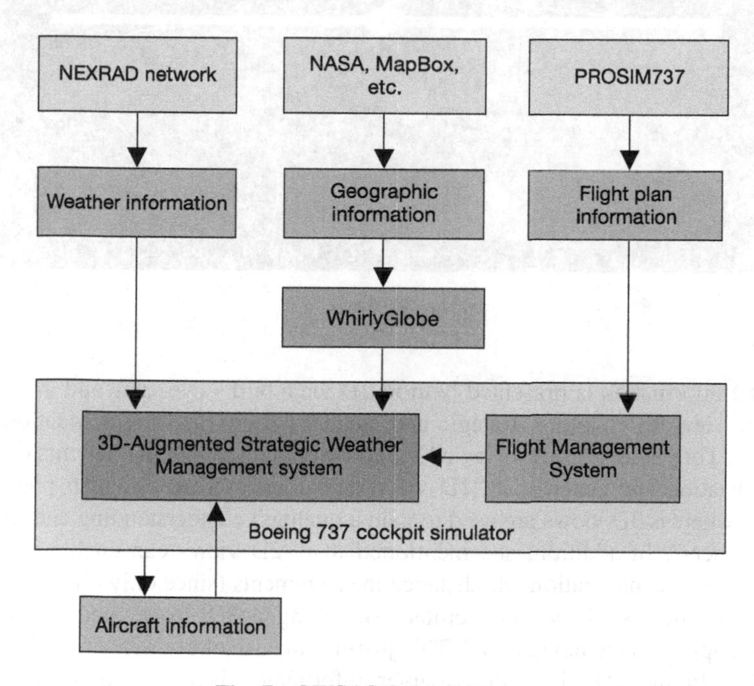

Fig. 7. OWSAS-3.2 architecture.

Content, Presentation and Interaction of OWSAS-3.2. Figure 8 shows an overview of OWSAS-3.2. Three key knowledge elicitation concepts (i.e. strategic weather information, 3D weather visualization, weather evolutions) were implemented in

OWSAS-3.2. WhirlyGlobe was selected as the new VGE to integrate maps, weather, flight path, airplane and camera information altogether. The maps information comes from NASA servers. The weather information is simulated based on radar archives and 3D radar images. The flight path information is generated via the PROSIM737 instruction station that is used to create flight scenarios for our B737 cockpit simulator and was transferred in OWSAS-3.2. The airplane information is collected from the simulator and was available in OWSAS-3 in real time via a third-party program that handled the data transfer (Boulnois et al. 2015).

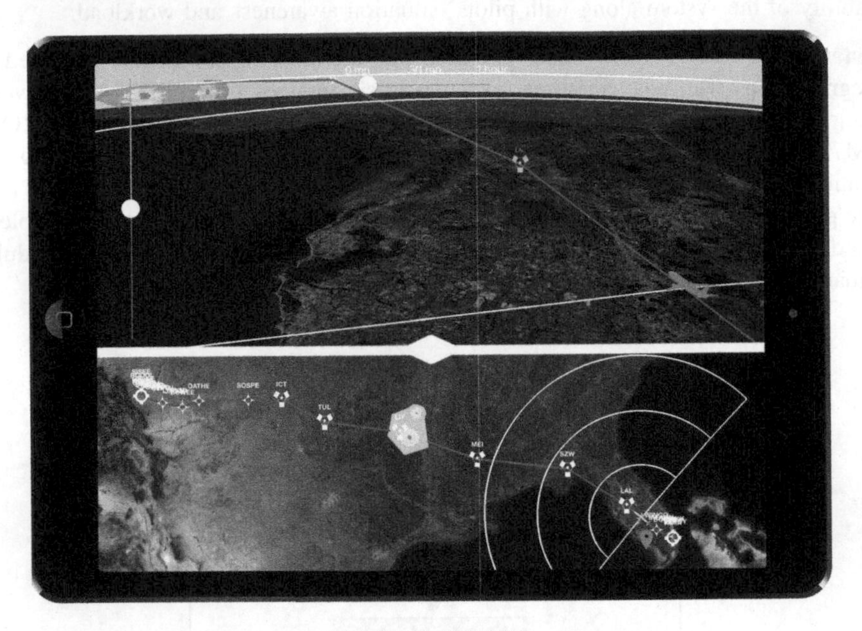

Fig. 8. OWSAS-3.2.

All the information is presented both in 2D via a bird's eye view and in 3D via an exocentric view for enabling strategic and tactical weather and flight situation awareness. Also, Tory et al. (2006) reviewed several works regarding the benefits of 2D and 3D visualization and stated that "2D views are often used to establish precise relationships, whereas 3D views are used to gain a qualitative understanding and to present ideas to others". In addition, she mentioned that "2D views can enable analysis of details and precise navigation and distance measurements (since only one dimension is ambiguous) whereas 3D views facilitate surveying a 3D space, understanding 3D shape, and approximate navigation". This justifies the use of both 2D and 3D views for OWSAS-3. In the 2D view, the weather information is represented with colored polygons (i.e. green, yellow and red). These polygons are then elevated in the 3D view to enable the 3D visualization. The flight plan information (i.e. set of navigation points) is represented by a set of white/gray symbols, aligned with FAA standards, linked with each other via a magenta line in both views. The coordinates and altitude of these information points are also presented to the user in a pop-up window, on click.

The airplane information is represented by a 3D model in the 3D view and an airplane symbol in the 2D view, both colored in gray. Distance/time information relative to the airplane is also represented by white 180° arcs around the airplane in both views. Finally, an orange pinpoint is used in the 2D view to inform the user on the location of the virtual camera (i.e. viewpoint of the 3D view).

The 3D view synchronizes with the 2D view when the pilot selects an area of interest in the 2D view (i.e. the 3D viewpoint is adjusted accordingly). The user also has the option to adjust the size of each 2D or 3D view dynamically by holding and sliding up and down the white views splitter in the middle of the screen. Also, a vertical slider is available on the left side of the 3D view allowing the user to change the viewpoint of the 3D view vertically only. A horizontal slider is integrated at the top of the 3D view to allow for visualization of weather forecast information (simulated also) up to one hour ahead. Finally, the rest of the interaction controls (e.g. zoom in/out, rotate, pan) are touch-based and are aligned with the native interaction controls that iPads provide to users.

Evaluation Design. The experiment was purely exploratory and was not designed to compare different groups of population.

Participants. Five pilots participated to the experiment. Three of them were experienced professional airline pilots with an average of 16,833 flight hours. The two other ones were novice private pilots with an average of 225 flights hours. All of them confirmed their ease with tablets with an average rating of 7.6/10 and four of them already used weather applications such as ForeFlight or MyRadar.

Scenario. One scenario was chosen for the experiment and consisted of a flight going from Miami to Denver. In the middle of the way, convective weather was developing and crossing the flight path of the aircraft (Fig. 9).

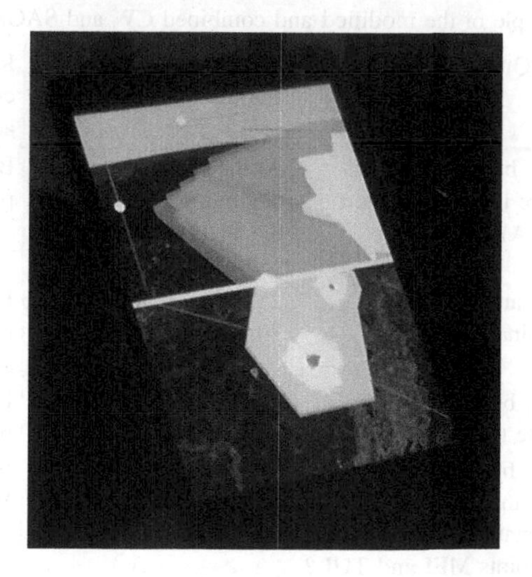

Fig. 9. Sample of the scenario proposed during the experiment.

Evaluation Methods. The various assessment methods used are available in Table 4 and were used for consolidating the proposed design options (i.e. compare expectations generated from a designer's perspective with the user's perspective) and for exploration purposes (i.e. collect information to improve the design).

Table 4. Evaluation methods used for the experiment

Criteria	Method	Purpose
Usability	System Usability Scale (SUS)	Exploration
	Cognitive Walkthrough (modified)	Exploration
Situation awareness	CC-SART	Exploration
	Situation Awareness Global Assessment Technique (SAGAT - modified)	Validation
Workload	NASA-Task Load Index (NASA-TLX - modified)	Exploration

The standard System Usability Scales (SUS) method was employed for measuring the usability of OWSAS (Brooke 1996). The thresholds of acceptability by Bangor et al. (2009) were used for interpreting SUS results.

The Cognitive Walkthrough (CW) is also a usability method that "encourages" the use of the system features via defined tasks (Wharton et al. 1994) and allows to discover potential design errors during their exploration, focusing on relevant cues and available actions after each task is performed. For reducing the intrusiveness during experiments, we observed how users performed the actions, instead of asking a series of questions as prescribed in the original method. Questions about the relevant cues were mapped with each SAGAT level, as described hereafter (Table 5).

Table 5. Sample of the modified and combined CW and SAGAT questions.

#	SA level/Action	Task/Question	Expected answer	Correct? (Y/N)	Relevant cues/Available actions	Used
1	Perception	In the bird's eye view: Where is Miami airport (KMIA) located?	At the beginning of the flight-path		BEV icon-pan/zoom	
2	Perception	What are KMIA coordinates?	25.7959° N, 80.2870° W		BEV icon and annotation-select	
3	Action	In the bird's eye view: Locate the waypoint MEI			BEV WP icon-pan/zoom	
4	Perception	In the bird's eye view: Is there any bad weather information between waypoints MEI and TUL?	Yes		BEV WP + weather colors-pan/zoom	

The Situation Awareness Global Assessment Technique (SAGAT) assesses the situation awareness of the pilots. This technique consists of freezing and blanking the system being evaluated and ask questions related to the 3 levels of Endsley's situation awareness model (i.e. perception, comprehension, projection) (Endsley 1988). However, we modified this technique and did not freeze nor blank the system before questions were asked to the pilots. Indeed, we aligned ourselves with Crutchfield et al. (n.d.) that stated that "there seems to be no reason to remove the situation to assess SA" and that "response time will capture SA differences" if pilots reexamine the display after the questions are asked. Moreover, this modified technique was combined with the CW because questions about relevant cues and available actions are aligned with SAGAT questions. The combination of these methods is shown in Table 5.

The Cognitive Compatibility - Situation Awareness Rating Technique (CC-SART) has been used to evaluate the system cognitive compatibility and is based on Rasmussen's model of human behavior (Rasmussen 1983) since cognitive compatibility "refers to ease of perceiving, thinking and doing, in line with past experience, training and expectations" (Taylor 1995). Three versions of this method are available. The 3D CC-SART was chosen since we wanted to relate situation awareness with a more generic mental model based on skills, rules and knowledge (Rasmussen 1983).

Finally, a modified version of the NASA-Task Load Index (NASA-TLX), that evaluates the workload of the pilots while using the system (Hart and Staveland 1988), was used by expanding the mental demand scale with a visual demand scale since OWSAS-3 is highly visual (Stephane 2013). The visual scale was defined as "How much visual activity is required to process the visual scene (2D vs 3D; 2D and 3D integration; cluttering effects; visual cues such as 3D perspectives, relative size, etc.)".

Equipment. A computer-based PowerPoint presentation was used to provide pilots with an overview of the experiment. The experiment itself was conducted on an iPad. OWSAS-3.2 was tested standalone and was not integrated in our B737 simulator. Video cameras were used to record the experiment, so we could observe the activity of the pilots while they were using the system and were able to discover potential emerging behaviors and properties.

Procedure. Pilots were welcomed in the cockpit design lab at the Human-Centered Design institute of Florida Institute of Technology although the system was tested standalone. They first filled out an informed consent form to make sure they agreed to participate and a background questionnaire about their flight qualifications and ease with tablets. A training system was developed (Fig. 10) so that the pilots could get familiar with the interaction capabilities provided by the VGE, as it would be the case when a new system is being integrated in an aircraft. The content of OWSAS-3.2 was not used for feeding the training system. Instead, basic shapes were used such as spheres or rectangles. The same presentation concepts were kept and were presented in a 2D bird's eye view and a 3D exocentric view. Finally, the interaction controls were the same since the same VGE was used. The pilots first discovered the system by themselves during a few minutes. They were also provided with an interaction controls guide to help them. They were then asked to perform basic tasks to make sure they understood the way the interaction controls work. We introduced them to OWSAS-3.2 system with PowerPoint slides, explained to them the functioning features were finally ready to start the experiment. The experiment was mainly driven by the tasks and questions (i.e. integration of Cognitive Walkthrough and SAGAT) that pilots respectively had to perform and answer. Pilots were asked to answer 12 SAGAT questions (i.e. four perception, six comprehension and two projection questions) and to perform three actions. The other evaluation sheets (i.e. SUS, CC-SART and NASA-TLX forms) were answered by the pilots after the experiment was done. An open-feedback session was included at the end of the experiment to collect additional information about their experience with the system.

Fig. 10. Training system.

Results. All pilots successfully went through the scenario and were able to use OWSAS' features and controls to perform every action. SAGAT perception and projection questions were successfully answered. Pilots' answers to SAGAT comprehension questions went beyond the safety expectations of the design team (e.g. enlarged safety margins reported by pilots, compared to the minimum ones expected by the design team). SUS results were very positive with an average score of 97/100. Based on Bangor et al. (2009), this score is acceptable (Fig. 11). System cognitive compatibility results (CC-SART) had also a good average (94/100). NASA-TLX results were very positive. However, OWSAS-3.2 was not tested in our B737 simulator at this time. Therefore, the results reflect only workload related to the use of OWSAS-3.2 standalone. Consequently, only the individual components were taken into account (especially the visual demand) to collect trends (Fig. 12). All the components were rated between 0 (low) and 100 (high) except for the performance that was rated between 0 (good) and 100 (poor). We can notice that the visual demand is balanced.

The first hypothesis of the design team is that the definition of the visual demand scale was not understood in the same way by all the pilots. While three pilots did not give any feedback when rating a low score for visual demand, two other pilots actually mentioned the following things: What could you do without it, right?", "Of course there is much visual demand, so visual demand is very high because you can't do this without visuals, you got to be able to look at the stuff, colors, terrain, lines, weather, so that is very high.". Finally, open discussions were very useful because pilots proposed a lot of ideas to improve OWSAS-3.2, such as the integration of altitude cues in the 3D view, a compass or a "play button" that loops the evolution of weather information.

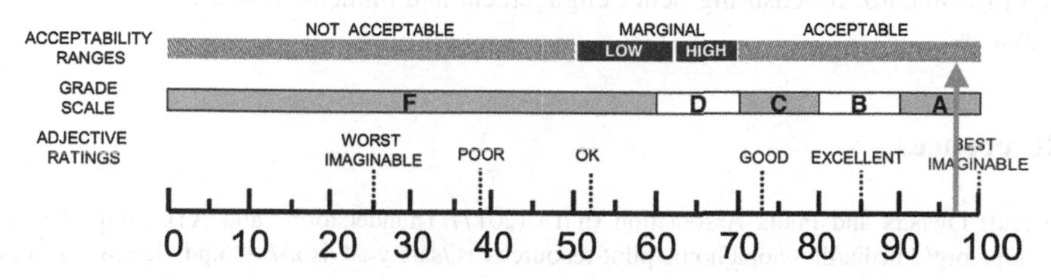

Fig. 11. SUS score interpretation based on Bangor et al. (2009).

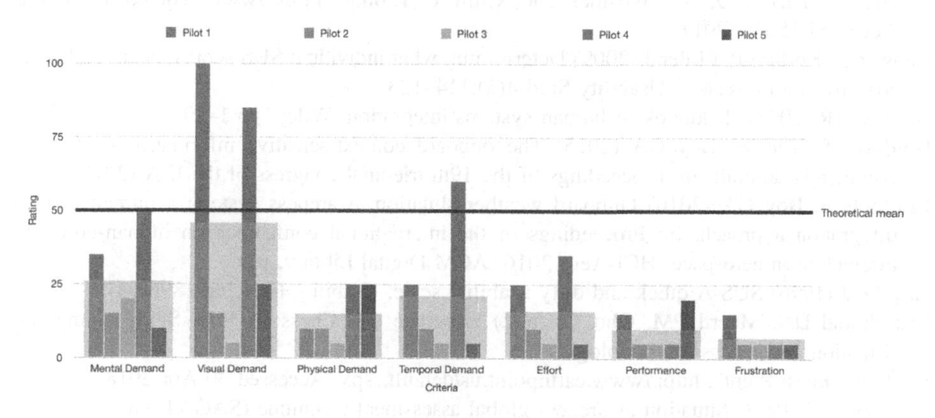

Fig. 12. NASA-TLX results by individual components.

4 Conclusion

This paper presented a four-year research for the Onboard Weather Situation Awareness System (OWSAS). It showed the tedious design process that, as Meinel and Leifer (2011) emphasized, in practice happens in a scrambled mode rather than a linear one, going back and forth through the five main design thinking stages. Participatory design was key for the design team and exploring the various design options with domain experts was crucial for making sure that the best design options were identified. Furthermore, insights into existing technologies enabled the design team to ensure that the visual 2D-3D frontend can be based on weather data structures employed in the current aeronautics infrastructure.

The results presented in the current paper, in continuity with results previously published, show first and foremost that there is a strong need to provide innovative 3D weather representations integrated with 2D existing ones. Furthermore, the evaluation results show that the current combination of content, presentation and interaction, based on requirements elicitation from expert pilots, is improved compared to previous OWSAS versions. Overall, the design team successfully applied HCD and DT principles.

Ongoing and future work focuses on the integration of OWSAS in the B737 cockpit simulator for ensuring better engagement and immersion in the proposed flight scenarios.

References

Aircraft Owners and Pilots Association (n.d.) (2017) Thunderstorms and ATC. https://www.aopa.org/-/media/files/aopa/home/pilot-resources/asi/safety-advisors/sa26.pdf?la=en. Accessed 09 Nov 2017

Anderson R (n.d.): EarthView. https://github.com/RossAnderson/EarthView. Accessed 09 Oct 2015

Ansari S (n.d.): NOAA's Weather and Climate Toolkit. https://www.ncdc.noaa.gov/wct/. Accessed 15 Feb 2017

Bangor A, Kortum P, Miller J (2009) Determining what individual SUS scores mean: adding an adjective rating scale. J Usability Stud 4(3):114–123

Booher HR (2003) Handbook of human systems integration. Wiley, pp 1–29

Boulnois S, Tan W, Boy GA (2015) The onboard context-sensitive information system for commercial aircraft. In: Proceedings of the 19th triennial congress of the IEA (2015)

Boulnois S, Boy GA (2016) Onboard weather situation awareness system: a human-systems integration approach. In: Proceedings of the international conference on human-computer interaction in aerospace, HCI-Aero 2016. ACM Digital Library, p 9

Brooke J (1996) SUS-A quick and dirty usability scale. Usability Eval Ind 189(194):4–7

Crutchfield DN, Moertl PM, Ohrt D. (n.d.) Expertise and Chess: A Pilot Study Comparing Situation Awareness Methodologies

E. Point "Earth Point". http://www.earthpoint.us/default.aspx. Accessed 30 Apr 2018

Endsley MR (1988) Situation awareness global assessment technique (SAGAT). In: Aerospace and electronics conference

FAA. FAQ: Weather Delay (n.d.). https://www.faa.gov/nextgen/programs/weather/faq/. Accessed 09 Nov 2017

FAA Operations & Performance Data (n.d.). https://aspm.faa.gov/opsnet/sys/Delays.asp. Accessed 09 Nov 2017

Hart SG, Staveland LE (1988) Development of NASA-TLX (Task Load Index): results of empirical and theoretical research. Adv Psychol 52:139–183

Globe3Mobile. (n.d.). http://glob3mobile.com. Accessed 08 Oct 2015

Grandin Q (2016) Onboard Weather Situation Awareness System. Florida Institute of Technology

Honeywell (n.d.). https://aerospace.honeywell.com/en/~/media/aerospace/files/brochures/n61-1486-000-000-weatherinformatioservice-bro.pdf. Accessed 10 Oct 2016

Kessinger C, Blackburn G, Rehak N, Ritter A, Milczewski K, Sievers K, McParland T (2015) Demonstration of a convective weather product into the flight deck. https://ams.confex.com/ams/95Annual/videogateway.cgi/id/30096?recordingid=30096. Accessed 10 Mar 2016

Lang JF (2015) Improvement on the Onboard Weather Situation Awareness System (OWSAS). Florida Institute of Technology

Laurain T (2014) Design of an onboard weather system. Florida Institute of Technology

Laurain T, Boy G, Stephane L (2015) Design of an on-board 3D weather situation awareness system. In: Proceedings of the 19th Triennial Congress of the IEA, vol. 9, p 14

Letondal C, Zimmerman C, Vinot JL, Conversy S (2015) 3D visualization to mitigate weather hazards in the flight deck: findings from a user study. In: IEEE symposium - 3D user interfaces (3DUI), pp 27–30

Meinel C, Leifer L (2011) Design thinking research. In: Plattner H, Meinel C, Leifer L (eds) Design thinking: understand, improve, apply. Springer, Heidelberg, pp xiii–xxi

MouseBird Consulting. WhirlyGlobe. http://mousebird.github.io/mousebird/. Accessed 01 May 2018

National Severe Storm Laboratory (n.d.). Thunderstorm Basics. http://www.nssl.noaa.gov/education/svrwx101/thunderstorms/. Accessed 09 Nov 2017

National Transportation Safety Board. (n.d.) (1980). Flight accidents occurring in the United States

Rasmussen J (1983) Skills, rules, and knowledge; signals, signs, and symbols, and other distinctions in human performance models. IEEE Trans Syst Man Cybern SMC-13:257–266

SketchUp (2018). https://www.sketchup.com. Accessed 30 Apr 2018

Stephane L (2013) Visual intelligence for crisis management. Ph.D. dissertation, Florida Institute of Technology. Proquest LLC

Stephane L (2014) Situated risk visualization in crisis management. In: Millot P (ed) Risk management in life-critical systems, Chap. 4, 1st edn. ISTE Ltd & John Wiley & Sons, Inc., pp 59–77

Taylor RM (1995) CC-SART: The development of an experiential measure of cognitive compatibility in system design. In: Report to TTCP UTP-7 human factors in aircraft environments, annual meeting, DCIEM, Toronto, 12–16 June 1995

Tory M, Kirkpatrick AE, Atkins MS, Moller T (2006) Visualization task performance with 2D, 3D, and combination displays. IEEE Trans. Vis Comput. Graph 12(1):2–13

Wharton C, Rieman J, Lewis C, Polson P (1994) The cognitive walkthrough method: a practitioner's guide. In: Nielsen J, Mack Rl (eds) Usability inspection methods. Wiley, pp 105–140

Wu SC, Luna R, Johnson WW (2013) Flight deck weather avoidance decision support: implementation and evaluation. In: Digital avionics systems conference (DASC) - IEEE/AIAA 32nd, p 5A2-1

Generative Games in Aviation

Vladimir Ponomarenko[1], Vitaliy Tretyakov[2], and Alexander Zakharov[3](✉)

[1] Flight Work Psychology, RAE, Moscow, Russia
[2] IAEA, St. Petersburg State University, St. Petersburg, Russia
4054489@mail.ru
[3] IAHPAA, EAAP, Rossiya Airlines, Moscow, Russia
za9162076979@gmail.com

Abstract. The generative games in aviation are used to increase readiness for actions in an extreme situation due to accelerated building and further maintenance of the IMAGE of a PROFESSIONAL, including the image of professional flight activity, which leads to increase in reliability of the activity.

An important element of the generative scenario in a crew/group is to add experience and put it into practice, i.e. to strengthen it. At the same time, different crews/groups creating the participants' experience resort to different scenarios, which can finally lead to absolutely unlike results including those that concern flight safety.

In the course of a game situation, pilots can see the interaction process, on the one hand, and they are engaged into it, on the other hand. Everyone resorts to their own scenario and can see what their actions lead to in the end of the game.

During the game, one could study how leadership skills are demonstrated and how situational understanding of any following action is formed by the group when a passenger in a game situation spontaneously assumes responsibility and says: "Landed…", i.e. the game generates a chain of the situations that requests decision-making.

In general, one could speak of development of skills recommended by the international community in course of games. The basis is made by of participants' interaction in a game situation together with resorting to one's own scenario.

Keywords: Professional pilot training · Generative games · Competencies International standards · Image of a professional · Ground & simulator training Human factors/CRM training · Safety

V. Ponomarenko—Academician RAE.
V. Tretyakov—IAEA Expert.
A. Zakharov—Academician IAHPAA, Full membership of EAAP, Leading Instructor "Rossiya Airlines".

S. Bagnara et al. (Eds.): IEA 2018, AISC 823, pp. 576–581, 2019.
https://doi.org/10.1007/978-3-319-96074-6_59

1 Introduction

Leading international airlines make acceptable level of flight safety the main priority and constantly introduce innovative experience, borrowing it sometimes from other industries. The international civil aviation community (ICAO, IATA) recommends to focus on the competencies development in flight crew training [3, 6, 15].

One of the competencies development methods recommended by the international aviation community is generative games [11, 12, 16], which have proved themselves in the energy sector as a valuable tool to improve the personnel reliability and are successfully used in Russia, CIS countries, France, Germany, USA [12, 14]. As practice has shown, game-based training has a positive effect in several situations: when retraining of flight personnel on the A-320, specifically in the process of mastering theoretical knowledge before training on the simulator; as part of human factor training and in joint pilot and flight crew training.

2 Airlines Pilot Training

The system of continuous flight personnel training of the of JSC "Rossiya Airlines" consists of ground training, simulator and in flight training and testing. According to current international recommendations it is aimed at maintaining 8 core competencies:

1. Application of Procedures.
2. Communication.
3. Aircraft Flight Path Management, automation.
4. Aircraft Flight Path Management, manual control.
5. Leadership and Teamwork.
6. Problem Solving and Decision Making.
7. Situation Awareness.
8. Workload Management.

Thus, 5 out of 8 competencies are related to non-technical (CRM Skills), which are difficult to develop through automated training. Therefore, we propose using generative games that form understanding through the knowledge consolidation and interaction. An important element of the crew generative scenario is the addition of their own experience and its practical implementation that leads to consolidation of knowledge. At the same time, group expertise is also developed, taking into account the individual characteristics of the participants. Different crews will play different scenarios and this leads to absolutely different results from the point of view of flight safety.

Generative games in aviation are used to increase preparedness to handle extreme situations by accelerating the formation and further maintenance of the PROFESSIONAL flight expertise, which increases reliability. Being professional aviator is inextricably linked with the spiritual principle. Pilot has to develop special personal qualities that will be crucial in the sky: humanity, responsibility, conscientiousness, self-criticism … everything that breeds truth, good, love, peace [8–10].

Therefore, a special role in the Airline is assigned to Instructors who are able to teach anyone. They are constantly and comprehensively improving and skillfully using innovative methods, spiritually uplifting the flight crew. When conducting ground and simulator training of the flight crew, the following stages are identified for the use of generative games:

1. Transition from theoretical knowledge to the cognitive skills formation in the course of the re-training program before training in simulators.
2. Forming the PROFESSIONAL IMAGE during aviation career while undergoing retraining or recurrent training.
3. Simulation of the actual situations that took place in the Airline, in order to understand the reasons for the ineffective interaction during the joint training of pilots and flight attendants (for example, the inspection situation)
4. Providing preparedness for non-standard and emergency situations during additional simulator training in the framework of corrective measures.

2.1 Ground Training

Several variants of generative games are used in the process of ground preparation of the flight crew:

- The game "Flight" and the card game "Limitations", which are based on the recommendation of "S7 Training";
- Game "SAFA Inspection" and the game "VIP on Board", designed by Flight & Technical Standard Service of "Rossiya Airlines".

Participants sit around the table playing board game. The facilitator conducts a short briefing, explaining the main purpose and rules of the game. Before the game starts, the participants choose from the proposed roles of the crew members, inspectors and passengers, depending on the scenario. The facilitator distributes cards assigning roles (black card for negative attitude and red card for positive attitude). The game begins with the first instruction given by the facilitator.

During the game session pilots participate in the interaction process and observe it at the same time. Everyone plays his/her role and sees the consequences of his/her actions at the end of the game.

In the process of discussion participants see how interaction takes place, when and for what reasons it becomes successful or unsuccessful. A collective experience is formed through the observation, which depends on the quantitative and qualitative composition of the group.

During the game the participants' emotional state gradually changes towards positive. Usually very skeptical at the beginning experienced instructor-pilots are very pleased with results at the end of the game and evaluate the game as a useful training method.

The game generates a chain of situations that require decision-making either by the Captain or the flight attendants. For example, the Captain may decide to land below the minimum altitude due to a passenger having a heart attack on board or a scandalous situation in the cabin that caused the flight attendants to tie the passenger to the seat. Thus, the decision-making skill is also developed in the game reality.

Conducting joint exercises helps to see the generating interaction when the Inspector appears on board the aircraft. Considering that both pilots and flight attendants work in their areas of responsibility, the game allows you to see errors made due to insufficient English knowledge and the shortcomings of the interaction between the flight and cabin crew.

In general, we can confirm about the recommended competencies development in the course of the game. The basis is the participants' interaction in the game situation, taking into account their own scenario.

Setup cards (bad/good) usage can be associated with linking person's behavior with a pure/impure Spirit. On the one hand, everyone can play bad card in order to understand if this attitude is comfortable (emotionally acceptable) for them, and on the other hand - to realize how this behavior looks from the outside and develop its own strategy of interaction.

2.2 Simulator Training

Simulator training being a kind of active training methods is aimed to train aircraft operations and flight procedures. In training, the simulator is a "mechanism" used by instructor in order to check the pilots professional level and to choose exercises that can increase this level. It is the Instructor's experience that allows to improve the flight crew expertise level through flight simulation.

Taking into account the international standard requirements [1, 2, 4, 5] for instructors of the Airline Companies, they must be classified in three categories:

- Ground CRMI,
- Simulator CRMI,
- Line CRMI.

This distinction also helps to increase the instructional level.

All instructors of "Rossiya Airline" received a special human factor training course, including competencies assessment during flight simulation.

The most interesting part was the practical part of the Course, which consisted of simulator training under the Line Oriented Flight Training (LOFT) scenario. During the simulator training session, both the instructor and the trained crew work were observed. The instructor "plays" for all the personnel with whom the crew works: a technician, a dispatcher, a senior flight attendant, etc.

25 simulator sessions were conducted and it was revealed that the A-320 instructors use the approved scenario of simulator training and rigidly stick to it during the training.

The scenario is as follows: traditional flight from Tel-Aviv to St. Petersburg (TLV-LED). There is a storm front at the departure airport that does not allow return and landing, so the crew after failure and inability to continue the flight in the RVSM zone, has a choice to either fly to the reserve airport (Larnaca) or return, provided the weather improves. Landing is with weight excess, conducted manually. If the crew decides to fly to the Russian border (Rostov), the instructor usually gives a new instruction "the passenger is feeling sick" and the plane continues to fly to the reserve airport.

The instructors of the B-737, B-747 and B-777, operated by the Airline, are more creative in organizing flight simulation and provide pilots with a route, including radio communication in Russian or English. For example, the instructor offers the crew at the preflight briefing to choose the flight route: Tel Aviv-Vnukovo (TLV-VKO) or Vnukovo-Mineralnye Vody (VKO-MRV). The first option involves the radio communication in English, the second - in Russian. In the example below the crew chooses radio communication in Russian, i.e. second route.

After conducting simulator training according to LOFT scenarios, the instructor's staff evaluates the flight crew members for 8 competencies.

3 Conclusion

Taking into account the domestic experience and international practice of conducting ground and simulator training for the flight crew, the existing flight training standards in the Airline, and also understanding the importance of professional pilots development in the flight environment, we can draw the following conclusions:

1. Generative games are a method of forming understanding based on the knowledge consolidation. They can be used by Instructors during both ground and simulator training.
2. The flight reliability increase through generative games is achieved due to the preparedness of crew members to act in extreme conditions and to accelerate the formation, as well as to further maintenance of the PROFESSIONALS IMAGE in aviation.
3. The participants emotional state changes during the game sessions. The recommended competencies are being developed based on the interaction according to participants own scenario.
4. Spiritual development, as the foundation for professional pilot, occurs in the game situation with the help of setup cards, through the behaviour understanding of a person with a pure/impure spirit.
5. Scenario generation during flight simulation depends on the instructor's professional level: a "strong" instructor spontaneously generates a simulation training scenario, without being tied to the template.
6. The crew member's professional development happens through the Instructors understanding of pilot capabilities and consolidation of the generated situation into the successfully acquired skill.
7. Knowledge generation through game-based training used as Instructor's technique will help to form pilot's "safe behavior" and maintain an acceptable flight safety level of the Airline.

References

1. Crew Resource Management (CRM) Training (2014) EASA
2. Crew Resource Management (CRM) Training (2006) Guidance for Flight Crew, CRM Instructors (CRMIS) and CRM Instructor-Examiners (CRMIES). CAA
3. Evidence-Based Training Implementation Guide (2014) IATA, Montreal-Geneva
4. Flight-crew human factors handbook. CAA (2014)
5. IOSA Standard Manual. IATA (2017)
6. Manual of Evidence-Based Training, 1st edn. ICAO, Montreal (2013)
7. Ponomarenko V (2016) Psychology of professional spirituality. 2nd edn., revised and enlarged, RAO, State Research Institute of the Ministry of Defense of the Russian Federation (aviation and space medicine), Moscow
8. Ponomarenko V (2016) Purpose and meaning of life in aviation. Kogito-Center, Moscow
9. Ponomarenko V (2013) Spiritual and moral foundations in the safety, vitality, philanthropy management system (Volume I)/Under the general editorship of Academician of the RAMS, Doctor of Medical Sciences in Aerospace Medicine, Honored Scientist of the Russian Federation A. Razumova. MNAPCHAK (International Public Academy of Human Factor in Aviation and Cosmonautics), Research Institute of Space Medicine and Ergonomics, SBHCI "Moscow Scientific and Practical Center for Medical Rehabilitation and Sports Medicine", Moscow
10. Ponomarenko V (2013) The unmade world is the spiritual creator of the aviator's personality (Volume II)/Under the general editorship of Academician of the RAMS, Doctor of Medical Sciences in the field of aerospace medicine, Honored Scientist of the Russian Federation A. Razumova. MNAPCHAK (International Public Academy of Human Factor in Aviation and Cosmonautics), Research Institute of Space Medicine and Ergonomics, SBHCI "Moscow Scientific and Practical Center for Medical Rehabilitation and Sports Medicine", Moscow
11. Ponomarenko V, Tretyakov V, Zakharov A (2017) Generative games as a method of aviation staff development. Saf Issues 2:3–9
12. Ponomarenko V, Tretyakov V, Zakharov A (2017) Generative games: the experience of implementation in the flight crew training. Hum Factor Probl Psychol Ergon 3:28–32
13. Tretyakov V (2016) Generative games. In: Psychology of transitions: modalities of game consciousness 2016, vol. 30. International Higher School of Practical Psychology, Riga, pp 30–31
14. Tretyakov V (2016) Generative games. Practical guidance for use. Publishing House "Humanitarian Center", Kharkov
15. Zakharov A (2016) Human factors and flight safety training: developing competencies. In: Anokhin A, Paderno P, Sergeev S (eds) The human factor in complex technological systems and environments (ERGO-2016). Interregional ergonomic association, Federal State educational establishment "PEIPK", Northern Star, St. Petersburg, pp 20–26
16. Zakharov A (2017) Use of game methods in the training of aviation personnel. In: Prokhorov A, Popova L, Bayanova L et al (eds) Materials of the russian psychological society congress 2017, vol. 1. Kazan University Publishing House, Kazan, pp 134–136

Ergonomics and Crisis Intervention in Aviation Accident Investigation

Michelle Aslanides[1,2,3](✉), Daniel Barafani[4], C. Mónica Gómez[5], Mariluz Novis[6], Maria da Conceição Pereira[7], Reynoso Humberto[4], and Pamela Suárez[4]

[1] Universidad Favaloro, Buenos Aires, Argentina
miaslanides@gmail.com
[2] Universidad Austral, Pilar, Argentina
[3] GIPSIA, Buenos Aires, Argentina
[4] Jiaac, Buenos Aires, Argentina
{dni,psuarez}@jiaac.gob.ar, mdhareynoso@hotmail.com
[5] GIPSIA, Doha, Qatar
mgomezcan@gmail.com
[6] GIPSIA, Madrid, Spain
mariluznovis@gmail.com
[7] GIPSIA, Recife, Brazil
concitapereira@gmail.com

Abstract. Since 2004 we have been analyzing and understanding aviation accident investigator's work, starting from France at the BEAD-air board, and following with JIAAC, in Argentina during the last two years. The problems they face having to investigate violation of rules was the first object of our research. The context of accident investigation is also difficult to understand systemic causality. It is prescribed and also sometimes impossible to achieve, so this requirement of the task impacts on their reliability and health through an assessment process based on the way they write the reports which doesn't really reflect their real analysis. Nowadays we are starting a new Project with JIAAC which combines an ergonomic approach with a psychological approach to be able to prevent the consequences of their stress during contexts that are exposing them to critical event such as death, to the families suffering or to media pressure. This produces a level of stress added to all the organizational and judicial aspects that interfere with their work. Our goal as a team is to prevent as much as possible these factors to intervene, and when it is not possible to prevent them, we have combined our intervention with the necessary psychological assistance to avoid further PTSD (Post-Traumatic Stress Disorder) symptoms. This work is an innovation in our aviation world, and we would be delighted to share our advances during 2018 IEA congress.

Keywords: Ergonomics · Psychology · Aviation · Accident investigation PTSD · Critical incident · Stress

GIPSIA: Grupo interdisciplinario e internacional para la promoción de la salud de los investigadores de accidentes aéreos.

S. Bagnara et al. (Eds.): IEA 2018, AISC 823, pp. 582–591, 2019.
https://doi.org/10.1007/978-3-319-96074-6_60

1 Introduction

Accident investigators are subject, due to the characteristics of their task, to situations that generate stress and anxiety due mainly to work conditions that require great physical and mental effort. ICAO itself recognizes in the Aviation Accident and Incident Investigation Manual, Doc 9756 AN/965 that: "Accidents can happen anywhere: airports, mountains, swamps, very thick forests, deserts, etc. Frequently it is necessary to pass penalties to reach the accident site in remote areas, so it is important that the investigators are in good physical shape. Likewise, the expected profile of the investigation requires high technical knowledge as well as experience. In the same document, Section 2.4.3. People say that:

"It is essential that every accident investigator has practical experience in aviation, as this will allow him to develop his investigative assets". The experience may have been acquired as a professional pilot, as an aeronautical engineer, or as an aircraft maintenance technician. Other specialized aviation sectors that could constitute a useful experience include Administration, operations, airworthiness, air traffic services, meteorology, and human factors.

Regarding the personal qualities that an investigator must have in Section 2.4.4. it is mentioned that:

"Apart from the technical knowledge, every accident investigator needs certain personal attributes, among them, integrity and impartiality to record the facts, be logical and perseverant in the surveys, which are sometimes carried out under unnerving conditions, and tact to deal with a great variety of people who have suffered the traumatic experience of an aviation accident."

We have seen through ergonomic intervention in their system that this is only a wish: in the real world these rules are not very easy to apply. In this context, our group GYPSIA is trying to help accident investigators to improve their working conditions through two different and complementary human factors approaches.

2 Goals of Our Team

The following paragraphs describe the work we have being doing since 2004 analyzing and understanding aviation accident investigator's work, starting from France at the BEAD-air[1] board, and following with JIAAC[2], in Argentina.

2.1 Ergonomic Approach: Investigator'S Difficulties Need to Find Solutions

Our work started by a requirement from the BEAD-air accident investigators who were very concerned with applying well their conceptual and methodological knowledge following James Reasons principles in the accident investigation process. This is

[1] BEAD-air: Bureau Enquetes Accident Défense "Air".

[2] Junta de Investigación de Accidentes de Aviación Civil.

mandatory, but it is not always easy to achieve in real accident investigation situations (Aslanides and Jollans 2006). Since 2016 we had the opportunity to assess JIAAC in Argentina, following the request of its president. Indeed, some differences between the perceived problems from the investigators perspective (in terms of systemic analysis difficulties and in terms of obstacles to include the systemic information in the public reports) and from the top management leading team. We were asked to understand why the investigators had such difficulties and how one could improve the quality of the public accident reports by acting on workers consciousness and skills.

It is hard to search for violation of rules in accident investigations
On the one hand, accident investigators have to face problems when investigating violation of rules. This was the first object of our research in BEAD-air (Aslanides et al. 2007) where we have stated our PhD Project on that basis, focusing our analysis on that problem, understanding why it was really practically impossible for them to accomplish that goal during accident investigations without having to infer too much the intentions of actors who have violated the rules, without having to avoid publishing their findings when they finally could Access that level of information.

It is hard to investigate the system from accident investigators perspective
The context of accident investigation is, indeed, a very hard one and other objects of investigation are also difficult to grasp, like the second one we got to better understand little by little: systemic causality. Indeed, this other object is complex for the accident investigators to analyze, to obtain information about, and also to write in the reports. As violation of rules, the difficult part of their work might be writing what has been analyzed with many difficulties. It is in a sense, an impossible goal to achieve, that has impacts on their reliability and health, being sometimes judged on the basis of the reports that don't really reflect their real analysis (Aslanides and Jollans 2006).

2.2 Psychological Approach to Crisis Intervention

Since 2016 we established the team GIPSIA, made of one aviation ergonomist and three aviation psychologists (co-authors in this paper) to integrate ergonomic approaches to safety, health and reliability with psychological approaches to distress prevention. We were trying to innovate and bring some improvements in health of aviation accident investigators especially after we saw the media reactions to the Lamia flight 2933 that was transporting Chapecoense soccer team from Brazil to Colombia. We made our first proposal at that moment, but due to lack of time, we never intervened in that context.

Since our offer was conveyed by JIAAC top management to Colombian authorities, and JIAAC authorities supported our approach, the project could be applied in Argentina at any time. Indeed, lately we have combined our intervention as ergonomists with the necessary psychological assistance to avoid further PTSD symptoms during 2017, after a plane crashed close to Buenos Aires, in the Delta del Paraná region, where it was very difficult to find the remains of the plane. Since it took a long time to find the plane, it took much more time than usual to start the investigation process. This added some pressure to the investigators and to the members of the management of JIAAC, who asked the psychologists of our team to give them the first support from a psychological intervention based on the Critical Incident Stress Management (CISM) perspective (Everly and Mitchell (2008), Mitchell 2012, 2013).

3 Methods

3.1 Ergonomics Approach

Based on the crossing of three types of descriptive techniques, in both fields we have proceeded to interview the investigators, we analyzed many of the regulatory documents that define the working context of investigators and, finally, we accessed to direct observation of their working situations to understand better their concerns. We have offered both BEAD-air and JIAAC all our techniques potential in order to show how we can proceed to detect all the critical working situations that these two organizations wanted us to prevent.

The BEAD-air intervention was more the consequence of a request made by the accident investigators themselves than by their management, even if afterwards the project was accepted by all of them. It started by an analysis of 70 accident reports asked by one of the researchers and experts that was at that time teaching all the human factors causality factors and models to the investigators, also the tutor of the ergonomics PhD work. We wanted at the beginning find out if violation of rules were mentioned as causes as a positive consequence of the human factors plan (Aslanides et al. 2007).

The JIAAC request in 2016 was similar, but regarding systemic approach, and came directly from the top management of that institution. The question was: why these aspects of the "real world" don't emerge in the reports that are written by investigators? We had many interviews and some verbal protocols where the accident investigators explained us how they proceeded while we read the reports with them. The JIAAC case consisted in understanding accident investigators perspective about the fact systemic analysis according to James Reason's model didn't appear as expected in the final version of the reports. This was considered as a weakness in accident investigators reliability, and it brought some discussions between investigators and their management. The ergonomist was asked to understand the root of these discussions that not always were leading to a good understanding. Individual interviews of all the accident investigators from Buenos Aires and from Córdoba were done in order to get all the input from all the diversity of investigators: the new hired ones, the more experienced, women, men, etc. The result of the diagnosis was written down in a report and given to the top manager of the JIAAC who took the decisions to prevent the difficult situations investigators had to deal with during all the phases of the investigation process.

3.2 Psychological Approach

In this sense, when we were in the middle of a proactive phase of the development of a multidisciplinary preventive project of multidisciplinary diagnosis-intervention, a critical event was presented which was the investigation of an accident considered to be greater, both due to the working conditions of the researchers and the Media impact.

This intervention was carried out through individual on-line interviews carried out by the psychologists of our team, adapting to the availability of the researchers as soon as they could leave the accident site. The intervention was based on the SAFER model that seeks to stabilize, normalize and contain, as well as offering coping measures. These interviews were confidential and in no case was the researcher's performance evaluated but simply the possible existence of symptoms or alterations of a cognitive or emotional type. Although the possibility of referral to a high level of care was offered, this help was not required by any of the researchers.

Three months afterwards of the individual intervention, a group interview was conducted with all the researchers through a CISD (crisis incident stress debriefing). This debriefing helps them to re-structure the critic event among all them, to normalize their behaviors and emotions as well as to reflect back the possible changes that could have happened in their health. Finally, we encourage them for effective coping to explore by themselves possible changes in their health in the next months.

Finally, a meeting was hold with the JIAC President and her assistants to summary our intervention and offer them different lines of intervention for the next future.

4 Results and Discussion

4.1 Bead-Air and JIAAC Results Related to the Ergonomic Intervention

Violations and systemic analysis can't be easily expressed in the reports even though they are analyzed by investigators

We found out the violations were not mentioned at all and we then wanted to check the causal factors that led to the lack of violations as causes in the reports (Aslanides et al. 2007). We found out many reasons in 9 cases that we then analyzed because of the concern from the accident investigators who were worried to be asked to detect violations in a context where they couldn't easily analyze that psychological object, nor reflect it easily in the reports (Aslanides and Jollans 2006).

Bead-air investigators verbal protocols and decision-making analysis shows it is almost impossible for accident investigators to understand violation of rules intention and mechanisms in the context of accident investigation if the analyst himself does not know the situations where these violations took place. It is difficult to identify and be certain about pilot or the technical worker's intention after the accident occurs since none of them will admit a conscious deviation from rules decision since they would have to face the legal processes. Sometimes the confident relationship that emerges between the analyst ergonomist and the investigator is enough to understand investigator's intention, but then these informations can't appear in the final report for the reasons we already exposed in other papers.

The second report, in Argentina, mentioned that investigators are stressed not only because of the traumatic factors (death, suffering of family of the victims, etc.) but mainly because of the way the assessment of their work is done. Indeed, it is a judgment concerning the way they work that only takes into account the final result of the process (the report) and doesn't take into account all the intermediate phases of the writing of the report, which also include other people deciding on the shape of the report itself. JIAAC Management took into account these findings and changed the way the management of the investigation took place.

If there are some intrinsic contextual obstacles to use all the terms and methods that are recommended by ICAO, accident investigators won't be always in a good position to achieve their goals, and therefore will always be stressed and feel bad at work, even if the management makes all the best efforts to make them feel well recognized. The gap between what has to be done and what is really possible to do in those contexts will always exist, it is only a question to recognize this gap. The PhD we are developing is, in this sense, a way to bring this recognition to the table. The PhD's scope is to bring together all of this information as arguments to explain that the boundaries of James Reason's model are exactly the ones that BEAD-air and IJAAC accident investigators perceive with regard to the violations of rules and systemic factors analysis. It is all about "the real" aspect of work that cannot be easily investigated during accident investigations because of all the legal pressure as we stated it in other publications (Aslanides 2007). Our main target is to demonstrate accident investigators not always have the means to approach the real work in the organizations they have to investigate and have less freedom to publish the analysis in the public reports, and to define a new set of prescribed tasks and methodological tools that will enable them to achieve the investigation in healthy and reliable conditions.

4.2 Crisis Intervention's Results

Results from the psychological intervention are confidential also but we will only say that they were carried out with no difficulties and very well accepted by the accident investigators. The interviews helped to restructure the event among all the accident investigators involved, to normalize and explain thoughts and emotions and to investigate possible changes that could have occurred at the level of physical and emotional health. Finally, some indications were given to explore for themselves possible changes in their health status in the coming months.

5 Perspectives and Conclusion

Our proposal for the future is to go on helping accident investigation boards (JIAAC, BEAD-air and others) carry out an adequate and effective assessment of physical and psychosocial risks and their consequences, as well as an adequate intervention to take care of the health of accident investigators.

This proposal includes a diagnostic phase and an intervention phase which will be done sometimes by the ergonomist, with his methods, sometimes by the psychologists, sometimes together. In every case we would try to identify working conditions that could affect the health of the investigators as well as the psychosocial risk factors associated with their activity.

Our idea is to develop the following lines of work:

- classify psychosocial risks by their level of frequency and severity to include them in risk matrices of a JIAAC safety management system (SMS) to mitigate their effects on investigators health.
  To direct actions towards the promotion of a culture of prevention and protection by preparing a report with proposals for intervention actions and practical guidelines. The intervention phase would consist in the experiencing preventive actions.
- Training and education of accident investigators in different aspects that improve their skills: (a) communication during dynamic and systematic crisis (b) CRM Crew Resource Management), (c) How to deal with the need to contact the family of victims, etc. (d) Training in crisis intervention to provide accident investigators with the psychological necessary skills to intervene in a critical event such as an accident investigation: (i) Training in conducting structured and semi-structured interviews with witnesses, victims, relatives, co-workers of victims of an accident, etc. (ii) Training to understand the impact of a traumatic event on the memory of witnesses. It is proven that witnessing a traumatic event has a decisive influence on memory. Invstigators must know this circumstance to make decisions about the veracity of the witnesses versions and about when it is convenient to interview them (iii) Training in the PEER program. The PEER training aims to equip PEER researchers with sufficient tools to help their peers cope with the physical, emotional or cognitive reactions that they may be suffering after experiencing a critical event and that require immediate action. This training consists of a series of modules already established for other professions such as air traffic controllers.
- Assistance and training in psychosocial management of crisis and emergencies adressed to managers of the investigation boards as direct or indirect intervinients in critical situations. Their behavior has a great impact in the overall management of a critial incident.
- CISM H24 Telephonic assistance program: individual intervention done using the telephone, offering continously the investigators the possibility to be heard concerning working situations.

6 Conclusions

We believe we have showed through our work it is possible to start an innovative approach of stress prevention in accident investigation field, combining ergonomic approaches with psychological ones to prevent the consequences of stress during inevitable contexts that are exposing accident investigators to stress factors related to the object of work itself (exposure to death, to suffering people like victims or their families, to their own suffering since sometimes they lose Friends or family members) and also to the organizational and judicial contextual factors that interfere with their work. It is possible to prevent as much as possible the factors that generate stress in accident investigators, and also try to prevent the consequences of stress once those factors can't be avoided. Both approaches at the same time are possible, and we believe, necessary. We believe this work is an innovation in aviation and in other work systems. We are delighted to share it with our community in IEA congress.

References

Amalberti R (1997) Human error in aviation. In: Soekkha H (ed) International aviation safety conference, IASC-1997, pp 91–108. Rotterdam, The Netherlands. Vsp 68, Utrech

Amalberti R (2001) The paradoxes of almost totally safe transportation systems. Saf Sci 37:109–126

Amalberti R, Auroy Y, Aslanides M (2004) Understanding violations and boundaries. The Canadian Healthcare Safety Symposium, Edmonton, Alberta, CA

Aslanides M, Valot C, Nyssen A-S, Amalberti R (2007) The evolution of error and violation descriptions in French Air Force accident reports: the impact of human factors education. Hum Factors Aerosp Saf 6(1):51–70

Aslanides M (2007) Ergonomía y aviación: Matrimonio de conveniencia Revista Aviador pp. 53–56

Aslanides M, Jollans JY et al (2006) Prevención mediante el control de los desvíos a las normas: caracterísiticas y limites del análisis de las violaciones en las investigaciones de accidentes. In: Primeras Jornadas latino-americanas de seguridad de vuelo y Factores Humanos, Aranjuez, Spain

Barriquault C, Amalberti, R (1999) L'influence des modèles de causalité sur l'analyse d'incidents de contrôle aérien. In: XXXIV Congrès de la SELF, Caen

Bourrier M (2000) "Le nucléaire à l'epreuve de l'organisation: Puf, coll. "Le travail humain" 1999, 294 p. 148 F. Nature Sciences Societes vol. 8, no. 1, p. 85

Bourrier M (2001) Organiser la fiabilité. In: Bourrier M (ed.) Risques collectifs et situations de crise. L'Harmattan, Paris

Carballeda G (1997) La contribution des ergonomes à l'analyse et à la transformation de l'organisation du travail: l'exemple d'une intervention relative à la maintenance dans une industrie de processus continu. Bordeaux Collection Thèses & Mémoires Bordeaux, Presses de l'imprimerie de l'Université de Bordeaux 2

Catino M (2006) Logiche dell'indagine: oltre la cultura delta colpa. Rass ital soc 47(1):7–36

Clot Y (1999) La fonction psychologique du travail. PUF, Paris

Critical incident stress management user implementation guidelines EATM M. Barbarino human factors management business division

CISM Programade Gestao de Incidentes Críticos 18 a 22 de fevereiro de 2008 Lisboa - Centro de Psicología da Forca Aérea

De la Garza C, Weill-Fassina A, Maggi B (1999) Modalités de réélaboration de règles: des moyens de compensation des perturbations dans la maintenance d'infrastructures ferroviaires. In: XXXIV Congres de la SELF, Caen

de Terssac G, Lompré N (1996) Pratiques organisationnelles dans les ensembles productifs: essai d'interprétation. L'ergonomie face aux changements technologiques et organisationnels du travail humain. J. C. Spérandio, Toulouse, pp 51–66, Octarès 1996

Everly GS Jr, Mitchell JT (2008) Integrative crisis intervention and disaster mental health. Chevron Publishing Corporation, Ellicot City

Helmreich R (2000) On error management: lessons from aviation. Br Med J 320:721–785

Hollnagel E (1993) Human reliability analysis, context, and control. Academic Press, London

Holloway CM, Johnson CW (2004) Distribution of causes in selected US aviation accident reports between 1996 and 2003. In: Welch N, Boyer A (eds) Proceedings of the 22nd international systems safety conference, international systems safety society

Johnson CW (2003) How will we get the data and what will we do with it? issues in the reporting of adverse healthcare events. Qual Saf Healthc 12:64–67

Everly GS Jr PhD, ABPP, FAPA, CCISM, The Johns Hopkins University, Maryland Trauma culpa y duelo - Hacia una psicoterapia integradora - Pau Pérez Sales 2da Edición

Johnson CW (2003) The failure of safety-critical systems: a handbook of accident and incident reporting. Glasgow University Press, Glasgow

Johnson CW, Holloway CM (2004) 'Systemic failures' and 'human error' in Canadian TSB aviation accident reports between 1996 and 2002. The impact of human factors education 69. In: Pritchett A, Jackson A (eds) HCI in Aerospace 2004. EURISCO, Toulouse, France

Johnson CW, Holloway CM (2003) A survey of causation in mishap logics. Reliab Eng Syst Saf J 80:271–291

Lawton R, Parker D (2002) Barriers to incident reporting in a healthcare system. BMJ Qual Saf Healthc 11:15–18

Leplat J (1998) L'analyse cognitive de l'erreur. Rev Eur Psychol Appl 49(1):31–41

Marx D (2001) Patient safety and the 'just culture': a primer for health care executives. In: Ahrq (ed) MERS: Medical event reporting system for transfusion medicine, Washington DC

Mauririo D, Reason J, Johnston N, Lee R (1995) Beyond aviation human factors. Ashgate-Avebury, Aldershot

Neboit M (1996) Erreur humaine et Prévention: Le point de vue de l'ergonome. L'erreur humaine: question de point de vue. In: Cambon de Lavalette B, Neboit M (eds) Marseille, pp 23-34, Octares 1996

Ombredane A, Faverge JM (1955) L'analyse du travail. PUF, Paris

Pereira MC (2012) Estudo preliminar sobre o impacto emocional em familiares de vitimas de accidentes aeronáuticos vol 3, no 3. Edição Especial - Psicologia Aplicada à Aviação

Pereira MC (2012) Emergencias e desastres: contribucoes da psicología, vol 3, no 3 (2012). Edição Especial - Psicologia Aplicada à Aviação

Pereira MC (2012) A avaliacao psicológica e o acidente aeronáutico: existe uma relacao? vol 3, no 3 (2012). Edição Especial - Psicologia Aplicada à Aviação

Perrow C (1984) Normal accidents, living with high-risks technologies. Basic Books, New York

Polet P, Vanderhaegen F et al (2003) Modelling border-line tolerated conditions of use (BTCU) and associated risks. Saf Sci 41(2–3):111–136

Poy M (2006) Aspectos funcionales de los riesgos y desvíos de las normas de seguridad en el trabajo. Un aporte a la comprensión de las relaciones entre actividad humana y seguridad. Tesis de Doctorado, Facultad de Ciencias Sociales - Universidad de Palermo, Buenos Aires

Queinnec Y, Marquie JC, Thon B (1991) Modèles, comportement et analyse du travail. Modèles en Analyse du travail. In: Amalberti, RDM, Theureau MJ (eds) Editions Mardaga, Liège

Rasmussen J (1997) Risk management in a dynamic society: a modelling problem. Saf Sci 27:183–213

Reason JT (1990) Human error. Cambridge University Press, New York

Reason JT (1997) Managing the risks of organizational accidents. Ashgate, Aldershot

Rochlin GI (1992) Defining high reliability organizations in practice: a taxonomic prologomena. In: Roberts KH (ed) New Challenges to understanding Organizations. Sage, Beverly Hills

Sarter N, Amalberti R (eds) (2000) Cognitive engineering in the aviation domain. Lawrence Erlbaum Associates, Hillsdale

Mitchell JT (2013) Care and feeding successful critical incident management team. Chevron Publishing, Ellicott City

Mitchell JT (2012) Critical incident stress. In: Figley Charles (ed) International encyclopedia of trauma. Sage Publications, New York

Sartcr N, Alexander H (2000) Error types and related error detection mechanisms in the aviation domain: analysis of aviation safety reporting system incident report. Int J Aviat Psychol 10:189–206

Strauch B (2002) Normal accidents: yesterday and today. In: Johnson CW (ed) Proceedings of the first workshop on the investigation and reporting of incidents and accidents (IRIA 2002). Department of Computing Science, University of Glasgow, Glasgow

Valot C Amalberti R (2001) Ergonomics in aviation. Le Travail Humain vol 64, editorial (3), no. 70, pp 193–196

Vaughan D (1999) Technologies à haut risques, organisations et culture: le cas de Challenger. In: Séminaire du Programme Risques Collectifs et Situation de crise, Paris

Westrum R (1995) Organisational dynamics and safety. In: McDonald N, Johnston N, Fuller R (eds) Application of psychology to the aviation system, pp 75–80. Avebury Aviation, Aldershot

Wioland L (1997) Etude des mécanismes de protection et de détection des erreurs, contribution à un modèle de sécurité écologique. Psychology of cognitive processes. Université Paris V, Paris

Woods D, Johansnesen L, Cook M, Sarter N (eds) (1994). Behind Human Error. CERSIAC SOAR 94–01. WPAFB, Dayton

The Effects of the Type of Rest Breaks on Return-to-Task Performance in Semi-automated Tasks with Varying Complexities

S. Zschernack(✉), M. Göbel, and Z. Hoyi

Rhodes University, Grahamstown 6140, South Africa
s.zschernack@ru.ac.za

Abstract. Automation in the aviation industry is acknowledged as a useful tool in reducing pilot workload. Different types of rest tasks are commonly prescribed fatigue countermeasures in the industrial setting and have been showed to elicit beneficial effects on prolonged human performance. Understanding the effects of different rest break activity and time out-of-the-loop during semi-automated flying on return to task performance has been adequately studied, thus highlighting its importance in the context of flight safety.

The present study requested participants to perform a tracking task in a laboratory where they changed from activity (30 min) to a break (2 vs. 30 min) and back to the activity (20 min). The task varied in the complexity of the activity (pure tracking vs. tracking plus memory plus rule-based decision making), the type of break (passive rest vs. actively supervising) and the duration of the break (2 min vs. 30 min). Performance was measured as effective response time in the tracking task and number of correct responses to secondary cognitive tasks.

Physiological measures included heart rate (HR), heart rate variability (HRV-time and frequency-domain), eye blink frequency and duration. The Karolinska Sleepiness Scale was used as a subjective measure.

The study concluded that active, administrative tasks, which allowed the operator to maintain some form of situational awareness by monitoring the automated system, achieved favourable effects of being more alert than the passive rest break of being disengaged from the system. The shorter duration of being out-of-the-loop from controlling the system proved to be more advantageous than the longer out-of-the-loop duration. In looking at the workload levels of arousal, the results suggest that the higher workload level is better at maintaining the alertness of operators.

Keywords: Return-to task performance · Rest break · Semi-automated tasks

S. Bagnara et al. (Eds.): IEA 2018, AISC 823, pp. 592–600, 2019.
https://doi.org/10.1007/978-3-319-96074-6_61

1 Background

Fig. 1. Fast food meal and packaging (cardboard packaging on the left and paper wrapper packaging on the right).

2 Methods

2.1 Experimental Design

This research project aims to understand the effects of food packaging on driving performance when eating while driving. This study compares two different types of packaging during driving, unpacking and driving and eating and driving in a simulator setting. It follows a repeated measure design, in which each participant performs all conditions.

There are two independent variables, namely type of packaging (cardboard and paper wrapper) and the activities with the packaging and meal consisting of individual food items such as a burger, chips and canned beverages (see Fig. 1). The two independent conditions namely the cardboard and paper wrapper packaging were compared. For each condition (cardboard and paper wrapper packaging) participants were required to perform four activities (pure driving 1, unpacking of the meal, eating of the meal and pure driving 2) while throughout driving dependent variables were measured. The pure driving 1 and 2 activities require participants to drive normally following the white line as accurately as possible presented on a screen in front of them. The second activity requires participants to unpack the meal, taking the burger and fried chips out of either the cardboard or paper wrapper packaging as well as open the cold beverage and putting the straw into it while driving. The third activity requires participants to consume the food while driving.

The order of the cardboard and paper wrapper conditions was randomised (half of the participants performed the test firstly with the cardboard packaging while the other of the participants began with the paper wrapper packaging), However the order of the activities was not, because eating before unpacking is not possible. Instead a second pure driving condition has been included to identify a possible learning effect that might mask the effect of packaging. Noteworthy, the study aims to look at the effects of the type of packaging on driving performance but does not compare the impact different types of foods have on driving performance.

In total twelve participants (4 males, 8 females) were recruited, their ages ranging from 20–30 years. Volunteering participants were recruited from students around Rhodes University and had to have a valid driver's license and a minimum of three years driving experience as anything less than 3 years driving experience can overestimate the results. Participants had to have good eye sight and no food allergies. Participants with visual impairment were required to wear contact lenses.

2.2 Dependent Variables

Tracking Deviation
Tracking deviation considers the driving ability of the participants'. This continuous tracking task reflected behavioural aspects of motorists (attention, and performance allocation) [7]. It highlighted the consistency or inconsistency to remain on the middle white line, with the tip of the yellow triangle (the front of the vehicle) as accurately as possible. Any distraction will result in a swerve away from the line demonstrating less vehicle control. The average deviation is calculated from the target line in meters and the accuracy of driving on the target, a percentage will be given. A low fidelity driving simulator was used to measure driving performance variable such as tracking deviation. A low-fidelity driving simulator was used to provide a controllable, repeatable environment which allows individuals to explore these potentially risky conditions in an ethical and safe context [14]. Previous studies have shown that a low fidelity simulator is suitable in measuring tracking deviation/lane keeping performance [15]. The driving simulator was presented by a curved road with a yellow triangle at the bottom of the screen indicating the vehicle. The speed on the simulator remained constant at 4 km/hr throughout the testing sessions.

Perceived Control
The perceived control is a subjective measure used to measure driver's estimation of perceived control as well as driving skills. A five point Likert scale was used with scores ranging from one being poor or worst driving ability to five being excellent or best driving ability.

2.3 Experimental Setup

The low fidelity driving simulator vehicle consisted of a light motor vehicle with a non-force feedback steering wheel on a metal dash board and a car seat which could be adjusted according to the anthropometry of the participant. Each participant sat behind the wheel of the simulator vehicle facing the white projector screen (1350 mm X 1030 mm) placed in front of them where the road scene of the simulator was projected. A tracking task using a simplified road scene was shown on the white projector. It required participants to use the steering wheel to direct the static yellow arrow representing the bonnet of a vehicle on the moving white line as accurately as possible (Fig. 2).

Fig. 2. The simulator software representing the visual scene that is projected on to the white projector screen

2.4 Procedure

The experimentation was conducted in the laboratory in the Department of Human Kinetics and Ergonomics at Rhodes University. After obtaining ethical clearance from the Human Kinetics and Ergonomics Ethics Committee participants were recruited.

Participants were required to attend an introduction and habituation session in the HKE department. For the testing session Participants were required to attend two testing sessions and were tested individually. For each condition (cardboard and paper wrapper packaging) participants were required to perform four activities (pure driving1, unpacking of meal, eating of meal and pure driving 2) while driving. The pure driving 1 and 2 activities required participants to drive normally following the white line as accurately as possible. The second activity required participants to unpack the meal, taking the burger and fried chips out of either the cardboard or paper wrapper packaging as well as open the cold beverage and putting the straw into it while driving. The third activity required participants to consume the food while driving. Participants did not unpack and eat the meal against time. Each driving activity lasted for 3-min with 3-min breaks in between, in order to set up for the next condition and for the participants to complete the perceived control questionnaire (Fig. 3).

Statistical analysis of data collected was performed using Microsoft Office Excel (Windows 7, 2010, USA) and STATISTICA 12.0 software programme. Two-factorial measures analyses of variances (ANOVAs) using the factors activity (eating, unpacking, pure driving) and packaging (cardboard, paper) were conducted to assess the differences between the conditions. A confidence level of $(p < 0.05)$ was used to show significance. Sex was considered as co-variate, but as no differences between male and females were found the results are being presented for the entire sample

Fig. 3. Participants during the unpacking (left) and eating activity (right)

3 Results

3.1 Tracking Deviation

The descriptive statistics of the tracking deviation is illustrated in Fig. 4 indicating the mean and standard deviation values for the different activities for each condition.

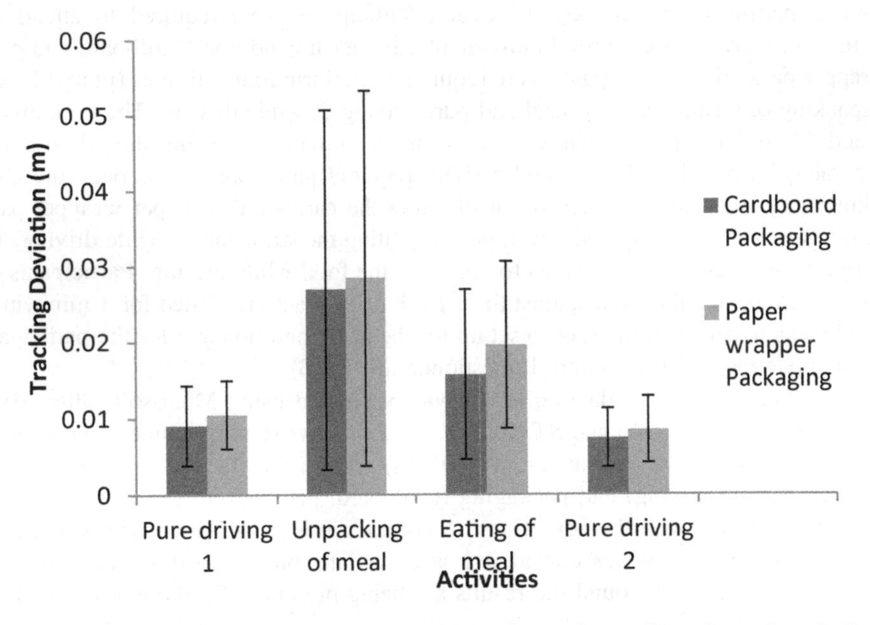

Fig. 4. Average tracking deviation for all activities for each condition (error bars indicate standard deviation, n = 12).

For the purpose of the study the higher the mean value for tracking deviation the worse the participants' vehicle controls. Participants performed worst in both packaging conditions during the unpacking of the meal activity (the mean value for driving deprivation tripled in comparison to pure driving 1 activity), followed by eating of the meal activity which doubled in mean value in comparison to pure driving 1 activity and then following this was the pure driving activities having better driving performance (Fig. 4).

Two-factorial measures analyses of variances (ANOVAs) using the factors activity (eating, unpacking, pure driving) and packaging (cardboard, paper) showed statistical significances between the activities ($p < 0.05$). There was no effect of packaging nor an interactional effect.

3.2 Perceived Control

For the purpose of the study the higher the value (ranging from 1–5) the better the participant's estimation of perceived control and driving skill, whereas the lower the value the worst the participant estimation of perceived control and driving skill. Participants perceived to have the least vehicle control during the unpacking of meal activity, followed by eating of the meal activity and then the pure driving 1 and 2 activities for both packaging conditions (Fig. 5).

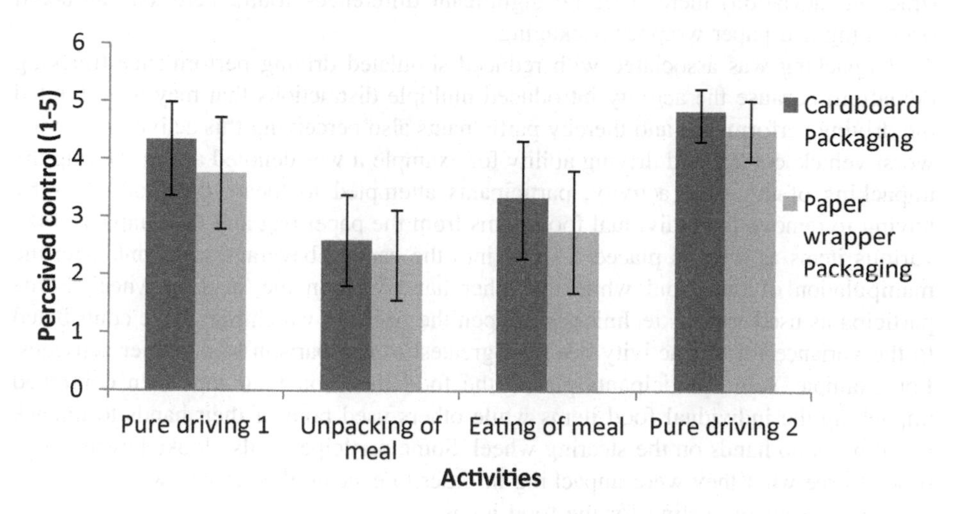

Fig. 5. Average score values and standard deviation (error bars) measured for all activities for each condition (rated between 1 = poor control and 5 = excellent control, n = 12).

Two-factorial measures analyses of variances (ANOVAs) using the factors activity (eating, unpacking, pure driving) and packaging (cardboard, paper) revealed statistical significances between the conditions ($p < 0.01$). There was no effect of packaging nor an interactional effect between the conditions and activities ($p > 0.05$).

4 Discussion

The task of eating and driving in a vehicle can result in visual and physical distraction thus contributing to increase risk of accidents [4]. With many individuals still eating and drinking behind the wheel and many fast food drive-thru's making it easier and accessible for people especially drivers to eat fast food while steering the wheel of a car, the task of driving has been normalized by our society and can contribute to reduced driving performance [5]. The main purpose of this study was to determine the effects of food packaging on driving performance when eating while driving and whether the packaging of food would make a difference in driving performance, thus improving driving safety.

The principle findings from this study suggest that food packaging does not influence driving performance. This means that driving is as dangerous with cardboard packaging as to paper wrapper packaging. Interestingly however, is that the study found that people perceived driving performance differently. The possible reasons for this are, participants perceived that the cardboard packaging was easier to handle while driving because they did not have to worry about the paper wrapper while driving. It was also perceived as easier to unpack while driving. However, participants perceived that they drove worst in paper wrapper condition but in fact they are not. This might have affected their confidence in driving, however in terms of driving performance (tracking deviation) there were no significant differences found between cardboard packaging and paper wrapper packaging.

Unpacking was associated with reduced simulated driving performance (tracking deviation) because the activity introduced multiple distractions that may have limited the driving performance and thereby participants also perceiving this activity to having worst vehicle control and driving ability for example it was denoted as fair. During the unpacking of the meal activity, participants attempted to locate the food items by having to remove the individual food items from the paper bag and then unpacked the various items as well as placed a straw into the canned beverage/soft drink with the manipulation of one hand while the other hand was on the steering wheel. Some participants used various techniques to open the package which may have contributed to the variance for this activity been the greatest in comparison to the other activities. For example, some participants placed the food items on their laps then continued unpacking the individual food items while others used both of their hands to unpack resulting in no hands on the steering wheel. Some participants also looked away a few times to see what they were unpacking in order to execute the activity whereas others did not look away, feeling for the food items.

During the eating of meal activity, it was also observed that during this activity participant kept the various food items with the use of one hand while the other hand was on the steering wheel. Participants also often looked away to consume the food to avoid spillage. Due to this, it may have also limited their driving performance (tracking deviation) and also participants perceiving this activity to having worst vehicle control and driving ability. Previous research has highlighted that eating of the meal contributes to incidences of inattention in lane change [3, 13]. Eating behind a wheel leads to increases in collision as well as lane position deviation [1]. A similar study also highlighted that distracted driving behaviour such as eating while driving appears to negatively impact driving measures of lane position control and reaction times [4]. The study also showed that those who were drinking a beverage behind the wheel were 18 percent more likely to have poor lane control. For example, the act of sipping on a soda beverage behind a wheel, participants found it difficult to remain in the center of the lane. This can be hazardous if a vehicle is attempting to pass you and you accidentally swerve into their lane [4]. These finding were consistent with the findings of the current study. Interestingly however in terms of subjective measure, previous research indicated that drivers perceived that the task of eating behind the wheel to be a more acceptable behaviour and are therefore less likely to identify the associated risks or modify eating behaviours behind the wheel [12, 13, 16). Therefore, current findings do not support previous works. Given the findings of the current study greater awareness on the impact of unpacking and eating while driving and driving safety are required.

The significant differences found between pure driving 1 and pure driving 2 activities for perceived control could be explained by the fact that participants felt more confident afterwards highlighting a possible learning effect considering that the unpacking and eating when they are already learning it increased the assuredly of these differences. However, this was not the case, as for performance while driving, pure driving 1 was not significantly different to pure driving 2 meaning that there was no possible learning effect that masked the effect of packaging. In terms of the effect of interactions, no differences were found for the interactional effect between the conditions and activities, meaning that the packaging type did not affect any possible changes between the activities. Therefore, driving performance was not made safer by putting the individual food items in different packaging types.

The study acknowledges that there were limitations in terms of equipment and protocol used in the study. In terms of equipment use, with the use of a low-fidelity simulator it may evoke unrealistic driving behaviour of drivers that eat and drink as it imitates an automated vehicle and does not imitate a manual vehicle making it easier for participants to complete. This may underestimate the behaviour of eating and drinking while driving on driving performance variables in a real scenario. The simulator software also does not represent road features such as traffic signs, road signs, billboards, stop streets, pedestrians and does not require participants to use both the brake and accelerator to stop and start a vehicle that take place in the real life setting

5 Conclusion

This study suggested that food packaging does not seem to influence driving performance when eating while driving, meaning that driving is as dangerous with cardboard packaging as to paper wrapper packaging. Interestingly however, is that the study found that people perceived driving performance differently. This might have affected their confidence in driving, but in terms of their driving performance which is tracking deviation it did not differ.

The unpacking of the meal activity highlighted the worst performance this may be explained by the fact that participants used various techniques to open the package which may have contributed to the variance for this activity been the greatest in comparison to the other activities. Following this was eating of the meal. These behaviours lead drivers neglecting to put both hands on the steering wheel which could have contributed to increased exposure and risk of accidents as well as participants perceiving these activities to having worst vehicle control and driving ability.

References

1. Alosco ML, Spitznagel MB, Fischer KH, Miller LA, Pillai V, Hughes J, Gunstad J (2012) Both texting and eating are associated with impaired simulated driving performance. Traffic Inj Prev 13(5):468–475
2. Beissmann T (2012) Eating, drinking while driving slows reaction times. http://www.caradvice.com.au/171581/eating-drinking-while-driving-slows-reaction-times-study. Accessed 4 Apr 2016
3. Campbell BN, Smith JD, Najm WG (2002) Examination of Crash Contributing Factors Using National Crash Databases. (Report No. DOT-VNTSC-NHTSA-02-07). Volpe national Transportation Systems Center: Cambridge, MA
4. Irwin C, Monement S, Desbrow B (2015) The influence of drinking, texting, and eating on simulated driving performance. Traffic Inj Prev 16(2):116–123
5. Jakle JA, Sculle KA (1999) Fast food: roadside restaurants in the automobile age. The John Hopkins University Press, United States of America
6. Locher J, Moritz O (2009) Eating while driving causes 80% of all car accidents, study show. http://www.nydailynews.com/new-york/eating-driving-80-car-accidents-study-shows-article-1.427796. Accessed 4 Mar 2016
7. Louw T (2013) An investigation into control mechanisms of driving performance resource depletion and effort regulation. Unpublished Master of Science thesis, Rhodes University, Grahamstown, South Africa
8. Lytx (2014) Lytx Data Finds Three Dangerous Activities You May Be Doing While Driving Every Day. http://www.lytx.com/press-releases/lytx-data-finds-three-dangerous-activities-you-may-be-doing-while-driving-every-day. Accessed 4 Apr 2016
9. Marsh K, Bugusu B (2007) Food packaging—roles, materials, and environmental issues. J Food Sci 72(3):39–55
10. Siu J (2013) Eating while driving deadlier than texting behind the wheel, study shows. http://www.autoguide.com/auto-news/2012/05/eating-while-driving-deadlier-than-texting-behind-thewheel-study-shows.html. Accessed 4 Mar 2016
11. Stutts JC, Reinfurt DW, Staplin L, Rodgman EA (2001) The role of driver distraction in traffic crashes. AAA Foundation for Traffic Safety, Washington, DC
12. Stutts J, Feaganes J, Reinfurt D, Rodgman E, Hamlett C, Gish K, Staplin L (2005) Driver's exposure to distractions in their natural driving environment. Accid Anal Prev 37(6):1093–1101
13. Tlhoaele K (2013) Understanding the multitask process of eating while driving and its potential effect on driving performance. Unpublished Honours thesis Rhodes University, Grahamstown, South Africa
14. Young K, Lee JD, Regan MA (2008) Defining driver distraction. In: Young K, Lee JD, Regan MA (eds) Driver distraction: theory, effects, and mitigation. CRC Press, Florida, USA, pp. 31–40
15. Young MS, Mahfoud JM, Walker GH, Jenkins DP, Stanton NA (2008) Crash dieting: the effects of eating and drinking on driving performance. Accid Anal Prev 40(1):142–148
16. White MP, Eiser JR, Harris PR (2004) Risk perceptions of mobile phone use while driving. Risk Anal 24(2):323–334

Author Index

S. Bagnara et al. (Eds.): IEA 2018, AISC 823, pp. 601–603, 2019.
https://doi.org/10.1007/978-3-319-96074-6

Printed in the United States
By Bookmasters